T0358971

Advanced Textbooks in Economics

Series Editors: C. J. Bliss *and* M.D. Intriligator

Currently Available:

OPTIMAL CONTROL THEORY
WITH ECONOMIC APPLICATIONS

ADVANCED TEXTBOOKS IN ECONOMICS

VOLUME 24

Editors:

C.J. BLISS

M.D. INTRILIGATOR

Advisory Editors:

W.A. BROCK

D.W. JORGENSON

A.P. KIRMAN

J.-J. LAFFONT

L. PHLIPS

J.-F. RICHARD

NORTH-HOLLAND
AMSTERDAM·LONDON·NEW YORK·TOKYO

OPTIMAL CONTROL THEORY WITH ECONOMIC APPLICATIONS

ATLE SEIERSTAD and KNUT SYDSÆTER

University of Oslo

NORTH-HOLLAND
AMSTERDAM · LONDON · NEW YORK · TOKYO

Elsevier
Radarweg 29, PO Box 211, 1000 AE Amsterdam, The Netherlands
The Boulevard, Langford Lane, Kidlington, Oxford OX5 1GB, UK

Published with a grant from the Norwegian Research Council for Science and the Humanities.

Library of Congress Cataloging in Publication Data

Seierstad, Atle.
    Optimal control theory with economic applications.

    (Advanced textbooks in economics ; v. 24)
    Bibliography: p.
    1. Control theory.   2. Economics--Mathematical
models.   I. Sydsaeter, Knut.   II. Title.   III. Series.
QA402.3.S4   1986   629.8'312   86-16504
ISBN 0-444-87923-4 (Elsevier Science Pub. Co.)

First printing: 1987
Second impression: 1993

ISBN: 0 444 87923 4

Working together to grow
libraries in developing countries

www.elsevier.com | www.bookaid.org | www.sabre.org

ELSEVIER   BOOK AID
International   Sabre Foundation

Transferred to Digital Printing 2007

INTRODUCTION TO THE SERIES

The aim of the series is to cover topics in economics, mathematical economics and econometrics, at a level suitable for graduate students or final year undergraduates specializing in economics. There is at any time much material that has become well established in journal papers and discussion series which still awaits a clear, self-contained treatment that can easily be mastered by students without considerable preparation or extra reading. Leading specialists will be invited to contribute volumes to fill such gaps. Primary emphasis will be placed on clarity, comprehensive coverage of sensibly defined areas, and insight into fundamentals, but original ideas will not be excluded. Certain volumes will therefore add to existing knowledge, while others will serve as a means of communicating both known and new ideas in a way that will inspire and attract students not already familiar with the subject matter concerned.

The Editors

CONTENTS

PREFACE

This book is an introduction to the theory of deterministic optimal control of systems governed by ordinary differential equations. It also includes an introduction to the classical calculus of variations.

The selection of topics has been influenced by needs arising in applications of optimal control theory to economics; hence the book will be of particular interest to students and researchers in this field. However, the theory, as well as the examples, will be of interest also to non-economists.

The theoretical sections concentrate on rendering existence results, necessary conditions and (direct) sufficient conditions. Control problems without state constraints as well as problems with mixed and/or pure state constraints have been treated. For problems without state constraints, results on the dependence of the optimal solution on parameters are given. Problems with jumps in the state variables are extensively discussed.

Topics like controllability and observability of control systems are not treated, and we only touch upon the dynamic programming approach and feedback controls. Moreover, we were unable to discuss problems of estimation and stochastic control, in particular the important topic of optimal control of large-scale models of the economy. Numerical solution procedures are not mentioned.

The book contains a rather comprehensive collection of theoretical results designed to cover a wide variety of control problems. A reader wanting an introduction to the field should not try to digest the entire catalogue of results at first reading. A suitable selection of sections for a reasonably elementary introduction is indicated subsequent to this preface.

An important aspect of this book is the large number of examples, in which the theory is applied to a wide variety of problems in economics. However, the type of applications found in this book, as well as their often more complicated counterparts in the literature, have a very modest purpose indeed. By simple – one or a few dimensional – models an attempt is made to illuminate some qualitative aspects or analytic points that might be important to keep in mind when turning to explanations of – and policy formulation in – a more complex reality.

Even for such types of applications, the models presented do not form a representative selection. Pedagogical considerations have been foremost in the selection of examples and exercises. Their primary purpose is to serve as suitable, not too lengthy, illustrations of how the theory can be applied. Nevertheless, the book does include simple models of economic growth in both closed and open economies, exploitation of both renewable and non-renewable resources, pollution control, behaviour of firms, and differential games.

Many of the theoretical results are stated without proofs, but with a high degree of rigour. We do prove almost all the results for sufficient conditions, since they are relatively easy and since the required concavity conditions prevail in many economic models. Proofs are also offered for many of the elementary results in the calculus of variations.

The intended level of rigour in the theoretical results and in the examples is roughly matching that prevailing in, say, general equilibrium theory and duality theory in economics or, for that matter, in mathematics generally.

Our emphasis on precision differentiates this book somewhat from the bulk of literature in this area. Actually, in control models appearing in the economic literature it is customary to be rather sloppy. For instance, people who would not dream of counting equations and unknowns in an argument for the existence of an equilibrium in economic models of general equilibrium frequently make much more heroic short-cuts in "solving" optimal control problems.

Most of the papers in this field refer to results in books like Arrow and Kurz (1970), Kamien and Schwartz (1981), and Sethi and Thompson (1981). These books are valuable and highly readable introductions to a rather complicated theory, but they are on many points intentionally non-rigorous. In our experience, people working in optimal control theory basing their understanding mainly on heuristic arguments and loosely formulated results often run into difficulties. In particular, this is true for problems with state space constraints. Even in many small-scale problems of this type, one has to be extremely careful in order to obtain a correct solution.

Colin W. Clark (1982) has commented on the problem in a favourable review of Kamien and Schwartz (1981):

"While pedagogically sound, the simple approach used in the text tends to underplay the difficulties that can occur – and often do – in practical application. The student should be warned that simplistic attempts to extend the theory outlined may not work; the applied optimal control literature already contains many blunders resulting from the failure to appreciate this fact".

In order to put our arguments for precision in a proper perspective, we want to stress that in our opinion pursuit of rigour can sometimes be highly misplaced. It often hampers imagination and stifles the development of new ideas. While engaged in explorative research one should often hold off demands for rigour during the early stages of investigation. Even publication of results lacking rigour might be advisable. However, by the time theoretical results begin to take on a more established role in a given body of knowledge, the call for rigour should be urgent.

There is a number of exercises throughout the text (with answers provided). The majority of the problems are quite easy and are designed to illustrate the use of the theorems in simple situations.

We have been working on this book and on related problems for many years. During this time we have received support and encouragement from a number of institutions and individuals. For some periods we have had grants from the Norwegian Council for Science and Humanities, The Norwegian Central Bank, Professor Keilhaus Minnefond, and the research fund of Christiania Bank og Kreditkasse.

A number of colleagues and students have given their reactions to parts of the manuscript. In particular we thank Kjell A. Brekke and Bruce Wolman for mathematical and linguistic corrections.

We also would like to thank the Department of Economics at the University of Oregon for offering Seierstad good working conditions in the fall of 1982, and the Department of Economics, University of California at Los Angeles for inviting Sydsæter to be a Visiting Scholar in the spring of 1983. A grant from the United States Educational Foundation which made Seierstad's visit to Oregon possible, is also gratefully acknowledged.

The Department of Economics at the University of Oslo has provided excellent facilities and assistance. We wish to thank its technical staff for enduring all our requests. In particular, our thanks go to Inger Johanne Kroksjø for drawing the figures and to Elin Begby who has expertly and patiently typed numerous versions of the manuscript.

Oslo, August 1985   Atle Seierstad and Knut Sydsæter

TOPICS AND NOTATION

A first course

The following list of topics in this book may serve as a first course in optimal control theory for economists:

Chapter 1, sections 1–8
Chapter 2, sections 1–7 (minus example 7), (section 8), section 9.
Chapter 3, sections 1, 2, 5 (until theorem 11), section 7 (until theorem 14), section 8.
Chapter 4, sections 1, 3, 4 (minus example 4)
Chapter 5, sections 1, 2.

Notation, terminology, and references

By and large we use standard notation and terminology, some of which is described in Appendices A and B, especially in A. 1 and B. 2 – B. 4. In particular, note that "w.r.t." is an abbreviation of "with respect to", and that "v.e." (virtually everywhere) means for all points, except a finite or countable number.

References are mainly given in the footnotes. The references do not necessarily give the original source of a theorem or a proof.

Solutions to the exercises are found at the end of the book. Though references are given for a few of the exercises, in most cases we have not traced the origin of the problems, even in cases where this is a simple task.

A reference to (5.2) is a reference to equation (2) in Chapter 5; a similar numbering is used for theorems and notes. Equation (A.1) is the first equation in Appendix A. The chapter number is omitted when referring to equations, theorems, and notes in the same chapter.

For ease of reference an index of theorems, notes, and examples is provided at the end of the book.

Some sections, subsections, notes and exercises are marked with an asterisk which indicates material of a more special character, and which can be omitted at first reading.

INTRODUCTION

In many areas of the empirical sciences such as physics, biology, and chemistry, as well as in economics, we study the time development of systems. Certain essential properties or characteristics of the system, called *the state of the system*, change over time. If there are n *state variables* we denote them by $x = (x_1, \ldots, x_n)$. The rate of change of the state variables w.r.t. time usually depends on many factors, including the actual values of the x_i's, and on certain parameters that can be controlled from the outside. These parameters are called *control variables* and are denoted by $u = (u_1, \ldots, u_r)$. We shall assume throughout this book that the laws governing the behaviour of the system through time are given by systems of ordinary differential equations.

The control variables can be freely chosen within certain bounds. If the system is in some state x^0 at time t_0, and if we choose a certain control function $u(t)$, then the system of differential equations will, usually, have a unique solution, $x(t)$. If there are a priori bounds on the values of the state variables, only those control functions will be *admissible* which give rise to state functions $x(t)$ satisfying the bounds. In general, there will be many admissible control functions, each giving rise to a specific evolution of the system. In an *optimal* control problem an optimality criterion is given which assigns a certain number (a "utility") to each evolution of the system. The problem is then to find an admissible control function which maximizes the optimality criterion in the class of all admissible control functions.

As we explained in the preface, we shall usually refrain from proving the basic necessary conditions of the optimal control problems which we study. This introductory section is designed for readers who want a preview of some of the problems that we are dealing with, an indication of important results and a rough description of some of the main ideas underlying the theory, including necessary conditions.

The tools available for solving optimal control problems are analogous to those used in static optimization theory. Before examining a simple control problem we digress for a moment with some remarks concerning static optimization.

Static optimization

In static optimization theory three main types of results are used in order to solve global optimization problems: (1) theorems on necessary conditions (typically: first-order conditions), (2) theorems on sufficient conditions (typically: first-order conditions plus suitable concavity/convexity conditions) and, finally, (3) existence theorems (typically: the Extreme Value Theorem).

To briefly illustrate with a simple optimization problem these three types of theorems, let f_0 be a continuously differentiable function of one variable defined on a closed, bounded interval $I = [u_0, u_1]$. We seek the solution to the problem

$$\max f_0(u) \quad \text{when } u \in I. \tag{1}$$

A necessary condition for $\bar{u} \in I$ to solve this problem is, in a somewhat unconventional formulation,

$$\max_{u \in I} \frac{df_0(\bar{u})}{du} \cdot u = \frac{df_0(\bar{u})}{du} \cdot \bar{u}, \tag{2}$$

or, equivalently,

$$\max_{u \in I} \frac{df_0(\bar{u})}{du} \cdot (u - \bar{u}) \leqslant 0. \tag{3}$$

Relying on some basic facts from calculus, this "maximum principle" can be proved very easily: Suppose $\bar{u} \in I$ solves problem (1). If $\bar{u}$ is an interior point of I, then $df_0(\bar{u})/du = 0$, and (3) is trivially satisfied. If $\bar{u} = u_0$, then $df_0(\bar{u})/du \leqslant 0$ (if $df_0(\bar{u})/du > 0$, $\bar{u} = u_0$ would not maximize f_0), and $u - \bar{u} = u - u_0 \geqslant 0$ for all $u \in I$. Hence (3) is satisfied. Finally, if $\bar{u} = u_1$, then $df_0(\bar{u})/du \geqslant 0$, and since $u - \bar{u} = u - u_1 \leqslant 0$ for all $u \in I$, (3) is again satisfied.

The first-order condition (3) will often be satisfied by non-optimal points in I (e.g. any stationary point of f_0 satisfies (3)). All points $\bar{u}$ which satisfy (3) are called *candidates* for optimality (w.r.t. the given necessary condition). Usually, the set of candidates will contain only a few points. If we compare the values of f_0 for each candidate, we can pick out from the set of all candidates the subset consisting of those candidates which give f_0 its highest value. Is this subset of "best" candidates the set of *optimal solutions*? In a more general framework we cannot be sure since the problem may fail to have an optimal solution. However, in the present case, by appealing to the Extreme Value Theorem (i.e. the Weierstrass theorem), we know that the problem has at least one solution, and the question above can be answered in the affirmative.

Above we used necessary conditions plus an existence theorem in order to solve problem (1). There is another method which is sometimes applicable. Suppose f_0 is a concave function. Then

$$f_0(u) - f_0(\bar{u}) \leqslant \frac{\mathrm{d}f_0(\bar{u})}{\mathrm{d}u} \cdot (u - \bar{u}) \qquad \text{for all } u \in I \tag{4}$$

by a well-known characterization of concave functions (see (B.1)). Suppose $\bar{u} \in [u_0, u_1]$ satisfies (3). Then from (4) it follows that $f_0(u) - f_0(\bar{u}) \leqslant 0$ for all $u \in [u_0, u_1]$. Hence $\bar{u}$ solves problem (1). Thus we observe that if f_0 is concave, the necessary condition (3) becomes sufficient. In this case, we need not compare the function values of different candidates. Note also that this argument is valid even if I is unbounded, open or half open. In the latter cases no general existence results are readily at hand.

Of course, the concavity requirement on f_0 is a rather severe restriction, and a lot of problems of the type (1) in which f_0 is not concave still have solutions which, in principle, can be found by using our first method.

A simple dynamic optimization problem

Consider first a very simple control problem:

$$\max_{u(t)} \int_{t_0}^{t_1} f_0(u(t), t) \mathrm{d}t, \qquad u(t) \in [u_0, u_1] = U, \tag{5}$$

where $u(t)$ is piecewise continuous, t_0, t_1, u_0, and u_1 are fixed numbers, and f_0 is a given real-valued function. The maximization problem is to find a piecewise continuous function $u(t)$ defined on $[t_0, t_1]$, with values in $[u_0, u_1]$, maximizing the integral in (5).

Let $u^*(t)$ be optimal in this problem. Then the following condition is necessary for the optimality of $u^*(t)$:

$$\max_{u \in U} f_0(u, s) = f_0(u^*(s), s) \qquad \text{at all points } s \text{ in } [t_0, t_1]. \tag{6}$$

This condition amounts to choosing at each instance of time s that value of $u^*(s) \in U$ which maximizes the integrand $f_0(u, s)$. It is obvious that this choice of u will give the integral its highest value.

Although this sounds rather convincing, let us give a more formal argument. Let u be an arbitrary point in U and let s be any continuity point of $u^*(t)$, $s \in (t_0, t_1]$. If δ is a small positive number, define the following "strong variation" of $u^*(t)$:

$$\begin{cases} u(t) = u & t \in [s - \delta, s] = I \\ u(t) = u^*(t) & \text{elsewhere} \end{cases} \tag{7}$$

and let

$$\Delta J(\delta) = \int_{t_0}^{t_1} f_0(u(t), t) dt - \int_{t_0}^{t_1} f_0(u^*(t), t) dt$$

$$= \int_{s - \delta}^{s} [f_0(u, t) - f_0(u^*(t), t)] dt. \tag{8}$$

By the optimality of $u^*(t)$, $\Delta J(\delta) \leqslant 0 = \Delta J(0)$ for all $\delta > 0$. Now, for small values of δ, we have the first-order approximation

$$\Delta J(\delta) = \Delta J(\delta) - \Delta J(0) \approx \Delta J'(0) \cdot \delta,$$

where $\Delta J'(0)$ is found by differentiating the last integral in (8) w.r.t. δ and putting $\delta = 0$. The result is $\Delta J'(0) = f_0(u, s) - f_0(u^*(s), s)$. It follows from the first-order approximation and $\Delta J(\delta) \leqslant 0$ that

$$f_0(u, s) - f_0(u^*(s), s) \leqslant 0. \tag{9}$$

(Expressed in an equivalent way, (9) follows from the fact that $\Delta J'(0) > 0$ contradicts the inequality $\Delta J(\delta) \leqslant 0$.) Evidently, (6) follows from (9).

A simple problem in which we have to plan ahead

Consider next a somewhat more complicated problem:

$$\max_{u(t)} \int_{t_0}^{t_1} f_0(x(t), u(t), t) dt, \tag{10a}$$

$$\dot{x}(t) = u(t), \qquad x(t_0) = x^0, \qquad x(t_1) \text{ free,} \tag{10b}$$

where, still, $u(t) \in [u_0, u_1] = U$ for all t. Here

$$x(t) = \int_{t_0}^{t} u(\tau) d\tau + x^0. \tag{11}$$

Thus to each control function $u(t)$, there corresponds a function $x(t)$. The problem is now to find a pair $(x^*(t), u^*(t))$ that maximizes the integral in (10a). Observe that when choosing values of $u^*(t)$ at each point of time in this problem we have to take the future into consideration: The values of $u^*(t)$ in the vicinity of any point s, influence the values of the corresponding solution $x^*(t)$ at all points of time after s. Since f_0 also contains x, this influence must be taken into account.

In order to derive necessary conditions for the optimality of $(x^*(t), u^*(t))$, let $u(t)$ be the perturbation of $u^*(t)$ given in (7) and let $x(t)$ be given by (11). Define

$$\Delta J = \int_{t_0}^{t_1} f_0 \, dt - \int_{t_0}^{t_1} f_0^* \, dt = \int_{s-\delta}^{s} (f_0 - f_0^*) dt + \int_{s}^{t_1} (f_0 - f_0^*) dt, \tag{12}$$

where $f_0 = f_0(x(t), u(t), t)$, $f_0^* = f_0(x^*(t), u^*(t), t)$. (Generally, the sign * means evaluation at $(x^*(t), u^*(t), t)$. In (12) we made use of the fact that for $t \leqslant s - \delta$, f_0 and f_0^* are equal.)

For $t > s$, $f_0 - f_0^* = f_0(x(t), u^*(t), t) - f_0(x^*(t), u^*(t), t) \simeq f_{0x}'^*(t) \Delta x(t)$ when δ is small, where

$$\Delta x(t) = x(t) - x^*(t) = \int_{t_0}^{t} (u(\tau) - u^*(\tau)) d\tau$$

$$= \int_{s-\delta}^{s} (u - u^*(\tau)) d\tau \approx [u - u^*(s)] \delta. \tag{13}$$

Thus, when δ is small

$$\Delta_2 J = \int_s^{t_1} (f_0 - f_0^*)dt \approx \int_s^{t_1} f_{0x}'^*(t)\Delta x(t)dt. \tag{14}$$

The difference $f_0(x(t), u(t), t) - f_0(x^*(t), u(t), t)$ is of the first order in δ, since $x(t) - x^*(t)$ is of the first order in δ. Hence, the difference $\int_I [f_0(x(t), u(t), t) - f_0(x^*(t), u(t), t)]dt$ is of the second order in δ, since $I = [s - \delta, s]$. Thus, in the first order

$$\Delta_1 J = \int_I (f_0 - f_0^*)dt = \int_I (f_0 - f_0(x^*(t), u(t), t))dt$$

$$+ \int_I (f_0(x^*(t), u(t), t) - f_0^*)dt \approx [f_0(x^*(s), u, s) - f_0^*(s)]\delta. \tag{15}$$

Thus

$$\Delta J = \Delta_1 J + \Delta_2 J \approx [f_0(x^*(s), u, s) - f_0^*(s)]\delta + [u - u^*(s)]\delta \int_s^{t_1} f_{0x}'^*(t)dt.$$

Again, the optimality of $u^*(t)$ requires $\Delta J \leq 0$. Defining $p(s) = \int_s^{t_1} f_{0x}'^*(t)dt$, we get:

$$f_0(x^*(s), u, s) - f_0(x^*(s), u^*(s), s) + p(s)[u - u^*(s)] \leq 0. \tag{16}$$

Finally, defining $H(x, u, p, t) = f_0(x, u, t) + pu$, (16) can be written as $H(x^*(s), u, p(s), s) \leq H(x^*(s), u^*(s), p(s), s)$. In other words, for all s,

$$\max_{u \in U} H(x^*(s), u, p(s), s) = H(x^*(s), u^*(s), p(s), s). \tag{17}$$

Note that $p(s)$ satisfies

$$\dot{p} = -f_{0x}'(x^*(s), u^*(s), s) =$$
$$-H_x'(x^*(s), u^*(s), p(s), s), \qquad \text{and with } p(t_1) = 0. \tag{18}$$

The necessary conditions for optimality in problem (10) consist of (17) and (18). The function H is usually called the *Hamiltonian*.

Let us compare condition (17) with condition (6) obtained for problem (5). In (6) we find the correct value of $u^*(s)$ simply by maximizing $f_0(u, s)$. For problem (10), (17) tells us instead to maximize $f_0(x^*(s), u, s) + p(s)u$ w.r.t. u. Here the function p takes care of the need described above to plan ahead.

A more general case

The treatment of the two preceding problems, exemplifies the type of arguments used in deriving necessary conditions in more complicated control problems. The reader will discover that a function $p(s)$ invariably turns up in necessary conditions serving the same forward-looking purpose.

We consider next a more complicated control problem in which the trivial differential equation $\dot{x} = u$ from problem (10) is replaced by a more complicated equation. (Some readers might prefer to study first Chapter 2, sections 1 and 2.) We study the problem

$$\max_{u(t)} \int_{t_0}^{t_1} f_0(x(t), u(t), t)\mathrm{d}t, \qquad \dot{x}(t) = f(x(t), u(t), t),$$

$$x(t_0) = x^0, \qquad x(t_1) \text{ free.} \tag{19}$$

Still x is a scalar, $u(t) \in [u_0, u_1] = U$ for all t and f_0 and f are real-valued functions. We assume that for each choice of piecewise continuous function $u(t)$ the differential equation for x has a continuous solution $x(t)$ with $x(t_0) = x^0$. The corresponding pair $(x(t), u(t))$ is then called *admissible*, and the problem is to find an admissible pair maximizing the integral in (19). An optimal pair is again denoted $(x^*(t), u^*(t))$. To find necessary conditions for this problem, we argue in the same manner as in problem (10). We define ΔJ, $\Delta_1 J$ and $\Delta_2 J$ by (12), (14) and (15) in the same way as before.

However, the calculation of $\Delta x(t) = x(t) - x^*(t)$ is different: Write $f = f(x(t), u(t), t)$, $f^* = f(x^*(t), u^*(t), t)$ and note that $x(t) = \int_{t_0}^t f\mathrm{d}\tau + x^0$, $x^*(t) = \int_{t_0}^t f^*\mathrm{d}\tau + x^0$. Thus for $t > s$,

$$\Delta x(t) = \int_{t_0}^t f\mathrm{d}\tau - \int_{t_0}^t f^*\mathrm{d}\tau = \int_{s-\delta}^s (f - f^*)\mathrm{d}\tau + \int_s^t (f - f^*)\mathrm{d}\tau. \tag{20}$$

Now, on $(s, t_1]$, $f(\tau) - f^*(\tau) = f(x(\tau), u^*(\tau), \tau) - f(x^*(\tau), u^*(\tau), \tau) \simeq f_x'^*(\tau)\Delta x(\tau)$.

Hence, for $t > s$,

$$\Delta x(t) \approx \int_{s-\delta}^{s} (f - f^*)d\tau + \int_{s}^{t} f_x'^*(\tau)\Delta x(\tau)d\tau. \tag{21}$$

For $t > s$, from (21), we get

$$\frac{d}{dt}\Delta x(t) = f_x'^*(t)\Delta x(t), \qquad \Delta x(s) = \int_{s-\delta}^{s} (f - f^*)d\tau \approx C\delta, \tag{22}$$

where $C = f(x^*(s), u, s) - f^*(s)$. The differential equation in (22) is a linear one. With initial condition $\Delta x(s) = C\delta$, the solution is $\Delta x(t) = C\delta \cdot \exp \int_{s}^{t} f_x'^*(\tau)d\tau$ (see (A.20)). Inserting $\Delta x(t)$ into (14), we get (using also (15)),

$$\Delta J = \Delta_1 J + \Delta_2 J \approx [f_0(x^*(s), u, s) - f_0^*(s)]\delta$$

$$+ \int_{s}^{t_1} f_{0x}'^*(t)C\delta \, \exp\left(\int_{s}^{t} f_x'^*(\tau)d\tau\right)dt$$

$$= [f_0(x^*(s), u, s) - f_0^*(s)]\delta$$

$$+ p(s)[f(x^*(s), u, s) - f^*(s)]\delta \tag{23}$$

where

$$p(s) = \int_{s}^{t_1} f_{0x}'^*(t) \, \exp\left(\int_{s}^{t} f_x'^*(\tau)d\tau\right)dt.$$

It is easily seen (cf. (A.20)) that $p(s)$ is the solution of the differential equation

$$\dot{p} = -f_{0x}'^*(t) - p(t)f_x'^*(t), \qquad p(t_1) = 0. \tag{24}$$

Define the Hamiltonian

$$H(x, u, p, t) = f_0(x, u, t) + pf(x, u, t). \tag{25}$$

By the first-order approximation in (23),

$$\Delta J \leqslant 0 \qquad \text{implies } H(x^*(s), u, p(s), s) \leqslant H(x^*(s), u^*(s), p(s), s).$$

In other words, we again get the maximum condition: For all $s \in [t_0, t_1]$:

$$\max_{u \in U} H(x^*(s), u, p(s), s) = H(x^*(s), u^*(s), p(s), s). \qquad (26)$$

Furthermore, (24) is equivalent to

$$\dot{p} = -H'_x(x^*(t), u^*(t), p(t), t), \qquad p(t_1) = 0. \qquad (27)$$

The necessary conditions for problem (19) consist of the two conditions (26) and (27). Thus, if $(x^*(t), u^*(t))$ is optimal, then (26) is satisfied, $p(s)$ being the function defined by (27). This result is referred to as the (*Pontryagin*) *Maximum Principle*.

The conditions (26) and (27) are not, in general, sufficient for the optimality of $(x^*(t), u^*(t))$. It turns out that by imposing appropriate concavity conditions on the Hamiltonian, (26) and (27) become sufficient as well (concavity of $H(x, u, p(t), t)$ w.r.t. (x, u) is sufficient).

We assumed in problem (19) that $x(t_1)$ is free. Proofs of necessary conditions in cases where $x(t_1)$ is restricted are more complicated. (See Chapter 2, section 11 and the references given to theorem 2.2.)

In optimal control problems arising in economics there are often additional restrictions of the type $h(x(t), u(t), t) \geqslant 0$. Proofs of necessary conditions for such problems are even harder.

Weak variations

We shall briefly discuss another type of variation of the control functions often used in deriving necessary conditions in optimal control problems.

Let us begin by considering problem (5). The following condition is necessary for the optimality of $u^*(t)$:

$$\max_{u \in [u_0, u_1]} \frac{\partial f_0(u^*(t), t)}{\partial u} \cdot (u - u^*(t)) \leqslant 0 \qquad \text{at all points } t \in [t_0, t_1]. \qquad (28)$$

This follows from (6) and the argument for (3), but let us sketch another proof.

Proof. Let $u(t) \in [u_0, u_1]$ for all t and define $v(t) = u(t) - u^*(t)$. Then

$u^*(t) + \alpha v(t) = \alpha \dot{u}(t) + (1 - \alpha)u^*(t) \in [u_0, u_1]$ for all $\alpha \in [0, 1]$. Define

$$V(\alpha) = \int_{t_0}^{t_1} f_0(u^*(t) + \alpha v(t), t) \mathrm{d}t. \tag{29}$$

Since $u^*(t)$ is optimal, $V(\alpha) \leq V(0)$ for all $\alpha \in [0, 1]$, and then $V'(0) \leq 0$. Differentiating (29) w.r.t. α and putting $\alpha = 0$ we obtain:

$$V'(0) = \int_{t_0}^{t_1} \frac{\partial f_0(u^*(t), t)}{\partial u} \cdot (u(t) - u^*(t)) \mathrm{d}t \leq 0. \tag{30}$$

Suppose (28) fails for some continuity point t' and some $u = u'$. Then $(\partial f_0(u^*(t), t)/\partial u) \cdot (u' - u^*(t)) > 0$ for all t in some interval $[a, t']$. Defining $u(t) = u'$ in $[a, t']$ and $u(t) = u^*(t)$ elsewhere, we get

$$V'(0) = \int_{a}^{t'} (\partial f_0(u^*(t), t)/\partial u)(u' - u^*(t)) \mathrm{d}t > 0,$$

which contradicts (30).　■

Suppose $f_0(u, t)$ is concave in u. Then, as in (4),

$$f_0(u(t), t) - f_0(u^*(t), t) \leq (\partial f_0(u^*(t), t)/\partial u)(u(t) - u^*(t))$$

for any $u(t)$, and it follows that condition (28) is sufficient for $u^*(t)$ to solve problem (5) in this case. (Without any assumption on f_0, (6) *is* a sufficient condition.)

The function $u^*(t) + \alpha v(t)$ in the proof of (28) is called *a weak variation* of $u^*(t)$. Both weak and strong variations are used for obtaining necessary conditions for optimality in dynamic optimization problems.

In fact, the above proof based on weak variations can also be generalized to the problems (10) and (19). In problem (10) it leads to the Euler equation (on integrated form) in the calculus of variations. Thus we see that in several problems useful conditions can be derived by using either strong or weak variations. In more complicated problems like those encountered in the later parts of this book, a simultaneous use of strong and weak variations are needed.

Necessary conditions — sufficient conditions — existence theorems

We end this introduction with a brief discussion of the way in which necessary conditions, sufficient conditions and existence theorems are generally applied in optimization problems.

Necessary conditions are used to single out a set of *candidates* for optimality. If there is more than one candidate, the best candidates can be found by calculating the values of the criterion for each candidate and comparing these values. Provided an existence theorem applies, we know that the set of best candidates is the set of *optimal solutions*. However, in the absence of a general existence theorem, we have no guarantee that the best candidates are optimal. In fact, even if the set of candidates is non-empty, no optimal solution need exist.

In cases where no general existence theorem applies it is still useful to find the set of best candidates. Knowing such candidates may enable us to prove that one or more of the candidates are optimal by arguments specific to the given problem.

Sufficient conditions are very convenient when they are applicable: A candidate satisfying them *is* optimal. However, the requirements needed (typically concavity conditions) are quite restrictive.

CALCULUS OF VARIATIONS

The classical calculus of variations is a forerunner to optimal control theory. Some of the main results in the calculus of variations were established already by Euler and Lagrange in the 18th century, and since then the subject has formed an important part of applied mathematics. In particular, it was gradually realized that the theory provided a unified view of many problems in physics.

   The calculus of variations can be developed as a special case of optimal control theory. Because of the difficulties involved in understanding the modern and more general theory, however, we feel that it is preferable to begin by presenting some important aspects of the classical calculus of variations. In fact, a number of important dynamic optimization problems in economics can be analysed in terms of the classical theory.

1. The simplest problem

We shall begin by introducing a problem from optimal growth theory that will serve as an example throughout this chapter.

Example 1   (How much should a nation save?)   Consider an economy developing over time where $K = K(t)$ denotes the capital stock, $C = C(t)$ consumption and $Y = Y(t)$ net national product at time t. Suppose that

$$Y = f(K) \quad \text{where } f'(K) > 0, \quad f''(K) \leqslant 0, \tag{1}$$

so that the net national product is a strictly increasing, concave function of the capital stock alone. For each t we have the relation

$$C(t) = f(K(t)) - \dot{K}(t), \tag{2}$$

which means that production, $f(K(t))$, is divided between consumption, $C(t)$, and investment, $\dot{K}(t)$. Moreover, let $K(0)=K_0$ be the historically given capital stock existing "today", $t=0$, and suppose that we are considering a fixed planning period $[0,T]$. Now, for each choice of investment function $\dot{K}(t)$ on $[0,T]$, the capital function is fully determined by $K(t)=K_0+\int_0^t \dot{K}(\tau)d\tau$ and (2) in turn determines $C(t)$. The question arises as to how the planning authorities are to choose the investment function. High consumption today is in itself preferable, but it leads (via (2)) to a low rate of investment which in turn results in a lower capital stock in the future thus reducing the possibilities for future consumption. One must somehow find a way to reconcile the conflict between providing for the present and taking care of the future. To this end, let us assume that the society has a utility function U, where $U(C)$ is the utility (flow) the country enjoys when the total consumption is C, and let us require that

$$U'(C)>0, \qquad U''(C)<0, \tag{3}$$

so that U is strictly increasing and strictly concave. (This assumption implies, in particular, that people at a high level of consumption derive less increase in satisfaction from a given increase in consumption than do people at a low level of consumption.)

Let us now introduce moreover a discount factor ρ and assume that the investment criterion is as follows: Choose $\dot{K}(t)$ for $t\in[0,T]$ such that the total discounted utility for the country in the period $[0,T]$ is as great as possible. Another way of formulating the problem is as follows: *Find the capital function $K=K(t)$, with $K(0)=K_0$, that maximizes*

$$\int_0^T U(C(t))e^{-\rho t}\,dt = \int_0^T U(f(K(t))-\dot{K}(t))e^{-\rho t}dt. \tag{4}$$

In order for the problem to have a solution some "terminal condition" on $K(T)$ is necessary, for example that $K(T)=K_T$ is given. (The study of this model was initiated by F. Ramsey (1928).)

The problem in example 1 is a special case of a problem that briefly formulated is the following (*usually called the simplest problem in the calculus of variations*)

$$\max \int_{t_0}^{t_1} F(t,x,\dot{x})dt \qquad \text{when } x(t_0)=x_0, x(t_1)=x_1, \tag{5}$$

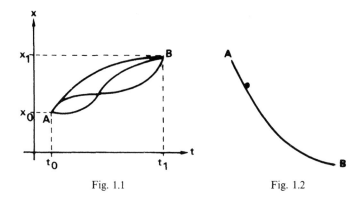

Fig. 1.1   Fig. 1.2

where F is a given "well-behaved" function of three variables and t_0, t_1, x_0 and x_1 are given numbers. More precisely: *Among all well-behaved functions $x(t)$ which satisfy $x(t_0)=x_0$ and $x(t_1)=x_1$, find one (if there exists any) such that the integral in (5) is as large as possible.* Geometrically the problem can be illustrated in this way, see fig. 1.1: The points $A=(t_0,x_0)$ and $B=(t_1,x_1)$ are given in the tx-plane. For each smooth curve that joins the points A and B, the integral in (5) has a definite value. Find the curve which makes the integral as large as possible.

So far we have asked for the maximization of the integral in (5). Since minimizing the integral of $F(t, x, \dot{x})$ leads to the same result as maximizing the integral of $-F(t, x, \dot{x})$, the relationship between the two problems is easily worked out, and conditions characterizing the maximization case is directly transferable to conditions characterizing the minimization case.

Let us briefly mention the problem that initiated the calculus of variations, the "*brachistochrone problem*". ("brachistochrone" means in Greek: shortest time). It is this: Given two points A and B in a vertical plane. The time required for a particle to slide along a curve from A to B under the influence solely of gravity will depend on the shape of the curve. Find that curve along which the particle uses the minimal time from A to B.

Ones first reaction to the problem might be that it is easy, the straight line joining A and B gives the solution. This is not correct. (However, the straight line between A and B solves another variational problem: Find the shortest curve joining A and B. See exercise 1.4.5.) By an ingenious method the Swiss mathematician J. Bernoulli solved the problem in 1696. The required curve is called the cycloid. Using elementary physics one can show that this problem reduces to that of minimizing an integral of the type appearing in (5).

Exercise 1.1.1 (a) The graphs of all the following functions pass through the points $(0,0)$ and $(1, e^2 - 1)$ in the tx- plane:

(i) $x = (e^2 - 1)t$,

(ii) $x = (e^2 - 1) \sin(\frac{1}{2}\pi t)$,

(iii) $x = e^{1+t} - e^{1-t}$,

(iv) $x = at^2 + (e^2 - 1 - a)t$.

Calculate in each case the value of $J(x) = \int_0^1 (x^2 + \dot{x}^2) dt$.

(b) Prove that the problem of maximizing $J(x)$ among all curves joining $(0,0)$ and $(1, e^2 - 1)$, has no solution. (Hint: What happens to $J(x)$ when x is given in (iv) above and $a \to \infty$?)

2. The Euler equation

In this section we consider the problem briefly formulated in (5) and we prove that a solution of it must satisfy a certain differential equation called the Euler equation.

Of course, some restrictions must be placed on the functions involved. We shall assume that the fixed function F is a C^2-function. It is more problematical to decide the restrictions to be placed on $x(t)$, since it is the variable in the problem. However, *some* conditions on $x(t)$ are necessary in order for the problem in (5) to make sense at all. A natural requirement is to assume that $x(t)$ is C^1, but to facilitate the proof we start by requiring more regularity. Let us call a C^2-function $x = x(t)$ that satisfies the boundary conditions $x(t_0) = x_0$, $x(t_1) = x_1$, *an admissible function* in problem (5). For each admissible function $x = x(t)$, $J(x)$ given by

$$J(x) = \int_{t_0}^{t_1} F(t, x(t), \dot{x}(t)) dt \qquad (6)$$

has a definite value. An admissible function $x^* = x^*(t)$ such that $J(x^*) \geqslant J(x)$ for all admissible functions $x = x(t)$, is called *an optimal solution*, or simply a *solution*, to problem (5).

Recall that a necessary condition for a differentiable function $f(x_1, \ldots, x_n)$ to have an extremum at an interior point in its domain of definition is that all the partial derivatives $\partial f / \partial x_i$ are equal to zero at the point. In deriving this result, we remember that we considered the behaviour of f only in a small neighbourhood of the point, and, in fact, it was sufficient to study the behaviour of f in the directions of the coordinate axes. The situation is analogous in the present case. In the calculus of variations, the Euler equation plays a role similar to the first-order condition in the static case. Moreover, in the proof of the Euler equation it suffices to compare the optimal solution of (5) with a very restricted class of admissible functions. With this in mind, let us begin the proof.

Suppose $x^* = x^*(t)$ *is* an optimal solution of the problem, and let $\mu(t)$ be any C^2-function which satisfies $\mu(t_0) = \mu(t_1) = 0$. For each real number α, let us define a new function $x(t)$ by

$$x(t) = x^*(t) + \alpha\mu(t).$$

(See fig. 1.3.) Note that if α is small, the function $x(t)$ is "near" the function $x^*(t)$. Clearly, $x(t)$ is admissible for all α, and since x^* is optimal, $J(x^*) \geqslant J(x^* + \alpha\mu)$ for all α. If the function $\mu(t)$ is kept fixed, $J(x^* + \alpha\mu)$

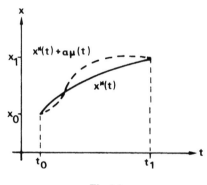

Fig. 1.3

becomes a function of α alone. Putting $I(\alpha) = J(x^* + \alpha\mu)$, we have

$$I(\alpha) = \int_{t_0}^{t_1} F(t, x^*(t) + \alpha\mu(t), \dot{x}^*(t) + \alpha\dot{\mu}(t)) \mathrm{d}t. \tag{7}$$

Here $I(0) = J(x^*)$, so $I(\alpha) \leqslant I(0)$ for all α. Hence the function I has a maximum at $\alpha = 0$. Since I is a differentiable function of α (see below), $I'(0) = 0$. This is the condition we use to deduce the required (Euler) differential equation, and it motivates our choice of comparison functions. Now, looking at (7), we see that to calculate $I'(0)$ we must differentiate the integral with respect to a parameter appearing in the integrand. The result is as follows[1]

$$I'(0) = \int_{t_0}^{t_1} [F'_2(t, x^*(t), \dot{x}^*(t)) \cdot \mu(t) + F'_3(t, x^*(t), \dot{x}^*(t)) \cdot \dot{\mu}(t)] dt,$$

or more briefly:

$$I'(0) = \int_{t_0}^{t_1} \left[\frac{\partial F}{\partial x} \mu(t) + \frac{\partial F}{\partial \dot{x}} \dot{\mu}(t) \right] dt. \tag{8}$$

We now integrate the second term by parts:

$$\int_{t_0}^{t_1} \frac{\partial F}{\partial \dot{x}} \dot{\mu}(t) dt = \left| _{t_0}^{t_1} \left(\frac{\partial F}{\partial \dot{x}} \right) \mu(t) - \int_{t_0}^{t_1} \frac{d}{dt} \left(\frac{\partial F}{\partial \dot{x}} \right) \mu(t) dt.$$

By making use of this result in (8) and rearranging, we get:

$$I'(0) = \int_{t_0}^{t_1} \left[\frac{\partial F}{\partial x} - \frac{d}{dt} \left(\frac{\partial F}{\partial \dot{x}} \right) \right] \mu(t) dt$$

$$+ \left(\frac{\partial F}{\partial \dot{x}} \right)_{t=t_1} \mu(t_1) - \left(\frac{\partial F}{\partial \dot{x}} \right)_{t=t_0} \mu(t_0). \tag{9}$$

Here $\mu(t_0) = \mu(t_1) = 0$, so the condition $I'(0) = 0$ reduces to

$$\int_{t_0}^{t_1} \left[\frac{\partial F}{\partial x} - \frac{d}{dt} \left(\frac{\partial F}{\partial \dot{x}} \right) \right] \mu(t) dt = 0. \tag{10}$$

In the argument leading to this result, $\mu(t)$ was a fixed function. However, (10) holds for *all* functions $\mu(t)$ which are C^2 on $[t_0, t_1]$ and which are 0 at t_0 and t_1. But then, as we shall prove at the end of this section, the term in brackets must vanish for all $t \in [t_0, t_1]$:

$$\frac{\partial F}{\partial x} - \frac{d}{dt} \left(\frac{\partial F}{\partial \dot{x}} \right) = 0. \tag{11}$$

[1]See e.g. Sydsæter (1981), example 4 p. 174.

This is *the Euler equation* discovered in 1744 by the Swiss mathematician Euler. We have proved that in order for $x^* = x^*(t)$ to solve the problem (5), $x^*(t)$ must satisfy (11). By a similar argument (or by noting that (11) is unchanged if F is changed to $-F$), we also find that (11) represents a necessary condition for the corresponding minimization problem.

If we look closely at the argument above, we realize that the admissible functions were required to be C^2-functions in order to ensure that (d/dt) $(\partial F/\partial \dot{x})$ exists and is continuous, thus allowing integration by parts. It turns out that by using a more involved argument,[2] one can show that equation (11) holds good even if we assume only that the admissible functions are C^1. The existence of $(d/dt)(\partial F/\partial \dot{x})$ is proved to be a consequence of the other conditions on the problem. All in all, we have the following result:

Theorem 1 Suppose F is a C^2-function of three variables. A necessary condition for $x^* = x^*(t)$ to maximize or minimize

$$J(x) = \int_{t_0}^{t_1} F(t, x, \dot{x}) dt,$$

among all C^1-functions $x(t)$ which satisfy the boundary conditions

$$x(t_0) = x_0, \qquad x(t_1) = x_1, \qquad (x_0, x_1 \text{ given numbers})$$

is that $x^*(t)$ is a solution to the differential equation

$$\frac{\partial F(t, x, \dot{x})}{\partial x} - \frac{d}{dt} \left(\frac{\partial F(t, x, \dot{x})}{\partial \dot{x}} \right) = 0. \quad \blacksquare \tag{12}$$

In equation (12) note that $(d/dt)(\partial F(t, x, \dot{x})/\partial \dot{x})$ denotes the total derivative of $\partial F/\partial \dot{x}$ with respect to t, where we must take into account that $\partial F/\partial \dot{x}$ is a function of three variables all depending on t. If we assume that $x = x(t)$ is C^2, we obtain (see exercise 1.2.4),

$$\frac{d}{dt} \left(\frac{\partial F(t, x, \dot{x})}{\partial \dot{x}} \right) = \frac{\partial^2 F}{\partial t \partial \dot{x}} + \frac{\partial^2 F}{\partial x \partial \dot{x}} \cdot \dot{x} + \frac{\partial^2 F}{\partial \dot{x} \partial \dot{x}} \cdot \ddot{x}. \tag{13}$$

By inserting this expression into (12) and rearranging, we obtain an alternative form of the Euler equation:

$$\frac{\partial^2 F}{\partial \dot{x} \partial \dot{x}} \cdot \ddot{x} + \frac{\partial^2 F}{\partial x \partial \dot{x}} \cdot \dot{x} + \frac{\partial^2 F}{\partial t \partial \dot{x}} - \frac{\partial F}{\partial x} = 0. \tag{14}$$

---

[2] See e.g. Gelfand and Fomin (1969).

Note that $\partial^2 F/\partial t \partial \dot{x}$ is here the same as $F''_{31}(t, x, \dot{x})$, entirely different from $(d/dt)(\partial F/\partial \dot{x})$. Now, (14) is an ordinary second-order differential equation (if $F''_{\dot{x}\dot{x}} \neq 0$). The solution of it will therefore, in general, depend on two constants. These constants are usually determined by the boundary conditions $x(t_0) = x_0, x(t_1) = x_1$.

Let us look at two examples to illustrate the theory.

Example 2   Find possible solutions of the Euler equation when

$$J(x) = \int_0^1 (x^2 + \dot{x}^2) dt, \qquad x(0) = 0, \qquad x(1) = e^2 - 1.$$

Here $F(t, x, \dot{x}) = x^2 + \dot{x}^2$, so $F'_x = 2x$, $F'_{\dot{x}} = 2\dot{x}$ and hence $(d/dt)(F'_{\dot{x}}) = 2\ddot{x}$. (Note that $F''_{\dot{x}t} = 0$.) The Euler equation (12) is therefore in this case $2x - 2\ddot{x} = 0$, or

$$\ddot{x} - x = 0.$$

(Check that you obtain the same result by using (14).) We have here arrived at a linear second-order differential equation with constant coefficients, and its solution is $x(t) = Ae^t + Be^{-t}$, where A and B are arbitrary constants. (See e.g. Sydsæter (1981), Chapter 6, in particular, section 2.) Since we want to have $x(0) = 0$, $x(1) = e^2 - 1$, we get the following equations for determining A and B: $0 = A + B$ and $e^2 - 1 = Ae + Be^{-1}$. It follows that $A = e$, $B = -e$, so that

$$x = x(t) = e^{1+t} - e^{1-t} \tag{15}$$

is the only solution of the Euler equation that satisfies the given boundary conditions. On the basis of theorem 1, what can be said about the function given in (15)? We know that *if there exists* an admissible function that maximizes or minimizes $J(x)$, then it must be the function given in (15). No other function can do it. We shall see later that this function actually solves the minimization problem. (See Example 12.) The corresponding maximization problem does not have any solution. This fact was established in exercise 1.1.1(b).

Example 3   Consider again the problem in example 1:

$$\max \int_0^T U(f(K(t)) - \dot{K}(t))e^{-\rho t} dt \tag{16a}$$

$$K(0) = K_0, \qquad K(T) = K_T. \tag{16b}$$

If we put $F(t, K, \dot{K}) = U(C)e^{-\rho t} = U(f(K) - \dot{K})e^{-\rho t}$, then

$$\frac{\partial F}{\partial K} = U'(C)f'(K)e^{-\rho t}, \qquad \frac{\partial F}{\partial \dot{K}} = U'(C)(-1)e^{-\rho t},$$

and the Euler equation is $\partial F/\partial K - (d/dt)(\partial F/\partial \dot{K}) = 0$, i.e.

$$U'(C)f'(K)e^{-\rho t} + \frac{d}{dt}(U'(C)e^{-\rho t}) = 0. \tag{17}$$

Here

$$\frac{d}{dt}(U'(C)e^{-\rho t}) = \left[\frac{d}{dt}U'(C)\right]e^{-\rho t} + U'(C)(-\rho)e^{-\rho t} \tag{18}$$

and

$$\frac{d}{dt}U'(C) = U''(C)\dot{C} = U''(C)(f'(K)\dot{K} - \ddot{K}). \tag{19}$$

From (17)–(19) we obtain by some algebraic manipulations

$$\ddot{K} - f'(K)\dot{K} + \frac{U'(C)}{U''(C)}(\rho - f'(K)) = 0. \tag{20}$$

In general, this is a quite complicated second-order differential equation, and explicit solutions cannot be obtained except in special cases. (Some such cases will be considered later. See e.g. exercise 1.2.5.)

Since $\dot{C} = f'(K)\dot{K} - \dot{K}$, we see from (20) that if $K = K(t)$ is a solution to problem (16), then the relative growth rate of C is (for $C \neq 0$)

$$\frac{\dot{C}}{C} = \frac{U'(C)}{CU''(C)}(\rho - f'(K)) = \frac{\rho - f'(K)}{\breve{\omega}}, \tag{21}$$

where $\breve{\omega} = El_C U'(C) = (C/U'(C))U''(C)$ is the elasticity of the marginal utility of consumption. Since we have assumed that $U' > 0$ and $U''(C) < 0$, we see that $\breve{\omega} < 0$. It follows that the relative growth rate of consumption is positive if and only if $f'(K)$ (the marginal productivity of capital) exceeds ρ (the discount factor). Note that $\dot{K}(0)$ is determined by the optimization, and hence so is $C(0) = f(K_0) - \dot{K}(0)$. In fact, there is in this problem a "trade off" between the value of $C(0)$ and the level of $\dot{C}(t)/C(t)$. If we are willing to settle for a low $C(0)$, then the relative growth rate of consumption might become high. On the other hand, an initial high value of $C(0)$ implies lower levels for $\dot{C}(t)/C(t)$.

Note 1 In example 3, and in some of the other examples and exercises in this chapter, we deliberately refrain from complete mathematical rigour. Firstly, the existence of a solution to (20) on $[0, T]$ is not shown. Secondly, a meaningful solution of problem (16) requires $C(t) > 0$. Thirdly, we have not specified the domain of U, frequently taken to be $(0, \infty)$ or $[0, \infty)$. In example 6.5 a precise set of conditions on a model closely related to that in example 3 is given, which secures the existence of an optimal solution.

The fundamental lemma in the calculus of variations.

For those who are interested, we now present a proof of the fact that if (10) is valid for all admissible $\mu(t)$, then (11) is valid. We need the following result:

The fundamental lemma   Suppose f is continuous on $[t_0, t_1]$ and suppose $\int_{t_0}^{t_1} f(t)\mu(t)\mathrm{d}t = 0$ for each C^2-function $\mu(t)$ which satisfies $\mu(t_0) = \mu(t_1) = 0$. Then $f(t) \equiv 0$ in $[t_0, t_1]$.

Proof.   We prove first that $f(t) \leqslant 0$ for all $t \in (t_0, t_1)$. Suppose on the contrary that $f(s) > 0$ for some $s \in (t_0, t_1)$. Then, by continuity of f, there must exist an interval (α, β) around s with $(\alpha, \beta) \subset (t_0, t_1)$ such that $f(t) > 0$ for all $t \in (\alpha, \beta)$. Define for each $t \in [t_0, t_1]$, $\mu(t)$ by (see fig. 1.4):

$$\mu(t) = \begin{cases} 0 & t \notin (\alpha, \beta) \\ (t-\alpha)^3(\beta-t)^3 & t \in (\alpha, \beta) \end{cases}.$$

Clearly, $\mu(t_0) = \mu(t_1) = 0$ and $\mu(t)$ is C^2 everywhere (at α and β one must use the definition of the derivative to prove this result). Now, $f(t)\mu(t)$ is strictly positive in (α, β), while $f(t)\mu(t) = 0$ outside of (α, β). Hence $\int_{t_0}^{t_1} f(t)\mu(t)\mathrm{d}t =$

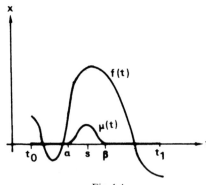

Fig. 1.4

$\int_\alpha^\beta f(t)\mu(t)dt > 0$. But this is a contradiction of the assumptions. In a similar way we can show that $f(t) \geqslant 0$ in (t_0, t_1). Consequently $f(t) = 0$ for all $t \in (t_0, t_1)$. By the continuity of f, $f(t_0)$ and $f(t_1)$ must be 0 as well.

In order to apply this result to the situation in (10) above, put $f(t) = F'_x - (d/dt)(F'_{\dot{x}})$. By our assumptions on F and $x = x(t)$, f is a continuous function on $[t_0, t_1]$, and by the lemma just proved, $f(t) = 0$ for all $t \in [t_0, t_1]$, so (11) follows. ∎

Exercise 1.2.1   Let $J(x) = \displaystyle\int_0^1 (t\dot{x} + \dot{x}^2)dt$.

(i)   Find the associated Euler equation and solve it.
(ii) Find the solution satisfying $x(0) = 1$, $x(1) = 0$.
(iii) Find the solution satisfying $x(0) = x_0$, $x(1) = x_1$.
(iv) Let x^* be the solution obtained in (iii) and define $V(x_0, x_1) = \int_0^1 (t\dot{x}^* + \dot{x}^{*2})dt$. ($V$ is called the optimal value function. See section 8.) Find an explicit expression for $V(x_0, x_1)$ and compute $\partial V/\partial x_0$ and $\partial V/\partial x_1$.

Exercise 1.2.2   Let $J(x) = \displaystyle\int_{t_0}^{t_1} (x^2 + \dot{x}^2 + 2xe^t)dt$.

Find the associated Euler equation and solve it.

Exercise 1.2.3   Let $J(x) = \displaystyle\int_1^2 \frac{\dot{x}^2}{t^2}dt$.

(i)   Find the solution of the Euler equation for which $x(1) = 1$, $x(2) = 2$.
(ii) Prove that the problem of maximizing $J(x)$ when $x(1) = 1$, $x(2) = 2$ has no (finite) solution. (Hint: Calculate $J(x)$ when $x = at^2 + (1 - 3a)t + 2a$.)
(iii) Does the result in (ii) imply that the solution in (i) minimizes $J(x)$?

Exercise 1.2.4   With suitable differentiability requirements placed on $x = x(t)$ and G, prove that

$$\frac{d}{dt}G(t, x, \dot{x}) = \frac{\partial G}{\partial t} + \dot{x}\frac{\partial G}{\partial x} + \ddot{x}\frac{\partial G}{\partial \dot{x}}.$$

Put $G(t, x, \dot{x}) = \partial F(t, x, \dot{x})/\partial \dot{x}$ and deduce equation (13).

Exercise 1.2.5 Consider problem (16) in example 3 in the case $U(C) = C^{1-v}/(1-v)$, $v \in (0, 1)$ and $f(K) = bK$, $b > 0$. Assume that $K_0 > 0$ and $K_T > 0$.

(i) Find the Euler equation in this case. Prove that the general solution for $b \neq a$, where $a = (b - \rho)/v$, is:

$$K(t) = Ae^{bt} + Be^{at}$$

(ii) Find the corresponding solution for $C(t)$.
(iii) Find the condition for a positive relative growth rate of consumption.
(iv) Suppose $K_0 e^{bT} > K_T$ and $a < b$. Find the solution through $(0, K_0)$ and (T, K_T), and show that $C(t) = bK(t) - \dot{K}(t) > 0$.

Exercise 1.2.6 In a dynamic model for the firm, H.Y. Wan encounters the problem of finding a function $x = x(t)$ that maximizes

$$\int_0^T [N(\dot{x}(t)) + \dot{x}(t)f(x(t))]e^{-rt}dt$$

where N and f are given functions, r and T are positive constants, and $x(0) = x_0$, $x(T) = x_T$. Prove that if $x = x(t)$ solves the problem,

$$\frac{d}{dt}N'(\dot{x}) = r(N'(\dot{x}) + f(x)).$$

Exercise 1.2.7 Let us generalize the problem in example 1. Let $Y(t) = f(K(t), t)$, and instead of $U(C)e^{-\rho t}$, put $U(C, t)$, so that the problem becomes

$$\max \int_0^T U(f(K, t) - \dot{K}, t)dt, \qquad K(0) = K_0, \qquad K(T) = K_T.$$

How does the Euler equation look like in this case? Find an expression for $\dot{C}/C$ corresponding to (21).

Exercise 1.2.8 A monopolist offers a certain commodity for sale on a market. If the production per unit of time is x, let $b(x)$ be the associated total cost. Suppose that the demand at time t for this commodity depends not only on the price $p(t)$, but also on $\dot{p}(t)$, so that the demand at time t is $D(p(t), \dot{p}(t))$, where D is a given function. If production is adjusted to the demand at each t, the total profit for the monopolist in the time interval $[0, T]$ is given by

$$\int_0^T [D(p, \dot{p})p - b(D(p, \dot{p}))]dt.$$

We suppose that $p(0)$ is given and that some condition on $p(T)$ is imposed. The problem for the monopolist is to find the price function $p(t)$ maximizing his profit.

(i)   Derive the Euler equation for this problem.
(ii) Put $b(x) = \alpha x^2 + \beta x + \gamma$, $x = D(p, \dot{p}) = Ap + B\dot{p} + C$ where α, β, γ, B, and C are positive constants, while A is negative. Find the Euler equation and its solution in this case.

3. Comments on the Euler equation; Legendre's necessary condition

A variational problem of the type considered in the previous section does not always have a solution. This fact was illustrated in example 2, where no solution to the maximization problem exists. It turns out to be difficult to give conditions that will determine whether a given variational problem has a solution. In section 9 we shall give one existence theorem. Often, however, we find that the geometric, physical or economic interpretation of the problem will strongly suggest whether a solution exists or not. If it *is* proved that the problem in (5) has a solution, then we know that it must satisfy the Euler equation. When there is only one function satisfying the Euler equation and the boundary conditions, it is *the* solution to the problem, provided it has a solution.

Concerning the problem of the existence of a function satisfying the Euler equation and the boundary conditions, we must note that we cannot appeal to the usual existence theorems for differential equations. In the present case we are interested in a solution to the differential equation defined on the whole of $[t_0, t_1]$ satisfying the boundary conditions $x(t_0) = x_0$ and $x(t_1) = x_1$. This is another type of boundary conditions than those involved in the usual existence theorem for $\ddot{x} = f(t, x, \dot{x})$ which guarantees only the existence of a solution when $x(t_0)$ and $\dot{x}(t_0)$ have given values. (Our present problem is in the theory of differential equations often called *a two-point boundary value problem.*)

In economic applications we can often bypass the problem of existence by relying on a theorem which tells us that if $F(t, x, \dot{x})$ is concave in x and $\dot{x}$, then a solution of the Euler equation satisfying the boundary conditions is necessarily optimal. (See section 6.)

We have formulated the variational problem as the one of finding the largest (or smallest) value of $J(x)$ when we search among *all* admissible functions $x = x(t)$. This *global* optimization problem is the one which is of primary interest in applications. However, if we return to the argument leading to the Euler equation, we see that it was a consequence of the condition $I'(0) = 0$ alone. But this is the condition for $I(\alpha)$ (given by (7)) to have a *local* extreme point at $\alpha = 0$. Hence we arrive at the Euler equation just by comparing the value of $J(x)$ at the optimal x^* with the value of $J(x)$ at $x^* + \alpha\mu$ where α is close to 0. If α is small, the function $x^* + \alpha\mu$ is "close to" the function x^*. The Euler equation therefore also represents a nec-essary condition for "local" optimality of $J(x)$. In many treatises of the calculus of variations properties of such "local" optima are studied exten-sively, for instance in Elsgolc (1962) and Gelfand and Fomin (1969).

It is possible to derive other necessary conditions for the solution to the problem in (5) in addition to the Euler equation. In order for $I(\alpha)$ given by (7) to have a maximum at $\alpha = 0$, it is necessary that $I''(0) \leqslant 0$, while $I''(0) \geqslant 0$ is necessary for a minimum. By using these facts, one can prove the following result (see also exercise 1.10.3):

Theorem 2 **(The Legendre's necessary condition)** Suppose that the con-ditions in theorem 1 are satisfied. A necessary condition for $J(x)$ given by (6) to have a maximum at $x^* = x^*(t)$ is that

$$F''_{\dot{x}\dot{x}}(t, x^*(t), \dot{x}^*(t)) \leqslant 0 \qquad \text{for all } t \in [t_0, t_1]. \tag{22}$$

The corresponding condition for a minimum is obtained by reversing the inequality in (22). ∎

The condition $I''(0) < 0$ (together with $I'(0) = 0$) is sufficient for $I(\alpha)$ to have a local maximum at $\alpha = 0$. Note, however, that it is *not* so that the corresponding condition $F''_{\dot{x}\dot{x}} < 0$ gives a sufficient condition for "local" solution to the corresponding variational problem.

Example 4 In the problem in example 2, $F(t, x, \dot{x}) = x^2 + \dot{x}^2$ so that $F''_{\dot{x}\dot{x}} = 2$. It follows from theorem 2 that the solution to the Euler equation we obtained cannot solve the maximization problem. This confirms our previous conclusion. (See the end of example 2.)

Example 5 For the problem in example 3, $F(t, K, \dot{K}) = U(C)e^{-\rho t}$, with $C = f(K) - \dot{K}$, so $F''_{\dot{K}\dot{K}} = U''(C)e^{-\rho t}$. Since we assumed $U''(C) < 0, F''_{\dot{K}\dot{K}} < 0$, and Legendre's necessary condition for a maximum is consequently satisfied in this case.

Constraints on the solution path

A number of dynamic optimization problems in economics are of the following type:

$$\max \int_{t_0}^{t_1} F(t, x, \dot{x}) dt$$

when

$$x(t_0) = x_0, \qquad x(t_1) = x_1, \qquad h(t, x(t), \dot{x}(t)) > 0. \tag{23}$$

If we compare with problem (5), the new feature is that the admissible function $x(t)$ for all $t \in [t_0, t_1]$ has to satisfy the inequality $h(t, x(t), \dot{x}(t)) > 0$, where h is a given C^1-function of three variables. (Problems more general than (23) will be studied extensively in Chapter 4. If we put $\dot{x} = u$, (23) is an optimal control problem.) The following result is quite obvious:

A C^1-function $x^*(t)$ which solves problem (23) must be a solution to the Euler equation.

In fact, from the continuity assumptions we see that $h(t, x(t), \dot{x}(t)) > 0$ for all $x(t)$ which are sufficiently close to $x^*(t)$, and the result follows because the Euler equation was derived by comparing $x^*(t)$ only with functions $x(t)$ close to $x^*(t)$.

Note that in problem (23) the function F needs only be defined and C^2 for triples $(t, x, \dot{x})$ which satisfy $h(t, x, \dot{x}) > 0$. This is an important fact which is utilized in many problems. For instance, in example 1 we did not discuss the domain of U, but very often (see for instance exercise 1.2.5) the U functions which are used are C^2 only for $C > 0$.

Exercise 1.3.1 Test the Legendre necessary condition on the problem in exercises 1.2.1, 1.2.2, and 1.2.3.

Exercise 1.3.2   Check whether the Legendre necessary condition is satisfied in the problem in exercise 1.2.8(ii).

4. Special cases

In general, the Euler equation is a very complicated nonlinear differential equation of the second order. Only in rather special cases can such equations be solved explicitly in the sense that the solution can be obtained in terms of elementary functions and their integrals. We consider now some standard cases in which the Euler equation can be more or less simplified. (F is the integrand in problem (5).)

(a) F does not depend on $\dot{x}$: $F = F(t, x)$

Since $F'_{\dot{x}} = 0$, the Euler equation reduces to the static optimization condition:

$$\frac{\partial F(t, x)}{\partial x} = 0, \qquad t \in [t_0, t_1]. \tag{24}$$

Equation (24) is not a differential equation so the solution does not involve any arbitrary constants. Hence, if we maximize $\int F dt$ subject to boundary conditions on $x(t_0)$ and/or $x(t_1)$, then the problem will usually have no solution. Equation (24) is, however, an interesting necessary condition for the problem of maximizing $\int F(t, x) dt$ with no conditions imposed on $x(t_0)$ or $x(t_1)$.

(b) F does not depend on x: $F = F(t, \dot{x})$

In this case $F'_x = 0$, so the Euler equation tells us that the time derivative of $F'_{\dot{x}}(t, \dot{x})$ is 0 for all $t \in [t_0, t_1]$. Hence

$$\frac{\partial F(t, \dot{x})}{\partial \dot{x}} = c \qquad \text{for some constant } c, \ t \in [t_0, t_1]. \tag{25}$$

This is a first-order differential equation. In principle, we solve it for $\dot{x}$ and integrate w.r.t. t to obtain the required function. (Examples of this case were given in exercises 1.2.1 and 1.2.3.)

(c) *F does not depend (explicitly) on t: $F = F(x, \dot{x})$.*

Variational problems of this type are frequently encountered in economics. (A case in point is our growth theory example (example 1), provided the discount rate ρ is 0.)

Using a little trick, we can simplify the Euler equation in this case. The trick consists of calculating the total derivative of $F(x, \dot{x}) - \dot{x}F'_{\dot{x}}(x, \dot{x})$ w.r.t. t (taking into account that x and $\dot{x}$ both depend on t) and see what happens. We get:

$$\frac{d}{dt}[F(x, \dot{x}) - \dot{x}F'_{\dot{x}}(x, \dot{x})] = F'_x \dot{x} + F'_{\dot{x}}\ddot{x} - \ddot{x}F'_{\dot{x}} - \dot{x}\frac{d}{dt}F'_{\dot{x}}$$

$$= \dot{x}\left[F'_x - \frac{d}{dt}F'_{\dot{x}}\right]. \tag{26}$$

It follows that for any $t \in [t_0, t_1]$, if either $\dot{x}(t) = 0$ or the Euler equation is satisfied, then the expression in (26) is 0, from which it follows that

$$F - \dot{x}F'_{\dot{x}} = C \qquad \text{for all } t \in [t_0, t_1] \tag{27}$$

for some constant C. Conversely, if (27) is satisfied, then for each $t \in [t_0, t_1]$, either $\dot{x}(t) = 0$ or the Euler equation is satisfied. Briefly formulated:

$$F - \dot{x}F'_{\dot{x}} = C \Leftrightarrow \dot{x} = 0 \qquad \text{or} \qquad F'_x - \frac{d}{dt}F'_{\dot{x}} = 0.$$

(27) is a first-order differential equation, and therefore, in general, easier to handle than the Euler equation. Because of the possibility that $\dot{x} = 0$, however, the Euler equation is not equivalent to (27). Every solution of the Euler equation is clearly a solution to (27) for some constant C. On the other hand, for each value of C, any solution to (27) that on no interval is constant, is a solution to the Euler equation.

Actually it is easy to prove that the Euler equation associated with $F = F(x, \dot{x})$ has a solution $x = k$ on an interval, only if $F'_x(x, \dot{x}) = 0$ in that interval. Hence, if $F'_x(x, \dot{x}) \neq 0$ in the whole of $[t_0, t_1]$, the Euler equation is equivalent to (27). Equation (27) is in this case often referred to as a *first integral* of the Euler equation.

Example 6 Consider example 1 with $\rho = 0$. Then the integrand is $F(t, K, \dot{K}) = U(f(K) - \dot{K})$, so $F'_{\dot{K}} = -U'(C)$ and equation (27) becomes

$$U(C) + \dot{K}U'(C) = c, \qquad (c \text{ constant}). \tag{28}$$

Since $F'_K = U'(C)f'(K) > 0$ by our assumption on f and U, (28) is equivalent to the Euler equation.

Note 2 In some cases it is easier to work with the original form of the Euler equation, even if (27) is a first integral. The problem in exercise 1.4.4 provides an example.

Exercise 1.4.1 Consider the problem

$$\min \int_0^1 (t+x)^4 \, dt, \qquad x(0)=0, \qquad x(1)=a.$$

Find the solution of the Euler equation for this problem and determine the value of a for which the solution is admissible. For that value of a, find the solution of the problem.

Exercise 1.4.2 Find the Euler equation and its solution where

(i) $J(x) = \int_{t_0}^{t_1} (t + tx + x^2) dt,$ (ii) $J(x) = \int_{t_0}^{t_1} \frac{\dot{x}^2}{t^3} dt,$

(iii) $J(x) = \int_{t_0}^{t_1} (x^2 + \dot{x}^2 - 2x \sin t) dt.$

Exercise 1.4.3 In a macroeconomic model by T. Haavelmo the following problem arises

$$\max_{K(t)} \int_{t_0}^{t_0+\theta} e^{-\rho(t-t_0)} [p\phi(K(t)) - q(\rho + \gamma)K(t)] dt,$$

where t_0, p, q, θ, ρ and γ are constants. Find the associated Euler equation.

Exercise 1.4.4 R.M. Goodwin encounters the problem of maximizing

$$\int_0^f \ln \left(\frac{y - \sigma \dot{y}}{l(t)} - \bar{z} \right) dt$$

where f, σ and $\bar{z}$ are constants, $l(t)$ is a positive, given function and $y = y(t)$ is the unknown function.

(i) Find the Euler equation.

(ii) Put $l(t) = l_0 e^{\alpha t}$ and solve the Euler equation if $\alpha\sigma \neq 1$.

Exercise 1.4.5 The length of the graph of a C^1-function $x = x(t)$ which connects the two points (t_0, x_0) and (t_1, x_1) is given by

$$L = \int_{t_0}^{t_1} (1 + \dot{x}^2)^{1/2}\, dt.$$

Prove that in order to minimize L, the graph of $x = x(t)$ must be a straight line.

Exercise 1.4.6 In a model by R. Solow on "optimal land use in a long, narrow city", the problem arises to find a function $y(x)$ that maximizes

$$J(y) = \left(\frac{2g}{A}\right)^{k+1} b \int_0^L \frac{(y(x))^{k+1}(A - y(x))^{k+1}}{(w - y'(x))^k}\, dx,$$

where g, A, k, b, L, and w are constants. Find a first integral of the Euler equation in this case.

Exercise 1.4.7 Suppose the integrand F in (5) is linear in $\dot{x}$: $F = f(t, x) + g(t, x)\dot{x}$. Prove that the Euler equation reduces to the static condition:

$$\frac{\partial f(t, x)}{\partial x} = \frac{\partial g(t, x)}{\partial t}. \tag{29}$$

(A solution $x(t)$ to (29) is called a *singular solution*. It will contain no constants of integration, so it will usually not satisfy the boundary conditions.)

5. Different types of terminal conditions

So far we have been studying variational problems of the form (5) in which the admissible functions were required to satisfy the boundary conditions $x(t_0) = x_0$, $x(t_1) = x_1$. This problem is one with *fixed* initial and terminal endpoints. In economic applications the initial conditions are usually historically given, but in many models it is natural to consider other types of terminal conditions. Here we shall examine three different types of them. Briefly formulated, our problem is as follows:

$$\max \int_{t_0}^{t_1} F(t, x, \dot{x})dt, \qquad x(t_0) = x_0, \tag{30}$$

when *one* of the following terminal conditions are satisfied,

$$x(t_1) \qquad\qquad \text{free } (t_1 \text{ fixed}) \tag{31a}$$

$$x(t_1) \geqslant x_1 \qquad\qquad (x_1 \text{ given number, } t_1 \text{ fixed}) \tag{31b}$$

$$x(t_1) = g(t_1), \qquad t_1 \text{ free, } g \text{ a given } C^1\text{-function.} \tag{31c}$$

When we study problem (30) with terminal condition, (31a), we seek a function $x^* = x^*(t)$ that maximizes the integral for those admissible functions which satisfy $x(t_0) = x_0$, and which have no restriction placed on $x(t)$ at $t = t_1$. We simply do not care what value $x(t)$ has at $t = t_1$.

Boundary conditions of the type (31b) are often encountered in economic problems. For a function $x(t)$ to be admissible in this case it has to attain at least the value x_1 at $t = t_1$.

In the case of condition (31c), we seek among all curves starting at the point (t_0, x_0) and ending at some point on the graph of $x = g(t)$, a curve which makes the integral as large as possible. This last boundary condition has, for example, been used in optimal growth theory in cases where $x = g(t)$, loosely speaking, represents a desirable growth path for the economy.

In the cases where (31a) or (31b) are the boundary conditions, the terminal time t_1 is fixed and the problems are referred to as *fixed-time problems*. In the corresponding problem with (31c) as the boundary condition, the terminal time t_1 might vary with the particular admissible function in question, and such problems are referred to as *variable-time problems*. Some admissible functions in the three types of problems are indicated in fig. 1.5.

A fundamental observation in each of the three problems is this: Suppose that $x^*(t)$ solves the problem. Then $x^*(t)$ has a definite value $x^*(t_1)$ at $t = t_1$. Now, if we compare the value of the integral in (30) for $x^* = x^*(t)$ with its value only for those admissible functions whose curves join (t_0, x_0) and $(t_1, x^*(t_1))$, then $x^*(t)$ must be optimal among these. *But then the Euler equation has to be satisfied*, by the same argument as in section 2. The

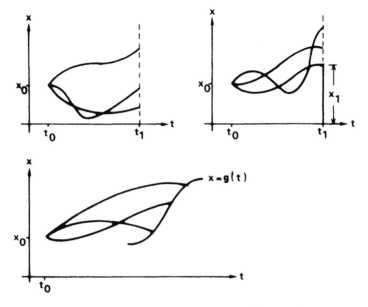

Fig. 1.5 The indicated curves represent some admissible functions in the three cases.

condition $x(t_0) = x_0$ will usually determine one of the constants involved in the general solution of the Euler equation. To determine the other constant, or, to put it another way, to determine where the optimal $x(t_1)$ is to end, we need a condition in each case that is called *a terminal condition* (for the right end) or *a transversality condition*. (The variational problem of finding the shortest path joining a fixed point P and a differentiable surface S, has an obvious solution: The straight line through P perpendicular (transverse) to S. In the calculus of variations and in optimal control theory, this condition of perpendicularity and its generalizations are called *transversality conditions*.)

We show below that the three terminal conditions are

$$\left(\frac{\partial F}{\partial \dot{x}}\right)_{t=t_1} = 0, \tag{32a}$$

$$\left(\frac{\partial F}{\partial \dot{x}}\right)_{t=t_1} \leqslant 0, \qquad (=0 \text{ if } x^*(t_1) > x_1) \tag{32b}$$

$$\left[F + (\dot{g} - \dot{x})\frac{\partial F}{\partial \dot{x}}\right]_{t=t_1} = 0. \tag{32c}$$

An economic interpretation of condition (32a) will be given in Note 13.

If in (31b) we consider the limiting case "$x_1 = -\infty$", then (32b) reduces to (32a), as it should. Moreover, if in (31c) we look at the limiting case in which "$\dot{g} = \infty$", so that the graph of g is vertical ($x = t_1$), then we see that (32c) can only be satisfied provided $(\partial F/\partial \dot{x})_{t=t_1} = 0$, which is condition (32a) again. Now, for those who are interested, let us look at the proofs.

Proof of (32a) We start out as in the proof of the Euler equation in section 2, let $x^*(t)$ be the optimal solution and compare the value of J at $x^*(t)$ with its value at $x^*(t) + \alpha\mu(t)$. But this time we require only $\mu(t_0) = 0$; $\mu(t_1)$ can take any value, since in this case it does not matter which value an admissible function takes at $t = t_1$. Defining $I(\alpha)$ as before by (7), the condition $I'(0) = 0$ must still be valid, so that the expression in (8) must be 0. Since the Euler equation must be satisfied for $x^* = x^*(t)$, and $\mu(t_0) = 0$, the expression in (9) is zero provided

$$\left(\frac{\partial F}{\partial \dot{x}}\right)_{t=t_1} \mu(t_1) = 0.$$

Since this must be valid for all admissible $\mu(t)$, the choice $\mu(t_1) \neq 0$ implies that (32a) must hold.

Proof of (32b) Suppose $x^* = x^*(t)$ is the solution of problem (30) with (31b) as the terminal condition. As usual, we compare the value of the integral in (30) for $x = x^*(t)$ with its value for $x = x^*(t) + \alpha\mu(t)$. In order for $x^* + \alpha\mu$ to be admissible in this problem, firstly $\mu(t_0)$ has to be 0, and secondly, we must have $x^*(t_1) + \alpha\mu(t_1) \geqslant x_1$. There are two cases to consider.

1. $x^*(t_1) > x_1$. If we let $\varepsilon = x^*(t_1) - x_1$, then $\varepsilon > 0$, and if we choose $|\mu(t_1)|$ and $|\alpha|$ so small that $|\mu(t_1)| \cdot |\alpha| < \varepsilon$, then $x^*(t_1) + \alpha\mu(t_1) > x_1$. Now, if we define $I(\alpha)$ as before by (7), $I(\alpha)$ will have a local maximum at $\alpha = 0$, so that $I'(0) = 0$, where $I'(0)$ is given by (9). Since the Euler equation is satisfied for $x^* = x^*(t)$ and $\mu(t_0) = 0$, we obtain as in the proof of (32a), $(\partial F/\partial \dot{x})_{t=t_1} \mu(t_1) = 0$. Since $\mu(t_1)$ can be chosen $\neq 0$, we conclude that $(\partial F/\partial \dot{x})_{t=t_1} = 0$.

2. $x^*(t_1) = x_1$. From the admissibility requirement $x^*(t_1) + \alpha\mu(t_1) \geqslant x_1$ we get in this case $\alpha\mu(t_1) \geqslant 0$. Choose $\mu(t)$ such that $\mu(t_0) = 0$ and $\mu(t_1) > 0$. Then $x^*(t) + \alpha\mu(t)$ is admissible for all $\alpha \geqslant 0$, and hence $I(0) \geqslant I(\alpha)$ for all $\alpha \geqslant 0$. But then $I'(0) \leqslant 0$, so $I'(0) = (\partial F/\partial \dot{x})_{t=t_1} \mu(t_1) \leqslant 0$. Since $\mu(t_1) > 0$, we conclude that $(\partial F/\partial \dot{x})_{t=t_1} \leqslant 0$. All in all, we have proved (32b).

Proof of (32c) (not rigorous) Suppose $x^*(t)$ is an admissible function solving

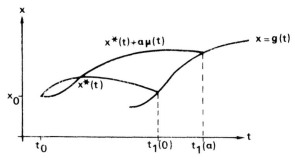

Fig. 1.6

the problem. Let t_1 be the smallest t for which $x^*(t) = g(t)$. (See fig. 1.6.) Let $\mu(t)$ be chosen such that $\mu(t_0) = 0$ and such that $x^*(t) + \alpha\mu(t)$ is an admissible function for all α. The smallest t for which $x^*(t) + \alpha\mu(t) = g(t)$ will, for fixed $\mu(t)$, depend on α. Denote it by $t_1(\alpha)$. (At this point we must think of $x^*(t)$ as a function defined also outside of $[t_0, t_1]$, but this technical point will not be explored here.) In particular, $t_1(0) = t_1$. Let us define

$$I(\alpha) = \int_{t_0}^{t_1(\alpha)} F(t, x^*(t) + \alpha\mu(t), \dot{x}^*(t) + \alpha\dot{\mu}(t)) dt. \tag{33}$$

For a fixed $\mu(t), I(\alpha)$ will have a maximum for $\alpha = 0$, so that $I'(0) = 0$. *Assuming* that $t_1(\alpha)$ is differentiable, we obtain by a familiar rule:[3]

$$I'(\alpha) = \int_{t_0}^{t_1(\alpha)} \left[\frac{\partial F}{\partial x} \mu(t) + \frac{\partial F}{\partial \dot{x}} \dot{\mu}(t) \right] dt + [F]_{t = t_1(\alpha)} \cdot t'_1(\alpha).$$

Here $\partial F/\partial x$, $\partial F/\partial \dot{x}$ and F are all evaluated at the point $(t, x^*(t) + \alpha\mu(t), \dot{x}^*(t) + \alpha\dot{\mu}(t))$. Integrating by parts we get

$$I'(\alpha) = \int_{t_0}^{t_1(\alpha)} \left[\frac{\partial F}{\partial x} - \frac{d}{dt} \frac{\partial F}{\partial \dot{x}} \right] \mu(t) dt + \left| \begin{matrix} t_1(\alpha) \\ \\ t_0 \end{matrix} \right. \left[\frac{\partial F}{\partial \dot{x}} \mu(t) \right] + [F]_{t = t_1(\alpha)} \cdot t'_1(\alpha).$$

By putting $\alpha = 0$ and making use of the fact that the Euler equation is

---

[3]See e.g. Sydsæter (1981), (4.9) p. 175.

satisfied for $x^* = x^*(t)$, we see that $I'(0) = 0$ reduces to

$$\left(\frac{\partial F}{\partial x}\right)_{t = t_1(0)} \mu(t_1(0)) + [F]_{t = t_1(0)} \cdot t_1'(0) = 0. \tag{34}$$

Now $x^*(t_1(\alpha)) + \alpha\mu(t_1(\alpha)) = g(t_1(\alpha))$. Differentiating w.r.t. α we obtain:

$$\dot{x}^*(t_1(\alpha)) \cdot t_1'(\alpha) + \mu(t_1(\alpha)) + \alpha\dot{\mu}(t_1(\alpha)) \cdot t_1'(\alpha) = \dot{g}(t_1(\alpha)) \cdot t_1'(\alpha).$$

$\dot{\alpha} = 0$ gives, in particular, $\dot{x}^*(t_1(0)) \cdot t_1'(0) + \mu(t_1(0)) = \dot{g}(t_1(0)) \cdot t_1'(0)$. Hence

$$[\dot{g}(t_1(0)) - \dot{x}^*(t_1(0))] t_1'(0) = \mu(t_1(0)). \tag{35}$$

Inserting (35) into (34) we get, since $t_1(0) = t_1$,

$$\left[[F]_{t = t_1} + (\dot{g}(t_1) - \dot{x}^*(t_1)) \left(\frac{\partial F}{\partial x}\right)_{t = t_1} \right] t_1'(0) = 0. \tag{36}$$

Now (35) is valid also for $\mu(t_1) \neq 0$, so $t_1'(0) \neq 0$. By cancelling $t_1'(0)$ and simplifying the notation, we see that we have obtained (32c). ∎

In summary we have arrived at the following result:

Necessary conditions for $x(t)$ to solve the problem (30) with (31a), (31b) or (31c) as the terminal conditions are:

(i) $x(t)$ must satisfy the Euler equation,
(ii) $x(t_0) = x_0$,
(iii) $x(t_1) \geqslant x_1$ in case (31b), $x(t_1) = g(t_1)$ in case (31c),
(iv) $x(t)$ must satisfy the transversality conditions (32a), (32b) or (32c), respectively.

Note 3   If we consider the problem of minimizing the integral in (30) with (31a), (31b) or (31c) as the terminal conditions, the corresponding (32a) and (32c) will be unchanged, whereas (32b) takes the form

$$\left(\frac{\partial F}{\partial x}\right)_{t = t_1} \geqslant 0, \qquad (=0 \text{ if } x(t_1) > x_1). \tag{37}$$

Note 4   Although the arguments for (32) given above assume (implicitly) that the admissible functions are C^2-functions, one can prove that they are valid even for C^1-functions.

Let us now look at some examples.

Example 7 Find the solution of the following problem provided it exists:

$$\min \int_0^1 (x^2 + \dot{x}^2)dt, \qquad x(0)=1, \qquad x(1) \text{ free.}$$

According to the theory above a solution to this problem must satisfy (i): the Euler equation, (ii): $x(0)=1$, (iii): $(F'_{\dot{x}})_{t=1}=0$. In example 2 we saw that the Euler equation in this case is $\ddot{x} - x = 0$, with the general solution $x(t) = Ae^t + Be^{-t}$. The condition $x(0)=1$ therefore implies $1 = A + B$. Moreover, $F'_{\dot{x}} = 2\dot{x}$, so condition (iii) reduces to $2\dot{x}(1)=0$. Here $\dot{x}(t) = Ae^t - Be^{-t}$, so $\dot{x}(1)=0$ implies $Ae - Be^{-1} = 0$ or $Ae^2 = B$. We therefore conclude that $A = (e^2 + 1)^{-1}$, $B = e^2(e^2 + 1)^{-1}$. Hence the only possible solution to the given problem is

$$x = Ae^t + Be^{-t} = \frac{1}{e^2 + 1}e^t + \frac{e^2}{e^2 + 1}e^{-t}.$$

(It is an immediate consequence of note 5 in the next section that this function actually solves the given problem.)

Example 8 Find the solution of the following problem provided it exists.

$$\min \int_0^1 (x^2 + \dot{x}^2)dt, \qquad x(0)=0, \qquad x(1) \geqslant 1.$$

A solution must satisfy (i): the Euler equation, (ii): $x(0)=0$, (iii): $x(1) \geqslant 1$, (iv): $(F'_{\dot{x}})_{t=1} \geqslant 0 (=0$ if $x(1)>1)$. Note that in (iv) we have $\geqslant 0$ since we are faced with a minimization problem. (See note 3 above.) As in example 7, the solution to the Euler equation is $x(t) = Ae^t + Be^{-t}$. Since by (ii) $x(0)=0$, we get $0 = A + B$ so $x(t) = A(e^t - e^{-t})$. Condition (iii) therefore implies

$$x(1) = A(e - e^{-1}) \geqslant 1. \tag{38}$$

It follows from (38) that $A > 0$. We see moreover that $\dot{x}(t) = A(e^t + e^{-t})$ and $F'_{\dot{x}} = 2\dot{x}$, so condition (iv) reduces to

$$\dot{x}(1) = A(e + e^{-1}) \geqslant 0, \qquad (=0 \text{ if } x(1)>1).$$

Since $A > 0$, we conclude that $x(1)>1$ is impossible. Hence $x(1)=1$, *i.e.*

$A(e - e^{-1}) = 1$. The only possible solution to the problem is therefore

$$x = A(e^t - e^{-t}) = \frac{e}{e^2 - 1} e^t - \frac{e}{e^2 - 1} e^{-t}.$$

(It is an immediate consequence of note 5 in the next section that this function actually solves the given problem.)

Example 9 Find the solution of the following problem provided it exists.

$$\min \int_0^{t_1} (x^2 + \dot{x}^2)dt, \qquad x(0) = 0, \qquad x(t_1) = g(t_1) = t_1 + 1, \qquad (t_1 > 0).$$

A solution to this problem must satisfy (i): the Euler equation, (ii): $x(0) = 0$, (iii): $x(t_1) = t_1 + 1$, (iv): $[F + (\dot{g} - \dot{x})F'_{\dot{x}}]_{t=t_1} = 0$.

As in Example 8, (i) and (ii) give us $x(t) = A(e^t - e^{-t})$ and $\dot{x}(t) = A(e^t + e^{-t})$. Hence (iii) implies

$$A(e^{t_1} - e^{-t_1}) = t_1 + 1. \tag{39}$$

Since $t_1 > 0$, it follows that $A \neq 0$. Moreover $F'_{\dot{x}} = 2\dot{x}$ and $\dot{g}(t) = 1$, so (iv) takes the form

$$A^2(e^{t_1} - e^{-t_1})^2 + A^2(e^{t_1} + e^{-t_1})^2 + (1 - A(e^{t_1} + e^{-t_1}))2A(e^{t_1} + e^{-t_1}) = 0.$$

It follows that $A = \frac{1}{2}(e^{t_1} + e^{-t_1})$, which inserted into (39) gives the following equation for determining t_1:

$$\frac{1}{2}(e^{t_1} + e^{-t_1})(e^{t_1} - e^{-t_1}) = t_1 + 1 \qquad \text{or} \qquad e^{2t_1} - e^{-2t_1} = 2t_1 + 2.$$

It can be shown that this equation has a single real solution t_1 (which lies between 0 and 1). But then A is determined and the only possible solution to the problem has been found.

Example 10 Consider the following variant of our growth problem

$$\max \int_0^T U(f(K) - \dot{K})e^{-\rho t}dt, \qquad K(0) = K_0, \qquad K(T) \text{ free.}$$

In addition to the Euler equation and $K(0) = K_0$, a solution to the problem must satisfy $(F'_{\dot{K}})_{t=T} = 0$ (since $K(T)$ is free). Here $F'_{\dot{K}} = -U'(C)e^{-\rho t}$ (see example 3), so the transversality condition tells us in this case that $U'(C(T))e^{-\rho T}=0$, i.e. $U'(C(T))=0$. We have assumed that $U'(C)>0$ for all C. *Hence we must conclude that no solution to the problem exists in this case.* (Even if we enlarge the class of admissible functions to those that are piecewise continuously differentiable we get the same conclusion. See example 19.) The result is intuitively understandable. Since we have no concern for future generations, it will pay to decrease K immediately before the end of the period, and there is no limit as to how fast it pays to decrease K.

Another terminal condition

We conclude this section by looking at a type of terminal condition which is a mixture of (31b) and (31c):

$$x(t_1)\geq x_1, \qquad t_1 \text{ free}, \qquad x_1 \text{ fixed}. \tag{40}$$

Suppose $x^*(t)$ solves problem (30) with (40) as the terminal condition, and let t_1^* be the optimal terminal time. Then $x^*(t)$ solves, in particular, problem (30) with $t_1=t_1^*$ and terminal condition $x(t_1^*)\geq x_1$. It follows that $x^*(t)$ has to satisfy the Euler equation and the transversality condition (see (32b)):

$$\left(\frac{\partial F}{\partial \dot{x}}\right)_{t=t_1^*} \leq 0 \ (=0 \text{ if } x^*(t_1^*)>x_1).$$

The function $x^*(t)$ solves also problem (30) with terminal condition $x(t_1)=x^*(t_1^*)$, t_1 free. This is a problem with terminal condition of type (31c) provided we put $g(t)=x^*(t_1^*)$ where $x^*(t_1^*)$ is now a constant. Hence $\dot{g}=0$, and (32c) reduces to the condition

$$(F-\dot{x}F'_{\dot{x}})_{t=t_1^*}=0.$$

We conclude that the correct transversality conditions corresponding to the terminal requirements (40) are

$$(F'_{\dot{x}})_{t=t_1} \leq 0 \ (=0 \text{ if } x^*(t_1)>x_1) \tag{41a}$$

$$(F-\dot{x}F'_{\dot{x}})_{t=t_1}=0. \tag{41b}$$

Example 11 (**Optimal extraction of a natural resource**) Suppose that at time $t=0$ there is a fixed amount $\bar{x}$ of some resource (say oil in a certain oil field) which is extractable. Let $u(t)$ be the rate of extraction. If T is the time at which extraction stops, then

$$\int_0^T u(t)dt \leqslant \bar{x}. \tag{42}$$

We assume that the world market price of the resource at time t is $q(t)$, so that sales revenue per unit of time at time t is $q(t)u(t)$. We assume further that the costs, C, per unit of time depends on $u(t)$ as well as on t, $C = C(u(t), t)$. The rate of profit at time t is then $\pi(u(t), t) = q(t)u(t) - C(u(t), t)$. The total discounted profit over the interval $[0, T]$, when the discount rate is r, is therefore

$$\int_0^T [q(t)u(t) - C(u(t), t)]e^{-rt}dt. \tag{43}$$

We consider the following problem: Find the time T and the rate of extraction $u(t)$ which maximize (43) subject to the constraint (42).

This problem can easily be transformed into a variational problem with terminal condition of the type (40). Define $x(t)$ as the remaining stock at time t,

$$x(t) = \bar{x} - \int_0^t u(\tau)d\tau. \tag{44}$$

Then $\dot{x}(t) = -u(t)$, $x(0) = \bar{x}$, and (42) reduces to the requirement $x(T) \geqslant 0$. Hence our problem is as follows:

$$\max_{x(t), T} \int_0^T [-q(t)\dot{x}(t) - C(-\dot{x}(t), t)]e^{-rt}dt,$$

$$x(0) = \bar{x}, \qquad x(T) \geqslant 0. \tag{45}$$

The integrand does not depend on x and according to (25) the Euler equation reduces to

$$[-q(t) + C'_u(-\dot{x}(t), t)]e^{-rt} = -c \tag{46}$$

for some constant $-c$. The transversality condition (41a) tells us that the left-hand side of (46) is $\leqslant 0$ at $t = T$. Hence $c \geqslant 0$, and (46) can be written

$$q(t) - \frac{\partial C(u(t), t)}{\partial u} = ce^{rt} \qquad (c \text{ non-negative constant}). \tag{47}$$

The left-hand side of this equation is the marginal profit $\partial \pi(u(t), t)/\partial u$. Thus (47) tells us that *in optimum the marginal profit must increase exponentially with a rate equal to the discount factor r*.

We have not used condition (41b). It is in this case

$$[-q(T)\dot{x}(T) - C(-\dot{x}(T), T)]e^{-rT} - \dot{x}(T)[-q(T) + C'_u(-\dot{x}(T), T)]e^{-rT} = 0.$$

Using $\dot{x}(T) = -u(T)$ we obtain:

$$\frac{u(T)}{C(u(T), T)} C'_u(u(T), T) = 1. \tag{48}$$

(48) tells us that in the optimal solution extraction stops at a time at which the elasticity of costs w.r.t. the rate of extraction is unity.

Note that we have no guarantee that the optimal rate of extraction is $\geqslant 0$ for all t. Thus in the present model it might be optimal in some periods to pump oil back into the oil field! In example 2.11 we study the problem above with the requirement $u(t) \geqslant 0$ imposed.

We have taken for granted that there is an optimal solution to the problem. Of course, this has to be shown. (The study of this type of model was initiated by Hotelling (1931).)

Exercise 1.5.1 Consider the problem

$$\min \int_0^1 (t\dot{x} + \dot{x}^2)dt, \qquad x(0) = 1$$

with (i) $x(1)$ free   or   (ii) $x(1) \geqslant 1$.

Find the possible solutions to these two problems (see exercise 1.2.1).

Exercise 1.5.2 Suppose that the problem

$$\min \int_1^{t_1} \frac{\dot{x}^2}{t^2} dt, \qquad x(1) = 0, \qquad x(t_1) = t_1^2$$

has a solution with $t_1 > 1$. Find it.

Exercise 1.5.3   Let $A(t)$ denote the total assets (wealth) of a certain person at time t, w his constant wage rate, and suppose that he can borrow or lend at a constant interest rate r so that his consumption at time t is given by

$$C(t) = rA(t) + w - \dot{A}(t). \tag{49}$$

Suppose that he plans his consumption from now, $t=0$, until his expected death date T so as to maximize

$$\int_0^T U(C(t))e^{-\rho t}\,dt, \tag{50}$$

where U is a utility function, $U' > 0$, $U'' < 0$, and ρ is a discount factor. His present assets $A(0)$ is equal to A_0 and assume that he wants to leave his heirs at least the amount A_T so that $A(T) \geq A_T$.

(i) What does the necessary conditions developed above tell us in this case? Show, in particular, that for $A(T)$ to be optimal one must have $A(T) = A_T$. (Is this intuitively understandable?)
(ii) Let $U(C) = a - e^{-bC}$ where a and b are positive constants. Solve the Euler equation in this case.

Exercise 1.5.4   In a growth model by S. Chakravarty the following problem arises:

$$\max \int_0^{t_1} \frac{1}{1-v}(bK - \dot{K})^{1-v}\,dt, \qquad K(0) = K_0, \qquad K(t_1) = kL_0 e^{nt_1},$$

where $0 < v < 1$ and b, K_0, k, L_0 and n are positive constants.

(i) Find the associated Euler equation and its general solution.
(ii) Assume that $K = K(t)$ solves the problem. Find three equations which determine the two constants in the general solution and t_1.

Exercise 1.5.5   If the boundary condition (31c) above is generalized to be $g(x(t_1), t_1) = 0$, show that the transversality condition (32c) is changed to

$$\left[g'_x F - (g'_t + \dot{x}g'_x)\frac{\partial F}{\partial \dot{x}} \right]_{t=t_1} = 0.$$

(Hint: Suppose $g(x, t) = 0$ defines x as a function of t around t_1 and use (32c).)

6. Sufficient conditions

The Euler equation is a first-order condition for "local" optimum analogous to the condition that the partial derivatives are 0 at a local extreme point in static optimization. In the static case the first-order conditions are sufficient for optimality provided the function is concave. We can prove a similar result for variational problems. Consider the problem

$$\max \int_{t_0}^{t_1} F(t, x, \dot{x}) dt, \qquad x(t_0) = x_0, \quad t_1 \text{ fixed,} \tag{51}$$

subject to *one* of the following terminal conditions:

$$\text{(i) } x(t_1) = x_1, \qquad \text{(ii) } x(t_1) \geqslant x_1, \qquad \text{(iii) } x(t_1) \text{ free.} \tag{52}$$

We assume that $F(t, x, \dot{x})$ is a C^2-function and that the admissible functions $x(t)$ are C^1-functions. We have proved previously that a necessary condition for an admissible function $x^*(t)$ to solve our problem is that it satisfies the Euler equation and the transversality conditions

$$\text{(ii) } \left(\frac{\partial F}{\partial \dot{x}}\right)_{t=t_1} \leqslant 0 (=0 \text{ if } x^*(t_1) > x_1), \qquad \text{(iii) } \left(\frac{\partial F}{\partial \dot{x}}\right)_{t=t_1} = 0 \tag{53}$$

provided the terminal conditions are (52ii) or (52iii). The next theorem tells us that these conditions are *sufficient* for optimality provided $F(t, x, \dot{x})$ is (jointly) concave in $(x, \dot{x})$.

Theorem 3 Suppose $F(t, x, \dot{x})$ is concave as a function of $(x, \dot{x})$ for each $t \in [t_0, t_1]$. If $x^* = x^*(t)$ satisfies the Euler equation, the boundary conditions and (in case the terminal condition is (52ii) or (52iii)) the transversality conditions (53ii) or (53iii), then $x^*(t)$ is globally maximal in the sense that if $x = x(t)$ is an arbitrary admissible function in one of the three problems, then

$$\int_{t_0}^{t_1} F(t, x^*, \dot{x}^*) dt \geqslant \int_{t_0}^{t_1} F(t, x, \dot{x}) dt. \quad \blacksquare$$

Proof. Since $F(t, x, \dot{x})$ is concave in $(x, \dot{x})$, then by a well-known result for concave differentiable functions (see (B.1)):

$$F(t, x, \dot{x}) - F(t, x^*, \dot{x}^*) \leqslant \frac{\partial F(t, x^*, \dot{x}^*)}{\partial x}(x - x^*)$$

$$+ \frac{\partial F(t, x^*, \dot{x}^*)}{\partial \dot{x}}(\dot{x} - \dot{x}^*). \tag{54}$$

Let us simplify our notation by letting $F = F(t, x, \dot{x})$ and using $*$ to denote that a function is evaluated at $(t, x^*, \dot{x}^*)$. From (54) we obtain

$$F^* - F \geqslant \frac{\partial F^*}{\partial x}(x^* - x) + \frac{\partial F^*}{\partial \dot{x}}(\dot{x}^* - \dot{x})$$

$$= \left[\frac{d}{dt}\left(\frac{\partial F^*}{\partial \dot{x}}\right)\right](x^* - x) + \frac{\partial F^*}{\partial \dot{x}}(\dot{x}^* - \dot{x}) = \frac{d}{dt}\left[\frac{\partial F^*}{\partial \dot{x}}(x^* - x)\right],$$

where the first equality is a consequence of the fact that x^* satisfies the Euler equation and the second equality is obtained by making use of the rule for differentiating a product. Since the resulting inequality is valid for all $t \in [t_0, t_1]$, we get:

$$\int_{t_0}^{t_1} (F^* - F)dt \geqslant \int_{t_0}^{t_1} \frac{d}{dt}\left[\frac{\partial F^*}{\partial \dot{x}}(x^* - x)\right]dt = \left|\frac{\partial F^*}{\partial \dot{x}}(x^* - x)\right|_{t_0}^{t_1}$$

$$= \left(\frac{\partial F^*}{\partial \dot{x}}\right)_{t=t_1}(x^*(t_1) - x(t_1)), \tag{55}$$

since "the contribution" from the lower limit is 0 as $x^*(t_0) = x(t_0)$. Let us put $d = (F_{\dot{x}}^{*\prime})_{t=t_1}(x^*(t_1) - x(t_1))$. It remains to be proved that is each of the three cases, $d \geqslant 0$.

Now, if the terminal condition is $x(t_1) = x_1$, then $d = 0$. If the terminal condition is $x(t_1) \geqslant x_1$ and $x^*(t_1) > x_1$, then from (53ii), $(F_{\dot{x}}^{*\prime})_{t=t_1} = 0$ and so $d = 0$. On the other hand, if $x(t_1) \geqslant x_1$ and $x^*(t_1) = x_1$, then from (53ii) $(F_{\dot{x}}^{*\prime})_{t=t_1} \leqslant 0$ and since $x^*(t_1) - x(t_1) \leqslant 0$ in this case, $d \geqslant 0$. Finally, if $x(t_1)$ is free we see from (53iii) that $d = 0$ again. Thus in all cases $d \geqslant 0$. ∎

Note 5 The corresponding theorem for the problem of minimizing the integral in (51), is clearly obtained by requiring $F(t, x, \dot{x})$ to be convex in $(x, \dot{x})$.

Note 6   Note that the theorem does not guarantee the existence of an admissible function satisfying the Euler equation and the appropriate transversality conditions. It just tells us that if such a function is found it must solve the problem. If there are several such solutions, all of them are optimal and, of course, they give the same value to the integral.

Note 7   Suppose we add to problem (51), (52) the constraint $h(t, x(t), \dot{x}(t)) > 0$. Then the Euler equation is still a necessary condition for optimality. (See the end of section 3.) If we require $F(t, x, \dot{x})$ to be concave in $(x, \dot{x})$ only for $(x, \dot{x})$ satisfying $h(t, x, \dot{x}) > 0$ (but for all $t \in [t_0, t_1]$), and if we require $h(t, x, \dot{x})$ to be quasi-concave in $(x, \dot{x})$, then the proof of theorem 3 goes through as before. (Quasi-concavity of h is imposed in order for the domain of F (as a function of $(x, \dot{x})$) to be convex.) In fact, for this sufficiency result to hold we might assume that the constraint is $h(t, x, \dot{x}) \geqslant 0$.

Example 12   Consider example 2. Here $F(t, x, \dot{x}) = x^2 + \dot{x}^2$ is clearly convex in x and $\dot{x}$. By using the result in note 5, we conclude that $x = e^{1+t} - e^{1-t}$ minimizes $J(x)$.

The condition that $F(t, x, \dot{x})$ is concave as a function of x and $\dot{x}$ does, of course, considerably restrict the class of functions F for which our theorem can be applied. However, in economic problems the concavity requirement is often a natural one, and the theorem is therefore a very important result. Let us consider our standard example from economic growth theory.

Example 13

$$\max \int_0^T U(f(K) - \dot{K}) e^{-\rho t} dt, \qquad K(0) = K_0, \qquad K(T) = K_T.$$

This is the problem in example 1. Since we assumed f to be concave, $f(K) + (-\dot{K})$ is the sum of two concave functions, hence concave in $(K, \dot{K})$. U is increasing and concave. Hence $U(f(K) - \dot{K}) e^{-\rho t}$ is concave in $(K, \dot{K})$.[4] From theorem 3 it follows that a solution to the Euler eq. ((20) in example 3) satisfying the boundary conditions is indeed optimal.

Exercise 1.6.1   Verify the optimality of the candidates found in exercises 1.2.1 and 1.2.3.

---

[4]See e.g. Sydsæter (1981), theorem 5.14 (iv), p. 254.

Exercise 1.6.2 Prove that the integrand in the variational problem in exercise 1.4.4 is concave in $(y, \dot{y})$ and thereby verify the optimality of the solution found in exercise 1.4.4 (ii). (You will need the result in note 7.)

Exercise 1.6.3 In a model by J.K. Sengupta the following problem arises:

$$\min \int_0^T (\alpha_1 \bar{Y}^2 + \alpha_2 G^2)dt, \qquad \dot{\bar{Y}} = r_1 \bar{Y} - r_2 G, \qquad \bar{Y}(0) = Y_0, \qquad \bar{Y}(T) \text{ free}$$

where α_1, α_2, r_1, r_2, T, and Y_0 are given positive constants.

(i) Formulate the problem as a variational problem with $\bar{Y} = \bar{Y}(t)$ as the unknown function and find the associated Euler equation.
(ii) Find the only solution to the equation that satisfies the boundary condition and the transversality condition.
(iii) Prove that the function obtained in (ii) solves the given problem.

Exercise 1.6.4 In connection with a problem on optimal timing of innovations M.I. Kamien and N.L. Schwartz encounter the following variational problem:

$$\min \int_0^T e^{-rt}(\dot{z}(t))^{1/a}dt, \qquad z(0) = 0, \qquad z(T) = A,$$

where r, a, T, and A are positive constants, $a < 1$. Solve this problem.

7. A more general optimality criterion

In many variational problems in economics it is natural to include in the maximization a function which measures the value or utility assigned to the terminal state. This leads to the following problem:

$$\max \left\{ \int_{t_0}^{t_1} F(t, x, \dot{x})dt + S(x(t_1)) \right\}, \qquad x(t_0) = x_0, \tag{56}$$

where S is a given function, while $x(t_1)$ is to be determined by the maximization. For example, in a planning problem, $S(x(t_1))$ might represent the value assigned to the terminal capital stock $x(t_1)$. In particular, if a stock of machines is left over, $S(x(t_1))$ might denote the *scrap value*.

Suppose that $x^* = x^*(t)$ is a solution to problem (56). Then, in particular, x^* is maximal compared with all admissible functions whose graphs join (t_0, x_0) and $(t_1, x^*(t_1))$, and all these functions give the same value to the scrap value function. It follows that x^* must satisfy the Euler eq. (12) in $[t_0, t_1]$. One of the constants in the general solution of this equation is determined by the initial condition $x^*(t_0) = x_0$. To find the other constant we must determine the appropriate transversality condition for the right end point in this case. If we assume that the function S is sufficiently smooth, the problem can be transformed into a well known one. In fact, if $x = x(t)$, then

$$S(x(t_1)) - S(x(t_0)) = \int_{t_0}^{t_1} \frac{\mathrm{d}}{\mathrm{d}t} S(x(t)) \mathrm{d}t = \int_{t_0}^{t_1} S'(x) \dot{x} \, \mathrm{d}t.$$

Now, for each admissible function $x(t)$ in problem (56), $S(x(t_0))$ has the constant value $S(x_0)$. Hence we see that problem (56) has a solution $x^*(t)$ if and only if the problem

$$\max \int_{t_0}^{t_1} [F(t, x, \dot{x}) + S'(x) \cdot \dot{x}] \mathrm{d}t, \qquad x(t_0) = x_0, \qquad x(t_1) \text{ free} \tag{57}$$

has the solution $x^*(t)$. Note that (57) is a standard variational problem with free right-hand side. The integrand is $F_1(t, x, \dot{x}) = F(t, x, \dot{x}) + S'(x)\dot{x}$, so $\partial F_1/\partial \dot{x} = \partial F/\partial \dot{x} + S'(x)$, so the transversality condition for problem (57) is $(\partial F_1/\partial \dot{x})_{t=t_1} = 0$, i.e.

$$(\partial F/\partial \dot{x})_{t=t_1} + S'(x(t_1)) = 0. \tag{58}$$

Thus we have the following result:

If $x^*(t)$ solves problem (56), then x^* satisfies the Euler equation (12) and the transversality condition (58).

In exercise 1.7.4 you are asked to prove that the Euler equation for F_1 reduces to the Euler equation for F.

Note 8 In order to use our previous theorem in connection with the problem in (57), we must assume that S is a C^3-function. By using another argument we can prove (58) by assuming S to be only a C^1-function, but this generalization is unimportant.

Also in the case of problem (56) it is easy to prove a global sufficiency result:

Suppose $F(t, x, \dot{x})$ is concave in $(x, \dot{x})$ and suppose S is a concave function. Then a solution $x^*(t)$ of the Euler equation that satisfies $x^*(t_0) = x_0$ and (58) is globally maximal.

Let us briefly indicate the proof. We assume that $x = x(t)$ is an admissible function so, in particular, $x(t_0) = x_0$, and we get (confer the proof of theorem 3):

$$\int_{t_0}^{t_1} F^* dt + S(x^*(t_1)) - \int_{t_0}^{t_1} F dt - S(x(t_1))$$

$$= \int_{t_0}^{t_1} (F^* - F) dt + S(x^*(t_1)) - S(x(t_1))$$

$$\geqslant \left(\frac{\partial F^*}{\partial \dot{x}}\right)_{t=t_1} (x^*(t_1) - x(t_1)) + S'(x^*(t_1)) \cdot (x^*(t_1) - x(t_1))$$

$$= \left[\left(\frac{\partial F^*}{\partial \dot{x}}\right)_{t=t_1} + S'(x^*(t_1))\right] \cdot (x^*(t_1) - x(t_1)) = 0.$$

Note 9   If we ask for the minimization of the integral in (56), then we get exactly the same necessary conditions. Moreover, if $F(t, x, \dot{x})$ is convex in $(x, \dot{x})$ and if S is a convex function, the necessary conditions are also sufficient.

Let us look at a simple example to see how the theory works.

Example 14

$$\min\left\{\int_0^1 (x^2 + \dot{x}^2) dt + (x(1))^2\right\}, \qquad x(0) = 1.$$

In this case the function S is given by $S(u) = u^2$. According to note 9 above, an optimal solution must satisfy (i): the Euler equation, (ii): $x(0) = 1$, (iii): $(\partial F / \partial \dot{x})_{t=1} + S'(x(1)) = 0$ where $F(t, x, \dot{x}) = x^2 + \dot{x}^2$.

The solution of the Euler equation satisfying $x(0) = 1$ is $x(t) = A(e^t - e^{-t}) + e^{-t}$ (see example 7). Moreover, $F'_{\dot{x}} = 2\dot{x}$ and $S'(u) = 2u$, so (iii) implies $2\dot{x}(1) + 2x(1) = 0$, or

$$2A(e + e^{-1}) - 2e^{-1} + 2A(e - e^{-1}) + 2e^{-1} = 0,$$

that is

$4Ae = 0.$

Hence $A = 0$ and it follows that the only possible solution to the problem is

$x(t) = e^{-t}.$

Since $F(t, x, \dot{x}) = x^2 + \dot{x}^2$ is convex in x and $\dot{x}$ and $S(u) = u^2$ is a convex function, $x(t) = e^{-t}$ is the solution of the problem.

Exercise 1.7.1 Solve the problem

$$\min\left\{\int_0^1 (t\dot{x} + \dot{x}^2)dt + (2x(1) + 3)\right\}, \qquad x(0) = 1.$$

Exercise 1.7.2 In a model of "individual savings and bequest behaviour", A.B. Atkinson (1971) studies the following problem

$$\max\left\{\int_0^T U(rA(t) + w - \dot{A}(t))e^{-\rho t}dt + e^{-\rho T}\phi(A(T))\right\}, \qquad A(0) = A_0.$$

The situation and meaning of the symbols are as in exercise 1.5.3, except that included in the maximization we here have a function ϕ measuring the utility from bequests made by the individual at his death. (No extra condition on $A(T)$.) We assume that $\phi' > 0$, $\phi'' < 0$. Give a set of sufficient conditions for $A(t)$ to solve this problem.

Exercise 1.7.3 Consider the problem

$$\max\left\{\int_0^T \frac{1}{1-v}(bK - \dot{K})^{1-v}dt + \alpha[K(T)]^\beta\right\}, \qquad K(0) = K_0,$$

where α, β, and v are positive constants.
(i) Find the transversality condition for the right endpoint.
(ii) Let $T = 10$, $v = \frac{1}{2}$, $b = 2$, $\alpha = 6$, $\beta = \frac{1}{2}$ and $K_0 = 10$ and find the solution to the problem in this case.

Exercise 1.7.4 Prove that the Euler equation associated with (57) reduces to equation (12).

8. Several unknown functions

In the variational problems discussed so far, there has been only *one* unknown function. Of course, in many situations variational problems arise in which several unknown functions are involved. Let us briefly consider the following problem

$$\max \int_{t_0}^{t_1} F(t, x(t), \dot{x}(t))dt, \qquad x(t_0) = x^0, \tag{59}$$

where $x(t) = (x_1(t), \ldots, x_n(t))$, $\dot{x}(t) = (\dot{x}_1(t), \ldots, \dot{x}_n(t))$, $x^0 = (x_1^0, \ldots, x_n^0)$, and where the following terminal conditions are imposed:

$$x_i(t_1) = x_i^1, \qquad i = 1, \ldots, l, \tag{60a}$$

$$x_i(t_1) \geqslant x_i^1, \qquad i = l+1, \ldots, m, \tag{60b}$$

$$x_i(t_1) \text{ free}, \qquad i = m+1, \ldots, n. \tag{60c}$$

Here $t_0, t_1; x_1^0, \ldots, x_n^0$ as well as $x_1^1, \ldots, x_m^1$ are given constants. We assume that F is a given C^2-function of $2n+1$ variables and we assume that the admissible functions $x_1(t), \ldots, x_n(t)$ are C^1-functions on $[t_0, t_1]$, satisfying the initial conditions and the terminal conditions (60).

Suppose $x^*(t) = (x_1^*(t), \ldots, x_n^*(t))$ solves the problem. Fix any i, $i = 1, \ldots, n$ and let $J(x_i(t))$ denote the value of the integral in (59) if we evaluate it at $(x_1^*(t), \ldots, x_{i-1}^*(t), x_i(t), x_{i+1}^*(t), \ldots, x_n^*(t))$, where $x_i(t)$ is any admissible function. By the optimality of $x^*(t)$, $J(x_i^*(t)) \geqslant J(x_i(t))$ for all admissible $x_i(t)$. Furthermore, $x_i^*(t)$ as well as $x_i(t)$ are equal to x_i^0 at $t = t_0$ and they both satisfy, according to the value of i, the terminal condition (60a), (60b) or (60c). But then $x_i^* = x_i^*(t)$ must satisfy the Euler equation associated with $J(x_i)$ and the appropriate terminal condition. Thus we have the following result:

Theorem 4 Necessary conditions for an admissible $(x_1^*(t), \ldots, x_n^*(t))$ to solve problem (59) with terminal condition (60), are that the Euler equations

$$\frac{\partial F}{\partial x_i} - \frac{d}{dt} \frac{\partial F}{\partial \dot{x}_i} = 0, \qquad i = 1, \ldots, n \tag{61}$$

are satisfied as well as the transversality conditions

$$x_i^*(t_1) = x_i^1, \qquad\qquad i = 1, \ldots, l, \qquad\qquad (62a)$$

$$(\partial F/\partial \dot{x}_i)_{t=t_1} \leqslant 0 \,(=0 \text{ if } x_i^*(t_1) > x_i^1), \qquad i = l+1, \ldots, m, \qquad\qquad (62b)$$

$$(\partial F/\partial \dot{x}_i)_{t=t_1} = 0, \qquad\qquad i = m+1, \ldots, n. \quad \blacksquare \qquad (62c)$$

Note 10   If the problem is to minimize the integral in (59) subject to the same conditions, the conclusions in theorem 4 are unchanged, except that the first inequality in (62b) is reversed.

In section 3 we formulated the Legendre's necessary condition for the simplest variational problem. A similar result is valid in the present circumstances.

Theorem 5   (The Legendre's necessary conditions)   Suppose that the conditions in connection with theorem 4 are satisfied. A necessary condition for the integral in (59) to have a maximum at $(x_1^*(t), \ldots, x_n^*(t))$ is that

$$\sum_{j=1}^n \sum_{i=1}^n F_{\dot{x}_i \dot{x}_j}''(t, x_1^*(t), \ldots, x_n^*(t), \dot{x}_1^*(t), \ldots, \dot{x}_n^*(t)) h_i h_j \leqslant 0 \qquad (63)$$

for all $t \in [t_0, t_1]$ and all numbers h_i, h_j; $i, j = 1, \ldots, n$. The corresponding condition for a minimum is obtained by reversing the inequality in (63). $\blacksquare$

For $n = 1$ the double sum in (63) reduces to $F_{\dot{x}_1 \dot{x}_1}''(t, x_1^*(t), \dot{x}_1^*(t)) h_1^2$. Hence we see that the inequality in (63) for $n = 1$ is exactly the corresponding one in theorem 2 (since $h_1^2 \geqslant 0$). A proof of theorem 5 is indicated in Exercise 1.10.3.

In an entirely similar way as in the one-variable case, we can also for this general situation prove a global sufficiency result.

Theorem 6   Consider problem (59), (60). Suppose $F(t, x_1, \ldots, x_n, \dot{x}_1, \ldots, \dot{x}_n)$ is concave in $x_1, \ldots, x_n, \dot{x}_1, \ldots, \dot{x}_n$ for each $t \in [t_0, t_1]$. If $(x_1^*(t), \ldots, x_n^*(t))$ is admissible, satisfies the Euler equations (61) and conditions (62), then $(x_1^*(t), \ldots, x_n^*(t))$ solves the problem in the sense that if $(x_1(t), \ldots, x_n(t))$ is an arbitrary admissible vector function, then

$$\int_{t_0}^{t_1} F(t, x_1^*, \ldots, x_n^*, \dot{x}_1^*, \ldots, \dot{x}_n^*) dt \geqslant \int_{t_0}^{t_1} F(t, x_1, \ldots, x_n, \dot{x}_1, \ldots, \dot{x}_n) dt. \quad \blacksquare$$

The proof of this result is a simple extension of the proof of theorem 3, so we skip it.

Note 11 Convexity of F w.r.t. $x_1, \ldots, x_n, \dot{x}_1, \ldots, \dot{x}_n$ for each $t \in [t_0, t_1]$ will secure global minimality in the corresponding minimization problem.

Example 15 Solve the problem

$$\min \int_0^{\pi/2} (\dot{x}_1^2 + \dot{x}_2^2 + 2x_1 x_2)dt, \qquad x_1(0) = x_2(0) = 0,$$

where (i) $x_1(\pi/2) = 1$, (ii) $x_2(\pi/2)$ free.

Here $F = \dot{x}_1^2 + \dot{x}_2^2 + 2x_1 x_2$ so the Euler equations are

$$\frac{\partial F}{\partial x_1} - \frac{d}{dt}\left(\frac{\partial F}{\partial \dot{x}_1}\right) = 2x_2 - \frac{d}{dt}(2\dot{x}_1) = 2x_2 - 2\ddot{x}_1 = 0 \tag{64}$$

$$\frac{\partial F}{\partial x_2} - \frac{d}{dt}\left(\frac{\partial F}{\partial \dot{x}_2}\right) = 2x_1 - \frac{d}{dt}(2\dot{x}_2) = 2x_1 - 2\ddot{x}_2 = 0. \tag{65}$$

In order to find the general solution to this system of differential equations we differentiate (64) twice w.r.t. t to obtain

$$\frac{d^4 x_1}{dt^4} - x_1 = 0.$$

The fundamental polynomial is $r^4 - 1 = (r-1)(r+1)(r-i)(r+i)$ so the general solution is

$$x_1 = Ae^t + Be^{-t} + C \sin t + D \cos t. \tag{66}$$

Since $x_2 = \ddot{x}_1$, we further obtain

$$x_2 = Ae^t + Be^{-t} - C \sin t - D \cos t. \tag{67}$$

By using $x_1(0) = x_2(0) = 0$ and $x_1(\pi/2) = 1$, we get $0 = A + B + D$, $0 = A + B - D$ and $1 = Ae^{\pi/2} + Be^{-\pi/2} + C$. The final equation needed to determine the four constants is obtained from the transversality condition associated with (ii):

$$\left(\frac{\partial F}{\partial \dot{x}_2}\right)_{t=\pi/2} = 2\dot{x}_2(\pi/2) = 2[Ae^{\pi/2} - Be^{-\pi/2} + D] = 0.$$

By solving the four equations we get $A = B = D = 0$, $C = 1$. We conclude that there is only one pair of functions, $x_1(t) = \sin t$, $x_2(t) = -\sin t$, which can solve the given problem.

Let us see what theorem 5 tells in this case. We find

$$\sum_{j=1}^{2} \sum_{i=1}^{2} \frac{\partial^2 F}{\partial \dot{x}_i \partial \dot{x}_j} h_i h_j = 2h_1^2 + 2h_2^2.$$

Since this expression is not $\leqslant 0$ for all h_1, h_2, we see from theorem 5 that the associated maximization problem has no solution. A closer examination must be undertaken to see if the pair of functions we have found actually solves our problem. Note that in this case F is not convex in $x_1, x_2, \dot{x}_1, \dot{x}_2$, so note 11 does not apply.

The optimal value function

Consider problem (59), (60). Suppose we calculate the value of the integral in (59) for all admissible vector functions $x(t)$. The supremum (possibly ∞) of the set of numbers obtained in this way depends on x^0, x^1, t_0 and t_1. We denote it by $V = V(x^0, x^1, t_0, t_1)$. (If $m = 0$ in (60), the terminal point is completely free and V does not contain x^1 as an argument.) $V(x^0, x^1, t_0, t_1)$ is called *the optimal value function*. If there is an optimal solution $x^*(t)$ for a given quadruple (x^0, x^1, t_0, t_1), then V is finite and equal to the integral in (59) evaluated at $x^*(t)$. One can prove the following interesting result.[5]

Theorem 7 (Differentiability of the optimal value function) Let $x^*(t) = (x_1^*(t), \ldots, x_n^*(t))$ be an admissible function which solves problem (59), (60) with $x^0 = \bar{x}^0$, $x^1 = \bar{x}^1$, $t_0 = \bar{t}_0$, $t_1 = \bar{t}_1$. Suppose $x^*(t)$ satisfies the Euler equations (61) and the transversality conditions (62). Suppose moreover that $F(t, x, \dot{x})$ is jointly concave in $(x, \dot{x})$. Then the optimal value function $V(x^0, x^1, \bar{t}_0, \bar{t}_1)$ is defined and is finite for all (x^0, x_1), and $(x^0, x^1) \rightarrow V(x^0, x^1, \bar{t}_0, \bar{t}_1)$ is differentiable at $(\bar{x}^0, \bar{x}^1)$, with

$$\frac{\partial V(\bar{x}^0, \bar{x}^1, \bar{t}_0, \bar{t}_1)}{\partial x_i^0} = -\frac{\partial F(\bar{t}_0, x^*(\bar{t}_0), \dot{x}^*(\bar{t}_0))}{\partial \dot{x}_i} \qquad i = 1, \ldots, n \tag{68}$$

$$\frac{\partial V(\bar{x}^0, \bar{x}^1, \bar{t}_0, \bar{t}_1)}{\partial x_i^1} = \frac{\partial F(\bar{t}_1, x^*(\bar{t}_1), \dot{x}^*(\bar{t}_1))}{\partial \dot{x}_i} \qquad i = 1, \ldots, m. \tag{69}$$

Suppose furthermore that $x^*(t)$ can be extended, as a solution of (61) to a somewhat larger interval (a, b) containing $[\bar{t}_0, \bar{t}_1]$. Assume also that $l = n$

---

[5]See exercises 1.8.8→1.8.10.

(fixed end). Then $V(x^0, x^1, t_0, t_1)$ is defined and finite in some neighbourhood of $(\bar{x}^0, \bar{x}^1, \bar{t}_0, \bar{t}_1)$ and V is differentiable at this point with partial derivatives (68), (69) and

$$\frac{\partial V(\bar{x}^0, \bar{x}^1, \bar{t}_0, \bar{t}_1)}{\partial t_0} = -H(\bar{t}_0, x^*(\bar{t}_0), \dot{x}^*(\bar{t}_0)) \tag{70}$$

$$\frac{\partial V(\bar{x}^0, \bar{x}^1, \bar{t}_0, \bar{t}_1)}{\partial t_1} = H(\bar{t}_1, x^*(\bar{t}_1), \dot{x}^*(\bar{t}_1)) \tag{71}$$

where

$$H(t, x, \dot{x}) = F(t, x, \dot{x}) - \sum_{i=1}^{n} \dot{x}_i \frac{\partial F(t, x, \dot{x})}{\partial \dot{x}_i}. \quad \blacksquare$$

Let us give an interpretation of one of the formulas, say, (68): If x_i^0 denotes the amount of some resource, then $-\partial F/\partial \dot{x}_i$, evaluated at $(\bar{t}_0, x^*(\bar{t}_0), \dot{x}^*(\bar{t}_0))$, measures, approximately, the change in the value of the optimal value function resulting from a unit increase in x_i^0.

Note 12 The extension property assumed in this theorem is automatically satisfied if $\det(F''_{\dot{x}\dot{x}}) \neq 0$ at $\bar{t}_0$ and $\bar{t}_1$.

Example 16 Consider the problem in exercise 1.2.5:

$$\max \int_0^T \frac{1}{1-v} (bK - \dot{K})^{1-v} e^{-\rho t} dt, \qquad K(0) = K_0, \qquad K(T) = K_T. \tag{72}$$

The solution is (see also example 13):

$$K^*(t) = A e^{bt} + B e^{at} \qquad \text{where } a = \frac{b - \rho}{v} < b, \tag{73}$$

and where

$$K_0 = A + B, \qquad K_T = A e^{bT} + B e^{aT}. \tag{74}$$

We want to test (68) in this case. We find $C^* = bK^* - \dot{K}^* = (b-a)Be^{at}$, and the optimal value function is

$$V(K_0, K_T, T) = \int_0^T \frac{1}{1-v} (b-a)^{1-v} B^{1-v} e^{a(1-v)t} e^{-\rho t} dt$$

$$=\frac{1}{1-v}(b-a)^{-v}(1-e^{-(b-a)T})B^{1-v}. \tag{75}$$

The equations in (74) define A and B as functions of K_0, K_T and T. Differentiating the system in (74) w.r.t. K_0 we obtain, in particular, $(1-e^{-(b-a)T})\partial B/\partial K_0 = 1$. Hence, from (75)

$$\frac{\partial V}{\partial K_0} = (b-a)^{-v}(1-e^{-(b-a)T})B^{-v}\frac{\partial B}{\partial K_0} = (b-a)^{-v}B^{-v}. \tag{76}$$

On the other hand (see example 3)

$$\frac{\partial F}{\partial \dot{K}} = -(C^*)^{-v}e^{-\rho t} = -(b-a)^{-v}B^{-v}e^{-avt}e^{-\rho t},$$

so

$$-\left(\frac{\partial F}{\partial \dot{K}}\right)_{t=0} = (b-a)^{-v}B^{-v}, \tag{77}$$

and we see from (76) and (77) that (68) is confirmed.

Note 13 Consider the variational problem ($x(t)$ a scalar)

$$\max \int_{t_0}^{t_1} F(t,x,\dot{x})dt, \qquad x(t_0)=x_0, \qquad x(t_1) \text{ free}, \tag{78}$$

and assume that $x^*(t)$ solves the problem. The corresponding fixed-endpoint problem with $x(t_1)=x_1$, where $x_1=x^*(t_1)$, has the same solution, $x^*(t)$. According to (68) the optimal value function for the latter problem, $V(x_0,x_1,t_0,t_1)$ satisfies

$$\frac{\partial V}{\partial x_1} = \frac{\partial F(t_1,x^*(t_1),\dot{x}^*(t_1))}{\partial \dot{x}}. \tag{79}$$

Recall that the transversality condition for the free right endpoint case tells us that the right-hand side of (79) is 0 (32a). This result is intuitively clear since we cannot increase the optimal value in the free-endpoint problem by varying the point at which $x^*(t_1)$ ends.

Free-time problems

Consider problem (59), (60) assuming that t_1 is not fixed, but free to vary in an interval $[T_1, T_2]$, $t_0 \leqslant T_1 < T_2$. Again it is obvious that the Euler equations (61) are necessary for optimality. A complete set of necessary conditions is given in this theorem:

Theorem 8 (Necessary conditions. Free final time problems) Consider problem (59), (60) with $t_1 \in [T_1, T_2]$, $t_0 \le T_1 < T_2$. If $(x_1^*(t), \ldots, x_n^*(t))$ defined in the interval $[t_0, t_1^*]$ solves the problem, then the Euler equations (61) and the transversality conditions (62) are satisfied for $t = t_1^*$. In addition the following condition holds if $t_1^* > t_0$:

$$\left[F - \sum_{i=1}^{n} \dot{x}_i \frac{\partial F}{\partial \dot{x}_i} \right]_{t = t_1^*} \begin{cases} \le 0 & \text{if } t_1^* = T_1 \\ = 0 & \text{if } t_1^* \in (T_1, T_2) \\ \ge 0 & \text{if } t_1^* = T_2. \end{cases} \tag{80}$$

Sufficient conditions are given in the next theorem. ∎

Theorem 9 (Sufficient conditions. Free final time problems.)[6] Consider problem (59), (60) with $t_1 \in [T_1, T_2]$. Assume that for each $T \in [T_1, T_2]$, there exist solutions $(x_1^T(t), \ldots, x_n^T(t)) = x^T(t)$ defined on $[t_0, T]$ and satisfying the Euler eqns. (61) and the transversality condition (62). Assume that the function $T \to x^T(T)$ is continuous and piecewise continuously differentiable and that the set $\{\dot{x}^T(t) : t_0 \le t \le T, T_1 \le T \le T_2\}$ is bounded. Assume further that F is concave in $(x, \dot{x})$ for each t. Finally, assume that there exists a $T^* \in [T_1, T_2]$ such that the function

$$d(T) = F(T, x^T(T), \dot{x}^T(T)) - \sum_{i=1}^{n} \dot{x}_i^T(T) \frac{\partial F(T, x^T(T), \dot{x}^T(T))}{\partial \dot{x}_i} \tag{81}$$

has the property that

$$d(T) \ge 0 \quad \text{for} \quad T \le T^* \quad \text{if} \quad T_1 < T^*$$
$$d(T) \le 0 \quad \text{for} \quad T \ge T^* \quad \text{if} \quad T_2 > T^*. \tag{82}$$

Then $(x_1^{T^*}(t), \ldots, x_n^{T^*}(t))$ is optimal.

If $T_1 = t_0$, it suffices to test the conditions in the theorem for $T > t_0$. ∎

Note 14 The solution $x^{T^*}(t)$ is optimal in the problem where $[T_1, T_2]$ is replaced by $[T_1, \infty)$, provided the conditions in the theorem hold for all intervals $[T_1, T_2]$, $T_2 > T^*$.

---

[6]Seierstad (1984a) and exercise 1.8.11.

Exercise 1.8.1 Find the solution to this problem:

$$\min \int_0^1 (\dot{x}_1^2 + \dot{x}_2^2 + \dot{x}_1\dot{x}_2)\,dt, \qquad x_1(0)=x_2(0)=0, \qquad x_1(1)=1, \; x_2(1) \geqslant 1.$$

Exercise 1.8.2 Find the Euler equations associated with

$$J(K,L) = \int_0^T (F(K,L) - \omega L - I)e^{-\delta t}\,dt,$$

where $I = \phi(\dot{K} - \mu K)$. Here $K = K(t)$ and $L = L(t)$ are the unknown functions, F and ϕ are given functions, while T, ω, δ, and μ are constants.

Exercise 1.8.3 In a multi-sectorial growth model J.K. Sengupta encounters the problem of finding functions $Y_1 = Y_1(t)$ and $Y_2 = Y_2(t)$ that solve the problem

$$\max \int_{t_0}^{t_1} (\ln C_1 + \ln C_2)\,dt, \qquad Y_1(t_0) = Y_1^0, \qquad Y_2(t_0) = Y_2^0,$$

where $C_1 = Y_1 - \lambda_1 g_1 h_1 \dot{Y}_1 - \lambda_1 g_2 h_2 \dot{Y}_2$, $C_2 = Y_2 - \lambda_2 g_1 h_1 \dot{Y}_1 - \lambda_2 g_2 h_2 \dot{Y}_2$, and some terminal conditions are imposed on $Y_1(t_1)$ and $Y_2(t_1)$. Find the Euler equations associated with this problem.

Exercise 1.8.4 Consider the problem

$$\max \left\{ \int_{t_0}^{t_1} F(t, x_1, \ldots, x_n, \dot{x}_1, \ldots, \dot{x}_n)\,dt + S(x_1(t_1), \ldots, x_n(t_1)) \right\},$$

where $x_i(t_0) = x_i^0$, $i = 1, \ldots, n$. We assume that S is differentiable w.r.t. all variables, and we have the same requirements on F as before.

(i) Prove that a solution $(x_1^*(t), \ldots, x_n^*(t))$ to the problem must satisfy (61).
(ii) Find the transversality condition for the right end.
(iii) Formulate and prove a global sufficiency theorem in this problem.

Exercise 1.8.5 In a study by A. Sandmo on investment and the rate of return, the following problem arises:

$$\max_{K,L} \int_0^\infty [p(t)F(K,L) - w(t)L - q(t)(\dot{K} + \delta K)]e^{-rt} dt$$

where the unknown functions $K = K(t)$ and $L = L(t)$ are subject to certain initial and terminal conditions. Assuming that the Euler eqns. (61) are valid for this infinite horizon problem (see section 10), prove that they reduce to

$$\dot{q}(t) = (\delta + r)q(t) - pF'_K(K,L) \quad \text{and} \quad w(t) = p(t)F'_L(K,L)$$

(Note that these equations are not differential equations. Cf. exercise 1.4.7.)

Exercise 1.8.6   Verify (68) and (69) in the case of exercise 1.2.1 (iv).

Exercise 1.8.7   Using theorem 9 solve the free-time problems:

(i) $\min \displaystyle\int_0^{t_1} (x^2 + \dot{x}^2)dt, \qquad x(0) = 0, \qquad x(t_1) = 1, \qquad t_1 \in [1, 2].$

(ii) $\min \displaystyle\int_0^{t_1} (x^2 + \dot{x}^2 + 3)dt, \qquad x(0) = 0, \qquad x(t_1) = 1, \qquad t_1 \in [1/2, \infty)$

Exercise 1.8.8*   Prove (69) in theorem 7 for the case in which $x(t)$ is a scalar function with terminal condition $x(t_1) = x^1$ or $x(t_1) \geqslant x^1$ in the following way:

(i)   Use (55) to show that $V(\bar{x}^0, x^1, \bar{t}_0, \bar{t}_1)$ is finite for all x^1 and that $[\partial F(t, x^*, \dot{x}^*)/\partial \dot{x}]_{t=\bar{t}_1}$ is a supergradient of $V(\bar{x}^0, x^1, \bar{t}_0, \bar{t}_1)$ w.r.t. x^1 at $\bar{x}^1$.

(ii)   Let $\tilde{x}(t) = x^*(t) + \alpha\mu(t)$, where $\alpha = x^1 - \bar{x}^1$, $\mu(\bar{t}_0) = 0$, $\mu(\bar{t}_1) = 1$, and define $L(x^1) = I(x^1 - \bar{x}^1) = \int_{\bar{t}_0}^{\bar{t}_1} F(t, \tilde{x}(t), \dot{\tilde{x}}(t))dt$. Prove that $L'(\bar{x}^1) = I'(0) = (\partial F^*/\partial \dot{x})_{t=\bar{t}_1}$. (See eq. (9).).

(iii)   Show that $V(\bar{x}_0, x^1, \bar{t}_0, \bar{t}_1) \geqslant L(x^1)$ for all x^1.

(iv)   Use (i)→(iii) to conclude that (69) is valid.

(v)   Generalize the argument in (i)→(iv) to the case where $x(t)$ is a vector.

Exercise 1.8.9*   The previous exercise contains a proof of (69). The rest of the results in theorem 7 are partially proved in this exercise and the next. Let $x(t)$ be a scalar function and consider problem (51), (52) and suppose

$x^*(t)$ satisfies the Euler equation on $[\bar{t}_0, \bar{t}_1]$. Extend the definition of $x^*(t)$ by letting $\dot{x}^*(t) \equiv \dot{x}^*(t_1)$ for $t > \bar{t}_1$. Define $\mu(t) = [x^*(\bar{t}_1) - x^*(\bar{t}_1 + \alpha)](t - \bar{t}_0)/(\bar{t}_1 + \alpha - \bar{t}_0)$ and define

$$K(\alpha) = \int_{\bar{t}_0}^{\bar{t}_1 + \alpha} F(t, x^*(t) + \mu(t), \dot{x}^*(t) + \dot{\mu}(t)) dt.$$

(i) Prove that

$$K'(0) = F(\bar{t}_1, x^*(\bar{t}_1), \dot{x}^*(\bar{t}_1)) - \frac{\dot{x}^*(\bar{t}_1)}{\bar{t}_1 - \bar{t}_0} \left[\int_{\bar{t}_0}^{\bar{t}_1} \frac{\partial F^*}{\partial x}(t - \bar{t}_0) dt + \int_{\bar{t}_0}^{\bar{t}_1} \frac{\partial F^*}{\partial \dot{x}} dt \right].$$

By partial integration of the latter integral,

$$\int_{\bar{t}_0}^{\bar{t}_1} (\partial F^*/\partial \dot{x}) dt = \left| \int_{\bar{t}_0}^{\bar{t}_1} (t - \bar{t}_0)(\partial F^*/\partial \dot{x}) - \int_{\bar{t}_0}^{\bar{t}_1} (t - \bar{t}_0)(d/dt)(\partial F^*/\partial \dot{x}) dt, \right.$$

and using the Euler equation, show that $K'(0)$ equals the right-hand side of (71) (for $n = 1$).

(ii) Since $x^*(\bar{t}_1 + \alpha) + \mu(\bar{t}_1 + \alpha) = x^*(\bar{t}_1)$, $V(\bar{x}^0, \bar{x}^1, \bar{t}_0, \bar{t}_1 + \alpha) \geqslant K(\alpha)$ (equality if $\alpha = 0$). Thus $K'(0)$ is a subderivative of V w.r.t. t_1 at $\bar{t}_1$.

Exercise 1.8.10* Consider the situation in theorem 6, assuming that $x(t)$ is a scalar function on $[\bar{t}_0, t_1]$, with $x(\bar{t}_0) = \bar{x}^0$, $x(t_1) = x^1$. Let t_1 be close to $\bar{t}_1$, $t_1 \in (a, b)$. Using a notation analogous to that in the proof of theorem 3, show by means of (55) that

$$\int_{\bar{t}_0}^{t_1} F \, dt - \int_{\bar{t}_0}^{\bar{t}_1} F^* \, dt = \int_{\bar{t}_0}^{t_1} (F - F^*) dt - \int_{t_1}^{\bar{t}_1} F^* \, dt$$

$$\leqslant \frac{\partial F^*}{\partial \dot{x}}(t_1)(x(t_1) - x^*(t_1)) - \int_{t_1}^{\bar{t}_1} F^* \, dt$$

$$= \frac{\partial F^*}{\partial \dot{x}}(t_1)(x(t_1) - x^*(\bar{t}_1)) - \frac{\partial F^*}{\partial \dot{x}}(t_1)(x^*(t_1) - x^*(\bar{t}_1)) - \int_{t_1}^{\bar{t}_1} F^* \, dt$$

$$= \frac{\partial F^*}{\partial \dot{x}}(t_1)(x(t_1) - x^*(\bar{t}_1)) - \frac{\partial F^*}{\partial \dot{x}}(t_1)\dot{x}^*(\bar{t}_1)(t_1 - \bar{t}_1)$$

$$+ F^*(\bar{t}_1)(t_1 - \bar{t}_1)$$

$$+ \text{a second-order term.}$$

Conclude that

$$V(\bar{x}^0, x^1, \bar{t}_0, t_1) - V(\bar{x}^0, \bar{x}^1, \bar{t}_0, \bar{t}_1) \leqslant (\partial F^*(t_1)/\partial \dot{x})(x(t_1) - x^*(\bar{t}_1))$$

$$+ H^*(\bar{t}_1)(t_1 - \bar{t}_1)$$

$$+ \text{a second-order term.}$$

Exercise 1.8.11* In the situation in theorem 9 define $W(T) = V(x^0, x^T(T), t_0, T)$ where V is the optimal value function.

(i)   Use theorem 7 to prove that for v.e. $T \in (T_1, T_2)$, $W'(T) = r(T) + d(T)$ where

$$r(T) = \sum_{i=1}^{n} \frac{\partial F(T, x^T(T), \dot{x}^T(T))}{\partial \dot{x}_i} \frac{dx_i^T(T)}{dT}. \qquad (83)$$

(ii) Using (62) $(t_1 = T)$, prove that for all $T \in (T_1, T_2)$ and $T' \in (T_1, T_2)$

$$\sum_{i=1}^{n} \frac{\partial F(T, x^T(T), \dot{x}^T(T))}{\partial \dot{x}_i} (x_i^{T'}(T') - x_i^T(T)) \leqslant 0.$$

Conclude that $r(T) = 0$ for v.e. $T \in (T_1, T_2)$.

(iii) Using (i) and (ii) and the continuity of $W(T)$, prove theorem 9, in the case where each $x^T(t)$ can be extended to a slightly larger interval than $[t_0, T]$.

9. Piecewise differentiable solutions; Corner conditions; Existence

In the proof of the Euler equation for the simplest variational problem in section 2, we assumed that the admissible functions were C^2-functions. However, we pointed out that the Euler equation holds good also if the admissible functions are C^1. Sometimes it is more natural to consider an even more general class of admissible functions, piecewise smooth functions, i.e. functions which are continuous with piecewise continuous derivatives of the first order. This is in a certain sense the most natural class of admissible functions to consider in connection with variational problems associated with the integral $\int F(t, x, \dot{x}) dt$. With suitable restrictions on F, a function $x = x(t)$ belonging to this class will give meaning to the integral, while further generalizations, for example to include functions that are only continuous, will be meaningless since $F(t, x, \dot{x})$ is not even defined when x is not differentiable.

Suppose now that we consider a standard variational problem of the form (5). Then it might be the case that the problem has no C^1-solution, but that it does have a piecewise smooth solution. The following example is a standard one.

Example 17

$$\min \int_{-1}^{+1} x^2(1 - \dot{x})^2 \, dt, \qquad x(-1) = 0, \qquad x(1) = 1.$$

If we denote the integral by $J(x)$, then $J(x) \geqslant 0$ for all x for which the integral is defined, since $x^2(1 - \dot{x})^2$ is always $\geqslant 0$. Now, if we put $x^*(t) = 0$ for $-1 \leqslant t \leqslant 0$, $x^*(t) = t$ for $0 < t \leqslant 1$, then $J(x^*) = 0$ so that x^* solves the minimization problem. $x^*(t)$ *is* continuous, and $\dot{x}^*(t) = 0$ for $-1 \leqslant t < 0$ and $\dot{x}^*(t) = 1$ for $0 < t \leqslant 1$, so that x^* has a piecewise continuous derivative, i.e. is piecewise smooth. $x^*(t)$ has *a corner* at $t = 0$.

Let us suppose that $x = x(t)$ is a smooth function. Then $x^2(1 - \dot{x})^2$ is a continuous function of t which is $\geqslant 0$ for all $t \in [-1, 1]$. In order for the integral to be 0, $x^2(t)(1 - \dot{x}(t))^2$ must be 0 for all $t \in [-1, 1]$. (See the proof of the fundamental lemma in the calculus of variations in section 2.) The only smooth functions that equate the integral to 0 are consequently $x = t$ and $x = 0$, but neither of these satisfy both boundary conditions. On the other hand it is clear that by using smooth functions we can get the integral as close to 0 as we wish (by using functions that "smoothly approximate" x^*). This argument shows that the given problem has no solution within the class of smooth functions.

Let us consider now the problem

$$\max \int_{t_0}^{t_1} F(t, x, \dot{x}) dt, \qquad x(t_0) = x_0, \tag{84}$$

with some terminal condition, for example $x(t_1) = x_1$, and assume that the admissible functions are piecewise smooth. Then the Euler equation will not, in general, be valid on all of $[t_0, t_1]$. However, one can prove that if x^* solves the problem, then there exists a constant c such that $x^* = x^*(t)$ in the interval $[t_0, t_1]$ satisfies the equation

$$\frac{\partial F}{\partial \dot{x}} = \int_{t_0}^{t} \frac{\partial F}{\partial x} \, d\tau + c. \tag{85}$$

(See Hestenes (1966), Chapter 2.) *This is called the integrated form of the Euler equation* since by differentiating in w.r.t. t (when that is allowed) one obtains the Euler equation as we know it. The equation in (85) is valid also when we are considering a solution to the problem of minimizing the integral in (84). For the special case considered in example 17, we easily see that (85) is satisfied with $c = 0$.

Differentiation of equation (85) w.r.t. t is allowed on an interval where $\dot{x}^*$ is continuous. Hence the Euler equation in its original form is valid between corners of the optimal solution. If corners exist, we must have conditions which tell us how the different parts of the graph of the solution are to be pieced together at the corners. These are given by the *Weierstrass–Erdmann corner conditions* which follow immediately from (85). They are:

$$\frac{\partial F}{\partial \dot{x}} \text{ is continuous ``along'' } x^* = x^*(t) \text{ for every } t. \tag{86}$$

In particular $F'_{\dot{x}}$ is continuous at corners.

Suppose for example that the optimal solution associated with (84) has one corner, say at $\bar{t} \in (t_0, t_1)$. Then the Euler equation is valid in $[t_0, \bar{t}]$ and in $[\bar{t}, t_1]$. On each of these intervals we get a second-order differential equation which must be satisfied by the optimal solution. If we solve these equations we get, in general, four constants to be determined. Two of them are determined by $x(t_0) = x_0$, $x(t_1) = x_1$. The third is determined by the requirement that $x^*(t)$ is continuous at $\bar{t}$ while the fourth is determined by condition (86), which at $t = \bar{t}$ is:

$$\lim_{t \to \bar{t}^-} \left(\frac{\partial F}{\partial \dot{x}} \right) = \lim_{t \to \bar{t}^+} \left(\frac{\partial F}{\partial \dot{x}} \right). \tag{87}$$

Of course, we obtain the same conditions if we minimize the integral in (84).

Example 18   In example 17, $F(t, x, \dot{x}) = x^2(1 - \dot{x})^2$ so that $\partial F / \partial \dot{x} = 2x^2(\dot{x} - 1)$.

For $x*(t)$ given in example 17, we get

$$\lim_{t \to 0^-} \left(\frac{\partial F}{\partial \dot{x}} \right) = 0, \qquad \lim_{t \to 0^+} \left(\frac{\partial F}{\partial \dot{x}} \right) = 0,$$

so (87) is satisfied in this case.

In this chapter we have obtained solutions to a number of variational problems. We have usually assumed that the admissible functions were once or twice continuously differentiable. If we extend the class of admissible functions in these problems to include piecewise smooth functions, will this result in other solutions to the problems with "improved" values of the integrals? Fortunately, this is very seldom the case. It turns out that the corners for $x = x(t)$ can occur only at points at which $F''_{\dot{x}\dot{x}}(t, x(t), \dot{x}(t))$ is equal to 0. Hence, if $F''_{\dot{x}\dot{x}}$ is different from 0 everywhere, a solution of the corresponding variational problem has no corners. Moreover, according to the Hilbert differentiability theorem, if $x*$ is a solution of the Euler equation without corners, then $x* = x*(t)$ is a C^m-function ($m \geq 2$) if $F(t, x, \dot{x})$ is a C^m-function w.r.t. all variables. For these results and for generalizations to the case where $x(t)$ is a vector function, we refer to Hestenes (1966), Chapter 2.

Example 19 In the growth problem in example 1 that we have considered repeatedly, $F(t, K, \dot{K}) = U(f(K) - \dot{K})e^{-\rho t}$, so $F''_{\dot{K}\dot{K}} = U''(f(K) - \dot{K})e^{-\rho t}$. Since we have assumed $U'' < 0$, we know that a solution to our problem will have no corners. Hence we do not improve the maximum by allowing $\dot{K}$ to be piecewise continuous in this case.

*An existence theorem**

We end this section by presenting an existence theorem for variational problems with several unknowns. The relationship between necessary conditions (here the Euler equations), sufficient conditions (here given by theorem 6) and existence theorems, are set out, in a more general framework, at the end of section 2 in Chapter 2.

Theorem 10[7] **(Existence of an optimal solution)** Consider problem (59), (60). Let F be C^2 and concave in $\dot{x} = (\dot{x}_1, \ldots, \dot{x}_n)$ for each (t, x) and assume that for each vector $p = (p_1, \ldots, p_n) \neq 0$ there exist positive constants q_p, r

---

[7]Rockafellar (1973) and Ioffe and Tihomirov (1979). For the last part of the theorem, see Seierstad (1985a). (The condition on F'_x is needed in order to show that the Euler equation on integrated form is satisfied. Then we can conclude that the solution is not only measurable, but is C^2.)

and s such that for all $x=(x_1,\ldots,x_n)$ and for all $\dot{x}=(\dot{x}_1,\ldots,\dot{x}_n)$,

$$F(t,x,\dot{x})+p\cdot\dot{x}\leqslant q_p+(r+s\|p\|)\|x\|, \tag{88}$$

where r and s are independent of p. Assume also that $\|F'_x\|$ is bounded by a constant that is independent of $(t,x,\dot{x})$. Assume, moreover, that

$$\det(F''_{\dot{x}\dot{x}}(t,x,\dot{x}))\neq 0 \text{ for all } (t,x,\dot{x}). \tag{89}$$

Then there exists an optimal C^2-solution $x(t)=(x_1(t),\ldots,x_n(t))$.

If $F(t,x_1,\ldots,x_n,\dot{x}_1,\ldots,\dot{x}_n)$ is non-decreasing w.r.t. each $x_1,\ldots,x_n$ and $l=0$ and $n=m$ (i.e. only inequality terminal constraints), then (88) can be replaced by the following condition: For all $p\in R^n$, $p\geqslant 0$, there exist constants q_p, r and s such that

$$F(t,x,\dot{x})+\sum_{j=1}^{n}p_j\dot{x}_j\leqslant q_p+(r+s\|p\|)(\max(0,x_1,\ldots,x_n)). \tag{90}$$

Note 15 Condition (88) holds if the following condition is satisfied for all $(t,x)\in[t_0,t_1]\times R^n$:

$$\frac{F(t,x,\dot{x})}{\|\dot{x}\|}\to-\infty \quad\text{as}\quad \|\dot{x}\|\to\infty \text{ uniformly in } (t,x). \tag{91}$$

Note 16 The boundedness condition on F'_x can be replaced by the condition that for any bounded set $B\subset R^n$, there exist constants a, K, and K' such that

$$\|F'_x(t,y,\dot{x})\|\leqslant K+K'|F(t,x,\dot{x})|$$
$$\text{for all } t,x,y,\dot{x}, \quad \|y-x\|<a, \quad x\in B.$$

Exercise 1.9.1 Check the condition $F''_{\dot{x}\dot{x}}\neq 0$ in some of the examples and exercises we have considered previously.

Exercise 1.9.2 Find a sufficient condition for solutions to the problem

$$\min\int_{t_0}^{t_1}(a\dot{x}^2+bx\dot{x}+cx^2)dt, \quad x(t_0)=x_0, \quad x(t_1)=x_1$$

not to have corners.

Exercise 1.9.3 Prove that the following problem has a solution by using the last part of theorem 10 and note 16

$$\max \int_0^T -e^{-(bK-\dot{K})}\mathrm{d}t, \qquad K(0)=K_0>0, \qquad K(T)\geqslant K_T>0.$$

10. Generalizations*

In this chapter we have considered some problems in the classical calculus of variations of interest to economists. We shall now offer some comments on different types of generalizations that are sometimes needed in economic applications.

Infinite horizon

One of the main areas for application of variational methods in economics is growth theory. In many of the planning models constructed one has introduced the fiction that the planning period is infinite, which leads to variational problems with unbounded domains of integration. Two quite serious mathematical problems arise: (1) What is the natural class of admissible functions? (One possibility: Piecewise smooth functions for which the integral converges.) (2) What are the correct transversality conditions at infinity? (It is extremely unlikely that your first guess is correct.)

For properly specified problems of this sort, the Euler equation is still a necessary condition for optimality. We shall take a closer look at these problems in Chapter 3. (See note 3.17.)

The integrand depends on higher-order derivatives

Consider the problem of finding a function $x(t)$ that maximizes or minimizes

$$\int_{t_0}^{t_1} F\left(t, x, \frac{\mathrm{d}x}{\mathrm{d}t}, \frac{\mathrm{d}^2x}{\mathrm{d}t^2}, \ldots, \frac{\mathrm{d}^nx}{\mathrm{d}t^n}\right)\mathrm{d}t, \tag{92}$$

where $x(t)$ and its first $n-1$ derivatives have preassigned values at t_0 and t_1

and F is a given function. One can show that a necessary condition for $x(t)$ to solve this problem is that the following *generalized Euler equation* is satisfied:

$$\frac{\partial F}{\partial x} - \frac{\mathrm{d}}{\mathrm{d}t}\left(\frac{\partial F}{\partial \dot{x}}\right) + \frac{\mathrm{d}^2}{\mathrm{d}t^2}\left(\frac{\partial F}{\partial \ddot{x}}\right) - \cdots + (-1)^n \frac{\mathrm{d}^n}{\mathrm{d}t^n}\left(\frac{\partial F}{\partial x^{(n)}}\right) = 0. \tag{93}$$

We shall not go into this problem, but refer to Gelfand and Fomin (1969) and Hestenes (1966).

The integrand depends on two variables

Another type of generalization that we shall briefly mention is the one in which the variable function depends on more than one variable. With suitable restrictions on F, we pose the problem of maximizing or minimizing

$$\iint_R F\left(t, s, x, \frac{\partial x}{\partial t}, \frac{\partial x}{\partial s}\right) \mathrm{d}t\,\mathrm{d}s, \tag{94}$$

where R is a closed domain in the plane and $x = x(t, s)$ is the unknown function. In addition, some restrictions on the behaviour of $x(t, s)$ on the boundary of R is usually imposed. One can show that a necessary condition for $x(t, s)$ to solve the problem is that it satisfies the following partial differential equation

$$\frac{\partial F}{\partial x} - \frac{\partial}{\partial t}\left(\frac{\partial F}{\partial x'_t}\right) - \frac{\partial}{\partial s}\left(\frac{\partial F}{\partial x'_s}\right) = 0. \tag{95}$$

We refer to Gelfand and Fomin (1969) for further details.

Variational problems with constraints

Generalizations in another direction than those mentioned above play a greater role in economics. We have in mind dynamic optimization problems in which we maximize or minimize an integral subject to constraints in the form of equations or inequalities. Although most problems of this kind are solvable by classical methods, optimal control theory is by now the natural

tool for attacking problems of this type. At this point we shall therefore end
our treatment of the classical theory.

Exercise 1.10.1 Let

$$J(x) = \int_0^{\pi/2} (\ddot{x}^2 - x^2 + t^2) dt,$$

$$x(0) = 1, \qquad \dot{x}(0) = 0, \qquad x\left(\frac{\pi}{2}\right) = 0, \qquad \dot{x}\left(\frac{\pi}{2}\right) = -1.$$

Find the associated Euler eq. (93) and its solution.

Exercise 1.10.2 Let

$$J(y) = \int_{-1}^{1} (\tfrac{1}{2}\mu y''^2 + \rho y) dx, \qquad y(-1) = 0,$$

$$y'(-1) = 0, \qquad y(1) = 0, \qquad y'(1) = 0$$

where μ and ρ are constants. Find the associated Euler equation and its
solution.

Exercise 1.10.3* Consider the standard variational problem

$$\max \int_{t_0}^{T} F(t, x, \dot{x}) dt, \qquad x(t_0) = x_0, \qquad x(T) = x_1 \in R \qquad (96)$$

where F is a C^2-function and the admissible functions, $x = x(t)$, are piecewise
smooth. Suppose $x^* = x^*(t)$ solves problem (96).
(a) Let u be an arbitrary real number, let t_1 be a point in $[t_0, T]$. Put
$a = u - \dot{x}^*(t_1)$, and define for $t \leqslant T$, $g(t) = x^*(t) + a(t - t_1)$, $\mu(t) = -a(t - t_0)$. Let
$t_1(\alpha)$ be the solution of the equation $x^*(t) + \alpha\mu(t) = g(t)$, $\alpha \geqslant 0$. Prove that
$t_1(\alpha) \leqslant t_1$ and $t_1'(0) < 0$.
(b) The function defined by

$$x(t) = \begin{cases} x^*(t) + \alpha\mu(t) & t \in [t_0, t_1(\alpha)] \\ g(t) & t \in (t_1(\alpha), t_1] \\ x^*(t) & t \in (t_1, T] \end{cases} \qquad (97)$$

is admissible in problem (96). Define the function J for $\alpha \geqslant 0$ by

$$J(\alpha) = I(\alpha) + \int_{t_1(\alpha)}^{t_1} F(t, g(t), \dot{g}(t)) dt + \int_{t_1}^{T} F(t, x^*(t), \dot{x}^*(t)) dt,$$

where $I(\alpha)$ is defined by (33) with $g(t)$ as defined in (a). Prove (see the proof of (36)) that

$$J'(0) = \left[[F]_{t=t_1} + (\dot{g}(t_1) - \dot{x}^*(t_1)) \left(\frac{\partial F}{\partial \dot{x}} \right)_{t=t_1} \right] t_1'(0)$$

$$- F(t_1, g(t_1), \dot{g}(t_1)) \cdot t_1'(0). \tag{98}$$

(c) By the optimality of $x^*(t)$, $J(\alpha) \leqslant J(0)$ for all $\alpha \geqslant 0$, so $J'(0) \leqslant 0$. Thus prove that for all t,

the function $u \to F(t, x^*(t), u) - u \dfrac{\partial F(t, x^*(t), \dot{x}^*(t))}{\partial \dot{x}}$

has a maximum at $u = \dot{x}^*(t)$. \hfill (99)

(This is the *Weierstrass necessary condition* for problem (96). In fact, by properly interpreting the symbols above, the arguments are valid also if $x(t)$ is a vector function.)

(d) Use the result in (c) to prove theorem 2.

CHAPTER 2

OPTIMAL CONTROL THEORY WITHOUT RESTRICTIONS ON THE STATE VARIABLES

In the previous chapter we studied certain problems in the classical calculus of variations. The theory we presented provides the background needed to understand a number of dynamic optimization problems appearing in the economic literature. However, in most of these problems there are features not easily incorporated into the classical framework as presented in Chapter 1. In particular, constraints in the form of inequalities are essential aspects in many of the problems discussed.

In this chapter we shall deal with the basic results for optimal control problems in which the values of the control variables are restricted. We present the Pontryagin Maximum Principle along with several sufficiency theorems and discuss the problem of existence of an optimal control. Here it is assumed that there are no restrictions on the state variables. Later, problems involving restricted state variables will be studied in Chapters 4, 5, and 6.

1. A sketch of the problem

Let us begin with a rough sketch of the type of economic problems that can be formulated as optimal control problems. In the process we introduce our notation.

Consider an economy evolving in time. At any time t, the economic system is in some *state*, which can be described by n real numbers

$$x_1(t), x_2(t), \ldots, x_n(t) \qquad \text{(state variables)}. \tag{1}$$

The amounts of capital goods in n different sectors of the economy might, for example, be suitable state variables.

It is often convenient to consider (1) as defining the coordinates of the vector $(x_1(t), \ldots, x_n(t))$ in R^n. As t varies, this vector occupies different positions in R^n, and we say that the system moves along a curve in R^n, or traces a path in R^n.

Let us assume now that the process going on in the economy (causing the $x_i(t)$'s to vary with t) can be controlled to a certain extent in the sense that there are a number of *control functions*

$$u_1(t), u_2(t), \ldots, u_r(t) \tag{2}$$

that influence the process. These control functions, or control variables, also called *decision variables* or *instruments* will typically be economic data such as tax rates, interest rates, the allocations of investments to different sectors etc. To proceed we have to know the laws governing the behaviour of the economy through time, in other words *the dynamics* of the system. We shall concentrate on the study of systems in which the development is determined by a system of differential equation in the form

$$\frac{dx_1(t)}{dt} = f_1(x_1(t), \ldots, x_n(t), u_1(t), \ldots, u_r(t), t)$$

$$\cdots\cdots\cdots\cdots\cdots\cdots\cdots\cdots\cdots\cdots\cdots\cdots\cdots\cdots$$

$$\frac{dx_n(t)}{dt} = f_n(x_1(t), \ldots, x_n(t), u_1(t), \ldots, u_r(t), t). \tag{3}$$

The functions $f_1, \ldots, f_n$ are given functions describing the dynamics of the economy. The assumption is thus that the rate of change of each state variable, in general, depends on all the state variables, all the control variables, and also explicitly on time t. The explicit dependence of the f_i-functions on t is necessary, for example, to allow for the laws underlying (3) to vary over time due to exogenous factors such as technological progress, growth in population, etc.

By using vector notation the system (3) can be described in a simple form. If we put

$$x(t) = (x_1(t), \ldots, x_n(t)), \qquad u(t) = (u_1(t), \ldots, u_r(t)), \qquad f = (f_1, \ldots, f_n),$$

then (3) is equivalent to

$$\frac{dx(t)}{dt} = f(x(t), u(t), t). \tag{4}$$

Suppose that the state of the system is known at time t_0, so that $x(t_0) = x^0$, where x^0 is a given vector in R^n. If we choose a certain control function $u(t) = (u_1(t), \ldots, u_r(t))$ defined for $t \geqslant t_0$, and insert it into (3), we obtain a system of n first-order differential equations with n unknown

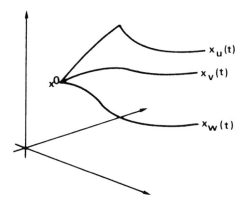

Fig. 2.1 Curves corresponding to the control functions u, v, w.

functions $x_1(t), \ldots, x_n(t)$. Since the initial point x^0 is given, the system (3) will, in general, have a unique solution $x(t) = (x_1(t), \ldots, x_n(t))$, geometrically represented by a curve in R^n. Since this solution is "a response" to the control function $u(t)$, it would have been appropriate to denote it by $x_u(t)$, but we usually drop the subscript u.

By suitable choices of the control function $u(t)$ many different evolutions of the economy can be achieved, each of them described by a curve in R^n. (See fig. 2.1.) However, it is unlikely that the possible evolutions will be equally desirable. We assume, then, as is usual in economic analysis, that the different alternative developments give different "utilities" that can be measured. More specifically, we shall associate with each control function $u(t)$ and its response $x(t)$ the number

$$J = \int_{t_0}^{t_1} f_0(x_1(t), \ldots, x_n(t), u_1(t), \ldots, u_r(t), t) \, dt$$

$$= \int_{t_0}^{t_1} f_0(x(t), u(t), t) \, dt, \tag{5}$$

where f_0 is a given function. Here t_1 is not necessarily fixed, and $x(t)$ might have some terminal condition on it at the end point t_1. The fundamental problem that we study in this chapter can now be formulated:

Among all control functions $u(t)$ that via (4) bring the system from the initial state x^0 to a final state satisfying the terminal conditions, find one (provided there exists any) such that J given by (5) is as large as possible.

Such a control function is called *an optimal control* and the associated path $x(t)$ is called *an optimal path*. J is often called *the criterion functional*.

In the next section we shall state this problem more precisely.

2. Precise assumptions; The main problems

Above we have indicated a rather general class of optimal control problems. In addition to the n state variables $x_1(t), \ldots, x_n(t)$ we introduced r control variables, $u_1(t), \ldots, u_r(t)$. The choice of these control functions determine the evolution of the system. So far we have not considered any restrictions to be placed on the control functions. However, in physical as well as in economic applications, the control functions usually cannot vary freely. Instead they are subject to constraints. If, for example, we study a multisectoral model and a control variable $u_1 = u_1(t)$ is the fraction of the total investment in the economy allocated to sector No. 1, then $u_1(t)$ can only vary between 0 and 1. Thus

$$0 \leqslant u_1(t) \leqslant 1$$

must be satisfied for all t. The domain of variation for $u_1(t)$ is thus the closed interval $[0, 1]$.

For our general problem we shall assume that $u(t) = (u_1(t), \ldots, u_r(t))$ is restricted to vary in a given set U in R^r, called *the control region*. U might be any given set in R^r, but of particular importance is the case in which U is a closed set in R^r. Actually, the fact that $u(t)$ is allowed to take values at the boundary of U is an important aspect of optimal control theory, since the classical calculus of variations cannot easily handle problems of this kind.

Another non-classical aspect of the new theory is the weak requirements placed on the control functions. In many applications an assumption of continuity is too restrictive. In the multisectoral model indicated above, one might, for example, wish to allow for the possibility that the optimal control of the process might involve sudden shifts in the control variables. As an illustration we might think of cases in which it is optimal to invest everything in sector No. 1 until $t = t'$, and thereafter invest nothing in this sector. Then $u_1(t)$ will be of the following form

$$u_1(t) = \begin{cases} 1 & t \in [t_0, t'] \\ 0 & t \in (t', t_1] \end{cases}.$$

Here $u_1(t)$ is a piecewise continuous function with a jump discontinuity at $t = t'$.

In general we assume that $u(t)$ is *piecewise continuous* in the sense that $u(t)$ has at most a finite number of discontinuities on each finite interval with finite jumps (i.e. one-sided limits) at each point of discontinuity. The values of $u(t)$ at the points of discontinuities will not be of any significance in the theory to follow, but let us decide that the value of $u(t)$ at a point of discontinuity t' is equal to the left-hand limit of $u(t)$ as t approaches t'. Let us require moreover that if $u(t)$ is defined in an interval $[t_0, t_1]$, then $u(t)$ is continuous at the end points of the interval.

The control functions we shall consider are thus piecewise continuous and take values in a fixed set U in R^r.

So far we have not stated the conditions to be placed on the functions $f_1, \ldots, f_n$ and f_0 involved in (3) and (5). It turns out that it suffices to impose the following assumptions:

$f_i(x, u, t)$ and $\partial f_i(x, u, t)/\partial x_j$ are continuous w.r.t. all the
$n + r + 1$ variables for $i = 0, 1, \ldots, n; j = 1, \ldots, n.$ (6)

(Differentiability of f_i w.r.t. $u_1, \ldots, u_r$ is, in particular, not required.) These regularity conditions are implicitly assumed in this book, unless otherwise noted.

Note that if $u(t)$ is discontinuous at τ, then $\dot{x}(\tau)$ does not necessarily exist, and the solution curve $x(t)$ has a kink at τ.

Two main problems

Two main classes of problems are examined in this chapter. First consider the problem formulated at the end of section 1. If we assume that the time interval $[t_0, t_1]$ is fixed (possibly with $t_1 = \infty$), then we are faced with a fixed initial and terminal time problem. Usually $x(t_0) = x^0$ is given, but there might be different types of boundary conditions at $t = t_1$. In section 4 the Pontryagin Maximum Principle is formulated for such a fixed time problem. This is a set of necessary conditions that an optimal control must satisfy. In the problem we have chosen boundary conditions at $t = t_1$ which are particularly useful in economic applications. In this case t_1 is finite. In chapter 3, section 7 we shall look at the corresponding infinite horizon problem.

The general Maximum Principle as stated in section 4 is a rather complicated result. Therefore in section 3 a simpler case is considered in which there are only one state variable and one control variable.

The second class of problems examined in this chapter is the variable terminal time problem. Usually in such problems, t_0 and $x(t_0) = x^0$ are given, and $x(t)$ is required to end up in a set T in R^n, but the time t_1 at which $x(t_1)$ "hits" the set T, may vary according to the control function used. The Maximum Principle for this case is presented in section 9.

In the two problems discussed above, the control region U is fixed, independent of t and x. Problems in which the range of the control functions depends on t and x are frequently encountered in economic applications. We shall deal with such problems in chapters 4, 5, and 6.

3. The Maximum Principle: Fixed time interval; One state and one control variable

In the previous section we formulated a rather general optimal control problem. Since the Maximum Principle for this general problem is quite complicated, we present first the special case in which there is only one state variable and one control variable.

The problem can be stated as such: Find a piecewise continuous control function $u(t)$ and an associated continuous and piecewise differentiable state vector $x(t)$, defined on the fixed time interval $[t_0, t_1]$, that will

$$\text{maximize} \int_{t_0}^{t_1} f_0(x(t), u(t), t) dt \tag{7}$$

subject to

$$\dot{x}(t) = f(x(t), u(t), t), \qquad x(t_0) = x^0 \qquad (x^0 \text{ a fixed number}), \tag{8}$$

one (and only one) of the following terminal conditions (x^1 fixed number)

$$\text{(a) } x(t_1) = x^1 \qquad \text{or} \qquad \text{(b) } x(t_1) \geq x^1 \qquad \text{or} \qquad \text{(c) } x(t_1) \text{ free,} \tag{9}$$

and the control variable restriction

$$u(t) \in U, \qquad U \text{ a given set on the real line.} \tag{10}$$

The problem is to find among all control functions $u(t)$ which via (8) bring $x(t)$ from the initial point x^0 to a point satisfying the given terminal condition, one which makes the integral in (7) as large as possible (provided such a control function exists).

Below we formulate a theorem which gives *necessary* conditions for a control function to solve this problem and thus to be an *optimal control*. Somewhat loosely analogous to our problem is the Lagrangian problem of maximizing a function subject to a constraint in the form of an equation. In our present case the problem is to maximize an integral subject to a constraint in the form of a differential equation. Associated with the constraint in the Lagrangian problem there is a constant, a Lagrangian multiplier. As to the problem above, the constraint is a differential equation on the interval $[t_0, t_1]$ and it can therefore be regarded as an infinite number of ordinary equations, one for each t. Hence, to the equation in (8) we associate a number $p(t)$ for each $t \in [t_0, t_1]$. The function $p(t)$ is usually called *an adjoint function* (or variable) associated with the differential equation. In addition, we associate a number p_0 to the function f_0 in (7).

In the Lagrangian multiplier theorem the Lagrangian function plays an important role. The comparable function in the present case is *the Hamiltonian function*, $H(x, u, p, t)$, defined by

$$H(x, u, p, t) = p_0 f_0(x, u, t) + p \cdot f(x, u, t). \tag{11}$$

The Maximum Principle transfers the problem of finding a $u(t)$ which maximizes the integral in (7) subject to the given constraints, to the problem of maximizing the Hamiltonian function w.r.t. $u \in U$. In addition it tells us how to determine the p-function. Precisely formulated, the *Pontryagin Maximum Principle* or *the Maximum Principle* (for our problem) is as follows.[1]

Theorem 1 (The Maximum Principle. Fixed time interval. $n = r = 1$) Let $u^*(t)$ be a piecewise continuous control defined on $[t_0, t_1]$ which solves problem $(7) \rightarrow (10)$ and let $x^*(t)$ be the associated optimal path. Then there exists a constant p_0 and a continuous and piecewise continuously differentiable function $p(t)$ such that for all $t \in [t_0, t_1]$,

$$(p_0, p(t)) \neq (0, 0) \tag{12}$$

$u^*(t)$ maximizes $H(x^*(t), u, p(t), t)$ for $u \in U$, that is,

$$H(x^*(t), u^*(t), p(t), t) \geq H(x^*(t), u, p(t), t) \text{ for all } u \in U. \tag{13}$$

---

[1] For references to proofs, see footnote 2, p. 85.

Except at the points of discontinuities of $u^*(t)$,

$$\dot{p}(t) = -\frac{\partial H^*}{\partial x} \qquad \text{where} \quad \frac{\partial H^*}{\partial x} = H'_x(x^*(t), u^*(t), p(t), t). \tag{14}$$

Furthermore,

$$p_0 = 1 \qquad \text{or} \qquad p_0 = 0. \tag{15}$$

Finally, to each of the terminal conditions (a), (b), (c) in (9) there corresponds a transversality condition:

$$p(t_1) \text{ no condition} \tag{16a}$$

$$p(t_1) \geqslant 0 (=0 \text{ if } x^*(t_1) > x^1) \tag{16b}$$

$$p(t_1) = 0. \quad \blacksquare \tag{16c}$$

Note 1 The inequality in (13) is valid for all $t \in [t_0, t_1]$. Recall our convention that at a point of discontinuity t' of u, $u(t')$ is equal to the left-hand limit of $u(t)$ when t approaches t'. Actually, the inequality in (13) is valid at t' also if we replace $u^*(t')$ by the right-hand limit of $u^*(t)$ at t'.

In the notes to the more general theorem 2 in the next section a few additional observations are supplied.

It is not at all evident how to apply this theorem. In fact, the manner in which it is used differs so significantly from one type of problem to another that no standard solution procedure can be devised. While presenting the more general theorem in the next section we shall indicate how "in principle" the theorem works and how it is able to single out one solution candidate (or a few).

Let us now consider two examples in which the Maximum Principle in each case produces a unique candidate for optimality. Our first example is a pure mathematical problem and it is almost as easy as can be.

Example 1

$$\max \int_0^1 x(t) \, dt \tag{17}$$

subject to

$$\dot{x}(t) = x(t) + u(t), \qquad x(0) = 0, \qquad x(1) \text{ free} \tag{18}$$

and a control variable restriction

$$-1 \leqslant u(t) \leqslant 1. \tag{19}$$

This is a problem covered by theorem 1. Let us see what the theorem tells us in this case.

Note first that the Hamiltonian is:

$$H = H(x, u, p, t) = p_0 x + p(x + u) = (p_0 + p)x + pu. \tag{20}$$

Suppose now that $(x^*(t), u^*(t))$ solves the problem. According to the theorem there must exist a constant p_0 and a continuous function $p = p(t)$ such that

$$(p_0, p(t)) \neq (0, 0) \qquad \text{for all } t \in [0, 1]. \tag{21}$$

Moreover, for each $t \in [0, 1]$,

$u^*(t)$ is that value of $u \in [-1, 1]$ which maximizes

$$H(x^*(t), u, p(t), t) = (p_0 + p(t)) x^*(t) + p(t) u. \tag{22}$$

The adjoint function $p(t)$ satisfies, according to (14), (except at the points of discontinuity of $u^*(t)$), the equation

$$\dot{p}(t) = -\frac{\partial H^*}{\partial x} = -p_0 - p(t). \tag{23}$$

Since $x(1)$ is free, (16c) implies

$$p(1) = 0. \tag{24}$$

From (21) we obtain in particular that p_0 and $p(1)$ cannot both be 0. Since $p(1) = 0$, $p_0 \neq 0$, and thus by (15), $p_0 = 1$. We have now put down all the information supplied by the Maximum Principle. To proceed, notice that the differential equation for $p(t)$ in (23) is particularly simple. In fact, it is a linear equation with constant coefficients so the solution is

$$p(t) = A e^{-t} - p_0 = A e^{-t} - 1. \tag{25}$$

To determine the constant A, we see that by (24) and (25), $0 = p(1) = A e^{-1} - 1$, so $A = e$ and therefore,

$$p(t) = e^{1-t} - 1. \tag{26}$$

Thus the adjoint function is fully determined. From (26) we see moreover that $\dot{p}(t) = -e^{1-t} < 0$ for all t. Since $p(1) = 0$, we conclude that $p(t) > 0$ in the interval $[0, 1)$.

Now let us turn to condition (22). Since only the term $p(t)u$ in the Hamiltonian depends on u, $u^*(t)$ is the value of $u \in [-1, 1]$ which maximizes $p(t)u$. When $t \in [0, 1)$, $p(t) > 0$, so in this case the maximum of $p(t)u$ is attained for $u = 1$. For $t = 1$, $p(t) = 0$ and (22) does not determine $u^*(1)$. The value of $u^*(t)$ at this single point is of no importance. However, we have previously decided to choose $u(t)$ as a continuous function at the endpoints of its domain of definition. Hence we must put $u^*(1) = 1$, so our proposal for an optimal control is this:

$$u^*(t) = 1, \qquad t \in [0, 1]. \tag{27}$$

The associated path $x^*(t)$ must satisfy (18). Solving this linear differential equation we get $x^*(t) = Be^t - 1$. By the initial condition $x^*(0) = 0$, so $B = 1$. Hence,

$$x^*(t) = e^t - 1. \tag{28}$$

We have now proved that *if* the given problem has a solution, the optimal control is given by (27) and the associated optimal path is given by (28). The corresponding value of the criterion functional is

$$\int_0^1 (e^t - 1)dt = e - 2. \tag{29}$$

We have illustrated above how the Maximum Principle works in a simple case. Actually, our effort was not necessary for solving the problem. From (18) and (19) we see that $u^*(t) \equiv 1$ produces the highest value of $x(t)$ for any $t \in [0, 1]$, and hence $u^*(t) \equiv 1$ must maximize $\int_0^1 x(t)dt$.

Our next example is a little more complicated and it is given an economic interpretation. It might be advisable to solve exercises 2.3.1 and 2.3.2 first.

Example 2   (Consumption versus investment)   Let U be a C^2-function of one variable with $U' > 0$ and $U'' < 0$ in $[0, \infty)$. Let $x(t)$ be a state variable, $u(t)$ a control variable and let T, x_0 and x_T be fixed positive numbers. Consider the following control problem:

$$\text{maximize} \int_0^T U(1 - u(t))dt, \tag{30}$$

$$\dot{x}(t) = u(t), \qquad x(0) = x_0, \qquad x(T) \geqslant x_T \tag{31}$$

$$u(t) \in [0, 1] \tag{32}$$

$$x_0 < x_T < x_0 + T. \tag{33}$$

(*Economic interpretation*: A country receives a constant stream of 1 unit of money as aid. Let $x(t)$ be the level of infrastructure at time t, and let $u(t)$ denote that part of the aid which is allocated to investment in infrastructure at time t. U is a utility function and $1 - u(t)$ is that part of the aid which is allocated to consumption. The planning period is $[0, T]$ and the assumption $x(T) \geqslant x_T$ means that the country aims at reaching at least the level x_T at the end of the planning period. The planning problem is to find that allocation of investment which maximizes the total utility.)

We again have a control problem of the type $(7) \rightarrow (10)$. In order to apply theorem 1, note first that the Hamiltonian is

$$H(x, u, p, t) = p_0 U(1 - u) + pu. \tag{34}$$

Assume that $(x^*(t), u^*(t))$ solves problem $(30) \rightarrow (33)$. According to theorem 1, there exist a constant p_0 and a continuous function $p = p(t)$, such that

$$(p_0, p(t)) \neq (0, 0) \qquad \text{for all } t \in [0, T], \tag{35}$$

and $p_0 = 0$ or $p_0 = 1$. Moreover,

> $u^*(t)$ is that value of $u \in [0, 1]$ which maximizes
$$H = H(x^*(t), u, p(t), t) = p_0 U(1 - u) + p(t)u. \tag{36}$$

The adjoint function $p(t)$ satisfies (except at points of discontinuity for $u^*(t)$),

$$\dot{p}(t) = -\frac{\partial H^*}{\partial x} = 0 \tag{37}$$

The transversality condition at $t = T$ is

$$p(T) \geqslant 0 \qquad (= 0 \text{ if } x^*(T) > x_T). \tag{38}$$

Having listed all the information supplied by the Maximum Principle, let us derive the implications of these conditions.

By (37) and (38) and the continuity of $p(t)$,

$$p(t) = c \geqslant 0 \qquad \text{(for some constant } c). \tag{39}$$

Consider next the maximum condition (36). We see that $H'_u = -p_0 U'(1 - u) + c$ and $H''_{uu} = p_0 U''(1 - u) \leqslant 0$, so the Hamiltonian is concave in

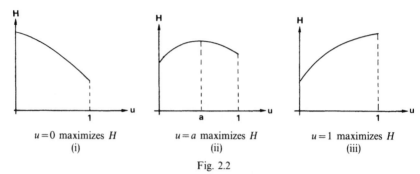

Fig. 2.2

u. There are three cases to consider, each illustrated in Fig. 2.2. If $u^*(t)=0$ then by (36), H'_u cannot be >0 for $u=0$, so $-p_0 U'(1)+c \leq 0$. If $u^*(t) \in (0,1)$, then $H'_u=0$ so $p_0 U'(1-u^*(t))=c$. Finally, if $u^*(t)=1$, then H'_u is ≥ 0 at $u=1$, so $-p_0 U'(0)+c \geq 0$. Briefly,

$$u^*(t)= \begin{cases} 0 & \Rightarrow c \leq p_0 U'(1) \\ \in (0,1) & \Rightarrow p_0 U'(1-u^*(t))=c \\ 1 & \Rightarrow p_0 U'(0) \leq c \end{cases} \tag{40}$$

We claim that $p_0=1$. Suppose on the contrary that $p_0=0$. Then using (35) and (39), we see that $H'_u=c>0$, so the Hamiltonian is maximized by $u=1$ for all t. Then by (31), $x(t)=t+x_0$, so $x(T)=T+x_0 > x_T$, by (33). But according to (38) this would imply $p(T)=c=0$, contradicting (35) for $t=T$.

Since $p_0=1$, $H=U(1-u)+cu$ is strictly concave and has a unique maximum point $\bar{u}$ which is independent of t. Hence $u^*(t)=\bar{u}$ for some choice of $\bar{u} \in [0,1]$. $\bar{u}=0$ is impossible since from (31) $x(t) \equiv x_0$, so $x(T)=x_0 \geq x_T$, contradicting (33). Since $\bar{u}>0$, $p(t)=c=0$ is impossible by (40). Hence from (38),

$$x^*(T)=x_T. \tag{41}$$

The optimality of allocating no aid to consumption can now be ruled out. In fact, if $\bar{u}=1$, then from (31), $x^*(t)=t+x_0$ and thus $x^*(T)=T+x_0$ which is inconsistent with (33) and (41). Thus $\bar{u} \in (0,1)$ and by (31), $x^*(t)=x_0+\bar{u}t$, so $x^*(T)=x_0+\bar{u}T$. Using (41) we see that

$$u^*(t)=\bar{u}=\frac{x_T-x_0}{T} \tag{42}$$

and, in turn,

$$x^*(t)=x_0+\frac{x_T-x_0}{T}t. \tag{43}$$

The Maximum Principle has supplied a unique proposal for an optimal pair $(x^*(t), u^*(t))$, with associated adjoint function (see (40))

$$p(t) \equiv c = U'\left(1 - \frac{x_T - x_0}{T}\right). \tag{44}$$

(It follows from a standard sufficiency theorem (see theorem 4) that we have found the optimal solution to the problem.)

The optimal value of the criterion functional will depend on the parameters x_0, x_T and T. If we denote it by $V = V(x_0, x_T, T)$, we obtain

$$V = \int_0^T U\left(1 - \frac{x_T - x_0}{T}\right) dt = U\left(1 - \frac{x_T - x_0}{T}\right) T, \tag{45}$$

and thus

$$\frac{\partial V}{\partial x_0} = U'\left(1 - \frac{x_T - x_0}{T}\right) = p(0),$$

$$\frac{\partial V}{\partial x_T} = -U'\left(1 - \frac{x_T - x_0}{T}\right) = -p(T). \tag{46}$$

Note that these results allow us to give price interpretations to the adjoint function. For instance, $-p(T)$ measures, approximately, the increase in the optimal total utility by increasing the terminal requirement on the level of infrastructure by one unit. In Chapter 3, section 5 we shall study general results which give price interpretations to the adjoint variables. See, in particular, example 3.7.

Exercise 2.3.1 In example 1, (i) $u(t) = \sin t$ and (ii) $u(t) = 1/2$ are both admissible controls. Compute the value, J, of the criterion functional in these cases and compare with the value found in (29) in the example.

Exercise 2.3.2 Consider the problem

$$\max \int_0^1 (x(t) + u(t)) dt, \qquad \dot{x}(t) = -x(t) + u(t) + t,$$

$$x(0)=1, \qquad x(1) \text{ free}, \qquad u(t) \in [0,1].$$

(i) Suppose $(x^*(t), u^*(t))$ solves the problem and write down all the conditions supplied by theorem 1.

(ii) Prove that $p_0 = 1$ (see example 1), and find the unique solution of the differential equation for $p(t)$ satisfying the transversality condition.

(iii) Show that the maximum condition tells us to choose $u^*(t)$ as that value of $u \in [0, 1]$ which maximizes $(1 + p(t))u$. Find $u^*(t)$.

(iv) Find the associated $x^*(t)$.

(v) Try to solve the problem without using the Maximum Principle. (Hint: Integrate the differential equation for $x(t)$, using $x(0)=1$.)

Exercise 2.3.3   Consider the problem

$$\max \int_0^T (1 - s(t))k(t)\, dt, \qquad \dot{k}(t) = s(t)k(t),$$

$$k(0) = k_0 > 0, \qquad k(T) \text{ free}$$

with $s(t) \in [0, 1]$, $T > 1$, s the control.

(i) Suppose $(k^*(t), s^*(t))$ solves the problem. Using theorem 1 show that $p_0 = 1$ and that

$$s^*(t) = 1 \qquad \text{if } p(t) > 1,$$

$$s^*(t) = 0 \qquad \text{if } p(t) < 1.$$

(ii) Show that $\dot{p}(t) = -1 + s^*(t)(1 - p(t))$ with $p(T)=0$, and that $\dot{p}(t) < 0$ in $[0, T]$. Conclude that in some interval $(t^*, T]$, $p(t) < 1$ with $p(t^*) = 1$. Find $p(t)$ on $(t^*, T]$ and show that $t^* = T - 1$.

(iii) Conclude that $s^*(t) = 1$ in $[0, T-1]$, $s^*(t) = 0$ in $(T-1, T]$.

(iv) Replace "$k(T)$ free" by $k(T) \geqslant k_T$ where $k_0 < k_T < k_0 e^T$, and find the unique candidate for optimality.

Exercise 2.3.4   In a model by K. Shell (1967) the following control problem is encountered:

$$\max \int_0^T (1 - s(t))e^{\rho t} f(k(t))e^{-\delta t}\, dt \qquad\qquad (47)$$

$$\dot{k}(t) = s(t)e^{\rho t} f(k(t)) - \lambda k(t), \qquad k(0) = k_0, \qquad k(T) \geqslant k_T,$$
$$0 \leqslant s(t) \leqslant 1. \tag{48}$$

Here $k(t)$ is the state variable (= the capital stock), $s(t)$ is the control variable (= the savings rate) and f a given function (= a production function). We assume that $f(k) > 0$ and that $\rho, \delta, \lambda, k_0$, and k_T are positive constants.

(i) Suppose $(k^*(t), s^*(t))$ solves the problem. Write down the differential equation for the adjoint function and the transversality condition.

(ii) What are the possible values taken by $s^*(t)$?

Exercise 2.3.5 Consider the model in example 2 for $x_T - x_0 = T$. Why is $u^*(t) \equiv 1$ the solution to the problem? Verify that the Maximum Principle is satisfied with $p(t) \equiv 1$ and $p_0 = 0$ in this case.

Exercise 2.3.6 Find the candidate(s) for optimality in the following problem by using theorem 1:

$$\max \int_0^T (1 - u(t))x(t)dt, \qquad \dot{x}(t) = u(t) \in [0, 1], \qquad x(0) = 0,$$

$$x(T) \geqslant x_1, \qquad x_1 < T/2.$$

Exercise 2.3.7 Consider the problem:

$$\max \int_0^T (1 - s(t))ak(t)dt, \qquad \dot{k}(t) = as(t)k(t), \qquad k(0) = k_0 > 0, \qquad k(T) \text{ free},$$

$$s(t) \in [0, 1], \qquad T > 1/a, \qquad s \text{ the control.} \tag{49}$$

(i) Note that in this model a depends on the time unit. Show that by a suitable choice of the time unit, we can obtain $a = 1$.

(ii) Find the solution candidate of (49) by using the results in exercise 2.3.3 and (i).

4. The Maximum Principle for problems with a fixed time interval

In the previous section we studied a fixed time optimal control problem with one state variable and one control variable. Let us consider now the general case in which there are n state variables and r control variables.

Briefly formulated our problem is as follows: Find a piecewise continuous control vector function $u(t) = (u_1(t), \ldots, u_r(t))$ and an associated continuous and piecewise differentiable state vector function $x(t) = (x_1(t), \ldots, x_n(t))$, defined on the fixed time interval $[t_0, t_1]$, that will

$$\text{maximize} \int_{t_0}^{t_1} f_0(x_1(t), \ldots, x_n(t), u_1(t), \ldots, u_r(t), t) \, dt,$$

$$(t_0, t_1 \text{ fixed}) \tag{50}$$

subject to the differential equations

$$\frac{dx_1(t)}{dt} = f_1(x_1(t), \ldots, x_n(t), u_1(t), \ldots, u_r(t), t)$$

$$\ldots\ldots\ldots\ldots\ldots\ldots\ldots\ldots\ldots\ldots\ldots\ldots\ldots\ldots\ldots\ldots$$

$$\frac{dx_n(t)}{dt} = f_n(x_1(t), \ldots, x_n(t), u_1(t), \ldots, u_r(t), t) \tag{51}$$

initial conditions

$$x_i(t_0) = x_i^0, \qquad i = 1, \ldots, n$$
$$(x^0 = (x_1^0, \ldots, x_n^0) \text{ a fixed point in } R^n) \tag{52}$$

terminal conditions

$$x_i(t_1) = x_i^1 \qquad i = 1, \ldots, l, \qquad\qquad (x_i^1 \text{ fixed}) \tag{53a}$$

$$x_i(t_1) \geqslant x_i^1 \qquad i = l+1, \ldots, m, \qquad (x_i^1 \text{ fixed}) \tag{53b}$$

$$x_i(t_1) \text{ free} \qquad i = m+1, \ldots, n, \qquad \blacksquare \tag{53c}$$

and control variable restriction

$$u(t) = (u_1(t), \ldots, u_r(t)) \in U \subseteq R^r \qquad (U \text{ a fixed set in } R^r). \tag{54}$$

In problem $(7) \rightarrow (10)$ the development of the system was governed by one differential equation describing the rate of change of the single state variable present. We associated with that differential equation an adjoint variable $p(t)$. In the present case there are n state variables and n differential equations. By analogy, we associate with each of these differential equations an adjoint function, numbered consecutively $p_1(t)$, $p_2(t), \ldots, p_n(t)$, and we form the *Hamiltonian function* $H(x, u, p, t)$ defined by

$$H(x, u, p, t) = p_0 f_0(x, u, t) + \sum_{i=1}^{n} p_i f_i(x, u, t)$$

$$= p_0 f_0(x, u, t) + p \cdot f(x, u, t). \tag{55}$$

Here $p = (p_1, \ldots, p_n)$, and $p \cdot f(x, u, t)$ denotes the scalar product of the vectors p and $f(x, u, t)$.

The *Pontryagin Maximum Principle* or the *Maximum Principle* for our problem is a direct generalization of Theorem 1[2] and it provides necessary conditions for optimality.

Theorem 2 (**The Maximum Principle. Fixed time interval**) Let $u^*(t)$ be a piecewise continuous control defined on $[t_0, t_1]$ which solves problem $(50) \rightarrow (54)$ and let $x^*(t)$ be the associated optimal path. Then there exist a constant p_0 and a continuous and piecewise continuously differentiable vector function $p(t) = (p_1(t), \ldots, p_n(t))$ such that for all $t \in [t_0, t_1]$,

$$(p_0, p_1(t), \ldots, p_n(t)) \neq (0, 0, \ldots, 0), \tag{56}$$

$u^*(t)$ maximizes $H(x^*(t), u, p(t), t)$ for $u \in U$, that is,

$$H(x^*(t), u^*(t), p(t), t) \geqslant H(x^*(t), u, p(t), t) \text{ for all } u \in U. \tag{57}$$

Except at the points of discontinuities of $u^*(t)$, for $i = 1, \ldots, n$:

$$\dot{p}_i(t) = -\frac{\partial H^*}{\partial x_i} \quad \text{where} \quad \frac{\partial H^*}{\partial x_i} = \frac{\partial H(x^*(t), u^*(t), p(t), t)}{\partial x_i} \tag{58}$$

Furthermore,

$$p_0 = 1 \text{ or } p_0 = 0, \tag{59}$$

and, finally, the following transversality conditions are satisfied,

$$
\begin{array}{lll}
p_i(t_1) \text{ no conditions} & \text{for } i = 1, \ldots, l, & \text{(60a)} \\
p_i(t_1) \geqslant 0 \, (=0 \text{ if } x_i^*(t_1) > x_i^1) & \text{for } i = l+1, \ldots, m, & \text{(60b)} \\
p_i(t_1) = 0 & \text{for } i = m+1, \ldots, n. \quad \blacksquare & \text{(60c)}
\end{array}
$$

---

[2]For proofs, see Pontryagin (1962b), Hestenes (1966), Lee and Markus (1967), Fleming and Rishel (1975) and Michel (1977).

Before we indicate how to go about using this theorem, let us note a few related facts.

Note 2 (a). The maximum condition (57) holds for left- and right-hand limits of $u^*(t)$, $t \in [t_0, t_1]$. (See also note 1.)

(b). If U is convex and H is strictly concave in u, then $u^*(t)$ is continuous.

Note 3 One can prove that $H(x^*(t), u^*(t), p(t), t)$ is continuous at all $t \in [t_0, t_1]$. Moreover, if $\partial f_i / \partial t$, $i = 0, 1, \ldots, n$ exist and are continuous, the following additional property can be obtained:

$$\frac{d}{dt} H(x^*(t), u^*(t), p(t), t) = \frac{\partial H(x^*(t), u^*(t), p(t), t)}{\partial t} \tag{61}$$

at all points of continuity of $u^*(t)$. (If $u^*(t)$ is differentiable and if $u^*(t)$ is an interior point of U, then (61) follows from the properties in theorem 2 in the following way (using vector notation):

$$\frac{d}{dt} H^* = \frac{\partial H^*}{\partial x} \cdot \dot{x}^* + \frac{\partial H^*}{\partial u} \cdot \dot{u}^* + \frac{\partial H^*}{\partial p} \cdot \dot{p} + \frac{\partial H^*}{\partial t}.$$

Since $\partial H^* / \partial x = -\dot{p}$, $\partial H^* / \partial p = f^* = \dot{x}^*$, and $\partial H^* / \partial u = 0$, the conclusion follows.)

Note 4 If some of the inequalities in (53b) are reversed, the corresponding inequalities in (60b) are reversed.

Note 5 Let us assume in problem $(50) \to (54)$ that the right endpoint is completely free, so that (53) reduces to the condition that $x_i(t_1)$ is free for $i = 1, \ldots, n$. Then by (60), $p_1(t_1) = \cdots = p_n(t_1) = 0$. From (56) with $t = t_1$, we conclude that $p_0 = 0$ is impossible. Hence from (59),

$$x_i(t_1) \text{ free}, \qquad i = 1, \ldots, n \Rightarrow p_0 = 1. \tag{62}$$

In the economic literature dealing with optimal control theory it is quite common to see, without any justification, the assumption $p_0 = 1$. Yet there are cases in which the Maximum Principle is satisfied with $p_0 = 0$. (See exercises 2.3.5, 2.4.4 and 2.9.9.) Incidentally, exercise 2.3.5 shows that the "abnormal" case $p_0 = 0$ is possible for problems with one state variable, contrary to a statement in Clark (1976), p. 108. We refer to problems in which $p_0 = 0$ as abnormal, since if $p_0 = 0$ the conditions in the Maximum Principle do not change if we replace the given function f_0 by *any* other function.

The problem is analogous to the corresponding problem in nonlinear programming. The numbers $p_0, p_1(t_1), \ldots, p_n(t_1)$ appear in the proof of the Maximum Principle as components of the normal of a separating hyperplane in much the same way as the multipliers $\lambda_0, \lambda_1, \ldots, \lambda_m$ do in the Fritz–John formulation of the Kuhn–Tucker Theorem. (See Takayama (1974), p. 106.) The Kuhn–Tucker constraint qualification implies that $\lambda_0 \neq 0$.

It *is* possible to find general criteria which guarantee that $p_0 = 1$ in optimal control problems. However, these criteria turn out not to be particularly useful.

Note 6* (Weaker regularity conditions.)
(a). Theorem 2 is still valid if we in (6) only assume of $\partial f_i / \partial x_j$ that they are defined and continuous for $(x, u, t) \in D$, where D is a neighbourhood in $E = R^n \times U \times [t_0, t_1]$ of the set $\{(x^*(t), u^*(t), t): t \in [t_0, t_1]\}$.
(b). The continuity and differentiability assumptions in theorem 2 and in (a) can be further relaxed. It is sufficient to make the following assumptions: There exists a finite or countable set of time points C in (t_0, t_1) such that the following properties hold: The functions f_i are *regular in E* in the sense that they are continuous as functions on E at any point $(x', u', t') \in E$ if $t' \notin C$, while if $t' \in C$, the limits of $f_i(x, u, t)$ exist both when $(x, u, t) \to (x', u', t'^-)$ and when $(x, u, t) \to (x', u', t'^+)$, $(x, u, t) \in E$. Finally, the functions $\partial f_i / \partial x_j$ are regular in D.

In this case, condition (57) is not satisfied for $t \in C$, nor do the properties in note 3 hold. Generalizations of this type are necessary, for instance, in the theory of differential games.

Note 7* Replace (54) by the condition that $u(t) \in U(t)$, where $U(t)$ for each t is a given set in R^r. Then the conditions in theorem 2 are still valid for U replaced by $U(t)$ in (57), provided $U(t)$ satisfies certain regularity conditions. In particular, if $U(t) = \{u : g_k(u, t) \geq 0, k = 1, \ldots, s\}$, then the following condition is sufficient: Define $I(u, t) = \{k : g_k(u, t) = 0\}$, and assume that the matrix $\{\partial g_k(u, t) / \partial u_j\}_{k \in I(u, t), j = 1, \ldots, r}$ has a rank equal to the number of elements in $I(u, t)$ for all (u, t) for which $I(u, t)$ is non-empty.[3]

---

[3] In the general case it suffices to assume that $U(t)$ is non-empty and contained in the closure of the set

$$\bigcup_n \bigcap_{\tau \in I_n} U(\tau),$$

where
$$I_n = [t - 1/n, \, t + 1/n] \cap [t_0, t_1].$$

The Maximum Principle is a rather complicated result. But, then, it provides *necessary* conditions for optimality in a very general optimization problem.

As mentioned in the previous section, there is no standard solution procedure that can be devised for optimal control problems. However, theorem 2 suggests a method that could be applied at least "in principle", if not in practice. We consider only the case $p_0 = 1$.

Note first that the terminal conditions (53) and the transversality conditions (60), taken together, amount essentially to n conditions on the pair $(x^*(t_1), p(t_1))$: (53a) and (60a) represents l conditions on the pair. (53b) and (60b) tells us that either $x_i(t_1) = x_1^i$ or $p_i(t_1) = 0$ for $i = l+1, \ldots, m$, thus representing $m - l$ conditions (disregarding for the moment the inequalities that have to be satisfied). Finally, (53c) and (60c) give us $n - m$ conditions. In all we thus have $l + m - l + n - m = n$ conditions on the pair $(x^*(t_1), p(t_1))$.

Now, let us assume that the maximization of $u \to H(x, u, p, t)$ for each (x, p, t) yields a unique maximum point u and that it leads to a "well-behaved" function $u = u(x, p, t)$. Insert this function for u into the vector differential equations $\dot{x} = f(x, u, t)$, $\dot{p} = -\partial H / \partial x$ (corresponding to (51) and (58)). Solve this system of $2n$ differential equations with initial conditions $x(t_0) = x^0$, $p(t_0) = p^0$, the latter being an unknown n-dimensional vector. The solutions $x(t, p^0)$, $p(t, p^0)$ both depend on p^0. The n parameters in p^0 are determined by the n conditions associated with (53) and (60). (The inequalities in (53b) and (60b), can in this setting be regarded as additional tests that the solution must pass.)

Heuristic as it may seem, the reasoning above is important because it indicates that the Maximum Principle contains enough conditions to single out just one or a few candidates for optimality. However, the procedure should not be taken as a guideline for the actual solving of control problems.

The examples studied in the previous section demonstrate that it is rather difficult to apply the Maximum Principle. Overviewing the difficulties, note firstly that a maximization of a function (the Hamiltonian) w.r.t. r variables $u_1, \ldots, u_r$ is involved, which sometimes lead to complicated nonlinear programming problems. Secondly, to solve the (usually) nonlinear differential equations in (51) and (58) is difficult, to say the least. The fact that the solutions of these equations are required to satisfy some initial conditions as well as some terminal conditions creates additional problems that can be formidable.

In spite of these difficulties, a vast number of papers published from 1964 onwards have applied the Maximum Principle to theoretical economic problems. As a matter of fact, in most of these papers explicit solutions have not been obtained; instead the principle has been used to deduce qualitative information about the optimal solutions.

To see how the principle works in a case with more than one state variable, we shall consider a two sector model.

Example 3 (**A two-sector model**)   Consider an economy consisting of two sectors where sector no. 1 produces investment goods, sector no. 2 produces consumption goods. Let

$x_i(t) =$ production in sector no. i per unit of time, $i = 1, 2$

and let $u(t)$ be the proportion of investments allocated to sector no. 1. We assume that $\dot{x}_1 = aux_1$ and $\dot{x}_2 = a(1 - u)x_1$, where a is a positive constant. Hence, the increase in production per unit of time in each sector is assumed to be proportional to investment allocated to the sector. By definition, $0 \leqslant u(t) \leqslant 1$, and if the planning period starts at $t = 0$, $x_1(0)$ and $x_2(0)$ are historically given.

In this situation a number of optimal control problems could be investigated. Let us, in particular, consider the problem of maximizing the total consumption in a given planning period $[0, T]$. Our precise problem is as follows:

$$\text{maximize} \int_0^T x_2(t) \mathrm{d}t \tag{63}$$

$$\dot{x}_1(t) = au(t)x_1(t), \qquad x_1(0) = x_1^0, \qquad x_1(T) \text{ is free}, \tag{64}$$

$$\dot{x}_2(t) = a(1 - u(t))x_1(t), \qquad x_2(0) = x_2^0, \qquad x_2(T) \text{ is free}, \tag{65}$$

$$0 \leqslant u(t) \leqslant 1. \tag{66}$$

Here a, x_1^0 and x_2^0 are positive constants.

Note first that the Hamiltonian in this case is

$$H = H(x, u, p, t) = p_0 x_2 + p_1 aux_1 + p_2 a(1 - u)x_1 \tag{67}$$

where p_1 and p_2 are the adjoint variables associated with (64) and (65), respectively.

Now, suppose $x^*(t) = (x_1^*(t), x_2^*(t))$ and $u^*(t)$ solve the problem. By (62), $p_0 = 1$. According to theorem 2, there exists a continuous vector function

$p(t) = (p_1(t), p_2(t))$ such that for each $t \in [0, T]$,

u*(t) is a value of $u \in [0, 1]$ which maximizes

$$H(x^*(t), u, p(t), t) = x_2^*(t) + p_1(t)aux_1^*(t) + p_2(t)a(1 - u)x_1^*(t). \qquad (68)$$

The adjoint functions $p_1 = p_1(t)$ and $p_2 = p_2(t)$ satisfy (except at the points of discontinuity for $u^*(t)$),

$$\dot{p}_1 = -\frac{\partial H^*}{\partial x_1} = -p_1 au^*(t) - p_2 a(1 - u^*(t)), \qquad (69)$$

$$\dot{p}_2 = -\frac{\partial H^*}{\partial x_2} = -1. \qquad (70)$$

Finally, since $x_1(T)$ and $x_2(T)$ are free,

$$p_1(T) = 0 \quad \text{and} \quad p_2(T) = 0. \qquad (71)$$

From (70) we get $p_2(t) = -t + C$, and by (71), $p_2(T) = -T + C = 0$, so

$$p_2(t) = T - t. \qquad (72)$$

Consider next (68). Collecting the terms in H which depend on u, we see that $u^*(t)$ must be chosen as a value of u which maximizes $a(p_1(t) - p_2(t))x_1^*(t)u$. Now, $x_1^*(0) = x_1^0 > 0$ and $\dot{x}_1^*(t) = au^*(t)x_1^*(t) \geqslant 0$, so $x_1^*(t) > 0$ for all t. Hence we see that (68) tells us to choose $u^*(t)$ in this way:

$$u^*(t) = \begin{cases} 1 & p_1(t) > p_2(t) \\ 0 & p_1(t) < p_2(t) \end{cases}. \qquad (73)$$

(When $p_1(t) = p_2(t)$, the value of $u^*(t)$ is not determined by (68). It turns out that $p_1(t) = p_2(t)$ at no more than two points (one of them being $t = T$) and the values we assign to $u^*(t)$ at these isolated points are of no importance.)

Note that by (69) and (71), $\dot{p}_1(T) = 0$. Since $\dot{p}_2(t) = -1$ and $p_1(T) = p_2(T) = 0$, it follows that $p_1(t) < p_2(t)$ in an interval to the left of T. Let t^* be the greatest value of t for which $p_1(t) \geqslant p_2(t) = T - t$. (Possibly, $t^* = 0$.) If we use (73), we see that $u^*(t) = 0$ in (t^*, T). Hence by (69), $\dot{p}_1(t) = -ap_2(t) = -a(T - t)$ for $t \in (t^*, T]$. By integration, it follows that $p_1(t) = \frac{1}{2}a(T - t)^2 + C_1$. Since $p_1(T) = 0$, it follows that $C_1 = 0$, so

$$p_1(t) = \frac{1}{2}a(T - t)^2, \quad t \in [t^*, T]. \qquad (74)$$

By the definition of t^*, $p_1(t^*) = p_2(t^*)$ if $t^* > 0$, and using (72) and (74) it follows that

$$t^* = T - 2/a \quad \text{(if } t^* > 0\text{)} \qquad (75)$$

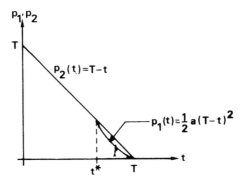

Fig. 2.3.

Suppose $T > 2/a$ so that $t^* > 0$. How does $p_1(t)$ behave in $[0, t^*]$? To examine this problem, note first that by (73) and (69)

$$\dot{p}_1(t) = \begin{cases} -a\, p_1(t) & \text{if} & p_1(t) > p_2(t) \\ -a\, p_2(t) & \text{if} & p_1(t) \leqslant p_2(t) \end{cases} \tag{76}$$

If $p_1(t) > p_2(t)$, then $-p_1(t) < -p_2(t)$, and hence $-ap_1(t) < -ap_2(t)$. If follows from (76) that whatever is the relation between $p_1(t)$ and $p_2(t)$, we have

$$\dot{p}_1(t) \leqslant -ap_2(t) = a(t - T).$$

In particular, for $t < t^*$, $\dot{p}_1(t) \leqslant a(t - T) < a(t^* - T) = -2$. Since $\dot{p}_2(t) = -1$ for all t and $p_1(t^*) = p_2(t^*)$, it follows that $p_1(t) > p_2(t)$ for $t < t^*$. Hence $u^*(t) = 1$ for $t \in [0, t^*]$. We conclude that if $T > 2/a$, the Maximum Principle suggests the following optimal control:

$$u^*(t) = \begin{cases} 1 & \text{in } [0, T - 2/a] \\ 0 & \text{in } (T - 2/a, T] \end{cases} \quad (T > 2/a). \tag{77}$$

From (69) it follows that $\dot{p}_1 = -ap_1$, i.e. $p_1(t) = Ce^{-at}$ in $[0, t^*]$ for some constant C. Since $p_1(t^*) = p_2(t^*) = T - t^* = 2/a$, we obtain

$$p_1(t) = \frac{2}{a} e^{-a(t - T + 2/a)} \quad \text{in } [0, T - 2/a]. \tag{78}$$

It is an easy task to find the associated functions $x_1^*(t)$ and $x_2^*(t)$, see exercise 2.4.1. The associated value of total consumption depends on the parameters x_1^0, x_2^0, T and a and it turns out to be

$$V(x_1^0, x_2^0, T, a) = \int_0^T x_2^*(t)dt = x_2^0 T + \frac{2}{a}x_1^0 e^{aT-2}. \tag{79}$$

Hence,

$$\frac{\partial V}{\partial x_1^0} = \frac{2}{a}e^{aT-2} = p_1(0), \qquad \frac{\partial V}{\partial x_2^0} = T = p_2(0), \tag{80}$$

where we used (78) and (72). Again we see special cases of general results which give interesting price interpretations to the adjoint variables. See also exercise 2.4.1.

For $T \leqslant 2/a, t^* = 0$, and the suggestion for an optimal control becomes:

$$u^*(t) = 0 \qquad \text{for all } t \in [0, T] \qquad (T \leqslant 2/a). \tag{81}$$

In this model the Maximum Principle provided us with only one suggestion for an optimal control (in each of the two cases $T > 2/a$, $T \leqslant 2/a$). In example 5 we prove that the optimal solution has been found.

Example 4   (Growth that pollutes)   Consider an economy in which, at any time t, the production $y(t)$ is proportional to the capital stock $k(t)$, $y(t) = ak(t)$, $a > 0$. A fraction $s(t)$ of $y(t)$ is invested, and the fraction $1 - s(t)$ of $y(t)$ is left for consumption. Let $z(t)$ be the accumulated waste at time t, and assume that the increase in waste per unit of time, $\dot{z}(t)$, is proportional to production $y(t)$, hence also proportional to the capital stock, say $\dot{z}(t) = ck(t)$, $c > 0$. The planning period is $[0, T]$, $k(0)$ and $z(0)$ are historically given, $k(T)$ is left free, while we require that $z(T) \leqslant z_T$, for some upper limit z_T of permissible waste.

Now, the benefits arising from consumption must be weighted against the disutility of the accumulated waste. We assume that the rate of net utility at any time t is $(1 - s(t))ak(t) - bz(t)$, where b is a positive constant. By changing the time unit, we may assume that $a = 1$. Moreover, by changing the physical units for measuring $z(t)$, we may assume $c = 1$. Our problem is then the following:

$$\max \int_0^T [(1 - s(t))k(t) - bz(t)]dt \tag{82}$$

subject to

$$\dot{k}(t) = s(t)k(t), \qquad k(0) = k_0, \qquad k(T) \text{ free} \tag{83}$$

$$\dot{z}(t) = k(t), \qquad z(0) = z_0, \qquad z(T) \leqslant z_T \tag{84}$$

$$s(t) \in [0, 1] \tag{85}$$

where T, b, k_0, z_0, and z_T are positive constants. In order to avoid too many cases to consider, we assume that

$$k_0 T < z_T - z_0, \qquad b < 1/2 \qquad \text{and} \qquad T > [1 - (1 - 2b)^{1/2}]/b. \tag{86}$$

In our problem there are two state variables, k and z, and one control variable, s. The Hamiltonian is

$$H = H(k, z, s, p_1, p_2, t) = p_0[(1 - s)k - bz] + p_1 sk + p_2 k \tag{87}$$

where p_1 and p_2 are the adjoint variables corresponding to k and z respectively.

Let us assume that $(k^*(t), z^*(t), s^*(t))$ solves our problem. The differential equations for the adjoint variables are

$$\dot{p}_1 = -\frac{\partial H^*}{\partial k} = -p_0(1 - s^*(t)) - p_1 s^*(t) - p_2 \tag{88}$$

$$\dot{p}_2 = -\frac{\partial H^*}{\partial z} = p_0 b. \tag{89}$$

Furthermore, from the transversality conditions, using note 4,

$$p_1(T) = 0 \tag{90}$$

$$p_2(T) \leqslant 0 \ (= 0 \text{ if } z^*(T) < z_T). \tag{91}$$

Suppose $p_0 = 0$. Then by (89), $p_2(t)$ is a constant. Since from (56), $(p_0, p_1(T), p_2(T)) \neq (0, 0, 0)$, it follows using (90) and (91) that $p_2(t) \equiv p_2(T) < 0$. Utilizing (A.20) (with $t_0 = T$), it follows that $p_1(t) < 0$ in all of $[0, T)$. From (83), $k(t) > 0$ for all t. To find a $s \in [0, 1]$ which maximizes the Hamiltonian, we maximize $p_1(t)k(t)s$ w.r.t. $s \in [0, 1]$. Since $p_1(t)k(t) < 0$, the optimal choice of s is $s = 0$. Thus $k(t) = k_0$, and $\dot{z}(t) = k_0$, i.e. $z(t) = z_0 + k_0 t$, and thus $z(T) = z_0 + k_0 T < z_T$ by (86). But this contradicts (91) since $p_2(T) < 0$. Hence $p_0 \neq 0$ and we can put $p_0 = 1$.

For $p_0 = 1$, $H = sk(p_1 - 1) + k - bz + p_2 k$. Since $k^*(t) > 0$ for all t, the maximum condition (57) gives us:

$$p_1(t) > 1 \Rightarrow s^*(t) = 1 \tag{92a}$$

$$p_1(t)=1 \Rightarrow s^*(t) \text{ is undetermined} \tag{92b}$$

$$p_1(t)<1 \Rightarrow s^*(t)=0 \tag{92c}$$

From (89) with $p_0=1$, $\dot{p}_2(t)=b$, so integrating and noting (91),

$$p_2(t)=b(t-T)+p_2(T), \qquad p_2(T)\leqslant 0. \tag{93}$$

Since $p_1(T)=0$, and $p_1(t)$ is continuous, $p_1(t)<1$ on some interval $(t',T]$. Let w be the smallest t' with this property. By (92c), $s^*(t)=0$ on $(w,T]$, so from (88) and (93), integrating and using (90), we obtain

$$p_1(t)=(1+p_2(T))(T-t)-\tfrac{1}{2}b(T-t)^2, \qquad t\in(w,T]. \tag{94}$$

The number w must be the largest root of the quadratic equation $p_1(t)=1$, so

$$w=T-\frac{1}{b}\left\{1+p_2(T)-[(1+p_2(T))^2-2b]^{1/2}\right\}. \tag{95}$$

(We must choose $-$ rather than $+$ in front of the root sign in order to obtain the largest root.)

Suppose that the radicand in (95) is <0, or that it is $\geqslant 0$ with $w\leqslant 0$. Then $p_1(t)<1$ in $(0,T]$, and from (92) we conclude that $s^*(t)\equiv 0$ in $[0,T]$. By (83) and (84), $k^*(t)=k_0$, $z^*(t)=k_0t+z_0$. In particular, by (86), $z^*(T)=k_0T+z_0<z_T$. From (91) we deduce that $p_2(T)=0$, and (95) reduces to

$$w=T-\frac{1}{b}[1-(1-2b)^{1/2}]. \tag{96}$$

By (86), $2b<1$, so the radicand in (96) is >0. By (86) again, $w>0$. We have obtained a contradiction and it follows that we have indeed that w is real and $w>0$. The behaviour of $p_1(t)$ given by (94) is then as indicated in fig. 2.4(i) or (ii).

Let us now consider what happens with $p_1(t)$ to the left of w. Note first that because of (92), equation (88) can be written in the form

$$\dot{p}_1(t)=-\max(1,p_1(t))-p_2(t). \tag{97}$$

We can discard the possibility in fig. 2.4(ii). In this case the function on the right-hand side of (94) satisfies (97) on all of $[0,T]$ (and (97) has no other solution on $[0,T]$). Then $p_1(t)<1$ except at a single point w, and therefore $s(t)\equiv 0$. But we saw above that $s(t)\equiv 0$ is impossible.

Since the behaviour in fig. 2.4(ii) is impossible, $\dot{p}_1(w)<0$ (see fig. 2.4(i)).

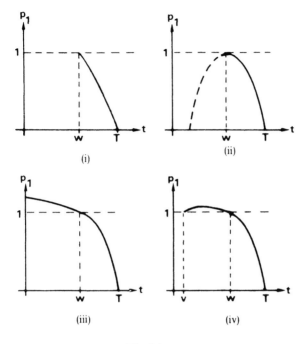

Fig. 2.4

From (97) we see that $\dot{p}_1(t)$ is continuous everywhere, and therefore $\dot{p}_1(t)<0$ for t to the left of w and close to w. Hence $p_1(t)>1$ on some interval (t',w). Le v be the smallest t' with this property. In the interval (v,w), $p_1(t)>1$, so by (97), $\dot{p}_1(t)=-p_1(t)-p_2(t)$. Making use of (93) and the fact that $p_1(w)=1$, we obtain after integrating the first-order differential equation ((95) is also used),

$$p_1(t)=he^{w-t}-bt+bT+b-p_2(T), \qquad t\in(v,w), \qquad (98)$$

where $h=[(1+p_2(T))^2-2b]^{1/2}-b$.

If $v\leqslant 0$, the situation is as indicated in fig. 2.4(iii) and $s^*(t)\equiv 1$ in $[0,w]$.

Consider next the case in which $v>0$. The situation is then as indicated in fig. 2.4(iv) with $p_1(v)=1$. We want to examine the behaviour of $p_1(t)$ to the left of v. Let us define the function g for all t by the formula on the right-hand side of (98). (Thus g and p_1 coincide on (v,w).) Then $g'(t)=-he^{w-t}-b$, $g''(t)=he^{w-t}$. If h were $\geqslant 0$, then $g'(t)$ would be <0 for all t. But then $v>0$

would be impossible. Hence h must be <0 in this case, i.e.

$$[(1+p_2(T))^2 - 2b]^{1/2} < b. \tag{99}$$

For $h<0$, $g''(t)<0$, so g is strictly concave. Since $g(v)=g(w)=1, g'(v)>0$ (see fig. 2.4(iv)). From (97) we see that $\dot{p}_1(t)$ is continuous everywhere, also at v. Hence $\dot{p}(v)=g'(v)>0$. But then $p_1(t)<1$ in some interval (y,v). Let y be the minimal number with this property. For all $t\in(y,v)$,

$$\dot{p}_1(t) = -\max(1, p_1(t)) - p_2(t) = -1 - p_2(t) > -1 - p_2(v)$$

since $p_2(t)$ is strictly increasing (see (93)). Since $-1-p_2(v)=\dot{p}_1(v)>0$, we see that $\dot{p}_1(t)>0$ for all $t\in(y,v)$ and $p_1(y)<1$. By the minimality of y we conclude that $y=0$. Thus $s^*(t)$ does not switch in the interval $[0,v)$.

We can now give a summary of our results. Having *assumed* existence of an optimal solution to our problem, we made use of the Maximum Principle to characterize the solution. We also proved that $s^*(t)$ has to switch at least once and at most twice. The relationship between the coefficients in our problem will determine which case prevails. Let us examine the different possibilities.

Case I (*The one switch case*) Here $s^*(t)=1$ in $[0,w]$, $s^*(t)=0$ in $(w,T]$. From (83) and (84) we obtain further

$$k^*(t) = \begin{cases} k_0 e^t & \text{in } [0,w] \\ k_0 e^w & \text{in } (w,T] \end{cases} \tag{100}$$

$$z^*(t) = \begin{cases} k_0 e^t + z_0 - k_0 & \text{in } [0,w] \\ k_0 e^w t + k_0 e^w(1-w) + z_0 - k_0 & \text{in } (w,T] \end{cases} \tag{101}$$

(a) With $p_2(T)=0$, w is given by (96). By (86) we see that w is well-defined (since $b<1/2$) and that $w>0$. Since $b>0$, it is obvious that $w<T$. The only thing we have to test in this case is that $z^*(T)\leqslant z_T$ (using (101)).
(b) Consider next the case $p_2(T)\neq0$. In this case $z^*(T)=z_T$. We now solve the equations (95) and $z^*(T)=z_T$ (see (101)) for the unknowns $p_2(T)$ and w. We then test if there is a solution with $p_2(T)<0$, $w\in(0,T)$.

Case II (*The two switch case*) Then $s^*(t)=0$ on $[0,v]$, 1 on $(v,w]$ *and* 0 on $(w,T]$.
From (83) and (84),

$$k^*(t) = \begin{cases} k_0 & \text{in } [0,v] \\ k_0 e^{t-v} & \text{in } (v,w] \\ k_0 e^{w-v} & \text{in } (w,T] \end{cases} \tag{102}$$

$$z^*(t) = \begin{cases} k_0 t + z_0 & \text{in } [0, v] \\ k_0 e^{t-v} + k_0(v-1) + z_0 & \text{in } (v, w] \\ k_0 e^{w-v}(t-w+1) + k_0 v - k_0 + z_0 & \text{in } (w, T] \end{cases} \tag{103}$$

(a) With $p_2(T) = 0$, w is given by (96) and $w \in (0, T)$. With these values of $p_2(T)$ and w, solve the equation $p_1(v) = 1$ (p_1 given by (98)) for v. Test if we have a solution $v \in (0, w)$ for which $z^*(T) \leqslant z_T$ (see (103)).

(b) Consider next the case $p_2(T) \neq 0$. Then $z^*(T) = z_T$. We now solve equations (95), $p_1(v) = 1$ (p_1 given by (98)) and $z^*(T) = z_T$ (see (103)) for $p_2(T)$, w and v. Test if we have a solution for which $p_2(T) < 0$, $w \in (0, T)$, $v \in (0, w)$.

We have thus described all possible cases that can occur. A candidate found in any of the cases I(a), (b), II(a), (b) passing the tests described, is a candidate for optimality. Whether or not any of these cases gives rise to a candidate depends on the values assigned to b, k_0, z_0, z_T and T. In exercise 2.4.3 we ask you to examine the model further. In particular, you are asked to prove that for any given set of values of the parameters, no more than one of the four cases has a solution.

Note, finally, that in spite of all our work, we still do not know if there is an optimal solution to our problem. For a proof of the existence of an optimal control, see exercise 2.8.4. Until such a proof is provided, we cannot be sure of finding a candidate in any of the cases I(a), (b), II(a), (b).

A pollution problem closely related to the problem above is studied in example 5.3.

Exercise 2.4.1 Consider example 3.

(i) For $T > 2/a$, find the functions $x_1^*(t)$ and $x_2^*(t)$ corresponding to the control function $u^*(t)$ given in (77).
(ii) Verify formula (79).
(iii) Prove that $\partial V / \partial T = H(x_1^*(T), x_2^*(T), u^*(T), p(T), T)$, where H is given in (67) (with $p_0 = 1$).

Exercise 2.4.2 Consider problem (63)→(66) in example 3. Replace the integral in (63) by

$$J = \int_0^T a(1 - u(t))x_1(t)\mathrm{d}t,$$

and find the unique solution candidate. (Compare with exercise 2.3.7.)

Exercise 2.4.3*   Consider example 4 again. Prove the following statements:

(i)   If $p_2(T)$ decreases from 0, then for each $t \in [0, T]$, $p_1(t)$ decreases. (Hint: $\dot{p}_1 = -\max(1, p_1) + b(T - t) - p_2(T)$. Apply the results in Appendix A.3.)

(ii) As $p_2(T)$ decreases from 0, w moves to the left, v moves to the right, $k^*(t)$ decreases and $z^*(t)$ decreases.

(iii) Case I(a) has at most one solution, and the same goes for case II(a)

(iv) If we find a solution $s^*(t)$, $k^*(t)$, $z^*(t)$ such that $z^*(T) = z_T$ and $p_2(T) < 0$ (in case I(b) and case II(b)), then case I(a) and case II(a) have no solutions. (Hint: Increase $p_2(T)$ to 0. What happens?). Moreover, case I(b) has at most one solution, and the same goes for case II(b).

(v)   At most one of the cases I(b) and II(b) has a solution. At most one of the cases I(a) and II(a) has a solution.

(vi) If (for a certain set of values of the parameters) there is a solution to our problem, then that solution is unique.

Exercise 2.4.4*   Prove that in the following cases the indicated controls are optimal and satisfy the Maximum Principle only for $p_0 = 0$.

(i) $\max \displaystyle\int_0^1 u\,dt, \qquad \dot{x} = (u - u^2)^2, \qquad x(0) = 0, \qquad x(1) = 0,$

$u \in [0, 2]$ $(u^*(t) \equiv 1$ is optimal)

(ii) $\max \displaystyle\int_0^1 \left(t - \frac{1}{2}\right) u\,dt, \qquad \dot{x} = u, \qquad x(0) = 0, \qquad x(1)$ free $\qquad \dot{y} = (x - tu)^2,$

$y(0) = 0, \qquad y(1) = 0, \qquad u \in (-\infty, \infty).$

(Any constant control is optimal. Note that $x - tu \equiv 0 \Rightarrow \dot{x} - u - t\dot{u} \equiv -t\dot{u} \equiv 0$.)

(iii) $\max \displaystyle\int_0^1 (u - 2v)\,dt, \qquad \dot{x} = (u - v)^2, \qquad x(0) = x(1) = 0, \qquad u \in [-1, 1],$

$v \in [-1, 1].$

$(u^*(t) \equiv v^*(t) \equiv -1$ is optimal.)

Exercise 2.4.5*   Assume that theorem 2 is known to hold for $l = n$ (fixed end case).

(i) Prove theorem 2 for the case $l=m \leqslant n$. (Hint: Let $(x^*(t), u^*(t))$ solve the problem and consider an auxiliary problem defined on $[t_0, t_1+1]$: Let f_0 and f be unchanged on $[t_0, t_1]$. Put $f_0=0$ and $f=v$ on $[t_1, t_1+1]$ where the control $v \in V = \{x \in R^n : x_i = 0 \text{ for } i=1, \dots, l\}$. We require that $x(t_0)=x^0$, $x(t_1+1)=x^*(t_1)$. Let $u^*(t) \in U$ be arbitrarily defined in $(t_1, t_1+1]$ and let $v^*(t) \equiv 0$. Show that $(u^*(t), v^*(t))$ is an optimal control in the new problem, and apply necessary conditions to derive (60) for $l=m \leqslant n$.)

(ii) Prove theorem 2 in the general case. (Hint: Put $x=y+z$ and consider the auxiliary problem:

$$\max \int_{t_0}^{t_1} f_0(y+z, u, t)dt, \qquad \dot{y}=f(y+z, u, t)-w, \qquad y(t_0)=x^0$$

$y_i(t_1)=x_i^1$, $i=1, \dots, m$, $y_i(t_1)$ free for $i>m$, $\dot{z}=w, z(t_0)=0$, $z(t_1)$ free; $w \in \{x:x_i=0, i \leqslant l \text{ and } i>m, x_i \geqslant 0 \text{ for } l+1 \leqslant i \leqslant m\}$. Here y and z are state variables and u, w are control variables. Choose for $l<i \leqslant m$, $w_i^* = 0$ if $x_i^*(t_1)=x_i^1$, $w_i^* > 0$ if $x_i^*(t_1)>x_i^1$ and $x_i(t_1)-\int_{t_0}^{t_1} w_i^*(t)dt=x_i^1$.)

Exercise 2.4.6 **(Current value Hamiltonians)** Consider problem (50)→(54) assuming that $f_0(x, u, t)=f^0(x, u, t)e^{-\delta t}$, where $\delta>0$ is a discount factor. Prove that in this case the necessary conditions can be rewritten in the following way: Define the current value Hamiltonian by $H^c(x, u, q, t)= p_0 f^0 + qf$. Then $(x^*(t), u^*(t))$ can only solve the problem if there exist a number $p_0, p_0=0$ or $p_0=1$, and a continuous and piecewise continuously differentiable function $q(t)$ such that $(p_0, q(t)) \neq (0,0)$ for all $t, u^*(t)$ maximizes $H^c(x^*(t), u, q(t), t)$ for all $t, q(t)$ satisfies v.e.

$$\dot{q}_i = -\partial H^c(x^*(t), u^*(t), q(t), t)/\partial x_i + \delta q_i$$

and $q(t_1)$ satisfies (60). (Hint: In theorem 2, replace $p(t)$ by $q(t)e^{-\delta t}$.)

Exercise 2.4.7 For the problem in exercise 2.3.4, write down the current value Hamiltonian, the differential equation for $q(t)$ and the transversality condition. What does the maximum condition (57) say in this case?

5. The Maximum Principle and the calculus of variations

Control theory can be regarded as an extension of the classical calculus of variations. In this section we show how the main problem of Chapter 1 can be formulated as an optimal control problem and see what the Maximum Principle tells us in this case.

Let us consider the problem studied in Chapter 1, section 8:

$$\max \int_{t_0}^{t_1} F(t, x_1, \ldots, x_n, \dot{x}_1, \ldots, \dot{x}_n) dt, \qquad x_i(t_0) = x_i^0, \qquad i = 1, \ldots, n \qquad (104)$$

subject to the boundary conditions (53a), (53b) and (53c). This problem is transformed to a control problem simply by letting $\dot{x}_1, \ldots, \dot{x}_n$ be the control functions. The given variational problem is thereby equivalent to the following control problem:

$$\max \int_{t_0}^{t_1} F(t, x_1, \ldots, x_n, u_1, \ldots, u_n) dt, \qquad x_i(t_0) = x_i^0, \qquad i = 1, \ldots, n \qquad (105)$$

where $\dot{x}_1 = u_1, \ldots, \dot{x}_n = u_n$ and the boundary conditions (53a), (53b) and (53c) are imposed.

In the original variational problem there were no restrictions on $\dot{x}_1, \ldots, \dot{x}_n$. Hence in our control problem there are no restrictions on the control functions, and the control region U is all of R^n. Note, moreover, that it suffices to assume that $x_1(t), \ldots, x_n(t)$ have piecewise continuous derivatives, since this assumption implies that $u_1(t), \ldots, u_n(t)$ are piecewise continuous, as is assumed in the Maximum Principle.

The control problem with which we are faced has a fixed time interval, and the system of differential equations is particularily simple. Assuming that an optimal solution exists, let us apply theorem 2. The Hamiltonian is

$$H = p_0 F(t, x_1, \ldots, x_n, u_1, \ldots, u_n) + \sum_{i=1}^{n} p_i u_i. \qquad (106)$$

According to the Maximum Principle, H as a function of $u_1, \ldots, u_n$ attains a maximum at the optimal control. Since U is all of R^n and H is differentiable w.r.t. $u_1, \ldots, u_n$, at the optimum we must have:

$$\frac{\partial H}{\partial u_i} = p_0 \frac{\partial F}{\partial u_i} + p_i = 0, \qquad i = 1, \ldots, n. \qquad (107)$$

The Maximum Principle tells us that the adjoint functions (associated with an optimal control) and p_0 cannot simultanuously be 0 for any t. From (107) we see that $p_0 = 0$ would imply $p_i = 0$ for all i. Hence $p_0 \neq 0$, so we can put $p_0 = 1$. The differential equations for the adjoint functions are

$$\dot{p}_i = -\frac{\partial H}{\partial x_i} = -p_0 \frac{\partial F}{\partial x_i} = -\frac{\partial F}{\partial x_i}. \tag{108}$$

Differentiating equation (107) w.r.t. t, we obtain

$$\frac{\mathrm{d}}{\mathrm{d}t}\left(\frac{\partial F}{\partial u_i} + p_i\right) = \frac{\mathrm{d}}{\mathrm{d}t}\left(\frac{\partial F}{\partial u_i}\right) + \dot{p}_i = 0.$$

Combining this result with (108), and replacing u_i with $\dot{x}_i$, we finally get

$$\frac{\partial F}{\partial x_i} - \frac{\mathrm{d}}{\mathrm{d}t}\left(\frac{\partial F}{\partial \dot{x}_i}\right) = 0, \qquad i = 1, \ldots, n. \tag{109}$$

These are precisely the Euler equations associated with our variational problem.

Since $p_0 = 1$ and $u_1 = \dot{x}_1, \ldots, u_n = \dot{x}_n$, we obtain from (107) and (108):

$$\frac{\partial F}{\partial \dot{x}_i} = -p_i(t) = -p_i(t_0) + \int_{t_0}^{t} \frac{\partial F}{\partial x_i} \mathrm{d}\tau, \tag{110}$$

which are the Euler equations in integrated form. (Cf. equation (85) of Chapter 1.) It follows immediately that the transversality conditions (60) in theorem 2 are identical to those obtained in theorem 4 of Chapter 1. Thus that theorem is reaffirmed.

Actually, the Maximum Principle reveals more information about (105) than we have related above. If we assume that F is a C^2-function, we obtain from (107), using $p_0 = 1$,

$$\frac{\partial^2 H}{\partial u_j \partial u_i} = \frac{\partial^2 F}{\partial u_j \partial u_i} \qquad i, j = 1, \ldots, n. \tag{111}$$

A necessary condition for H to have a maximum is that

$$\sum_{i=1}^{n} \sum_{j=1}^{n} \frac{\partial^2 H}{\partial u_j \partial u_i} h_i h_j \leqslant 0 \qquad \text{for all } h_i, h_j$$

(see e.g. Sydsæter (1981), theorem 5.8. p. 230). If we use (111), we see that the Legendre necessary condition in theorem 5 of Chapter 1 follows.

According to the Maximum Principle, the adjoint function $p_i(t)$ is continuous for all.$t \in [t_0, t_1]$. Because of (110) we see that the Weierstrass–Erdmann corner condition (86) of Chapter 1 follows.

Note also that the maximum condition (57) reduces to the Weierstrass necessary condition stated in exercise 1.10.3.

We proved above that the Maximum Principle applied to a standard variational problem leads to the Euler equation. With proper concavity conditions on F we can easily establish the converse.

Theorem 3* Consider the control problem

$$\max \int_{t_0}^{t_1} F(t, x, u)dt, \qquad \dot{x} = u, \qquad x(t_0) = x^0,$$

$x(t_1)$ satisfies (53). $\hspace{4cm}$ (112)

Assume that $F(t, x, u)$ is C^2 and concave in $u = (u_1, \ldots, u_n)$ and let $x^*(t) = (x_1^*(t), \ldots, x_n^*(t))$ be a solution of the Euler equations (109) on an interval I contained in $[t_0, t_1]$. Then $u^*(t) = \dot{x}^*(t)$ maximizes the Hamiltonian $H = F(t, x^*(t), u) + \sum_{i=1}^{n} p_i(t)u_i$ for $u \in R^n$, $t \in I$, with $p_i(t)$ defined by

$$p_i(t) = -\frac{\partial F(T, x^*(T), u^*(T))}{\partial u_i} - \int_T^t \frac{\partial F(\tau, x^*(\tau), u^*(\tau))}{\partial x_i} d\tau \hspace{2cm} (113)$$

where T is any fixed point in I. $\blacksquare$

Proof $H(x, u, p, t) = F(t, x, u) + pu$ is concave in u. Hence $u^*(t)$ maximizes $H(x^*(t), u, p(t), t)$ for $u \in R^n$ provided for all $i = 1, \ldots, n$

$$\frac{\partial H(x^*(t), u^*(t), p(t), t)}{\partial u_i} = \frac{\partial F(t, x^*(t), u^*(t))}{\partial u_i} + p_i(t) = 0.$$

By integrating the Euler equation (109), from T to t, we obtain:

$$\frac{\partial F(t, x^*(t), u^*(t))}{\partial u_i} = \frac{\partial F(T, x^*(T), u^*(T))}{\partial u_i}$$

$$+ \int_T^t \frac{\partial F(\tau, x^*(\tau), u^*(\tau))}{\partial x_i} d\tau$$

The conclusion follows. $\blacksquare$

Exercise 2.5.1   In an article by S. Strøm the following problem arises:

$$\max \int_0^T \{U(x(t)) - b(x(t)) - gz(t)\} dt$$

$$\dot{z}(t) = ax(t), \qquad z(0) = z_0, \qquad z(T) \text{ free.}$$

Here $U(x)$ measures the utility enjoyed by the society when producing the amount x, $b(x)$ denotes the total cost associated with the production and $z(t)$ is the amount of waste in the nature at time t. We assume that $U' > 0$, $U'' < 0$, $b' > 0$, $b'' > 0$. The state variable in this case is $z(t)$, while $x(t)$ is the control variable, $x(t) \in (-\infty, \infty)$. a and g are positive constants.

(i)   What is the Hamiltonian associated with the problem?
(ii)   Let $p(t)$ be the adjoint function associated with $z(t)$. Find the differential equation for $p(t)$.
(iii)   What does the transversality conditions say in this case? Prove that $p_0 = 1$.
(iv)   Using (ii) and (iii), prove that $p(t) = g(t - T)$, $t \in [0, T]$.
(v)   Prove that in order for $x(t)$ to solve the problem, it must satisfy

$$U'(x(t)) = b'(x(t)) + ag(T - t). \tag{114}$$

(vi)   Prove that if $x(t)$ satisfies (114), then $x(t)$ is strictly increasing. (Hint: Differentiate (114) w.r.t. t.)
(vii)   Transform the problem above to a variational problem by eliminating $x(t)$ and find the associated Euler equation. Give a new proof of (114).

6. Sufficient conditions for fixed time problems

As previously emphasized, the Pontryagin Maximum Principle gives necessary conditions for optimality. If the principle proposes a certain number of solution candidates, we know there are no other that can possibly solve the problem. However, the Maximum Principle cannot by itself tell us whether a given candidate is optimal or not, nor does it tell us whether or not an optimal solution exists.

The analogy with the calculus of variations suggests that provided we impose suitable concavity/convexity conditions on the functions involved, a solution candidate satisfying the necessary conditions will indeed be optimal.

Let us consider the problem in connection with theorem 2. Suppose we find a control function $u^*(t)$ with associated path $x^*(t)$, a constant p_0 and functions $p_1(t),\ldots,p_n(t)$ such that all the conditions in theorem 2 are satisfied. Is $x^*(t)$ an optimal path? If $u(t) \in U$ is any piecewise continuous control and $x(t)$ is a continuously differentiable function such that (51), (52) and (53) are satisfied, we call $(x(t), u(t))$ *an admissible pair*. Then $(x^*(t), u^*(t))$ is optimal provided for *all* admissible pairs $(x(t), u(t))$,

$$\Delta = \int_{t_0}^{t_1} f_0(x^*(t), u^*(t), t) dt - \int_{t_0}^{t_1} f_0(x(t), u(t), t) dt \geqslant 0. \tag{115}$$

Suppose $p_0 = 1$. Then using (55), $f_0(x, u, t) = H(x, u, p, t) - p \cdot f(x, u, t)$, so if we put $x^* = x^*(t)$, $u^* = u^*(t)$, etc., and make use of the fact that $\dot{x}^* = f(x^*, u^*, t)$, $\dot{x} = f(x, u, t)$, we get

$$\Delta = \int_{t_0}^{t_1} [H(x^*, u^*, p, t) - H(x, u, p, t)] dt + \int_{t_0}^{t_1} p \cdot (\dot{x} - \dot{x}^*) dt. \tag{116}$$

Suppose U is a convex set, and suppose that $H(x, u, p, t)$ is concave as a function of x and u (jointly), with $p = p(t)$ as the adjoint vector function supplied by the Maximum Principle, $x \in R^n$, $u \in U$, $t \in [t_0, t_1]$. Furthermore, suppose $f_0, \ldots, f_n$ fulfil our previous regularity condition and in addition that $\partial f_i / \partial u_j$ exist and are continuous. If we simplify our notation and put $H^* = H(x^*, u^*, p, t)$, $H = H(x, u, p, t)$, etc., and if we use a standard result on concave functions (see (B.1)), we obtain:

$$H - H^* \leqslant \frac{\partial H^*}{\partial x} \cdot (x - x^*) + \frac{\partial H^*}{\partial u} \cdot (u - u^*) \tag{117}$$

In vector notation (58) tells us that $\partial H^* / \partial x = -\dot{p}$, so it follows that

$$\Delta \geqslant \int_{t_0}^{t_1} [\dot{p} \cdot (x - x^*) + p \cdot (\dot{x} - \dot{x}^*)] dt + \int_{t_0}^{t_1} \frac{\partial H^*}{\partial u} \cdot (u^* - u) dt. \tag{118}$$

From the concavity condition on H it follows, in particular, that $H(x^*(t), u, p, t)$ is concave in u for each $t \in [t_0, t_1]$. Then, according to exercise 2.6.5 (iii),

$$\frac{\partial H^*}{\partial u} \cdot (u^* - u) \geqslant 0 \text{ for all } u \in U \text{ if and only if}$$

the maximum condition (57) is satisfied. $\tag{119}$

Hence we see that the second integral in (118) is ≥ 0, so

$$\Delta \geq \int_{t_0}^{t_1} [\dot{p} \cdot (x - x^*) + p \cdot (\dot{x} - \dot{x}^*)] dt. \tag{120}$$

Now, $\dot{p} \cdot (x - x^*) + p \cdot (\dot{x} - \dot{x}^*) = (d/dt)[p \cdot (x - x^*)]$, so

$$\Delta \geq \int_{t_0}^{t_1} \frac{d}{dt} [p \cdot (x - x^*)] dt = \Big|_{t_0}^{t_1} p \cdot (x - x^*)$$

$$= p(t_1) \cdot (x(t_1) - x^*(t_1)), \tag{121}$$

since the contribution from the lower limit is 0 as $x(t_0) - x^*(t_0) = 0$. Using (53) and (60) we see easily that

$$p(t_1) \cdot (x(t_1) - x^*(t_1)) = \sum_{i=1}^{n} p_i(t_1)(x_i(t_1) - x_i^*(t_1)) \geq 0.$$

Consequently, $\Delta \geq 0$ and $(x^*(t), u^*(t))$ is optimal. If $H(x, u, p(t), t)$ is strictly concave in (x, u), then the inequality (117) becomes a strict inequality if either $x \neq x^*$ or $u \neq u^*$.[4] Then $\Delta > 0$. This argument shows that any admissible pair $(x(t), u(t))$ which is not identically equal to $(x^*(t), u^*(t))$ is nonoptimal. Hence we have proved the following result:

Theorem 4 (**The Mangasarian sufficiency theorem**) Let $(x^*(t), u^*(t))$ be an admissible pair in problem $(50) \rightarrow (54)$. Suppose U is convex and that $\partial f_i / \partial u_j$ exist and are continuous. If there exists a continuous and piecewise continuously differentiable function $p(t) = (p_1(t), \ldots, p_n(t))$ such that the following conditions are satisfied with $p_0 = 1$,

$$\dot{p}_i(t) = -\frac{\partial H^*}{\partial x_i}, \qquad \text{v.e.} \qquad i = 1, \ldots, n \tag{122}$$

$$\sum_{j=1}^{r} \frac{\partial H^*}{\partial u_j} (u_j^*(t) - u_j) \geq 0 \text{ for all } u \in U \text{ for all } t \tag{123}$$

$$\begin{aligned}
& p_i(t_1) \text{ no conditions} && i = 1, \ldots, l \\
& p_i(t_1) \geq 0 \ (= 0 \text{ if } x_i^*(t_1) > x_i^1) && i = l+1, \ldots, m \\
& p_i(t_1) = 0 && i = m+1, \ldots, n
\end{aligned} \tag{124}$$

---

[4]See the text below (B.1).

$H(x, u, p(t), t)$ is concave in (x, u) for all t,   (125)

then $(x^*(t),\ u^*(t))$ solves problem $(50) \rightarrow (54)$.

If $H(x,\ u,\ p(t),\ t)$ is strictly concave in $(x,\ u)$, then $(x^*(t),\ u^*(t))$ is the unique solution to the problem. ∎

Let us consider the problem in example 1. Here the pair $u^* = 1$, $x^* = e^t - 1$ satisfies all the conditions in theorem 2 with $p_0 = 1$. The Hamiltonian is

$$H(x, u, p, t) = x + p(x + u)$$

which is linear and hence concave in x and u, and the control region $U = [-1,\ 1]$ is convex. According to theorem 4, the pair $u^*(t) = 1$, $x^*(t) = e^t - 1$ solves the problem.

Note 8   Let $A(t)$ be a convex set in R^{n+r} for each $t \in [t_0,\ t_1]$, and let us impose the additional restriction $(x(t),\ u(t)) \in A(t)$ in the problem above. Assume that we replace (123) with

$$\sum_{j=1}^{r} \frac{\partial H^*}{\partial u_j} (u_j^*(t) - u_j) \geq 0 \qquad \text{for all } (x, u) \in A(t) \cap (R^n \times U) \qquad (126)$$

and instead of (125) we require that H is concave in $(x,\ u)$ for $(x,\ u) \in A(t) \cap (R^n \times U)$. Then theorem 4 is still valid. (The proof is an easy modification of the proof above.)

Note 9*   If the Hamiltonian H in theorem 4 is not strictly concave in $(x,\ u)$, the solution to our control problem is not necessarily unique. Suppose, however, that we find one solution $(x^*(t),\ u^*(t))$ together with some $p(t)$ which satisfy all the conditions in theorem 4. Then for *this same* $p(t)$ all other optimal solutions $(\bar{x}(t),\ \bar{u}(t))$ satisfy the sufficient conditions. This property can often be used to prove that a unique optimal solution exists. In the case $U = R^r$ note that if $H(x,\ u,\ p(t),\ t)$ has continuous second-order partial derivatives w.r.t. x and u in a neighbourhood of $(x^*(t),\ u^*(t^+))$ for each t, and the matrix $[H_{u_i u_j}''(x^*(t),\ u^*(t^+),\ p(t),\ t)]$ is negative-definite, then the solution $(x^*(t),\ u^*(t))$ is unique.[5] (Then, in fact, $u^*(t)$ is also continuous.)

[5] Seierstad (1975).

The Arrow concavity condition

In the proof of theorem 4, the concavity of H w.r.t. x and u was an essential assumption. Note that if $f_0(x, u, t)$ and $f_1(x, u, t), \ldots, f_n(x, u, t)$ are all concave in (x, u), then the Hamiltonian is concave in (x, u) provided $p_i \geqslant 0$, $i = 1, \ldots, n$. If $f_0(x, u, t)$ is concave and $f_1(x, u, t), \ldots, f_n(x, u, t)$ are all linear in x, u, then H is always concave in (x, u).[6]

It turns out that in a number of interesting control problems in economics the Hamiltonian function is not concave in (x, u). (See, for instance, example 6.)

Arrow (see Arrow and Kurz (1970)) has proposed an alternative condition which is weaker than concavity of H w.r.t. (x, u). Define

$$\hat{H}(x, p, t) = \max_{u \in U} H(x, u, p, t), \tag{127}$$

assuming that the maximum value exists. The Arrow condition is as follows:

$$\hat{H}(x, p, t) \text{ is concave in } x \text{ for fixed values of } p \text{ and } t. \tag{128}$$

The following result can be established:

Theorem 5 (**The Arrow sufficiency theorem**)   Let $(x^*(t), u^*(t))$ be an admissible pair in problem $(50) \rightarrow (54)$. If there exists a continuous and piecewise continuously differentiable function $p(t) = (p_1(t), \ldots, p_n(t))$ such that the following conditions are satisfied with $p_0 = 1$,

$$\dot{p}_i(t) = -\frac{\partial H^*}{\partial x_i}, \qquad \text{v.e.} \qquad i = 1, \ldots, n \tag{129}$$

$$H(x^*(t), u, p(t), t) \leqslant H(x^*(t), u^*(t), p(t), t)$$
$$\text{for all } u \in U \text{ and all } t \tag{130}$$

$$\begin{aligned}
& p_i(t_1) \text{ no conditions,} && i = 1, \ldots, l \\
& p_i(t_1) \geqslant 0 \, (= 0 \text{ if } x_i^*(t_1) > x_i^1) && i = l+1, \ldots, m \\
& p_i(t_1) = 0, && i = m+1, \ldots, n
\end{aligned} \tag{131}$$

$$\hat{H}(x, p(t), t) = \max_{u \in U} H(x, u, p(t), t) \text{ exists and is concave in } x \text{ for all } t, \tag{132}$$

then $(x^*(t), u^*(t))$ solves problem $(50) \rightarrow (54)$.

---

[6]Sydsaeter (1981), Theorem 5.14.

If $\hat{H}(x, p(t), t)$ is strictly concave in x for all t, then $x^*(t)$ is unique (but $u^*(t)$ is not necessarily unique). ■

Proof. Suppose $(x, u) = (x(t), u(t))$ is an arbitrary admissible pair. We have to prove the inequality in (115). Suppose we are able to establish, for all x and all $t \in [t_0, t_1]$, the inequality $(p = p(t))$:

$$H(x^*, u^*, p, t) - H(x, u, p, t) \geqslant \dot{p} \cdot (x - x^*). \tag{133}$$

Then using (116), the inequality in (120) follows, and $\Delta \geqslant 0$ is derived in the same way as in the proof of theorem 4.

In order to prove (133), note first that from the definition of $\hat{H}$, $H(x^*, u^*, p, t) = \hat{H}(x^*, p, t)$ and $H(x, u, p, t) \leqslant \hat{H}(x, p, t)$. Hence,

$$H(x^*, u^*, p, t) - H(x, u, p, t) \geqslant \hat{H}(x^*, p, t) - \hat{H}(x, p, t). \tag{134}$$

Thus we see that it suffices to prove that for all $t \in [t_0, t_1]$ and all x,

$$\hat{H}(x, p, t) - \hat{H}(x^*, p, t) \leqslant -\dot{p} \cdot (x - x^*), \tag{135}$$

i.e. that $-\dot{p}$ is a *supergradient* for $\hat{H}(x, p, t)$ at x^*. The first step in a proof of (135) is to establish by a standard separating hyperplane argument (Rockafellar (1970a), Chapter 5, section 23) the existence of a supergradient a for $\hat{H}(x, p, t)$ at x^* (see (B.2)):

$$\hat{H}(x, p, t) - \hat{H}(x^*, p, t) \leqslant a \cdot (x - x^*) \qquad \text{for all } x \in R^n. \tag{136}$$

(If we knew, a priori, that $\hat{H}$ were differentiable, then we could simply put $a = \partial \hat{H}/\partial x$.) From (134) and (136) we see that (putting $u = u^*$)

$$H(x, u^*, p, t) - H(x^*, u^*, p, t) \leqslant a \cdot (x - x^*) \quad \text{for all } x \in R^n. \tag{137}$$

Define for a fixed $t \in [t_0, t_1]$, $g(x) = H(x, u^*, p, t) - H(x^*, u^*, p, t) - a \cdot (x - x^*)$. It follows from (137) that $g(x) \leqslant 0$ for all x. Since $g(x^*) = 0$, x^* maximizes g. Hence $g'(x^*) = 0$, and this equation gives $a = \partial H^*/\partial x = -\dot{p}$, using (129). (133) now follows from (137). ■

Theorem 5 is proved in Seierstad and Sydsæter (1977). (See note 1 to theorem 5, p. 373.) Theorem 5 *is* more general than theorem 4, since concavity of $H(x, u, p, t)$ w.r.t. (x, u) implies concavity of $\hat{H}(x, p, t)$ w.r.t. x. (See exercise 2.6.4.)

Note 10   It follows from the proof above that $\hat{H}$ is, in fact, differentiable w.r.t. x at x^*, and $\partial \hat{H}/\partial x = -\dot{p}$. (The proof shows that a in (136) is unique. By concavity, $\partial \hat{H}/\partial x$ exists.)

Example 5 Consider the two-sector model in example 3. The Hamiltonian is, for $p_0 = 1$,

$$H(x, u, p, t) = x_2 + ap_1 ux_1 + ap_2(1 - u)x_1$$

which is not concave in x_1, x_2, and u (jointly). $\hat{H}$ defined by (127) is in this case (make use of (73) in example 3),

$$\hat{H}(x, p, t) = \begin{cases} ap_1 x_1 + x_2 & p_1(t) > p_2(t) \\ ap_2 x_1 + x_2 & p_1(t) \leqslant p_2(t). \end{cases}$$

For each $t \in [0, T]$, $\hat{H}(x, p, t)$ is a linear, and hence concave function of x_1 and x_2. Since the triple $(x_1^*(t), x_2^*(t), u^*(t))$ satisfies all the conditions in the Maximum Principle (in each of the two cases $T > 2/a$, $T \leqslant 2/a$), it is optimal.

Some observations related to theorem 5 are gathered in the following notes.

Note 11 Let $A(t)$ be a convex set in R^n for each $t \in [t_0, t_1]$ and let us impose the additional restriction $x(t) \in A(t)$ for all $t \in [t_0, t_1]$ in problem (50)→(54). Then theorem 5 still holds provided $x^*(t)$ is an interior point of $A(t)$ for all $t \in [t_0, t_1]$, and provided $\hat{H}(x, p(t), t)$ is concave (only) for $x \in A(t)$ for each $t \in [t_0, t_1]$. (This result can be proved by the same kind of arguments as those needed in the proof of theorem 5 (cf. the set A in Appendix 3 in Seierstad & Sydsæter (1977).)

Note 12* The notes 8 and 11 as well as the theorems 4 and 5 are valid also for the weaker regularity conditions of note 6. In fact, in note 11 the functions f_i need only be defined and be regular in the set $\{(x, u, t): x \in A(t), u \in U, t \in [t_0, t_1]\}$. In note 8 the functions f_i need only be defined and be regular in the set $\{(x, u, t): (x, u) \in A(t), t \in [t_0, t_1]\}$, D can be a neighbourhood relative to this set, and $\partial f_i / \partial u_k$ and $\partial f_i / \partial x_j$ can be required to exist and be regular in D only.

Note 13* For cases in which $\hat{H}(x, p(t), t)$ fails to be strictly concave in x, the following theorem can sometimes be used to obtain uniqueness.[7]

Theorem 6* Suppose $u^*(t)$ is a piecewise continuous control with associated path $x^*(t)$ which satisfies all the conditions in theorem 5 for some adjoint

---

[7]Seierstad (1985c). Note that (139) implies $\dot{p} = -\partial H(\bar{x}(t), \bar{u}(t), p(t), t)/\partial x$, and $(H(x, \bar{u}(t), p(t), t) + \dot{p}x \leqslant \bar{H}(x, p(t), t) + \dot{p}x \leqslant H^*(t) + \dot{p}x^*$ for any x.)

function $p(t)$ and for $p_0 = 1$. Let $(\bar{x}(t), \bar{u}(t))$ be any optimal solution. Then the following conditions are satisfied for all $t \in [t_0, t_1]$:

$$H(\bar{x}(t), \bar{u}(t), p(t), t) = \max_{u \in U} H(\bar{x}(t), u, p(t), t). \tag{138}$$

Except at points of discontinuity of $u^*(t)$,

$$H(x^*(t), u^*(t), p(t), t) + \dot{p}(t)x^*(t)$$

$$= H(\bar{x}(t), \bar{u}(t), p(t), t) + \dot{p}(t)\bar{x}(t). \tag{139}$$

Finally,

$$p(t_1)x^*(t_1) = p(t_1)\bar{x}(t_1). \quad \blacksquare \tag{140}$$

In the economic literature using optimal control theory theorems 4 and 5 are often referred to in cases where it is impossible to obtain explicit solutions for $x^*(t)$ and $u^*(t)$. Note that it is then essential *to prove the existence* of a pair $(x^*(t), u^*(t))$ satisfying all the conditions in theorem 4 or 5, in order to be sure that an optimal solution exists. A vague reference to the effect that the proper concavity conditions are satisfied is inadequate. In example 7 in the next section we shall consider a model illustrating our point. (See also Seierstad and Sydsæter (1983).)

A useful result*

In some control models the optimal control $u^*(t)$ turns out to be an interior point of the control region U for all t in some subinterval I of $[t_0, t_1]$. Then for $t \in I$, the maximum condition implies the first-order condition $\partial H^*/\partial u = 0$. (The * indicates, as usual, that the gradient vector $\partial H/\partial u$ is evaluated along the optimal path.) On the other hand, if $H(x^*(t), u, p(t), t)$ is concave in u, the first-order condition is sufficient for $u^*(t)$ to maximize the Hamiltonian. Suppose $\partial H^*/\partial u = 0$ at some point in I and $(d/dt)(\partial H^*/\partial u) \equiv 0$ in I, assuming $u^*(t)$ to be a C^1- function on I. Then surely, $\partial H^*/\partial u \equiv 0$ in I. Note that when calculating the total derivative $(d/dt)(\partial H^*/\partial u)$, the resulting expression will contain $\dot{u}$ as well as $\dot{x}$ and $\dot{p}$. The latter two derivatives can be replaced by $f(x, u, t)$ and $-\partial H^*/\partial x$, respectively. Hence $(d/dt)(\partial H^*/\partial u) = 0$ becomes a differential equation satisfied by $u^*(t)$. This equation is sometimes used in the process of constructing the optimal solution. For easy reference we formulate the main point above in a theorem.

Theorem 7 Suppose $(x^*(t), u^*(t))$ is an admissible pair in problem

$(50) \to (54)$ and assume that $u^*(t)$ is a C^1-function satisfying on a subinterval I of $[t_0, t_1]$ the equations

$$\frac{d}{dt} \frac{\partial H(x^*(t), u^*(t), p(t), t)}{\partial u_j} = 0, \qquad j = 1, \ldots, r. \tag{141}$$

Suppose, moreover, that $\partial H / \partial u_j$ evaluated at some $t' \in I$ is 0 for all $j = 1, \ldots, r$. If $H(x^*(t), u, p(t), t)$ is concave in u, then for each $t \in I$, $u^*(t)$ maximizes this function for $u \in U$. ∎

Note 14 The type of sufficient conditions studied above is most frequently used in economic applications. There are, however, some sufficiency results of another type in the literature.

In the calculus of variations there are (global) sufficiency conditions based on the assumption that the optimal solution can be imbedded in a set (field) of extremals (solutions of the Euler equation) passing through all points around the optimal solution. The Weierstrass necessary condition (see exercise 1.10.3) are then assumed to hold for all extremals.

Closely related sufficient conditions for optimal control problems are indicated in exercise 2.11.2. They are based on the dynamic programming approach in which one studies a partial differential equation having the above mentioned set of extremals as characteristics.

Exercise 2.6.1 Prove that the solution candidate found in exercise 2.3.2 solves the problem.

Exercise 2.6.2 Use theorem 5 to show that the solution given in exercise 2.3.3 (iv) is optimal. Use theorem 6 (in fact (138)) to show that the optimal solution is unique.

Exercise 2.6.3 A "drastic" example in which theorem 4 fails, but theorem 5 can be used to prove sufficiency is the following:

$$\max \int_0^1 3u(t)dt, \qquad \dot{x}(t) = (u(t))^3,$$

$$x(0) = 0, \qquad x(1) \leqslant 0, \qquad u(t) \in [-2, \infty).$$

Prove that $u^* = 1$ in $[0, 8/9]$, $u^* = -2$ in $(8/9, 1]$ (with $p_0 = 1$ and $p(t) \equiv -1$) is optimal by using theorem 5. (The optimal solution is not unique.) Graph the Hamiltonian $H = 3u - u^3$ in the interval $[-2, \infty)$ and explain why theorem 4 cannot be used.

Exercise 2.6.4 Let S and U be convex sets in R^n and R^r, respectively, and let $F(x, u)$ be a real-valued concave function of (x, u), $x \in S$, $u \in U$. Define

$$f(x) = \max_{u \in U} F(x, u) \qquad (142)$$

where we assume that the maximum value exists for each $x \in S$. Prove that f is concave in S.
(Hint: Let $x_1, x_2 \in S$, $\lambda \in [0, 1]$ and choose $u_1, u_2 \in U$ such that $f(x_1) = F(x_1, u_1)$, $f(x_2) = F(x_2, u_2)$.)
  Let B be a convex set in $R^n \times R^r$ and define $U_x = \{u : (x, u) \in B\}$. Prove again that f defined in (142) is concave for $U = U_x$.

Exercise 2.6.5.

(i) Let f be a C^1-function in a set A and let S be a convex set in the interior of A. Let $\bar{x} \in S$. Prove that

$$f(x) \leqslant f(\bar{x}) \text{ for all } x \in S \Rightarrow \sum_{i=1}^{n} f_i'(\bar{x})(\bar{x}_i - x_i) \geqslant 0 \text{ for all } x \in S.$$

  (Hint: Consider $g(t) = f(\bar{x}) - f(tx + (1 - t)\bar{x})$ for $t \in [0, 1]$.)
(ii) What does the result in (i) tell us in the case $n = 1$? Illustrate.
(iii) Prove that the implication in (i) can be reversed if f is concave in S.

Exercise 2.6.6 Consider the general control problem (50)→(54). For each (x, t) define the set $N_0(x, U, t)$ in R^{n+1} by

$$N_0(x, U, t) = \{(f_0(x, u, t), f(x, u, t)), u \in U\}. \qquad (143)$$

Suppose $N_0(x, U, t)$ has the following property.

  If $x_1, x_2 \in R^n$, $\lambda \in [0, 1]$, $c_1 \in N_0(x_1, U, t)$ and $c_2 \in N_0(x_2, U, t)$, then there exists a $c_3 \in N_0(\lambda x_1 + (1 - \lambda)x_2, U, t)$ such that
  $$\lambda c_1 + (1 - \lambda)c_2 \leqslant c_3. \qquad (144)$$

Prove that if $p(t) \geqslant 0$ and if $p_0 = 1$, then $\hat{H}(x, p(t), t)$ defined in (127) is concave in x.

7. Some applications of the Arrow sufficiency theorem

In this section we consider two optimal control problems which are closely related to the Ramsey model studied in several examples of Chapter 1 (examples 1, 3, 6, 10, 13). Our main concern is to demonstrate how theorem 5 should properly be used to prove optimality.

The common starting point for both problems is the following model, where the variables have the same meaning as in example 1 of Chapter 1:

$$\max \int_0^T U(f(K) - \dot{K})dt, \qquad K(0) = K_0 > 0,$$

$$K(T) \text{ free}, \qquad \dot{K} \geqslant 0, \qquad f(K) - \dot{K} \geqslant 0. \tag{145}$$

Compared with the problem in example 1 of Chapter 1, note that there is no discounting here. On the other hand, we assume that we cannot eat capital, $(\dot{K} \geqslant 0)$. Moreover, we assume that a negative consumption rate is not permissible.

In the form presented above we are faced with a calculus of variations problem subject to constraint. There are several ways of transforming the problem into an optimal control problem. The obvious choice is to put $\dot{K} = u$ as a control variable. However, in that case $f(K) - \dot{K} \geqslant 0$ takes the form $f(K) - u \geqslant 0$, a type of constraint which is, so far in our theory, not permissible. Let us introduce instead the savings rate as a control variable, and put $\dot{K}(t) = u(t)f(K(t))$. Then the two latter constraints in (145) are taken care of by the requirement $0 \leqslant u(t) \leqslant 1$. Thus our problem above is equivalent to the following control problem:

$$\max \int_0^T U[(1 - u(t))f(K(t))]dt \tag{146}$$

$$\dot{K}(t) = u(t)f(K(t)), \qquad K(0) = K_0 > 0, \qquad K(T) \text{ free} \tag{147}$$

$$0 \leqslant u(t) \leqslant 1. \tag{148}$$

In our first example the utility function U and the production function f are particularly simple.

Example 6   (Consumption versus investment)

$$\max \int_0^T [1 - e^{-(1 - u(t))K(t)}]dt, \qquad T > 1 \tag{149}$$

$$\dot{K}(t) = u(t)K(t), \qquad K(0) = K_0 > 0, \qquad K(T) \quad \text{free} \qquad (K_0 \text{ fixed}) \quad (150)$$

$$0 \leqslant u(t) \leqslant 1. \tag{151}$$

Thus we see that the utility function is $U(C) = 1 - e^{-C}$, and the production function is simply $f(K) = K$.

As a preliminary exercise let us consider two extreme policies. If $u(t) \equiv 1$ (all is invested), then the integral in (149) becomes $\int_0^T (1 - 1) dt = 0$. On the other hand, if $u(t) \equiv 0$ (all is consumed), then by (150), $K(t) \equiv K_0$, and thus $\int_0^T (1 - e^{-K_0}) dt = T(1 - e^{-K_0})$. Since K_0 and T are both positive, $T(1 - e^{-K_0}) > 0$. Hence, of the two extreme policies considered, to consume everything all the time is better. There is, as we shall see, in the general case, a better policy to be pursued. Intuitively, it is natural to expect that a policy in which we have an initial period with high investments and a final period with low investments might be better than the two extreme policies examined above. Let us now try to make use of our machinery.

Our strategy for solving the problem is as follows: We consider the Maximum Principle for our problem (theorem 2) and we try to find a pair $(K^*(t), u^*(t))$ which satisfies all the conditions in the Maximum Principle. (We are not concerned with finding all pairs satisfying the necessary conditions.) We succeed in finding such a pair, and proceed by proving that the sufficiency conditions in theorem 5 are satisfied for $(K^*(t), u^*(t))$. We conclude that $(K^*(t), u^*(t))$ really solves our problem.

Note first that since $K(T)$ is free, by (62), $p_0 = 1$. The Hamiltonian is

$$H(K, u, p, t) = 1 - e^{-(1 - u)K} + puK. \tag{152}$$

Suppose $(K^*(t), u^*(t))$ solves our problem. Then $p(t)$ satisfies

$$\dot{p} = -\frac{\partial H^*}{\partial K} = -(1 - u^*)e^{-(1 - u^*)K^*} - pu^*, \qquad p(T) = 0. \tag{153}$$

Furthermore, for each $t \in [0, T]$, $u^*(t)$ is a value of $u \in [0, 1]$ which maximizes

$$\phi(u) = H(K^*(t), u, p(t), t) = 1 - e^{-(1 - u)K^*(t)} + p(t)uK^*(t). \tag{154}$$

Note that $\phi'(u) = -K^*(t)e^{-(1 - u)K^*(t)} + p(t)K^*(t)$ and $\phi''(u) = -(K^*(t))^2 e^{-(1 - u)K^*(t)} < 0$. From (150) and (151), $K^*(t) > 0$ for all t, so $\phi(u)$ is a strictly concave function on $[0, 1]$. We conclude that the following properties hold:

$$u = u^*(t) = 0 \qquad \text{maximizes } \phi(u) \Leftrightarrow \phi'(0) \leq 0 \Leftrightarrow p(t) \leq e^{-K^*(t)} \tag{155a}$$

$$u = u^*(t) \in (0, 1) \text{ maximizes } \phi(u) \Leftrightarrow \phi'(u^*(t)) = 0 \Leftrightarrow$$

$$p(t) = e^{-(1 - u^*(t))K^*(t)} \tag{155b}$$

$$u = u^*(t) = 1 \qquad \text{maximizes } \phi(u) \Leftrightarrow \phi'(1) \geq 0 \Leftrightarrow p(t) \geq 1. \tag{155c}$$

We know that $p(T) = 0$. The continuous function $g(s) = e^{-K^*(s)} - p(s)$ defined on $[0, T]$ therefore satisfies $g(T) > 0$. Let t^* be the smallest number t such that $g(s) > 0$ for all $s \in (t, T]$. Possibly $t^* = 0$.

If $t^* > 0$, then $g(t^*) = 0$ and $g(t) > 0$ for all $t \in (t^*, T]$. From (155a) we conclude that $u^*(t) \equiv 0$ and thus $\dot{p}(t) = -e^{-K^*(t)} < 0$ in $(t^*, T]$. Since $p(T) = 0$, $p(t) > 0$ in $[t^*, T)$. By (153) and using (A.20) with $t_0 = t^*$, we see that

$$p(t) > 0 \quad \text{and} \quad \dot{p}(t) < 0 \quad \text{in } [0, T). \tag{156}$$

Since $u^*(t) = 0$ in $(t^*, T]$ we see from (150) that $K^*(t) \equiv K_T$ in $(t^*, T]$, where K_T is some constant to be determined. (We shall see that K_T is determined, in the end, by the initial condition $K^*(0) = K_0$.) From (153) we obtain $\dot{p}(t) = -e^{-K^*(t)} = -e^{-K_T}$ in $(t^*, T]$ and thus $p(t) = -e^{-K_T}t + C$. Since $0 = p(T) = -e^{-K_T}T + C$, $C = e^{-K_T}T$, so $p(t) = e^{-K_T}(T - t)$. In particular, $p(t^*) = e^{-K_T}(T - t^*)$, and since $p(t^*) = e^{-K^*(t^*)} = e^{-K_T}$ as well, it follows that $e^{-K_T} = e^{-K_T}(T - t^*)$. Hence

$$t^* = T - 1. \tag{157}$$

This is our proposal for t^*. It is > 0 as $T > 1$. Since $\dot{K}^*(t) \geq 0$, $K_T \geq K_0 > 0$, thus $p(t^*) = e^{-K_T} < 1$. From (156) it follows that p is strictly decreasing on $[0, T]$. Define t_* as 0 if $p(t) < 1$ for all $t \leq t^*$ and define t_* by $p(t_*) = 1$ otherwise. In the interval (t_*, t^*) we have $p(t) < 1$, so from (155c), $u^*(t) < 1$. We guess that $u^*(t) > 0$ in (t_*, t^*). Then, from (155b), $p(t) = e^{-(1 - u^*(t))K^*(t)}$ in (t_*, t^*).

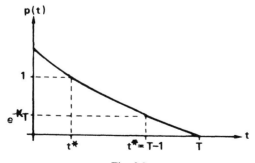

Fig. 2.5

Inserting this expression for $p(t)$ into (153) we obtain

$$\dot{p}(t) = -p(t)$$

that is

$$p(t) = p(t^*)e^{-(t-t^*)} = e^{-K_T}e^{-(t-t^*)} = e^{-K_T + t^* - t}.$$

If $t_* > 0$, then $1 = p(t_*) = e^{-K_T + t^* - t_*}$, and thus $-K_T + t^* - t_* = 0$. Hence

$$t_* = t^* - K_T = T - 1 - K_T \qquad \text{if } t_* > 0. \tag{158}$$

Furthermore, for $t \in [t_*, t^*]$,

$$e^{-(1 - u^*(t))K^*(t)} = e^{-K_T + t^* - t} \qquad \text{or}$$

$$u^*(t)K^*(t) - K^*(t) = -K_T + t^* - t. \tag{159}$$

By (150), $\dot{K}^*(t) = u^*(t)K^*(t)$, so

$$\dot{K}^*(t) - K^*(t) = -t + t^* - K_T.$$

The solution to this linear differential equation is (since $K^*(t^*) = K_T$),

$$K^*(t) = -e^{t-t^*} + t + K_T - t^* + 1 \qquad (t \in [t_*, t^*]) \tag{160}$$

In particular, we see by using (158) that if $t_* > 0$, then

$$K^*(t_*) = -e^{t_* - t^*} + t_* + K_T - t^* + 1 = -e^{-K_T} + 1. \tag{161}$$

Since p is strictly decreasing and $p(t_*) = 1$, we see from (155c) that $u^*(t) \equiv 1$ in $[0, t_*)$. From (153) it follows that $\dot{p}(t) = -p(t)$ and thus, as in the interval (t_*, t^*),

$$p(t) = e^{T - 1 - K_T - t} \qquad t \in [0, t_*]. \tag{162}$$

Moreover, $\dot{K}^*(t) = K^*(t)$, with $K^*(0) = K_0$, so $K^*(t) = K_0 e^t$. Thus $K^*(t_*) = K_0 e^{t_*}$. Combining this result with (161), we obtain, using (158):

$$K_0 e^{T - 1 - K_T} = -e^{-K_T} + 1 \qquad \text{if } t_* > 0.$$

Solving this equation for K_T, we obtain

$$K_T = \ln(K_0 e^{T - 1} + 1) \qquad \text{if } t_* > 0. \tag{163}$$

From (158) and (163) we see that $t_* > 0 \Rightarrow T - 1 > \ln(K_0 e^{T - 1} + 1) \Leftrightarrow e^{T - 1} > K_0 e^{T - 1} + 1 \Leftrightarrow K_0 < 1 - e^{1 - T}$. Note, moreover, that from (159), using (157), (160) and (163), we find the following suggestion for an optimal control in the interval (t_*, t^*) when $t_* > 0$:

$$u^*(t) = \frac{K^*(t) - K_T + t^* - t}{K^*(t)} = \frac{1 - e^{t - T + 1}}{-e^{t - T + 1} + t + \ln(K_0 e^{T - 1} + 1) + 2 - T}. \tag{164}$$

(It is a useful exercise to check that $u^*(t)$ defined by (164) belongs to $(0, 1)$ when $t \in (t_*, t^*)$ and that $u^*(t_*) = 1$, $u^*(t^*) = 0$.)

We are ready to suggest the following optimal control. To spare some words, we shall from now on assume that $K_0 < 1 - e^{1-T}$. Our suggestion is:

$$u^*(t) = \begin{cases} 1 & t \in [0, t_*] \\ u^*(t) \text{ given in (164)} & t \in (t_*, t^*] \\ 0 & t \in (t^*, T] \end{cases} \tag{165}$$

where $t^* = T - 1$, $t_* = T - 1 - \ln (K_0 e^{T-1} + 1)$.
The associated path $K^*(t)$ is given by

$$K^*(t) = \begin{cases} K_0 e^t & t \in [0, t_*] \\ -e^{t-T+1} + t + \ln (K_0 e^{T-1} + 1) + 2 - T & t \in (t_*, t^*] \\ \ln (K_0 e^{T-1} + 1) & t \in (t^*, T] \end{cases} \tag{166}$$

and the adjoint function $p(t)$ is

$$p(t) = \begin{cases} \dfrac{e^{-t+T-1}}{K_0 e^{T-1} + 1} & t \in [0, t^*] \\[3mm] \dfrac{T - t}{K_0 e^{T-1} + 1} & t \in (t^*, T]. \end{cases} \tag{167}$$

We want to prove the optimality by using theorem 5. It is then logically necessary first to make sure that our candidates $u^*(t)$ and $K^*(t)$ (with associated $p(t)$ given by (167)) really satisfy *all* conditions in the Maximum Principle (with $p_0 = 1$). In fact, some of these verifications are rather trivial, and some of them are obtained just by observing that the implications used can be reversed.

The assumptions on the parameters are: $T > 1$, $0 < K_0 < 1 - e^{1-T}$. With $t_* = T - 1 - \ln(K_0 e^{T-1} + 1)$, $t^* = T - 1$, we first make sure that $0 < t_* < t^* < T$. In the text below equation (163), $0 < t_*$ is established. The inequality $t_* < t^*$ is equivalent to $\ln (K_0 e^{1-T} + 1) > 0$ which is satisfied since $K_0 > 0$. Finally, $t^* < T$ is obvious. We proceed by noting:

(a) In each of the three time intervals we verify that $\dot{K}^*(t) = u^*(t)K^*(t)$, and we see also that $K^*(t)$ is continuous at t_* and t^*. Since $K^*(0) = K_0$ is obvious, (150) *is* satisfied.

(b) From (165) and the remark below (164) we see that $u^*(t) \in [0, 1]$, so (151) is valid.

We look now at the Maximum Principle, theorem 2. Our $u^*(t)$ given by (165) is piecewise continuous, and the function $p(t)$ given by (167) is continuous (with the value $(K_c e^{T-1}+1)^{-1}$ at t^*). Moreover,

(c) the maximum condition (57) is satisfied. On $[0, t_*)$ and $(t^*, T]$ this is easy to establish. On (t_*, t^*) it follows from (164), (166) and (167) that

$$p(t) = e^{-(1-u^*(t))K^*(t)}. \tag{168}$$

(Actually, $u^*(t)$ was defined by (168) in (t_*, t^*).) If we look back at the expression found for $\phi'(u)$ below (154), we see that $\phi'(u^*(t)) = 0$. Since $\phi(u)$ is concave, we conclude that $u^*(t)$ defined in (t_*, t^*) in (165) really maximizes $\phi(u)$.)

(d) We verify next that $p(t)$ satisfies (153) for $u = u^*(t)$, $K = K^*(t)$, $p = p(t)$. In the two intervals $[0, t_*]$ and $[t^*, T]$ this is trivial. In the interval (t_*, t^*) we verify that (153) is satisfied when $u^*(t)$, $K^*(t)$ and $p(t)$ are inserted.

Since $p_0 = 1$, (59) is taken care of. Finally, (60) reduces to the condition $p(T) = 0$, which is obviously satisfied.

We have proved now that the pair $(K^*(t), u^*(t))$ defined by (166) and (165) satisfies all the conditions in the Maximum Principle. We shall prove that we have found the optimal solution to our problem by using note 11 to theorem 5.

We let $A = [0, \infty)$ and define for all $K \in A$,

$$\hat{H}(K, p(t), t) = \max_{u \in [0,1]} H(K, u, p(t), t)$$

$$= \max_{u \in [0,1]} (1 - e^{-(1-u)K} + p(t)uK). \tag{169}$$

We must prove that for all $t \in [0, T]$, $\hat{H}(K, p(t), t)$ is a concave function for $K \geq 0$. Here $p(t)$ is given by (167).

Consider first those t for which $p(t) \geq 1$. As in (155c) we see that $u = 1$ maximizes H, and therefore

$$\hat{H}(K, p(t), t) = p(t)K \qquad \text{for } p(t) \geq 1, \tag{170}$$

which is linear, and hence concave as a function of K.

Consider next those t for which $p(t) < 1$. If $p(t) \leq e^{-K}$, we see as in (155a) that $u = 0$ maximizes H for $u \in [0, 1]$, so

$$\hat{H}(K, p(t), t) = 1 - e^{-K} \qquad \text{for } p(t) \leq e^{-K}, \qquad \text{i.e. for } K \leq -\ln p(t). \tag{171}$$

If $e^{-K} < p(t) < 1$, then as in (155), H is maximized by some $u \in (0, 1)$ for which

$e^{-(1-u)K} = p(t)$, i.e. $uK = \ln p(t) + K$. Hence,

$$\hat{H}(K, p(t), t) = 1 - p(t) + p(t) \ln p(t) + p(t)K \qquad \text{for } K > -\ln p. \qquad (172)$$

Combining (171) and (172) we see that for those t for which $p(t) < 1$,

$$\hat{H}(K, p(t), t) = \begin{cases} 1 - e^{-K} & 0 \leqslant K \leqslant -\ln p(t) \\ 1 - p(t) + p(t) \ln p(t) + p(t)K & -\ln p(t) < K \end{cases} \qquad (173)$$

It remains to prove that $\hat{H}(K, p(t), t)$ defined in (173) is a concave function of K. We see at once that $\hat{H}$ is concave on each of the two subintervals. Moreover, $\hat{H}$ is continuous at $K = -\ln p(t)$ with the value $1 - p(t)$. In order to be sure that $\hat{H}$ is concave for all $K \geqslant 0$, we have to examine the left-hand and the right-hand derivatives of $\hat{H}$ at $K = -\ln p(t)$. Now, $\partial \hat{H}/\partial K = e^{-K}$ for $0 < K < -\ln p(t)$ and $\partial \hat{H}/\partial K = p(t)$ for $K > -\ln p(t)$. Since $e^{-K} = p(t)$ for $K = -\ln p(t)$, it follows that $\hat{H}$ is differentiable at $K = -\ln p(t)$ with derivative $p(t)$. $\hat{H}$ is graphed in fig. 2.6 and is clearly concave in K.

For all $t \in [0, T]$ (whether $p(t) \geqslant 1$ or $p(t) < 1$) we have proved that $\hat{H}$ is a concave function of K for $K \geqslant 0$. Thus we can conclude that provided $K_0 < 1 - e^{1-T}$, $(K^*(t), u^*(t))$ given by (166) and (165) solves problem (149)–(151).

The value of the integral in (149), when we evaluate it along the optimal solution, will depend on K_0 and T. Denoting it by $V(K_0, T)$ and letting

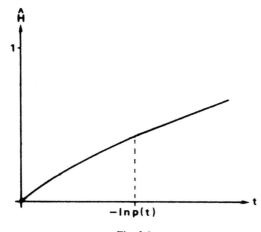

Fig. 2.6

$R = K_0 e^{T-1} + 1$, we find

$$V(K_0, T) = \int_0^{t_*} 0\, dt + \int_{t_*}^{t^*} (1 - p(t))\, dt + \int_{t^*}^{T} (1 - e^{-K^*(t)})\, dt$$

$$= \int_{t_*}^{t^*} \left(1 - \frac{1}{R} e^{-t+T-1}\right) dt + \int_{t^*}^{T} (1 - e^{-\ln R})\, dt$$

$$= t^* - t_* + \frac{1}{R}(e^{-t^*+T-1} - e^{-t_*+T-1}) + \left(1 - \frac{1}{R}\right)(T - t^*) = \ln R,$$

where we made use of $t^* = T - 1$ and $t_* = T - 1 - \ln R$. Thus our result is:

$$V(K_0, T) = \ln (K_0 e^{T-1} + 1) \qquad (0 < K_0 < 1 - e^{1-T}). \tag{174}$$

Note, in particular, that

$$\frac{\partial V}{\partial K_0} = \frac{e^{T-1}}{K_0 e^{T-1} + 1} = p(0), \qquad \text{(using (167))}. \tag{175}$$

Thus $p(0)$ measures (approximately) the increase in maximal utility by increasing the initial stock by one unit. For general results in this direction, see theorem 9 in Chapter 3.

Note that the switching point $t^* = T - 1$ is independent of K_0, while $t_* = T - 1 - \ln (K_0 e^{T-1} + 1)$ does depend on K_0. If the initial capital stock K_0 increases, we see that the time t_* (at which we stop to save everything) decreases, while V given by (174) increases. If T increases, t^*, t_* and V all increase $(dt_*/dT = (K_0 e^{T-1} + 1)^{-1} > 0)$.

We have examined above the case in which $K_0 < 1 - e^{1-T}$. If $K_0 \geq 1 - e^{1-T}$, then $t_* = 0$ and we encourage the reader in exercise 2.7.2 to work out the solution in this case. The case in which $T = \infty$ is studied in example 3.13.

In the previous example the application of theorem 5 was straightforward. We obtained explicit formulas for $K^*(t)$, $u^*(t)$, and $p(t)$, and the verification of all the conditions in the theorem was quite easy. The next example is more complicated due to the fact that the utility function and the production function are not specified. Hence we are unable to find explicit formulas for $K^*(t)$, $u^*(t)$, and $p(t)$.

Readers who have been exposed to much more complicated control problems "solved" in the literature might be astonished by the length and the tediousness of the arguments we give below. However, as we have previously emphasised, when using sufficient conditions it is essential to prove the existence of a pair $(K^*(t), u^*(t))$ which satisfies *all* the relevant conditions. To prove such existence we have to rely on the qualitative theory of differential equations reviewed in Appendix A.

Example 7 (Consumption versus investment) We study problem (146)–(148) which is, briefly formulated,

$$\max \int_0^T U[(1-u)f(K)]dt, \qquad \dot{K} = uf(K),$$

$$K(0) = K_0 > 0, \qquad K(T) \quad \text{free}, \qquad u \in [0, 1]. \tag{176}$$

The utility function U is defined and C^3 in $(0, \infty)$. If U is defined also at 0, assume that it is continuous there. Assume, moreover,

$$U' > 0 \text{ and } U'' < 0 \text{ on } (0, \infty), \qquad U'(C) \to \infty \text{ as } C \to 0^+. \tag{177}$$

The production function f is defined and C^2 in $(0, \infty)$ and

$$f''(K) < 0 \text{ on } (0, \infty), \qquad f'(\bar{K}) = 0 \text{ for some } \bar{K} > K_0,$$
$$f(K) > 0 \text{ on } (0, \bar{K}], \quad T > 1/f'(K_0). \tag{178}$$

Thus f is strictly concave and has a maximum at the "bliss-level" $\bar{K}$. Examples of utility and production functions satisfying the requirements in (177) and (178) are graphed in figs. 2.7 and 2.8.

We shall solve our problem by using the Arrow sufficiency theorem

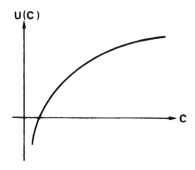

Fig. 2.7

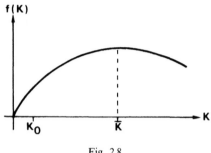

Fig. 2.8

(theorem 5). The Hamiltonian with $p_0 = 1$ is

$$H = H(K, u, p, t) = U(C) + puf(K) \qquad \text{where } C = f(K) - uf(K), \tag{179}$$

and the differential equation for $p = p(t)$ is

$$\dot{p} = -\frac{\partial H}{\partial K} = -f'(K)[(1 - u)U'(C) + pu]. \tag{180}$$

Since $K(T)$ is free,

$$p(T) = 0. \tag{181}$$

We want to find a pair $(K^*(t), u^*(t))$ which satisfies all the conditions in theorem 5. Since we appeal to sufficient conditions, we shall make use of the method of "informed guessing" to obtain a candidate for an optimal solution and then test if all conditions in theorem 5 are satisfied for our candidate.

Our first guess is that $u^*(t) = 1$ is not optimal at any point t. (If U is not defined at $C = 0$, $u^*(t) = 1$ is not even admissible. In the opposite case, our guess can be motivated as follows. If we decrease $u^*(t)$ slightly below 1 on a short time interval, then $C(t)$ increases slightly above 0. Since $U'(0^+) = \infty$, this increase in $C(t)$ implies a direct and substantial increase in the criterion functional which (probably) more than counteract the negative change caused by the change in $K(t)$ at later points of time.)

Next, we guess that it is not worth while at any time to accumulate more capital than $\bar{K}$, the bliss-level of capital, and that the optimal capital stock at T is less than $\bar{K}$:

$$K^*(t) < \bar{K} \qquad \text{for all } t. \tag{182}$$

It follows from (178) and (182) that for all $t \in [0, T]$,

$$f'(K^*(t)) \geqslant 0. \tag{183}$$

From the Hamiltonian we find

$$\frac{\partial H}{\partial u} = -f(K)U'(C) + pf(K) = f(K)(p - U'(C)) \tag{184}$$

and

$$\frac{\partial^2 H}{\partial u^2} = (f(K))^2 U''(C) \leqslant 0. \tag{185}$$

From (185) we see that H is a concave function of u. Thus in order for $u^*(t) = 0$ to maximize H it is necessary and sufficient that the derivative of $H(K^*(t), u, p(t), t)$ w.r.t. u is $\leqslant 0$ at $u = 0$, i.e. $f(K^*(t))(p(t) - U'[f(K^*(t))]) \leqslant 0$. We are going to propose a $K^*(t)$ which is > 0 and satisfies (182). Thus by (178), $f(K^*(t)) > 0$. Hence

$$u^*(t) = 0 \qquad \text{maximizes } H \Leftrightarrow p(t) \leqslant U'[f(K^*(t))]. \tag{186}$$

Having assumed away $u^*(t) = 1$, we see from (186) that $u^*(t) \in (0, 1) \Leftrightarrow p(t) > U'[f(K^*(t))]$. Moreover, since the derivative of H w.r.t. u must be 0 at an interior maximum,

$$u^*(t) \in (0, 1) \Rightarrow U'[(1 - u^*(t))f(K^*(t))] = p(t). \tag{187}$$

If we look back on our original problem, it is reasonable to assume that it is optimal to have $u^*(t) = 0$ in some interval $[t^*, T]$. From (176) it follows that in this case:

$$K^*(t) = K(T) = K_T \quad \text{for some constant } K_T \geqslant K_0 \text{ on } [t^*, T], \tag{188}$$

and from (180),

$$\dot{p}(t) = -f'(K_T)U'[f(K_T)], \qquad t \in [t^*, T]$$

Integrating and using (181), we find

$$p(t) = -f'(K_T)U'[f(K_T)](t - T), \qquad t \in [t^*, T], \tag{189}$$

and, in particular,

$$p(t^*) = -f'(K_T)U'[f(K_T)](t^* - T). \tag{190}$$

Let us choose t^* as the smallest value of t' for which $p(t) \leqslant U'[f(K^*(t))]$ in

$[t', T]$. Then

$$p(t^*) = U'[f(K^*(t^*))] = U'[f(K_T)], \qquad \text{and}$$
$$u^*(t) = 0 \text{ in } (t^*, T] \tag{191}$$

From (182), (190) and (191) we easily find (since $K^*(T) = K_T < \bar{K}, f'(K_T) > 0$)

$$t^* = T - \frac{1}{f'(K_T)}. \tag{192}$$

Note that if the marginal productivity $f'(K_T)$ is high, t^* is close to T, as is to be expected. We see moreover that if K_T increases, then $f'(K_T)$ decreases (since $f'' < 0$) and thus t^* decreases.

Since we have assumed that $u^*(t) \neq 1$ at all t, it is natural to expect that $u^*(t) \in (0, 1)$ in $[0, t^*)$, and this shall be our guess. In this case, from (187) and (180) we find

$$\frac{\mathrm{d}}{\mathrm{d}t} U'(C) = -f'(K)U'(C).$$

If we calculate the derivative on the left-hand side of this equation and rearrange, we have the following differential equation:

$$\ddot{K} = f'(K)\dot{K} + \frac{U'[f(K) - \dot{K}]}{U''[f(K) - \dot{K}]} f'(K). \tag{193}$$

(Note that the second-order differential equation (193) is the same as the Euler equation for the variational problem in example 3 of Chapter 1 for $\rho = 0$.)

In the interval $[0, t^*)$, where we have assumed $u^*(t) \in (0, 1)$, we see from (187) and (180) that the adjoint function $p(t)$ must satisfy the following differential equation:

$$\dot{p} = -f'(K)p. \tag{194}$$

Equation (191) gives the value of p at t^*.

We have now reached the following situation: In the interval $[t^*, T]$, $u^*(t) \equiv 0$, $K^*(t) \equiv K_T$ and $p(t)$ given by (189) satisfy, for all values of $K_T \in (0, \bar{K})$, all the relevant conditions in the Maximum Principle with $p_0 = 1$, (the initial condition being disregarded.) To test the maximum condition, we use (186). In fact, $p(t) \leqslant U'[f(K^*(t))] \Leftrightarrow -f'(K_T)U'[f(K_T)](t - T) \leqslant U'[f(K_T)] \Leftrightarrow -f'(K_T)(t - T) \leqslant 1 \Leftrightarrow t \geqslant T - 1/f'(K_T) = t^*$.)

In the interval $[0, t^*)$ we have proposed $u^*(t) \in (0, 1)$, with $u^*(t)$ defined implicitly in (187). For this choice of $u^*(t)$ to be feasible, the Maximum

Principle implies that $K^*(t)$ and $p(t)$ must satisfy (193) and (194). Our proposal for an optimal path satisfies $\dot{K}^*(t^{*+})=0$. By note 2(b), $\dot{K}^*(t^{*-})=0$ also.

Let a pair of solutions to (193) and (194) satisfying the conditions $K(t^*)=K_T>0$, $\dot{K}(t^*)=0$ and $p(t^*)=U'[f(K_T)]$ be denoted by $K(t;K_T)$, $p(t;K_T)$.

To solve our problem we proceed according to the following program:

(A) Prove the existence of a number $K_T^0<\bar{K}$ such that there is a solution $K(t;K_T^0)$ to (193) defined on $[0, t^*)$ which satisfies $K(0;K_T^0)=K_0$. The separable differential equation (194) then "inherits" a solution $p(t;K_T^0)$ on $[0, t^*)$.

(B) Verify that $K(t;K_T^0)$, $p(t;K_T^0)$ and $u^*(t)$ defined by

$$\dot{K}(t;K_T^0)=u^*(t)f(K(t;K_T^0)) \tag{195}$$

satisfy the Maximum Principle on $[0,t^*)$, with $p_0=1$.

Then define the triple $(K^*(t), p(t), u^*(t))$ on $[0, T]$ in the following way: In $[0, t^*)$, $t^*=T-1/f'(K_T^0)$, put $K^*(t)=K(t; K_T^0)$, $p(t)=p(t; K_T^0)$ and let $u^*(t)$ be defined by (195). In the interval $[t^*, T]$, put $K^*(t)=K_T^0$, let $p(t)$ be defined by (189) with $K_T=K_T^0$ and let $u^*(t)=0$. Then it follows that $(K^*(t), p(t), u^*(t))$ satisfies all conditions in the Maximum Principle with $p_0=1$. If we can prove that

(C) $\hat{H}(K, p(t), t)$ (see (127)) is concave in K for each $t\in[0, T]$,

then from theorem 5 we can finally conclude that $(K^*(t), p(t), u^*(t))$ solves our problem.

To prove (A) we have to study the solutions to (193) backwards from t^* with $K_T\in[K_0, \bar{K})$. It is quite easy to establish the existence of a solution $K(t; K_T)$ on some small interval to the left of t^*. In fact, (193) has the form $\ddot{K}=F(t, K, \dot{K})$ and F is continuous in t, K, and $\dot{K}$ and F is C^1 w.r.t. K and $\dot{K}$ in a neighbourhood of $(t^*, K_T, 0)$. By the usual existence and uniqueness theorem for ordinary differential equations (see Theorem A.10, (A), (C)), there exists a unique solution $K(t; K_T)$ of (193) in some interval $(t^*-\delta, t^*]$, such that $K(t^*; K_T)=K_T$ and $\dot{K}(t^*; K_T)=0$. It is easy to see that $\delta>0$ can be choosen so small that for $t\in(t^*-\delta, t^*)$,

$$K(t; K_T)>K_0/2, \tag{196a}$$

$$\dot{K}(t; K_T)>0, \tag{196b}$$

$$f(K(t; K_T))-\dot{K}(t; K_T)>0. \tag{196c}$$

Property (196a) follows from the continuity of K. As to (196b), note that

$\ddot{K}(t^*; K_T) = U'[f(K_T)] \, f'(K_T)/U''[f(K_T)] < 0$. Since $\ddot{K}$ (by (193)) is a continuous function of t, and $\dot{K}(t^*; K_T) = 0$, we can choose δ so small that (196b) is satisfied. Furthermore, $g(t) = f(K(t; K_T)) - \dot{K}(t; K_T)$ is a continuous function of t and $g(t^*) = f(K_T) > 0$. We conclude that (196c) is also satisfied for a suitable choice of δ.

Let $(t_*, t^*]$ be the largest half-open interval in $(-\infty, t^*]$ in which there is a solution $K(t; K_T)$ to (193) (with $K(t^*; K_T) = K_T$ and $\dot{K}(t^*; K_T) = 0$) such that (196) is valid. (It is convenient to study solutions to (193) in $(-\infty, t^*]$, although we shall be interested, in the end, only in functions which are defined in $[0, t^*]$.) We shall need the following inequality:

For some $a > 0$, $\quad f(K(t; K_T)) - \dot{K}(t; K_T) > a$ for all

$t \in (t'', t^*]$, $t'' = \max(0, t_*)$, $\quad$ for all $K_T \in [K_0, \bar{K}]$. $\qquad$ (197)

To prove (197), let $g(t)$ be the backwards solution to the separable differential equation $\dot{g} = -f'(K_0/2)g$, $g(t^*) = U'[f(K_T)]$. The solution to this equation will be compared to the solution $p(t; K_T)$ to (194). Now,

$$g(t) = U'[f(K_T)]\exp(f'(K_0/2)(t^* - t))$$

$$p(t; K_T) = U'[f(K_T)] \exp\left(\int_t^{t^*} f'(K(\tau; K_T))\,d\tau\right).$$

Since $f'(K_0/2) > f'(K(t; K_T))$ for all $t \in (t'', t^*]$, and since $U' > 0$, we see that $p(t; K_T) \leqslant g(t) \leqslant g(0)$, the latter inequality following from the fact that g is decreasing on $(0, t^*]$. Now, from fig. 2.9 (which is based on the properties of U' in (177) and on (187) we see that there exists an $a > 0$, such that

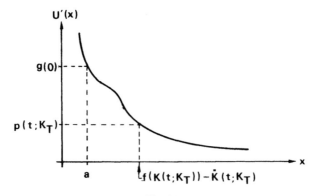

Fig. 2.9

$U'(a) = g(0)$, and such that (197) holds. To proceed we need the following result. Let $K_T \in [K_0, \bar{K}]$. Then

> there exists a maximal closed interval $[t', t^*] \subset [0, t^*]$ such
> that $K(t; K_T)$ satisfies (193) in this interval and such that
> $K(t; K_T)$ satisfies (196) in (t', t^*). If $t' > 0$, $K(t'; K_T) = K_0/2$.   (198)

In order to prove (198) we shall rely on a result in the theory of differential equations. Let the sets E and A in R^2 and R^3, respectively, be defined by

$$E = \{(K, \dot{K}): K > K_0/2, \dot{K} > 0, f(K) - \dot{K} > a/2\}, \qquad A = E \times R. \qquad (199)$$

We apply Theorem A.11 and note A.7 to equation (193), (see (A.41), (A.42)), and we conclude that one of the following three cases must occur:

(i)   There exists a solution $K(t; K_T)$ on all of $(-\infty, t^*)$ lying in A.
(ii)   There exists a solution $K(t; K_T)$ in A on the interval $(t_*, t^*]$, $t_* > -\infty$, and $K(t_*^+; K_T)$ belongs to the boundary of the set E. Thus, either (a) $K(t_*^+; K_T) = K_0/2$, (b) $\dot{K}(t_*^+; K_T) = 0$ or (c) $f(K(t_*^+; K_T)) - \dot{K}(t_*^+; K_T) = a/2$.

(iii)   There exists a solution $K(t; K_T)$ in A on $(t_*, t^*]$ with $t_* > -\infty$ and $\overline{\lim} \max\{|K(t; K_T)|, |\dot{K}(t; K_T)|\} = \infty$ when $t \to t_*^+$.

Case (iii) cannot occur. From $\dot{K}(t; K_T) > 0$ we get $0 \leqslant K(t; K_T) \leqslant K(t^*; K_T) = K_T$ for all $t \in (t_*, t^*]$, so $K(t; K_T)$ is bounded. Moreover, using (197), $0 < \dot{K}(t; K_T) < f(K(t; K_T)) - a \leqslant f(K(t^*; K_T)) - a = f(K_T) - a$. Thus $\dot{K}(t; K_T)$ is bounded as well.

Depending on the value of K_T, cases (i) and (ii) can both occur. In the case of (ii), if $t_* \geqslant 0$ we claim that $K(t_*; K_T) = K_0/2$. To prove this, it suffices to prove that (b) and (c) in (ii) cannot occur. (b) is impossible because if we insert $\dot{K}(t_*; K_T) = 0$ into (193), then $\ddot{K}(t_*; K_T) < 0$ and thus $\dot{K}(t; K_T) < 0$ for $t > t_*$ and t close to t_*, a contradiction. Since $f(K(t; K_T)) - \dot{K}(t; K_T) \geqslant a$, by continuity of f, K and $\dot{K}$, (c) is impossible as well.

To obtain (198) in the case of (i), we put $t' = 0$, while in the case of (ii) we put $t' = \max(0, t_*)$.

The types of behaviour of $K(t; K_T)$ shown in fig. 2.10 are now possible. Note that $K(t; \bar{K}) \equiv \bar{K}$ is a solution to (193) since $f'(\bar{K}) = 0$. Consider moreover $K(t; K_0)$. If t' is 0 in this case, then $K(t; K_0)$ is a solution on $[0, t^*]$ and $K(0; K_0) < K_0$ since $\dot{K}(t; K_0) > 0$ on $(0, t^*)$ (See curve (1) in fig. 2.10.) If $t' > 0$, $K(t; K_0)$ behaves like curve (2) in fig. 2.10. It is therefore reasonable to expect that if we decrease K_T from $\bar{K}$ towards K_0, the solution $K(t; K_T)$ moves downwards and that the initial point $K(0; K_T)$ eventually reaches K_0 for some value K_T^0.

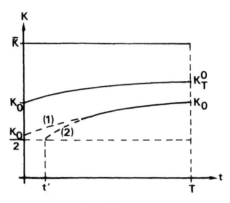

Fig. 2.10

Technically we argue in the following way: Define $K_T^0 = \inf \Gamma$, where $\Gamma = \{K_T: K_T \in [K_0, \bar{K}], K(t; K_T)$ exists on $[0, t^*)$, lies in A, and $K(0; K_T) \geq K_0\}$. $\bar{K} \in \Gamma$, so $\Gamma \neq \phi$. Γ is moreover bounded below, so K_T^0 is well defined. Now, for the solution $K(t; K_T^0)$ the number t' is (I) $t'=0$ or (II) $t'>0$. We claim that (II) cannot occur. In fact by the continuous dependence of the solution on the terminal conditions (see theorem A.8), there is a $\delta>0$ such that if $|K_T - K_T^0| < \delta$, then $|K(t; K_T) - K(t, K_T^0)| < K_0/4$ for $t=t'$. Now, $K(t; K_T) \geq K_0$ for all t and all $K_T > K_T^0$, and therefore, in particular, for $K_T^0 < K_T < K_T^0 + \delta$, $K(t'; K_T) - K(t'; K_T^0) \geq K_0 - K_0/2 = K_0/2$. This result contradicts the inequality $|K(t'; K_T) - K(t', K_T^0)| < K_0/4$. Thus $t'=0$ for $K_T = K_T^0$.

Let us show that $K(0; K_T^0) > K_0$ is impossible. If this was the case, there would exist solutions $K(t; K_T)$ on $[0, t^*)$ also for K_T slightly smaller than K_T^0 and for K_T near enough K_T^0, $K_T < K_T^0$, $K(0; K_T) > K_0$, since $K(0; K_T)$ depends continuously on K_T, contradicting the definition of K_T^0. Hence $K(0; K_T^0) \leq K_0$.

Neither is $K(0; K_T^0) < K_0$ possible. If this was the case, then since $K(0; K_T)$ depends continuously on K_T for K_T near K_T^0, $K(0; K_T) < K_0$ also for K_T slightly greater than K_T^0, contradicting the definition of K_T^0.

If we put $K^*(t) = K(t; K_T^0)$, we have the following properties satisfied on $[0, t^*]$: $K^*(0) = K_0$, $K^*(t^*) = K_T^0$, $\dot{K}^*(t^*) = 0$, $K^*(t) > 0$, $\dot{K}^*(t) > 0$ for $t < t^*$ and $f(K^*(t)) - \dot{K}^*(t) > 0$. Equation (195) is now $\dot{K}^*(t) = u^*(t) f(K^*(t))$, and we see that it determines $u^*(t) \in (0, 1)$.

Consequently we have now established the existence of an admissible pair $(K^*(t), u^*(t))$ on the whole interval $[0, T]$. On $[t^*, T]$ we found an adjoint function $p(t)$ (given by (189) with K_T replaced by K_T^0), and we proved that all the conditions in the Maximum Principle were satisfied on $(t^*, T]$ with $p_0 = 1$. In particular, $p(t^*) = U'[f(K_T^0)]$. In order to prove that all the conditions in the Maximum Principle are satisfied with $p_0 = 1$ on $[0, t^*]$ as well, it remains to prove that the maximum condition is satisfied with an adjoint function $p(t)$ satisfying $p(t^*) = U'[f(K_T^0)]$ and (194). To this end note that $\dot{K} = \dot{K}^*(t)$, according to theorem 3, maximizes the Hamiltonian $U[f(K^*(t)) - \dot{K}] + q(t)\dot{K}$ for $\dot{K} \in (-\infty, \infty)$ and $t \in [0, t^*]$ provided $q(t)$ is defined by:

$$q(t) = U'[f(K_T^0)] - \int_{t^*}^{t} U'[f(K^*(\tau)) - \dot{K}^*(\tau)]f'(K^*(\tau))\mathrm{d}\tau. \qquad (200)$$

In particular, $q(t^*) = U'[f(K_T^0)]$, and $\dot{q}(t) = - U'[f(K^*(t)) - \dot{K}^*(t)]f'(K^*(t))$. Hence $q(t) = U'[f(K^*(t)) - \dot{K}^*(t)]$. (Differentiate the latter expression using that (193) implies $(\mathrm{d}/\mathrm{d}t)U'(C) = -f'(K)U'(C)$, and note that $q(t^*)$ has the correct value since $K^*(t^*) = K_T^0$ and $\dot{K}^*(t^*) = 0$.) It follows that $q(t)$ satisfies (194) and we conclude that $q(t) = p(t)$ is the required adjoint function. We see immediately that $u^*(t) = \dot{K}^*(t)/f(K^*(t))$ maximizes the Hamiltonian $H(K^*(t), u, p(t), t)$ given in (179).

From the discussion following (B) we see that our problem is completely solved if we can prove (C), i.e. the concavity of $\hat{H}(K, p(t), t)$ w.r.t. $K \in [0, \infty)$. Note that

$$\hat{H}(K, p(t), t) = \max_{u \in [0, 1]} [U[f(K) - uf(K)] + puf(K)]$$

$$= \max_{\dot{K} \in [0, f(K)]} [U[f(K) - \dot{K}] + p\dot{K}].$$

We make use of the result in exercise 2.6.4. In fact, $U[f(K) - \dot{K}] + p\dot{K}$ is concave in K, $\dot{K}$ (see e.g. theorem 5.14 in Sydsæter (1981).) Moreover, if $K_1 \geqslant 0$, $K_2 \geqslant 0$, $0 \leqslant \dot{K}_1 \leqslant f(K_1)$, $0 \leqslant \dot{K}_2 \leqslant f(K_2)$ and $\lambda \in [0, 1]$, then

$$0 \leqslant \lambda \dot{K}_1 + (1 - \lambda)\dot{K}_2 \leqslant \lambda f(K_1) + (1 - \lambda)f(K_2) \leqslant f(\lambda K_1 + (1 - \lambda)K_2)$$

since f is concave. Thus $\lambda \dot{K}_1 + (1 - \lambda)\dot{K}_2 \in [0, f(\lambda K_1 + (1 - \lambda)K_2)]$, and we conclude that $\hat{H}(K, p(t), t)$ is concave in K.

Exercise 2.7.1   Consider the problem

$$\max_{u} \int_0^T (1-u)K^a dt, \qquad \dot{K} = uK^a,$$

$$K(0) = K_0 > 0, \qquad K(T) \text{ free}, \qquad u \in [0, 1] \tag{201}$$

where a is a constant, $0 < a < 1$.

(i) Prove that if $aK_0^{a-1}T \leqslant 1$, then $u^*(t) \equiv 0$, $K^*(t) \equiv K_0$, $p(t) = -aK_0^{a-1}(t - T)$ solves the problem.

(ii) Prove that if $aK_0^{a-1}T > 1$, then with $t^* = aT - K_0^{1-a}$, the optimal solution is

$$u^*(t) = \begin{cases} 1 & t \in [0, t^*] \\ 0 & t \in (t^*, T] \end{cases} \tag{202}$$

$$K^*(t) = \begin{cases} [(1-a)t + K_0^{1-a}]^{1/(1-a)} & t \in [0, t^*] \\ [a(1-a)T + aK_0^{1-a}]^{1/(1-a)} & t \in (t^*, T] \end{cases} \tag{203}$$

$$p(t) = \begin{cases} \left[\dfrac{a(1-a)T + aK_0^{1-a}}{(1-a)t + K_0^{1-a}} \right]^{a/(1-a)} & t \in [0, t^*] \\ \\ \dfrac{1}{(1-a)T + K_0^{1-a}} (T-t) & t \in (t^*, T] \end{cases} \tag{204}$$

Exercise 2.7.2   Consider example 6 and assume that $K_0 \geqslant 1 - e^{1-T}$. Find the solution to the problem in this case.

Exercise 2.7.3   Assume in example 7 that $f'(K) > 0$ for all $K \geqslant 0$, so that f has no bliss level.

(i) Let $A = f'(K_0)$, $B = f(K_0)$, and consider the differential equation $\dot{z} = Az + B$, $z(0) = K_0$. Prove that $\dot{z} \geqslant \dot{K}(t)$ for any solution $K(t) = K(t; K_T)$ for which $K(0) \geqslant K_0$.

(ii) Show that $K(t; z(T)) > K_0$ for all $t \geqslant 0$.

(iii) Show that the solution even in this case is exactly as in example 7. (The arguments connected with fig. 2.10 still hold.)

Exercise 2.7.4*   Use theorem 6 to show that the solution in example 6 is unique.

(Hint: Prove that the maximization of $H(K, u, p(t), t)$ gives a unique locally Lipschitzian solution $u(K)$. Insert $u(K)$ into $\dot{K} = uK$ and use theorem A.5.)

8. Existence theorems

It can well happen that an optimal control problem has no solution. For instance, in most optimal control problems in economics it is easy to formulate targets which are entirely unreachable. These are trivial cases of non-existence of an optimal control. However, even for problems in which admissible control functions exist, we are not automatically guaranteed the existence of an optimal control. As an example of a "practical" optimal control problem of this sort, think of a cooking plate which can be switched on and off, and assume that in some time period we want to keep the temperature as close to 100°C as possible (for instance by minimizing the time integral of the square deviation from 100°C). If we start out with a temperature of 100°C, it is obvious that there is no limit to how often we should switch the plate on and off. No "optimal control" exists. However, if we formulate the problem more realistically taking into account the costs of switching on and off, then intuitively an optimal control exists.[8]

In the literature on applications of optimal control theory, it is often argued that the "nature of the problem" (based on, say physical or economic considerations) implies that an optimal solution exists. Such heuristic arguments might give interesting clues to the existence problem, but they do not *replace* a mathematical proof of existence. In fact, a mathematical proof of existence is often a partial test for the appropriateness of the mathematical model used.

The Filippov–Cesari theorem

The first existence theorem we shall present is the standard one in the literature. Consider the general control problem (50)–(54). For each (x, t) define the set $N(x, U, t)$ in R^{n+1} by

$$N(x, U, t) = \{(f_0(x, u, t) + \gamma, f_1(x, u, t), \ldots, f_n(x, u, t)): \gamma \leqslant 0, u \in U \}. \quad (205)$$

An essential assumption in the next theorem is that the set $N(x, U, t)$ be

---

[8]Seierstad (1985c).

convex. This convexity requirement implies that our system has the following property: If we are at position x at time t and can "drive" in two directions $\dot{x}_1$ and $\dot{x}_2$, then we can drive in any direction $\dot{x}$ which is a convex combination of $\dot{x}_1$ and $\dot{x}_2$, and obtain an instantanuous "gain" (in terms of the value of f_0) not lower than the same convex combination of the gains associated with $\dot{x}_1$ and $\dot{x}_2$.

Theorem 8   (Filippov–Cesari. Existence of an optimal control)   Consider the standard optimal control problem (50)–(54). Assume that:

There exists an admissible pair $(x(t), u(t))$.   (206)

$N(x, U, t)$ is convex for each (x, t).   (207)

U is closed and bounded.   (208)

There exists a number b such that $\|x(t)\| \leqslant b$

for all $t \in [t_0, t_1]$ and all admissible pairs $(x(t), u(t))$.   (209)

Then there exists an optimal pair $(x^*(t), u^*(t))$ (where $u^*(t)$ is measurable).[9]

∎

---

[9]Cesari (1983), Chapter 9. We do not use measurable control functions anywhere in the text except for the statement of theorems on existence. Formally, when using existence theorems jointly with necessary conditions, the latter ones ought to be formulated in such a way that they apply even to measurable control functions. The following changes are then necessary: An admissible pair $(x(t), u(t))$ in problem (50)–(54) consists of, by definition, a bounded measurable function $u(t)$ and an absolutely continuous function $x(t)$ such that (51) holds a.e. In theorem 2 the words "continuous and piecewise continuously differentiable function $p(t)$" has to be changed to "absolutely continuous function $p(t)$", and (57) and (58) hold a.e. The regularity conditions needed for the measurable version of the necessary conditions are as follows: f_i and $\partial f_i/\partial x$ $(i=0, 1, \ldots, n)$ are: (1) separately measurable in t, (2) separately continuous in x and in u, (3) bounded on bounded subsets of R^{n+r+1}. The functions $\partial f_i/\partial x$ need only be defined in a neighbourhood of $(x^*(t), t)$ for all $t \in [t_0, t_1]$ and all $u \in U$.

Only for one model (see footnote 12 below) have we commented on the insignificant changes that normally are needed when we apply necessary conditions to a problem and allow for measurable controls.

If the set $N(x, U, t)$ is not convex, optimal solutions might fail to exist. In such situations sometimes the concept of *a relaxed (or generalized) problem* is introduced: Let $P^{n+2} = \{a: a \in R^{n+2}, a \geqslant 0, \Sigma_{i=1}^{n+2} a_i = 1\}$. Let V be the set of vectors $(u^1, \ldots, u^{n+2}, a)$ where $u^i \in U$ and $a \in P^{n+2}$. Define the function $\tilde{F}(x, v, t) = (F_0(x, v, t), F(x, v, t))$ by $\tilde{F}(x, v, t) = \Sigma_{i=1}^{n+2} a_i \tilde{f}(x, u^i, t)$ where $\tilde{f} = (f_0, f)$. In our standard control problem (50)→(54) replace f_0 by F_0 and f by F, and let V be the control region rather than U. Problem (50)→(54) modified in this way is called *the relaxed problem*. In the relaxed problem $N(x, V, t)$ is always convex. Thus, if the original system satisfies (211) and all the conditions in theorem 8, except convexity of $N(x, U, t)$, then an optimal pair in the relaxed problem always exists.

A fundamental property of an admissible *relaxed* pair $(x'(t), v'(t))$ is that there always exists a pair $(x(t), u(t))$ satisfying (51), (52), (54) but not necessarily the terminal conditions (53), such that the difference $x(t) - x'(t)$ can be made uniformly as small as we wish, and such that the

Involved in this theorem is a technical, mathematical problem indicated by the word "measurable". In fact, the theorem does not guarantee the existence of a piecewise continuous optimal control, it only ensures the existence of a measurable optimal control. A measurable function might be less well-behaved than a piecewise continuous function. (It may have more discontinuity points.) However, the risks involved in assuming that the optimal control, whose existence is ensured by theorem 8, is piecewise continuous are small indeed.

A generalization of theorem 8 to the case where U is unbounded is given in theorem 6.19. For an generalization to nonclosed sets U, see note 6.25.

Before we consider some examples, let us note a few useful facts related to the theorem.

Note 15   Condition (207) can be dropped if all the f_i-functions are linear in x, i.e.

$$f_i(x, u, t) = \sum_{j=1}^{n} a_{ij}(t)x_j + g_i(u, t), \qquad i = 0, 1, \ldots, n \tag{210}$$

where $a_{ij}(t)$ and $g_i(u, t)$ are continuous functions (Neustadt, 1963)).

Note 16   Condition (209) is implied by the following condition:

There exist piecewise continuous functions $a(t)$ and $b(t)$ such that $|| f(x, u, t)|| \leqslant a(t)||x|| + b(t)$ for all (x, t), $u \in U$. $\tag{211}$

Note 17*. The regularity conditions in (6) (which are implicitly assumed throughout this book) can be replaced by the weaker conditions that $f_i(x, u, t)$, $i = 0, 1, \ldots, n$ are defined for $u \in U$, $x \in R^n$, $t \in [t_0, t_1]$ and continuous in (x, u) and t separately. For each fixed (x, u), $t \to f_i(x, u, t)$ can even have a finite number of discontinuity points.

Example 8   Let us prove the existence of an optimal control in example 1.

---

criterion functional has approximately the same value. Also, if $x^r(t)$ is an optimal relaxed solution which satisfies the necessary conditions for $p_0 = 1$, but not for $p_0 = 0$, then there even exist original *admissible* pairs having the above-mentioned approximating properties.

Since existence is so easily obtained for relaxed problems, and because of the approximation properties of relaxed pairs, several authors have argued that relaxed systems are the proper objects to study. (See e.g. Young (1969), Warga (1972)). In control problems in economics we often have enough convexity for ordinary existence theorems and even sufficiency theorems to apply, so a study of relaxed problems is less important.

The set defined in (205) is in this case

$$N(x, U, t) = \{(x + \gamma, x + u), \qquad \gamma \leqslant 0, \qquad u \in [-1, 1]\}.$$

This set is indicated in Fig. 2.11. Note that for all choices of x and t, $N(x, U, t)$ is a convex set. Moreover, the control region $U = [-1, 1]$ is closed and bounded. Since $f(x, u, t) = x + u$ and $|u| \leqslant 1$, $|f(x, u, t)| = |x + u| \leqslant |x| + |u| \leqslant |x| + 1$. Thus condition (211) in note 16 is satisfied. We know in this case that there exist admissible pairs $(x(t), u(t))$. According to theorem 8 (with (209) replaced by (211)) we conclude that the problem has an optimal solution.

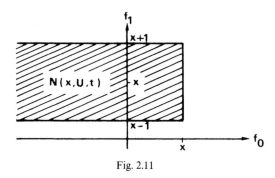

Fig. 2.11

Example 9   (Consumption versus investment)   Consider the problem

$$\max \int_0^T U[(1 - u(t))f(K(t))]e^{-\rho t}dt \tag{212}$$

$$\dot{K}(t) = u(t)f(K(t)) - \delta K(t), \qquad K(0) = K_0 > 0, \qquad K(T) \geqslant K_T > 0 \tag{213}$$

$$u(t) \in [0, 1]. \tag{214}$$

We assume that $U' > 0$, $U'' \leqslant 0$, and the constants T, ρ, and δ are positive.

This is clearly a variant of the Ramsey model. Compared with (146)–(148), we have here included a discount factor and moreover assumed that there is capital depreciation.

We claim that condition (207) is satisfied in this case. Keep (K, t) fixed and let y_1, y_2 be two arbitrary points in $N(K, U, t)$, i.e.

$$y_1 = (U[(1 - u_1)f(K)]e^{-\rho t} + \gamma_1, u_1 f(K) - \delta K) \qquad \text{for some } \gamma_1 \leqslant 0, u_1 \in U$$

$$y_2 = (U[(1 - u_2)f(K)]e^{-\rho t} + \gamma_2, u_2 f(K) - \delta K) \qquad \text{for some } \gamma_2 \leqslant 0, u_2 \in U.$$

Let $\lambda \in [0, 1]$. We want to prove that $y_3 = \lambda y_1 + (1 - \lambda)y_2 \in N(K, U, t)$. Put $\lambda y_1 + (1 - \lambda)y_2 = (z_1, z_2)$. Then the first component z_1 is

$$z_1 = \lambda U[(1 - u_1)f(K)]e^{-\rho t} + \lambda \gamma_1 + (1 - \lambda) U[(1 - u_2)f(K)]e^{-\rho t} + (1 - \lambda)\gamma_2$$
$$= \{\lambda U[(1 - u_1)f(K)] + (1 - \lambda)U[(1 - u_2)f(K)]\}e^{-\rho t} + \lambda \gamma_1 + (1 - \lambda)\gamma_2.$$

Now, $U'' \leqslant 0$ so U is concave. Therefore,

$$\lambda U[(1 - u_1)f(K)] + (1 - \lambda)U[(1 - u_2)f(K)]$$
$$\leqslant U[\lambda(1 - u_1)f(K) + (1 - \lambda)(1 - u_2)f(K)]$$
$$= U[f(K) - (\lambda u_1 + (1 - \lambda)u_2)f(K)]$$
$$= U[(1 - u_3)f(K)],$$

if we put $u_3 = \lambda u_1 + (1 - \lambda)u_2$. Then $u_3 \in [0, 1]$. Using this result we see from the last inequality that

$$z_1 \leqslant U[(1 - u_3)f(K)]e^{-\rho t} + \lambda \gamma_1 + (1 - \lambda)\gamma_2. \tag{215}$$

Define $\gamma_3 = z_1 - U[(1 - u_3)f(K)]e^{-\rho t}$. Then from (215),

$$\gamma_3 \leqslant \lambda \gamma_1 + (1 - \lambda)\gamma_2 \leqslant 0 \qquad \text{since } \gamma_1 \leqslant 0 \text{ and } \gamma_2 \leqslant 0.$$

As to the second component z_2,

$$z_2 = \lambda u_1 f(K) - \lambda \delta K + (1 - \lambda)u_2 f(K) - (1 - \lambda)\delta K$$
$$= (\lambda u_1 + (1 - \lambda)u_2)f(K) - \delta K = u_3 f(K) - \delta K.$$

Piecing all this together, we see that we have found a $u_3 \in [0, 1]$ and a $\gamma_3 \leqslant 0$ such that $\lambda y_1 + (1 - \lambda)y_2 = (U[(1 - u_3)f(K)]e^{-\rho t} + \gamma_3, \ u_3 f(K) - \delta K)$. Hence $\lambda y_1 + (1 - \lambda)y_2 \in N(K, U, t)$. Thus we have proved that $N(K, U, t)$ is convex.

The model (212)–(214) above generalizes several of the models considered in this book. See exercises 2.8.1 and 2.8.2.

*Another existence theorem**

In the next theorem a modification of the Filippov–Cesari existence theorem is given which deals with a case in which there are no equality terminal constraints.

Theorem 9 (Existence of an optimal control in a monotonous

case.) Consider the standard optimal control problem (50)–(54), but assume that $l=0$ in (53), so that there are no equality terminal constraints. Then theorem 8 is valid if we replace (207) by

$f_i(x, u, t)$ is non-decreasing in x and concave in u
    for $i=0, 1, \ldots, n$. $\qquad\qquad\qquad\qquad\qquad\qquad\qquad$ (216a)

U is convex $\qquad\qquad\qquad\qquad\qquad\qquad\qquad\qquad\qquad\qquad$ (216b)

and provided we replace (209) by (211). ∎

Note 18 If $f_i(x, u, t)=g_i(x, u, t)+a_i(t)x_i$, $i=1,\ldots, n$ then condition (216a) can for $i=1,\ldots, n$ be replaced by the condition that $g_i(x,u,t)$ is non-decreasing in x and concave in u, with g_i and a_i as continuous functions.[10]

In many economic models (216) (or the condition in note 18) is satisfied. (For one case, see example 9.)

Note also that there are cases in which theorem 9 applies, but where theorem 8 fails (and vice versa).

Note 19 Condition (216) can be given a slightly weaker form, (which is equivalent to (216) if U is convex). Define

$$\hat{N}(x, U, t)=\{(f_0(x,u,t)+\gamma_0, f_1(x,u,t)+\gamma_1, \ldots, f_n(x,u,t)+$$
$$+\gamma_n : \gamma_1 \leqslant 0, \ldots, \gamma_n \leqslant 0,\ u\in U\}.$$

Then (216) can be replaced by

$f_i(x, u, t)$ is non-decreasing in x for $i=0,1,\ldots, n$. $\qquad\qquad$ (217a)

$\hat{N}(x, U, t)$ is convex for all (x, t), where $t\in[t_0, t_1]$. $\qquad\qquad$ (217b)

When "everything is convex/concave"*

The last theorem we present in this section deals with problems in which there is "a lot of" convexity/concavity". It sums up properties not only related to existence but to necessary conditions and sufficient conditions as well.[11]

---

[10]Seierstad (1985b). A proof can be obtained by applying theorem 8 to a problem where $\dot{x}=f$ is replaced by $\dot{x}=f+\gamma$, $\gamma\leqslant 0$, γ a control variable in R^n. Cf. the set $\hat{N}(x, U, t)$ of note 19.
[11]Seierstad (1985b).

Theorem 10   Consider our standard optimal control problem (50)–(54) and assume that there are no equality constraints in (53) (i.e. $l=0$). Let us assume that

$$f_i(x, u, t) \text{ is non-decreasing in } x, \ i=0, 1, \ldots, n \tag{218}$$

$$f_i(x, u, t) \text{ is concave in } (x, u), \ i=0, 1, \ldots, n. \tag{219}$$

$$U \text{ is convex.} \tag{220}$$

There exist piecewise continuous functions $a(t)$ and $b(t)$ such that

$$\|f(x, u, t)\| \leq b(t) + a(t)\|x\| \text{ for all } x, t \text{ and all } u \in U. \tag{221}$$

Then:

(a) If U is closed and bounded and if there exists an admissible pair, then there exists an optimal solution, $(x^*(t),\ u^*(t))$. The control $u^*(t)$ is unique if $f_0(x, u, t)$ is strictly concave in u for each x, or if $f_0(x, u, t)$ is strictly increasing in some variable x_j, with $f_j(x, u, t)$ strictly concave in u.

(b) If there is an admissible pair $(x(t), u(t))$ such that $x(t_1)$ strictly satisfies all the inequalities in (53), then the conditions in theorem 2 with $p_0 = 1$, are necessary as well as sufficient for optimality.

Note 20   If $f_i(x,\ u,\ t) = g_i(x,\ u,\ t) + a_i(t)x_i, \ i=1, \ldots, \ n$, then condition (218) can be replaced by the condition that $g_i(x,\ u,\ t)$ is non-decreasing in x.

Note 21   Conditions (219) and (220) can be replaced by:

With N defined in (205), if $x_1,\ x_2 \in R^n$, $\lambda \in (0,\ 1)$ and $c_1 \in N(x_1, U, t)$, $c_2 \in N(x_2, U, t)$, then there exists a $c_3 \in N(\lambda x_1 + (1-\lambda)x_2, U, t)$ such that

$$\lambda c_1 + (1-\lambda)c_2 \leq c_3. \tag{222}$$

Uniqueness of $u^*(t)$ holds provided the inequality in (222) is strict in the first component, for $c_1 \neq c_2$. Uniqueness is also assured if the inequality in (222) is strict in some coordinate $i \ (i=1, 2, \ldots, n)$ for $c_1 \neq c_2$, provided also $f_0(x, u, t)$ is strictly increasing in x_i.

Note 22   If we know a priori that all admissible functions $x(t)$ belong to a fixed closed convex set A in R^n, then it suffices to assume that (218), (219), (221) and (222) are satisfied for $x \in A$, $u \in U$. Similarly, in theorem 8 it suffices to assume (207) for $x \in A$, and in theorem 9 it suffices to assume (216a) for

$x \in A$ (and $x \in A$ in (217a, 217b). Furthermore, (211) need only hold for $x \in A$ in order for (209) to be valid.

Note 23 In many economic models it is natural to assume that $0 \leqslant \dot{x}_i \leqslant f_i(x(t), u(t), t)$ where $f_i(x, u, t) \geqslant 0$ for all (x, u, t), $i = 1, \ldots, n$. Note that this type of models can be rewritten as $\dot{x}_i = f_i(x(t), u(t), t) \cdot v_i$, where v_i is a control in $[0, 1]$. If U is convex and f_i, $i = 0, \ldots, n$ are concave in u, then $N(x, U', t)$ is convex, where $U' = \{(u, v) : u \in U, \ 0 \leqslant v_i \leqslant 1\}$.

Note 24 Observe that all the results in section 8 hold if we only assume the regularity conditions in note 6.

Example 10 (Consumption versus investment) Consider problem (176) in example 7, replacing the terminal condition by

$$K(T) \geqslant K_T', \qquad \text{where } K_T' \text{ is a given number,} \tag{223}$$

$$K_0 < K_T' < \tilde{K}(T), \tag{224}$$

and where $\tilde{K}(t)$ is the solution to $\dot{K} = f(K)$, $K(0) = K_0 > 0$. We assume that f is defined on $[0, \infty)$ with $f(0) = 0$, $f' > 0$, and $f'' < 0$. (Theorem A.4 guarantees the existence of $\tilde{K}(t)$ on $[0, T]$: Put $A = (K_0/2, \infty) \times (-\infty, \infty)$. Condition (A.10) is satisfied since $f(K) - f(K_0) \leqslant f'(K_0)(K - K_0)$, by concavity of f. Put $A'' = [K_0, \infty) \times (-\infty, \infty)$. Since $\dot{K} = f(K) \geqslant 0$ in A, $K(t) \geqslant K_0$ for all t, so (A.11) is satisfied for $t_1 = \infty$.) Note that $\tilde{K}(t)$ for each t is the highest value of $K(t)$ obtainable.

We assume that the utility function U is continuous in $[0, \infty)$ (0 included), and C^2 on $(0, \infty)$ with

$$U' > 0, \ U'' < 0 \text{ on } (0, \infty), \ U'(C) \to \infty \text{ and } CU'(C) \to 0$$
$$\text{as } C \to 0^+. \tag{225}$$

(For instance, $U(C) = C^a$, $a \in (0, 1)$, satisfies these conditions.)

The latter condition in (225) is imposed in order to ensure that the conditions in note 6(a) are satisfied, for $K > K_0/2$. Note, in particular, that since $CU'(C) \to 0$ as $C \to 0^+$, the partial derivative $\partial U/\partial K = U'[(1-u)f(K)] \cdot (1-u)f'(K) = U'[C] \cdot C \cdot f'(K)/f(K)$ approaches 0 as $u \to 1^-$, while $\partial U/\partial K = 0$ for $u = 1$, so $\partial U/\partial K$ is continuous also at $u = 1$.

We shall solve this problem by using theorem 10 and notes 21 and 22. Let the set A in note 22 be $A = [K_0/2, \infty)$. Note that properties (218) and (221) are satisfied for $K \in A$. The set N in note 21 is in the present case with $U_1 = [0, 1]$, $N(K, U_1, t) = \{(U[(1-u)f(K)] + \gamma, \ uf(K)) : u \in [0, 1], \ \gamma \leqslant 0\} =$

$\{(U[f(K)-\dot{K}]+\gamma,\ \dot{K}):0\leqslant\dot{K}\leqslant f(K),\ \gamma\leqslant 0\}$. Let K_1, $K_2\in A$ and put $c_1=(c_1^1,\ c_1^2)=(U[f(K_1)-\dot{K}_1]+\gamma_1,\ \dot{K}_1)$, $c_2=(c_2^1,\ c_2^2)=(U[f(K_2)-\dot{K}_2]+\gamma_2,\ \dot{K}_2)$, where $\gamma_1\leqslant 0,\gamma_2\leqslant 0$. Let $\lambda\in(0,\ 1)$ and define $K_3=\lambda K_1+(1-\lambda)K_2$, $\dot{K}_3=\lambda\dot{K}_1+(1-\lambda)\dot{K}_2$. If we define $c_3=(c_3^1,\ c_3^2)=(U[f(K_3)-\dot{K}_3],\ \dot{K}_3)$, then using $U'>0$, $U''<0$, and $f''<0$, we find (by the same type of arguments as in example 9) that $c_3^1>\lambda c_1^1+(1-\lambda)c_2^1$, and, moreover, $c_3^2=\dot{K}_3=\lambda\dot{K}_1+(1-\lambda)\dot{K}_2$, by definition, such that $0\leqslant\dot{K}_3\leqslant f(K_3)$ and $c_3\in N(K_3,\ U_1,\ t)$.

The set $U_1=[0,\ 1]$ is closed and bounded, $(K,\ u)\equiv(\tilde{K}(t),\ 1)$ is admissible and $f_0(K,\ u)=U[(1-u)f(K)]$ is strictly concave in u for each $K\in A$. We conclude from (a) in the theorem and notes 21, 22 that there exists an optimal solution $(K^*(t),\ u^*(t))$, where $u^*(t)$ is unique.

The admissible pair $(\tilde{K}(t),\ 1)$ has the property that $\tilde{K}(T)>K_T'$. We deduce from (b) in the theorem that the conditions in theorem 2 with $p_0=1$ are necessary as well as sufficient for optimality.

In order to find the optimal pair $(K^*(t),\ u^*(t))$ we use the necessary conditions with $p_0=1$.

Note first that $u^*(t)$ is never 1. In fact, if $u\to 1$, then $C=(1-u)f(K)\to 0$ and $H_u'=-f(K)U'(C)+p(t)\to-\infty$, so H cannot be maximized by $u=1$. Thus $u^*(t)\in[0,\ 1)$ for all t.[12]

Now, $u^*(t)\equiv 0$ is impossible, since it would imply $K^*(t)\equiv K_0$, which contradicts $K^*(T)\geqslant K_T'>K_0$. If we define $t^*=\inf\{t:u^*(s)\equiv 0\text{ on }(t,\ T]\}$, then $t^*>0$, but, possibly, $t^*=T$. In any case there exist points $t'<t^*$ and arbitrarily close to t^* where $u^*(t')\in(0,1)$ and $\partial H^*/\partial u=0$ at t', i.e. $-U'[C^*(t')]\cdot f(K^*(t'))+p(t')f(K^*(t'))=0$. Hence $0=p(t')-U'[C^*(t')]<p(t')-U'[f(K^*(t'))]$ since $U''<0$ and $C^*(t')<f(K^*(t'))$. Thus $p(t')>U'[f(K^*(t'))]$. By continuity, the latter inequality holds for all t in some small interval $(\alpha,\ \beta)$ around t', and since (186) holds, $u^*(t)>0$ in $(\alpha,\ \beta)$. Hence $u^*(t)\in(0,\ 1)$ in $(\alpha,\ \beta)$ and thus the Euler equation (193) is satisfied. (According to the Hilbert differentiability theorem (see Chapter 1, section 9), $K^*(t)$ is C^2.)

---

[12] Let us indicate how the rest of the arguments in this example must be modified if we allow $u^*(t)$ to be only measurable.

First, we obtain $u^*(t)\in[0,1)$ for a.e.t, since (57) holds only a.e. Next, if we define t^* as the smallest point such that $u^*(t)=0$ a.e. in $(t^*,\ T]$, (such a point t^* exists), there exist points $t'<t^*$ arbitrarily near t^* at which (57) holds (since it holds a.e.). In fact, $\partial H^*/\partial u=0$ a.e. in (α,β). For $K^*(t)$ inserted into $\partial H/\partial u$ imagine that we solve the equation $\partial H/\partial u=0$ for each t in (α,β) with u as unknown. Then we obtain a C^1 function $\bar{u}(t)$, satisfying $\partial H/\partial u=0$ for all $t\in(\alpha,\beta)$. Since $u^*(t)$ satisfies $\partial H/\partial u=0$ a.e. in (α,β), $u^*(t)=\bar{u}(t)$ a.e., and we may assume that $u^*(t)\equiv\bar{u}(t)$. (The definition of $u^*(t)$ on a set of measure zero does not matter.) Assuming that α in fact equals t_*, $u^*(t)$ is continuously differentiable on (t_*,t') and the Euler equation is satisfied on all (t_*,t'). Finally, all the subsequent arguments in the example goes through without change.

Let t_* be the smallest t such that $u^*(t) \in (0, 1)$ in (t_*, t'). Then the Euler equation is satisfied on (t_*, t'). We claim that $t_* = 0$. By contradiction, assume that $t_* > 0$. Then $u^*(t_*^+) = 0$ and $\dot{K}(t_*^+) = 0$. By (193), $\ddot{K}(t_*^+) < 0$, hence $\dot{K}(t) < 0$ for t slightly larger than t_*, a contradiction.

Thus $t_* = 0$. Letting $t' \to t^{*-}$, we obtain that $u^*(t) \in (0, 1)$ and that the Euler equation is satisfied for all $t \in [0, t^*]$.

We conclude from the analysis above that an optimal pair $(K^*(t), u^*(t))$ must be of the following type: On some interval $[0, t^*)$ (perhaps $t^* = T$), $u^*(t) \in (0, 1)$ with $K^*(t)$ satisfying the Euler equation (193). On $(t^*, T]$, $u^*(t) \equiv 0$. Furthermore, one of the following three cases, (a), (b), and (c), prevails:

(a) $t^* = T$ and the solution $K^*(t)$ of the Euler equation on $[0, T]$ satisfies the boundary conditions $K^*(0) = K_0$, $K^*(T) = K_T'$. (Since $\partial H^*/\partial u = 0$ near T implies $p(t) = U'[(1 - u^*(t))K^*(t)] \geqslant U'[f(K^*(t))]$ for t near T, we obtain $p(T) \geqslant U'[f(K^*(T))] > 0$. By the transversality conditions, $p(T) > 0$ implies $K^*(T) = K_T'$.)

(b) $t^* < T$, and $K^*(t)$ satisfies the Euler equation on $[0, t^*]$ with terminal conditions $K^*(t^*) = K_T'$, $\dot{K}^*(t^*) = 0$. (By note 2(b), we know that $u^*(t)$ is continuous at t^*, so $\dot{K}^*(t^{*-}) = u^*(t^{*-})K^*(t^*) = 0$.) If we denote this solution by $K(t; t^*)$, t^* is a parameter to be determined by the condition $K(0; t^*) = K_0$.

(c) $t^* < T$, and $K^*(t)$ satisfies the Euler equation on $[0, t^*]$ with terminal conditions $K^*(t^*) = K' > K_T'$ and $\dot{K}^*(t^*) = 0$. If we denote this solution by $K(t; t^*, K')$, the parameters t^* and K' are determined by the conditions $K(0; t^*, K') = K_0$ and $t^* = T - 1/f'(K')$. (For the latter condition, see (192).)

By theorem 10 and the preceeding arguments we have indeed established the existence of a unique optimal pair $(K^*(t), u^*(t))$ which is of one of the types (a), (b), and (c) above. Thus the combined use of existence results and necessary conditions make it possible to conclude that one of the cases does take place. Existence results replace here the use of the theory of differential equations and the two point boundary arguments of example 7.

Exercise 2.8.1 Prove the existence of a (measurable) optimal control in exercise 2.3.2 and in example 3 by using theorem 8 and note 16. Draw a picture of $N(x, U, t)$ in both cases.

Exercise 2.8.2 Prove the existence of a (measurable) optimal control in exercise 2.3.3 (iv) by using theorem 9.

Exercise 2.8.3 Consider the problem

$$\max \int_0^1 (1-u)x^2 dt \qquad \dot{x}=ux,$$

$$x(0)=x_0>0, \qquad x(1) \text{ free}, \qquad u\in[0,1]$$

(a) Prove the existence of an optimal solution.
(b) Using necessary conditions, solve the problem.

Exercise 2.8.4 Prove the existence of a (measurable) optimal control in example 4 by using theorem 8.

Exercise 2.8.5 Consider the problem

$$\max \int_0^T (x^2-4x)\,dt, \qquad \dot{x}=u, \qquad x(0)=0, \qquad x(T) \text{ free}, \qquad u\in[0,1].$$

(i) Find all pairs (x,u) which satisfy the necessary conditions.
   (Hint: $T=4$ is a borderline case.)
(ii) Find the optimal solution.

Exercise 2.8.6 Consider the problem

$$\max \int_0^1 (x_2^2-x_1^2)\,dt$$

$$\dot{x}_1(t)=u(t), \qquad x_1(0)=0, \qquad x_1(1) \text{ free}$$
$$\dot{x}_2(t)=[u(t)]^2, \qquad x_2(0)=0, \qquad x_2(1) \text{ free}$$
$$u(t)\in[-1,1].$$

(i) Prove that the necessary conditions (theorem 2) have a unique suggestion for an optimal solution: $u^*(t)=x_1^*(t)=x_2^*(t)=p_1(t)=p_2(t)=0$, $p_0=1$.
(ii) Explain why there is no optimal control in this problem.
   (Hint: What happens if we switch $u(t)$ rapidly between -1 and 1?).

(This model tells us that even if there is a unique solution candidate to the necessary conditions, so that the Hamiltonian is globally maximal, the problem may fail to have an optimal solution. The method proposed in

Bensoussan, Hurst and Näslund (1974), p. 33, and used in their book, is therefore incorrect.)

Exercise 2.8.7   Consider the problem

$$\max \int_0^1 (1-u)(x-1)\mathrm{d}t \qquad \dot{x} = ux^2,$$

$$x(0) = x_0 > 0, \qquad x(1) \text{ free}, \qquad u \in [0, 1].$$

Assume $x_0 < 1$.

(a) Prove that an optimal solution exists. ((209) holds even if (211) fails.)
(b) Prove that the necessary conditions have only one solution.
   (Hint: Define $q(t) = p(t)(x^*(t))^2 - x^*(t) + 1$. Note that $u^*(t) = 1$ if $q(t) > 0$, $u^*(t) = 0$ if $q(t) < 0$. Show that $\dot{q}(t) < 0$.)

Exercise 2.8.8.   Assume in example 10 that $u \in (-\infty, 1]$. Assume also that $\lim U(c) < \infty$ when $c \to \infty$. In example 6.5 it is shown that an optimal solution in this problem exists. Use necessary conditions to show that it satisfies the Euler equation on all $[0, T]$.

Exercise 2.8.9   Prove that $N(x, U, t)$ is convex if U is convex, f_0 is concave in u and f is of the form $f(x, u, t) = g(x, t)u + h(x, t)$.
(Hint: See example 9.)

9. Variable final time problems

So far we have been considering optimal control problems with a fixed time interval. In some control problems arising in economics the final time t_1 is not fixed, but is a variable which is determined by the optimization problem. An important case is the *minimal time problem* in which we want to steer a system from its initial state to a desired state so that the time of transition is as small as possible.

   The general control problem with variable time was formulated in section 2. We shall consider, in particular, problem $(50) \to (54)$, with t_1 free $(t_1 \in (t_0, \infty))$. Thus, among all admissible control functions $u(t)$ which in a time interval $[t_0, t_1]$ via (51) bring the system from x° to a point which satisfies the boundary conditions (53), find one (provided there exists any) such that the integral in (50) is maximal. *In contrast to the situation in* section 4, the time t_1 is not a priori fixed since the different admissible control functions are allowed to be defined on different time intervals.

In this section we briefly consider some basic facts concerning variable time problems.

Necessary conditions.

The Maximum Principle for the present problem is as follows:

Theorem 11   **(The Maximum Principle. Variable final time)**   Let $u^*(t)$ be a piecewise continuous control defined on $[t_0, t_1^*]$ which solves problem $(50) \rightarrow (54)$ with t_1 free $(t_1 \in (t_0, \infty))$ and let $x^*(t)$ be the associated optimal path. Then all the conditions in theorem 2 are satisfied on $[t_0, t_1^*]$ and, in addition,

$$H(x^*(t_1^*), u^*(t_1^*), p(t_1^*), t_1^*) = 0. \quad \blacksquare \tag{226}$$

Compared with a fixed final time problem we have in the present case an additional unknown, t_1^*. Fortunately, we have also an extra condition, (226).

One natural way of solving a free final time problem is for any $t_1 > t_0$ to solve first the corresponding problem with t_1 fixed. Denote the solution to this problem $(x_{t_1}(t), u_{t_1}(t))$, with associated adjoint function $p_{t_1}(t)$. Then the solution to the free final time problem is obtained by considering t_1 to be an unknown parameter. Condition (226) tells us to determine t_1 by the condition $F(t_1) \equiv H(x_{t_1}(t_1), u_{t_1}(t_1), p_{t_1}(t_1), t_1) = 0$. The latter condition is natural in view of a result stated in Chapter 3: Let V be the value of the criterion with $x_{t_1}(t), u_{t_1}(t)$ inserted. Under certain conditions (see theorem 3.9), $\partial V / \partial t_1 = F(t_1)$, so $F(t_1^*) = 0$ is a necessary condition for t_1^* to be optimal. (These comments also shed light on the conditions in the following note.)

Note 25   (a) In theorem 11, we have no restrictions on the interval of definition of the admissible functions $[t_0, t_1]$, except $t_1 > t_0$. Suppose T_1, T_2 are fixed numbers, $t_0 \leqslant T_1 < T_2$, and suppose we require $t_1 \in [T_1, T_2]$. Then theorem 11 is still valid provided $t_1^* \in (T_1, T_2)$. If $t_1^* = T_1$, then the equality in (226) is replaced by $\leqslant$; if $t_1^* = T_2$, then the equality is replaced by $\geqslant$.[13] (If

---

[13] Hestenes (1966), Chapter V or Neustadt (1976), Chapter V. It turns out that if $u^*(t)$ is only measurable, $H(x^*(t_1^*), u(t_1^*), p(t_1^*), t_1^*)$ in (226) must be replaced by $\sup_{u \in U} H(x^*(t_1^*), u, p(t_1^*), t_1^*)$ (which *is* finite).

$T_1 = t_0$ and $t_1^* = t_0$, theorem 11 and the condition in this note make no sense. The following conditions are then necessary: There exist a number p_0, $p_0 = 0$ or $p_0 = 1$, and a vector $p(t_1^*)$ with $(p_0, p(t_1^*)) \neq (0, 0)$ such that $p(t_1^*)$ satisfies (60) and $\sup_{u \in U} H(x^*(t_1^*), u, p(t_1^*), t_1^*) \leq 0$.)

(b)* Theorem 11 and this note also hold for the weakened regularity assumptions of note 6, provided $t_1^* \notin C$. Consider next the case $t_1^* \in C$. If $t_1^* \in (T_1, T_2)$, any optimal control satisfies the fixed final time maximum principle on $[t_0, t_1^*]$ and the inequalities $H(x^*(t_1^*), u^*(t_1^*), p_1(t_1^*), t_1^{*+}) \leq 0$ and $H(x^*(t_1^*), u^*(t_1^*), p_2(t_1^*), t_1^{*-}) \geq 0$ for two (perhaps) different adjoint functions $p_1(t)$ and $p_2(t)$. If $t_1^* = T_1$, resp. $t_1^* = T_2$, only $p_1(t)$ (resp. $p_2(t)$) and the first (last) inequality is relevant.[14]

The minimal time problem

We have already referred to the minimal time problem. In fact, if we let $f_0(x, u, t) \equiv -1$, then

$$\int_{t_0}^{t_1} f_0(x, u, t)\, dt = \int_{t_0}^{t_1} (-1)\, dt = -(t_1 - t_0).$$

Since maximizing $-(t_1 - t_0)$ implies minimizing $t_1 - t_0$, we realize that with our choice of f_0 the problem of theorem 11 is precisely one of finding the control function which in minimal time leads the system from x^0 to a state where (53) is satisfied.

In the minimal time problem, $H = -p_0 + p \cdot f(x, u, t)$. Thus (226) reduces to $p(t_1^*) \cdot f(x^*(t_1^*), u^*(t_1^*), t_1^*) = p_0$. Since $(p_0, p(t)) \neq (0, 0)$ always, we conclude that in the minimal time problem $p(t_1^*) \neq 0$ if $t_1^* > t_0$. In fact, $p(t) \neq 0$ for all t.

Note that in the minimal time problem, p_0 does not enter into the differential equations (51) and (58), and, essentially, it drops out of the maximum condition (57) as well. Thus when solving minimal time problems, to begin with, only the fixed time necessary conditions are used. These conditions only determine $p(t)$ up to a positive scalar. (If $p(t)$ satisfies the necessary conditions, so does $ap(t)$, $a > 0$.) Next, letting $p(t)$ be some choice of adjoint function, then (226) essentially amounts to the following necessary condition: $p(t_1) f(x^*(t_1), u^*(t_1), t_1) \geq 0$. This is the final test a

---

[14]Neustadt (1976), Chapter V contains results implying these assertions.

solution has to pass to be acceptable as a candidate for optimality. (If the left-hand side of the last inequality equals 0, then (226) is satisfied with $p_0 = 0$. If the inequality is strict, then by multiplying $p(t)$ by a scalar $a > 0$, it is always possible to obtain that (226) is satisfied with $p_0 = 1$.)

An existence theorem

The Filippov–Cesari existence theorem can be modified to cover the variable time problem provided we know that the terminal time t_1 belongs to a bounded interval. The result is as follows:

Theorem 12 (Existence. Free final time problem)[15] Consider problem (50)→(54) but assume that t_1 is free to vary in $[T_1, T_2]$, $t_0 \leqslant T_1 < T_2$. Suppose that the conditions in theorem 8 are satisfied on $[t_0, T_2]$. Then there exists an optimal (measurable) control. If all f_i are linear in x i.e. (210) holds, then convexity of $N(x, U, t)$ is not needed. ∎

Sufficient conditions

For variable time problems it is hard to find sufficient conditions of any practical value, due to an inherent lack of convexity properties in such problems. There have been a few proposals in the literature (Mereau and Powers (1976), Petersson and Zalkin (1978)) which sometimes work, but most often do not. The following theorem proved in Seierstad (1984a) seems more promising.

Theorem 13 (Sufficient conditions. Free final time problems) Consider problem (50)→(54) with $t_1 \in [T_1, T_2]$, $t_0 \leqslant T_1 < T_2$. Suppose that for each $T \in [T_1, T_2]$ there exists an admissible pair $(x_T(t), u_T(t))$ defined on $[t_0, T]$, with associated adjoint function $p_T(t)$ which satisfies all the conditions in the Arrow sufficiency theorem 5. Assume further that $u_T(t)$ belongs to a fixed, bounded subset U' of U for all t and T. $x_T(T)$ is continuous in T and $\{p_T(T) : T \in [T_1, T_2]\}$ is bounded. Assume finally that the function

$$F(T) = H(x_T(T), u_T(T), p_T(T), T) \qquad (p_0 = 1) \tag{227}$$

---

[15]Cesari (1983), Chapter 9.

has the property that there exists a $T^* \in [T_1, T_2]$ such that

$$F(T) \geq 0 \quad \text{for} \quad T \leq T^* \text{ if } T_1 < T^*$$
$$F(T) \leq 0 \quad \text{for} \quad T \geq T^* \text{ if } T_2 > T^* \tag{228}$$

Then the pair $(x_{T^*}(t), u_{T^*}(t))$ defined on $[t_0, T^*]$ solves problem $(50) \to (54)$ with $t_1 \in [T_1, T_2]$. The pair is unique if (228) is valid also when all the inequalities in (228) are strict and $\hat{H}(x, p_{T^*}(t), t)$ is strictly concave in x for all $t \in [t_0, T^*]$. When $T_1 = t_0$ it is only necessary to test the conditions of the theorem for $T > T_1$. ∎

It follows from the remarks subsequent to theorem 11 that under certain conditions, $\partial V / \partial T = F(T)$. This relationship makes condition (228) natural.

Note 26   Suppose t_1 is required to belong to an interval $[T_1, \infty)$, $t_0 \leq T_1$. If the triple $(x_{T^*}^*(t), u_{T^*}^*(t), T^*)$ satisfies the conditions in theorem 13 for all intervals $[T_1, T_2]$ containing T^*, then the triple is optimal.

Note 27   Assume that (228) is not satisfied for any T^*, but suppose there exists a partition of $[T_1, T_2]$ into subintervals $[a_{i-1}, a_i]$ and points $T_i^* \in [a_{i-1}, a_i]$ such that (228) holds in each subinterval for $T^* = T_i^*$. Then one of the points T_i^* is optimal. The correct one is obtained by comparing the values of the criterion associated with the T_i^*'s.

Before we start to consider some examples, let us briefly mention a sufficiency result which deals with minimal time problems.

Consider our standard problem $(50) \to (54)$. An admissible path $x(t)$ must satisfy the terminal condition (53). Let us define the *target set* T by:

$$T = \{x \in R^n, x_i = x_i^1, i = 1, \ldots, l, \ x_i \geq x_i^1, i = l+1, \ldots, m$$
$$x_i \in R, i = m+1, \ldots, n\}. \tag{229}$$

Thus, if $x(t)$ is admissible, $x(t_1) \in T$. The sufficiency theorem we have in mind is the following result.[16]

Theorem 14*   (Sufficient conditions in minimal time problems)   Consider problem $(50) \to (54)$ with t_1 free and $f_0 \equiv -1$ (i.e. t_1 is to be minimized). Assume that the admissible pair $(x^*(t), u^*(t))$ defined in the interval $[t_0, t_1^*]$, $t_1^* > t_0$, satisfies all the necessary conditions of theorem 11 with $p_0 = 1$ for some adjoint function $p(t)$. Assume that $\hat{H}(x, p(t), t)$ defined in (127) is

---

[16]Seierstad (1984a).

concave in $x \in R^n$ for all $t \in [t_0, t_1^*]$. Assume, finally, that for each a in the target set T, and each $t' \in [t_0, t_1^*)$, there exists a control $u^1(t)$ defined on $[t', t_1^*)$ and a solution $x^1(t)$ of (51) for $u(t) = u^1(t)$, with $x^1(t') = a$, such that $p(t_1^*)\dot{x}^1(t) \geqslant 0$ in (t', t_1^*) with strict inequality for at least one t. Then $(x^*(t), u^*(t))$ is optimal. ■

Note 28*   The existence of the control function $u^1(t)$ with the stated properties is secured if U is compact, (211) is valid and if there exists a closed, convex cone C in R^n such that for each $a \in T$, $b \in C$, $t \in [t_0, t_1^*]$, there is a $u \in U$ such that $f(x, u, t) \in C$ and $p(t_1^*)f(x, u, t) > 0$ with $x = a + b$.

Examples

We shall study three examples in this section. The first one is a generalization of a model studied previously.

Example 11   (**Optimal extraction of a natural resource**)   We consider the resource extraction problem of Example 11 of Chapter 1, and this time we assume that the rate of extraction $u(t)$ is $\geqslant 0$. The problem is therefore

$$\max_{u(t),\, T} \int_0^T [q(t)u(t) - C(u(t), t)]e^{-rt}\, dt \tag{230}$$

$$\dot{x}(t) = -u(t), \qquad x(0) = \bar{x} > 0, \qquad x(T) \geqslant 0, \qquad u(t) \geqslant 0. \tag{231}$$

Recall that $x(t)$ is the stock remaining at time t, $q(t)$ is the (given) world market price at time t and $C(u, t)$ is the cost per unit of time at time t.

The Hamiltonian function is

$$H(x, u, p, t) = p_0[q(t)u - C(u, t)]e^{-rt} - pu. \tag{232}$$

Suppose $u^*(t)$, $x^*(t)$, both defined over some interval $[0, T^*]$, solve our problem. Then there exists a continuous adjoint function $p(t)$ such that for all $t \in [0, T^*]$,

$$(p_0, p(t)) \neq (0, 0), \tag{233}$$

$$u^*(t) \text{ maximizes } p_0[q(t)u - C(u, t)]e^{-rt} - pu \quad \text{for } u \geqslant 0. \tag{234}$$

Except at the points of discontinuities of $u^*(t)$,

$$\dot{p}(t) = -\frac{\partial H}{\partial x} = 0. \tag{235}$$

Moreover, $p_0 = 1$ or $p_0 = 0$, and

$$p(T^*) \geqslant 0 \ (=0 \text{ if } x^*(T^*) > 0). \tag{236}$$

Finally, (see condition (226) in theorem 11),

$$p_0[q(T^*)u^*(T^*) - C(u^*(T^*), T^*)]e^{-rT^*} = p(T^*)u^*(T^*). \tag{237}$$

(If $T^* = 0$, the conditions must be modified according to note 25(a).)

From (235) we see that $p(t) = \bar{p}$ for some constant $\bar{p}$, and by (236), $\bar{p} \geqslant 0$ ($=0$ if $x^*(T^*) > 0$).

Suppose $p_0 = 0$. Then from (233), $p(t) = \bar{p} \neq 0$ and it follows that $\bar{p} > 0$ and then $x^*(T^*) = 0$. On the other hand, with $p_0 = 0$ and $\bar{p} > 0$, it follows from (234) that $u^*(t) \equiv 0$, and then by (231), $x^*(T^*) = \bar{x} > 0$, a contradiction. Thus $p_0 = 1$.

If we look at (234), we see that if $u^*(t) = 0$ is optimal, then $\partial H / \partial u \leqslant 0$. On the other hand, if $u^*(t) > 0$ maximizes H, then $\partial H / \partial u = 0$. Hence from (234),

$$\left[q(t) - \frac{\partial C(u^*(t), t)}{\partial u} \right] e^{-rt} - \bar{p} \leqslant 0 \ (=0 \text{ if } u^*(t) > 0). \tag{238}$$

(If $C(u, t)$ is convex in u, then H is concave in u, and (238) becomes necessary as well as sufficient for $u^*(t)$ to maximize H for $u \geqslant 0$.)

If $u^*(t) > 0$, we see from (238) that

$$q(t) - \frac{\partial C(u^*(t), t)}{\partial u} = \bar{p}e^{rt}. \tag{239}$$

This is equation (47) in example 11 of chapter 1 with $c = \bar{p}$.

An optimal solution with $T^* > 0$ must satisfy conditions (233) to (238). However, without further information about $q(t)$ and $C(u, t)$, it is difficult to derive much information about $u^*(t)$. In fact, we cannot even be sure that an optimal control exists.

Let us sketch the solution in a special case. Let $q(t) = q$, a positive constant, $C = C(u) = au^2 + bu + c$, a, b, c positive constants with $q > b$. Assume furthermore that $d = \max_{u \geqslant 0} (qu - au^2 - bu - c) > 0$. This condition implies that production is profitable at least for some u. The maximum is attained by $u_0 = (q - b)/2a$, and $d = (q - b)^2/4a - c$. (The condition $d > 0$ implies in this case that $T^* > 0$: If $T^* = 0$, $x^*(T^*) = \bar{x} > 0$, so $p(T^*) = 0$ and $p_0 = 1$. The inequality $d > 0$ then contradicts $\max_{u \geqslant 0} (qu - au^2 - bu - c)e^{-rt} \leqslant 0$.)

Condition (238) is here

$$(q - 2au^*(t) - b)e^{-rt} - \bar{p} \leqslant 0 \qquad (=0 \text{ if } u^*(t) > 0) \tag{240}$$

and, in particular, if $u^*(t) > 0$,

$$u^*(t) = \frac{q - b - \bar{p}e^{rt}}{2a}. \tag{241}$$

Next, observe that if $u^*(t') > 0$ for some t', then $u^*(t) > 0$ for all $t < t'$, since the inequalities $q - b > q - 2au^*(t') - b = \bar{p}e^{rt'} > \bar{p}e^{rt}$ imply that (240) cannot be satisfied for $u^*(t) = 0$. Thus, $u^*(t)$ is > 0 on some interval $[0, t'')$, and $u^*(t)$ is the decreasing function of t given in (241).

Assuming that t'' has been chosen as large as possible (with $u^*(t) > 0$ on $[0, t'')$), we might have an interval $(t'', T]$ at which $u^*(t) = 0$. (An argument similar to the one above gives that if $u^*(t) = 0$ for some t, then later on it also sticks to this value.) However, condition (237) fails if $u^*(T^*) = 0$ (the left-hand side becomes negative). Hence, $u^*(t) > 0$ on all $[0, T^*]$ and is given by the formula in (241).

In the present case condition (237) is

$$[qu^*(T^*) - a[u^*(T^*)]^2 - bu^*(T^*) - c]e^{-rT^*} - \bar{p}u^*(T^*) = 0. \tag{242}$$

Suppose $\bar{p} = 0$. Then (242) reduces to $de^{-rT^*} = 0$, a contradiction. Hence $\bar{p} > 0$ and $x^*(T^*) = 0$. Thus $\int_0^{T^*} u^*(t)dt = \bar{x}$, i.e.

$$\frac{1}{2a}\left[(q - b)T^* - \frac{\bar{p}}{r}(e^{rT^*} - 1)\right] = \bar{x}. \tag{243}$$

Inserting the expression for $u^*(T^*)$ in (241) into (242) and rearranging, we obtain:

$$(\bar{p}e^{rT^*})^2 - 2(q - b)\bar{p}e^{rT^*} + (q - b)^2 - 4ac = 0. \tag{244}$$

The two unknowns $\bar{p}$ and T^* are now determined by (243) and (244). Note that (244) is a quadratic equation in $\bar{p}e^{rT^*}$ with the solution $\bar{p}e^{rT^*} = q - b \pm 2(ac)^{1/2}$. Since $u^*(T^*) > 0$ from (241), we see that the $+$ sign is not possible. Hence

$$\bar{p}e^{rT^*} = q - b - 2(ac)^{1/2}. \tag{245}$$

(This expression for $\bar{p}$ is > 0 since $d > 0$.) Inserting the expression for $\bar{p}e^{rT^*}$ into (243) and solving for $\bar{p}$, we obtain

$$\bar{p} = 2ar\bar{x} + q - b - 2(ac)^{1/2} - r(q - b)T^*. \tag{246}$$

Note that $\bar{p}$ decreases as T^* increases. Using (245) we find from (243) that T^* satisfies the equation

$$r(q - b)T^* - [q - b - 2(ac)^{1/2}] + [q - b - 2(ac)^{1/2}]e^{-rT^*} - 2ar\bar{x} = 0. \tag{247}$$

Denoting the left-hand side of (247) by $g(T^*)$, we find $g(0)<0$, $g(T^*)\to\infty$ as $T^*\to\infty$, and $g'(T^*)=r(q-b)(1-e^{-rT^*})+2r(ac)^{1/2}e^{-rT^*}>0$. Hence (247) has a unique positive solution T^*.

Let us sum up our argument so far: By means of necessary conditions we have proved that if there exists an optimal control, then that control is given by (241) on $[0, T^*]$ where T^* is the unique solution to (247).

Since the control region $U=[0, \infty)$ is not bounded and since T^* is not restricted to a bounded interval, the existence of an optimal control is not guaranteed by theorem 12. Yet we shall construct an existence argument based on theorem 12. Parallel arguments can often be used in similar situations.

Consider an auxiliary problem where $u\in[0, A]$ and $T^*\in[0, B]$ are restrictions we add to the other restrictions in the problem. The constants A and B are sufficiently large numbers. In particular, A is chosen so large that $C'(A)>q$, and B is chosen $>\bar{x}/\delta$, where δ is a fixed positive number such that $qu-au^2-bu-c<0$ for $u\in[0,\delta)$. (Note that the expression is <0 for $u=0$.)

We claim that the unique candidate presented above for problem (230), (231) is the only solution of the necessary conditions in the auxiliary problem as well. It suffices to show that no candidate in the auxiliary problem can satisfy $u^*(t)=A$ or $T^*=B$. If this is shown, then a candidate in the auxiliary problem must satisfy exactly the same conditions as a candidate in problem (230), (231).

The proof that $p_0=1$ is the same as before. Note next that the possibility that $u^*(t)=A$ for some t can be ruled out: If $u^*(t)=A$, then $H'_u\geqslant0$ for $u=A$, which contradicts $C'(A)>q$ and $\bar{p}\geqslant0$.

Also in the auxiliary problem we see that $u^*(t)>0$ is given by (241) on some interval $[0, t'')$ with $u^*(t)=0$ on $[t'', T]$. Moreover, $u^*(T^*)=0$ is impossible both when $T^*<B$ and $T^*=B$. In the latter case, (237) with $=$ replaced by $\geqslant0$ is contradicted. (See note 25 (a).) Hence $t''=T^*$ and $u^*(T^*)>0$. We now show that $T^*=B$ is impossible. Since $u^*(t)$ is non-increasing, $u^*(t)\geqslant u^*(T)$ for all t, and it follows that $\int_0^{T^*} u^*(t)dt=\bar{x}\geqslant\int_0^{T^*} u^*(T^*)dt=T^*u^*(T^*)$. Hence $T^*\leqslant\bar{x}/u^*(T^*)$. Then $\bar{x}/\delta<\bar{x}/u^*(T^*)$, so $u^*(T^*)<\delta$. The latter condition makes the left-hand side of (242)<0, while the necessary conditions in note 25(a) requires it to be $\geqslant0$. This contradiction allows us to conclude that $T^*<B$.

We have proved that for sufficiently large values of A and B any optimal solution must be of the type (241) with $\bar{p}$ and T^* given by (246) and (247). In *any* auxiliary problem theorem 12 (use note 15) is easily seen to guarantee the existence of an optimal solution. Thus the candidate

$(u^*(t), T^*)$ already found is *the* optimal solution in any auxiliary problem where A and B are sufficiently large. But then this pair $(u^*(t), T^*)$ must be optimal also in the original problem since any admissible pair $(u(t), T)$ in the original problem is also an admissible pair in an auxiliary problem for some choice of A and B sufficiently large, and $(u^*(t), T^*)$ is optimal in this problem.

Example 12 **(Oil extraction using imported technology)** An oil-producing country seeks to increase its export of oil. The required technology has to be imported. We let $x(t)$ denote the value of the oil production at time t, and let $K(t)$ be the total invested capital. The flow of imports is denoted by $y(t) = \dot{K}(t)$, and we assume that $x(t) = \alpha K(t)$ for some positive constant α, such that

$$\dot{x}(t) = \alpha y(t). \tag{248}$$

Suppose there is a certain inertia in the economy and in the world market for the required machinery, so that the country cannot control the flow of imports directly, but is able to control $\dot{y}(t)$. In particular, let us assume that

$$\dot{y}(t) = u(t) \tag{249}$$

with

$$u_0 \leqslant u(t) \leqslant u_1, \tag{250}$$

where u_0 and u_1 are fixed numbers, u_0 negative and u_1 positive.

The problem we shall consider is to find that policy which in minimal time, brings the value of oil production, $x(t)$, to a prescribed level $\bar{x}$ whereupon the flow of imports is terminated.

In mathematical terms, the problem is such: Find a control function $u(t)$, constrained by (250), which leads the system, whose evolution is determined by (248) and (249), from the initial point $(x(0), y(0))$ to the point $(\bar{x}, 0)$ in minimal time.

We realize at once that this is a minimal time problem of the type covered by theorem 11 if we put $f_0 = -1$. (The mathematical problem is closely related to example 1 in section 5 in Pontryagin (1962b).)

If we denote the adjoint functions associated with (248) and (249) by $p = p(t)$ and $q = q(t)$ respectively, the Hamiltonian is

$$H = -p_0 + \alpha p y + q u. \tag{251}$$

The adjoint functions must satisfy

$$\dot{p} = -\frac{\partial H}{\partial x} = 0, \qquad \dot{q} = -\frac{\partial H}{\partial y} = -\alpha p, \tag{252}$$

and $(p(t), q(t)) \neq (0,0)$ for all t. Hence for some constants A and B,

$$p(t) = A, \qquad q(t) = -\alpha A t + B, \qquad (A, B) \neq (0, 0). \tag{253}$$

Evidently, $q(t) = 0$ for at most one single $t \geqslant 0$.

According to the Maximum Principle, if $u(t)$ is optimal, then for each t $u(t)$ must maximize H subject to (250). In the Hamiltonian H, the term $qu = q(t)u$ is the only one depending on u. Hence, if $q(t) > 0$, then H is maximal if $u(t) = u_1$, while H is maximal if $u(t) = u_0$ provided $q(t) < 0$. Making use of (253), we find the following proposal for an optimal choice for $u(t)$:

$$u(t) = \begin{cases} u_1 & \text{if } -\alpha A t + B > 0 \\ u_0 & \text{if } -\alpha A t + B < 0. \end{cases} \tag{254}$$

Thus the optimal $u(t)$ is piecewise constant with at most one shift of value, occurring at a value of t for which $\alpha A t = B$.

The corresponding functions $x(t)$ and $y(t)$ are found from (248) and (249). *On an interval where $u(t) = u_1 > 0$*, (249) gives us (b a constant)

$$\dot{y}(t) = u_1 \qquad \text{that is } y(t) = u_1 t + b. \tag{255}$$

From (248) we get in turn, (a is a new constant)

$$\dot{x}(t) = \alpha(u_1 t + b) \text{ that is } x(t) = \tfrac{1}{2}\alpha u_1 t^2 + \alpha b t + a. \tag{256}$$

In order to study the relationship between x and y, we eliminate t from the expressions for $x(t)$ and $y(t)$. We get

$$x = \frac{1}{2}\frac{\alpha}{u_1}(y-b)^2 + \frac{\alpha b}{u_1}(y-b) + a = \frac{1}{2}\frac{\alpha}{u_1}y^2 + \left(a - \frac{1}{2}\frac{\alpha b^2}{u_1}\right). \tag{257}$$

The graph of (257) in the xy plane is a parabola with vertex at $(a - \tfrac{1}{2}\alpha b^2/u_1$, $0)$ and axis along the positive x-axis. For each point (x_1, y_1) in the plane there is one and only one such parabola passing through the point, namely the one for which

$$a - \frac{1}{2}\frac{\alpha b^2}{u_1} = x_1 - \frac{1}{2}\frac{\alpha}{u_1}y_1^2.$$

From (255) we see that $\dot{y}(t) > 0$, so y increases as t increases.

On an interval where $u(t)=u_0<0$, we find in a similar manner as above that the relationship between x and y is

$$x=\frac{1}{2}\frac{\alpha}{u_0}y^2+\left(a-\frac{1}{2}\frac{\alpha b^2}{u_0}\right).\tag{258}$$

The graph of this relation is a parabola with vertex at $(a-\frac{1}{2}\alpha b^2/u_0, 0)$, but the axis is now along the negative x-axis, since $u_0<0$ implies $\frac{1}{2}\alpha/u_0<0$. Again, there is one and only one such parabola through each point in the xy plane. In the present case, $\dot{y}(t)=u_0<0$, so $y(t)$ decreases as t increases.

The objective as formulated above, was to reach the point $(\bar{x}, 0)$ in minimal time from an initial point $(x(0), y(0))$. The Maximum Principle implies that an optimal control is piecewise constant, with values u_0 or u_1, and with, at most, one switch of value. In order to reach the point $(\bar{x}, 0)$ *with* $u(t)=u_1$ *in the final phase*, the system has to follow the unique parabola (257) passing through $(\bar{x}, 0)$. Since $y(t)$ increases with t in this case, the system must, in the final phase, follow the part of the graph of the parabola (257) denoted by C_1 in fig. 2.12. The other possibility of behaviour in the final phase is that the system hits the target $(\bar{x}, 0)$ under the influence of the control $u(t)=u_0$. Then the system has to follow the parabola (258) passing through $(\bar{x}, 0)$. In this case $y(t)$ decreases as t increases, and the relevant arc is denoted by C_2 in fig. 2.12. (Fig. 2.12 is drawn with $|u_0|>|u_1|$.)

Depending on the initial position $(x(0), y(0))$ of the system, the Maximum Principle will yield different suggestions for the optimal steering of the system towards the target. We shall examine the different possibilites, but

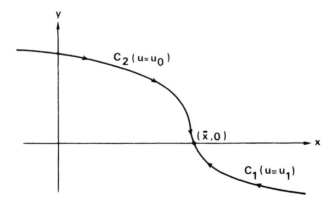

Fig. 2.12

we assume that $x(0) < \bar{x}$ (i.e. that the initial value of oil production is below the desired target).

I. Assume that $(x(0), y(0))$ happens to be the point P_0 on the arc C_2 in fig. 2.13(i). By choosing $u = u_0$, the system moves along C_2 until it reaches $(\bar{x}, 0)$. This is the only optimal path possible. If we were to shift the value of u from u_0 to u_1, at the point P_1, then the system would have to follow an arc of a parabola which does not pass through the target $(\bar{x}, 0)$, since additional shifts in the value of u are not permitted.

II. Consider the case in which the initial point P_0 lies above the parabola C_2 (fig. 2.13(ii)). We then put $u = u_0$ first, and let the system move along the corresponding parabola until it meets C_1. At that point P_1 we switch to $u = u_1$ and follow C_1 until the system hits the target $(\bar{x}, 0)$. This is the only path consistent with the requirements in the Maximum Principle. For instance, if we start out from P_0 with $u = u_1$, then we cannot reach $(\bar{x}, 0)$ without switching the value of u at least twice.

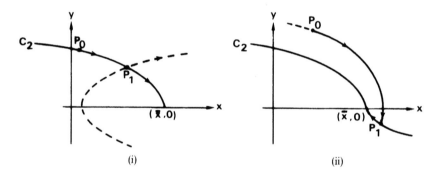

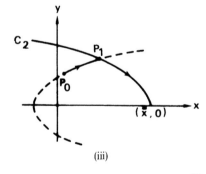

Fig. 2.13

III. We study next the case in which P_0 lies below the arc C_2, as indicated in fig. 2.13(iii). It is easy to see that the only possible optimal solution in this case is to start out with $u=u_1$ and follow the corresponding parabola until it meets C_2 at P_1. Then we switch to $u=u_0$ and follow C_2 until we reach $(\bar{x}, 0)$.

The analysis above has shown us that the Maximum Principle in each case produces a unique trajectory along which the system moves from P_0 to $(\bar{x}, 0)$. Since the Maximum Principle represents *necessary* conditions for optimality, we can conclude that if there exist optimal solutions to our problem, then the trajectories we have found are indeed optimal.

Existence of an optimal control follows from theorem 12. We might also use theorem 14. To this end, consider case III, see fig. 2.13(iii). Since we have a switch at some point in $(0, t_1^*)$, (t_1^* the minimal final time), both A and B are $\neq 0$. From $t=0$ we start out with $u(t)=u_1$, so, by (254), $B>0$. As t increases, $-\alpha At + B$ has to decrease in order for $u(t)$ to switch from u_1 to u_0. But then αA has to be positive. It follow from (253) that $p(t_1^*)>0$, $q(t_1^*)<0$. The target set $T=\{(\bar{x}, 0)\}$. Thus the a vector in the theorem is unique, $a=(\bar{x}, 0)$. Let $t' \in [0, t_1^*)$. Choose $u^1(t)=0$ in $[t', t_1^*-\varepsilon)$ for some positive number ε, $u^1(t)=u_0$ in $[t_1^*-\varepsilon, t_1^*)$. Then, in $[t', t_1^*-\varepsilon)$, $y^1(t)=0$, in $[t_1^*-\varepsilon, t_1^*)$, $y^1(t)=u_0[t-(t_1^*-\varepsilon)]=u_0[(t-t_1^*)+\varepsilon] \geqslant u_0\varepsilon$. Now $p(t_1^*)\dot{x}^1(t)+q(t_1^*)\dot{y}^1(t)=A\alpha y^1(t)+(-\alpha At_1^*+B)u^1(t)=0$ in $[t', t_1^*-\varepsilon)$ and is $\geqslant A\alpha u_0\varepsilon+(-\alpha At_1^*+B)u_0$ >0 if ε is small, $t>t_1^*-\varepsilon$. It remains to test the concavity of $\hat{H}$. But $\hat{H}$ is in this case linear in (x, y). Hence theorem 14 secures optimality.

The reader is encouraged to prove optimality in the cases I and II as well.

Our next example is a variable time problem from economic growth theory. We consider a model in which it is necessary to strike a balance between two conflicting goals: to attain high consumption and to reach a certain level in real capital as fast as possible.

Example 13 (**Consumption versus a rapid attainment of a given production capacity**) Consider the following problem:

$$\max \int_0^T \{-q+(1-u(t))x(t)\}dt, \qquad T\in[0, T_2] \tag{259}$$

$$\dot{x}(t)=u(t)x(t)+a, \qquad x(0)=0, \qquad x(T)\geqslant x_1 \tag{260}$$

$$u(t)\in[0, 1]. \tag{261}$$

We think of $x(t)$ as the production capacity for some country at time t, $u(t)x(t)$ is the fraction of production used for investments, a is a constant basic trend below which the increase in capacity cannot fall (due to technical progress), and $(1-u(t))x(t)$ is left for consumption. The optimality criterion in (259) implies that we put some emphasis on high consumption and some emphasis on attaining fast the level x_1.

In problem (259)→(261), q, a, x_1 and T_2 are positive parameters, and we shall assume that

$$\ln(1+x_1/a) < T_2, \tag{262a}$$

$$ae^{T_2-1} < q, \tag{262b}$$

$$x_1 + a > q, \tag{262c}$$

$$x_1 < aT_2 \tag{262d}$$

((262) is satisfied, for instance, if $q=1.5$, $a=1$, $x_1=1$, $T_2=1.1$.)

Note that we increase $x(t)$ most rapidly by choosing $u(t)\equiv 1$. Then $x(t)=a(e^t-1)$. Let T_1 be the solution of $x(t)=a(e^t-1)=x_1$, i.e.

$$T_1 = \ln(1+x_1/a). \tag{263}$$

We see that it is impossible for $x(t)$ to attain the required level x_1 earlier than at T_1, and we might as well replace the interval $[0, T_2]$ in (259) by $[T_1, T_2]$.

We shall solve the problem by appealing to theorem 13. Let T be any number in $[T_1, T_2]$ and let $(x_T(t), u_T(t))$ denote a solution candidate for problem (259)→(261) with T fixed *and* $x_T(T)\geqslant x_1$, with corresponding adjoint function $p_T(t)$. By studying the necessary conditions with $p_0=1$, it is easy to see that $u_T(t)$ is 0 or 1 according as $p_T(t)$ is <1 or >1. Moreover, $\dot{p}_T(t) = -\max(1, p_T(t))$, so $p_T(t)$ is strictly decreasing. We conclude that $u_T(t)$ switches from 1 to 0 at most once in $[0, T]$. Let the switching point, which depends on T, be denoted by $\hat{T}(T)$.

We establish first the following three facts:
(I)   If $T=T_1$, $\hat{T}(T_1)=T_1$.
(II) If $T>T_1$, $\hat{T}(T)<T$.
(III) If $\hat{T}(T)=0$, $T \geqslant x_1/a > T_1$

(I) follows from the fact that if $u_T(t)$ switches from 1 to 0 before T_1, $x_T(t)$ will never reach the level x_1. To prove (II), suppose on the contrary that $\hat{T}(T)=T$, with $T>T_1$. Then $p_T(t)\geqslant 1$ in $[0, T]$ so $p_T(T)>0$ and hence $x_T(T)=x_1$.

On   the   other   hand,   since $u_T(t)\equiv 1$, $x_T(t)=a(e^t-1)$,   so

$x_T(T) = a(e^T - 1) > a(e^{T_1} - 1) = x_1$, a contradiction. To prove (III), note that if $\hat{T}(T) = 0$, $u_T(t) \equiv 0$, $x_T(t) = at$, so $x_T(T) = aT$ and $x_T(T) \geqslant x_1$ implies $T \geqslant x_1/a$. Note that $x_1/a > T_1 = \ln(1 + x_1/a)$ since $\ln(1 + z) < z$ for all $z > 0$.

For $\hat{T}(T) \in (0, T)$ it follows easily that

$$u_T(t) = 1, \qquad x_T(t) = a(e^t - 1), \qquad p_T(t) = e^{\hat{T}(T) - t} \qquad \text{in } [0, \hat{T}(T)]$$
$$u_T(t) = 0, \qquad x_T(t) = a(e^{\hat{T}(T)} - \hat{T}(T) + t - 1), \qquad p_T(t) = \hat{T}(T) - t + 1$$
$$\text{in } (\hat{T}(T), T] \tag{264}$$

and $\hat{T}(T)$ has to be determined by the condition $x_T(T) \geqslant x_1$, i.e.

$$e^{\hat{T}(T)} - \hat{T}(T) \geqslant x_1/a + 1 - T \tag{265}$$

and the condition $p_T(T) \geqslant 0$ ($= 0$ if $x_T(T) > x_1$), i.e.

$$\hat{T}(T) - T + 1 \geqslant 0 \qquad (= 0 \text{ if } e^{\hat{T}(T)} - \hat{T}(T) > x_1/a + 1 - T). \tag{266}$$

For $T = T_1$, we saw that $\hat{T}(T_1) = T_1$. In this case $u_T(t) = 1$, $x_T(t) = a(e^t - 1)$ and we put $p_T(t) = e^{T - t}$. (Any $p_T(t) = Ae^{-t}$ with $A \geqslant e^T$ satisfies the necessary conditions.)

Let T belong to $(T_1, x_1/a)$. Then, according to (II) and (III), $\hat{T}(T) \in (0, T)$. Then $p_T(T) = \hat{T}(T) - T + 1 > 1 - T > 0$ at least as long as T is less than 1. Thus, for $T \in (T_1, \min(1, x_1/a))$, $\hat{T}(T)$ is determined by (265) with $\geqslant$ replaced by $=$:

$$e^{\hat{T}(T)} - \hat{T}(T) = x_1/a + 1 - T. \tag{267}$$

Note that if T increases from T_1, $x_1/a + 1 - T$ decreases, and from fig. 2.14

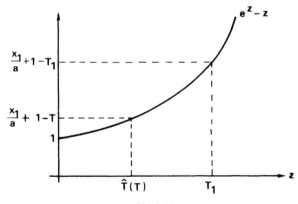

Fig. 2.14

we see that $\hat{T}(T)$ decreases. In fact, we obtain a solution $\hat{T}(T)>0$ as long as $x_1/a+1-T>1$, i.e. $T<x_1/a$. However, equation (267) can only define the optimal switching point as long as $p_T(T)\geqslant 0$, i.e. $\hat{T}(T)\geqslant T-1$.

It turns out that it is convenient to consider the two cases $x_1/a\leqslant 1$ and $x_1/a>1$ separately.

The case $x_1/a\leqslant 1$

If $T<x_1/a$, $T<1$, so $p_T(T)=\hat{T}(T)-T+1>0$, and $\hat{T}(T)$ *is defined by* (267):

(A) For $T\in[T_1,x_1/a)$, $\hat{T}(T)$ is defined by (267), $u_T(t)$, $x_T(t)$, and $p_T(t)$ are defined in (264).

Suppose next that $T\in[x_1/a,1]$. Since in case (A), $\hat{T}(T)\to 0$ as $T\to x_1/a$, it is natural to explore the possibility that $\hat{T}(T)=0$. In that case $u_T(t)\equiv 0$ and $p_T(t)\leqslant 1$. Moreover, $x_T(t)=at$, so $x_T(T)=aT>x_1$ for $T>x_1/a$, and thus $p_T(T)=0$. Since $\dot{p}_T(t)=-1$, $p_T(t)=T-t\leqslant 1$ for all t, which requires $T\leqslant 1$. We therefore arrive at the following suggestion:

(B) For $T\in[x_1/a,1]$, $\hat{T}(T)=0$, $u_T(t)\equiv 0$, $x_T(t)\equiv at$, $p_T(t)\equiv T-t$.

Suppose $T>1$ (which requires $T_2>1$). Then $T>x_1/a$, and according to our discussion of fig. 2.14, $\hat{T}(T)$ is no longer determined by (267), so (265) is satisfied with $>$. By (266) we then have $\hat{T}(T)=T-1$. Thus we suggest:

(C) For $T\in[1,T_2]$, $\hat{T}(T)=T-1$, $u_T(t)$, $x_T(t)$, and $p_T(t)$ are defined in (264).

The behaviour of $\hat{T}(T)$ as T increases from T_1 to T_2 may seem somewhat surprising, but it can be explained in the following way: When T is small, $\hat{T}(T)$ has to be >0 in order for $x_T(T)$ to reach the level x_1. When T becomes somewhat larger, $x_T(T)$ can reach above this level even with $\hat{T}(T)=0$. Only if T is >1 does it pay to save, i.e. $\hat{T}(T)>0$.

The case $x_1/a>1$

Again, as T increases from T_1, $x_T(T)=x_1$ and $\hat{T}(T)$ is given by (267). This goes on as long as $p_T(T)=\hat{T}(T)-T+1\geqslant 0$. Here $p_T(T)=0$ for $\hat{T}(T)=T-1$. Putting $\hat{T}(T)=T-1$ into (267) and solving for T, we find the solution $T'=1+\ln(x_1/a)$. ($T'>T_1$ for $x_1/a>1$. In fact, if $g(u)=1+\ln u-\ln(1+u)$, $u>0$, $g'(u)=1/u(1+u)>0$, and $g(u)=0\Leftrightarrow u=1/(e-1)$, so $g(u)>0$ for $u>1/(e-1)\approx 0.58$.) Hence

(D) For $T \leqslant T' = 1 + \ln(x_1/a)$, $\hat{T}(T)$ is the solution to (267). For $T > T'$, $\hat{T}(T) = T - 1$. In both cases $u_T(t)$, $x_T(t)$, and $p_T(t)$ are given in (264).

We have finally come up with a proposal for a solution to problem (259)–(261) for any fixed $T \in [T_1, T_2]$. It is easy to check that the fixed final time sufficient conditions are satisfied. Most of these checks are immediate. We restrict our attention to testing that $p_T(T) \geqslant 0$ in all cases. For case (A) we showed that $p_T(T) > 0$, for cases (B) and (C), $p_T(T)$ is clearly $\geqslant 0$. Consider, finally, case (D). If $T \leqslant 1$, then $p_T(T)$ is clearly $\geqslant 0$. If $T \in (1, 1 + \ln(x_1/a)]$, then $e^{T-1} \leqslant x_1/a$, so $e^{T-1} - (T-1) \leqslant x_1/a - T + 1 = e^{\hat{T}(T)} - \hat{T}(T)$. Since $e^z - z$ is increasing for $z > 0$, $\hat{T}(T) \geqslant T - 1$, and hence $p_T(T) \geqslant 0$. If $T > 1 + \ln(x_1/a)$, $p_T(T) = \hat{T}(T) - (T-1) = 0$.

Since $\hat{H}(x, p_T(t), t) = -q + ap(t) + \max(1, p_T(t))x$ is linear, the Arrow concavity condition is satisfied and all the pairs $(x_T(t), u_T(t))$ specified above satisfy the conditions in theorem 5.

Note next that $u_T(t)$ for all t belongs to a bounded set, as does $p_T(T)$ for all $T \in [T_1, T_2]$. For $x_1/a \leqslant 1$, $x_T(T) = x_1$ in $[T_1, x_1/a]$, $x_T(T) = aT$ in $(x_1/a, 1]$, and $x_T(T) = ae^{T-1}$ in $(1, T_2]$. For $x_1/a > 1$, $x_T(T) = x_1$ in $[T_1, T']$, $x_T(T) = ae^{T-1}$ in $(T', T]$ where $T' = 1 + \ln(x_1/a)$. In both cases we see that $x_T(T)$ is continuous.

In the present case, $F(T)$ as defined in (227) is

$$F(T) = -q + (1 - u_T(T))x_T(T) + p_T(T)u_T(T)x_T(T) + ap_T(T). \tag{268}$$

For $x_1/a \leqslant 1$, $F(T) = -q + x_1 + a(\hat{T}(T) - T + 1) = -q + ae^{\hat{T}(T)}$ in $[T_1, x_1/a]$, $F(T) = -q + aT$ in $(x_1/a, 1]$, and $F(T) = -q + ae^{T-1}$ in $(1, T_2]$. $F(T)$ is decreasing in $[T_1, x_1/a]$, increasing in $(x_1/a, T_2]$. $F(T_1^+) = -q + ae^{T_1} = -q + a + x_1 > 0$ by (262c). If $T_2 > 1$, $F(T_2) = -q + ae^{T_2-1} < 0$ by (262b). If $T_2 \leqslant 1$, $F(T_2) = -q + aT_2 < 0$, since by (262b) $T_2 < 1 + \ln(q/a) \leqslant q/a$.

Note that from (268) $\hat{T}(x_1/a) = 0$, so $F((x_1/a)^-) = -q + a$, while $F((x_1/a)^+) = -q + x_1$. Thus $F((x_1/a)^-) \geqslant F((x_1/a)^+)$, and for $x_1/a < 1$, F is discontinuous at x_1/a. In the cases $a > q$ and $a < q$ the behaviour of $F(T)$ is illustrated in fig. 2.15 (a) and (b). If $a \geqslant q$, we put $T^* = x_1/a$. Then (228) holds for $T \neq T^*$. For $T = T^*$ see below. If $a < q$, then we see that there exists a unique point $T^* \in (T_1, x_1/a)$ such that (228) is satisfied. In fact, since $0 = F(T^*) = -q + ae^{\hat{T}(T^*)}$, $\hat{T}(T^*) = \ln(q/a)$ and from (267)

$$T^* = \frac{x_1 - q}{a} + \ln\frac{q}{a} + 1. \tag{269}$$

For $x_1/a > 1$, $F(T) = -q + ae^{\hat{T}(T)}$ in $[T_1, T']$ and $F(T) = -q + ae^{T-1}$ in $(T', T_2]$, $T' = 1 + \ln(x_1/a)$. Note that by using (267) we see that $\hat{T}(T') = T' - 1$

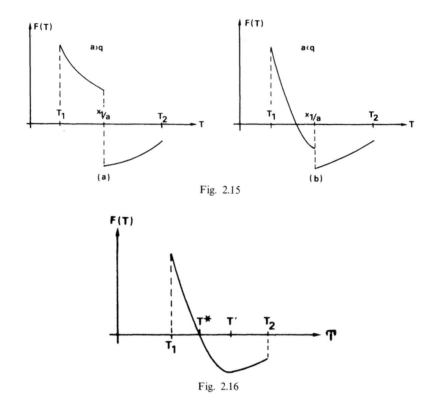

Fig. 2.15

Fig. 2.16

and F is continuous also at T'. Moreover, $F(T)$ decreases in $[T_1, T']$, while $F(T)$ increases in $(T', T_2]$. Again $F(T_1^+)>0$ and $F(T_2)<0$. Hence there is a unique $T^* \in (T_1, T')$ which satisfies (228). (See fig. 2.16.) Again T^* is given by (269).

We have now found an optimal triple $(x_{T^*}(t), u_{T^*}(t), T^*)$ in all cases.

(In the case $x_1/a \leqslant 1$ and $a > q$ (see fig. 2.15(a)), $F(T^*) = F(x_1/a) = -q + x_1 < 0$. This seems to contradict (226). However, there exists another adjoint function $p(t)$ satisfying all the conditions in theorem 11, such that, in particular, the corresponding F-function is 0 at T^*. In fact, $p(t) = q/a - t$.)

Synthesizing an optimal control

In the previous examples in this chapter we obtained optimal controls and associated optimal paths as functions of time. These time functions were dependent on the initial conditions. In an actual economy there will always be "disturbances" of different types which will bring the system out of a computed optimal path. If we continue to implement the optimal control initially computed, the economy will usually develop in a non-optimal fashion and fail drastically to hit the desired target.

In example 12 we presented the solution to the problem in a form which makes it possible to tackle the problem of disturbances. We intentionally expressed the optimal control as a function of the state of the system: For any point in the plane we know the optimal choice of $u(t)$ at that point. Thus, whatever was the previous history of the system up to a certain point, we know how to go from there to the target in minimal time. In this case we say that we have *synthesized* the optimal control (or that we have a *closed loop* or a *feedback control*).

However, we have not in any example discussed what would be a sensible next step: To try to insure against adverse effects of such disturbances. This would amount to introducing disturbances as stochastic variables in our models. This important field in applied control theory is beyond the scope of this book, and as scattered references to a vast literature we mention the books by Chow (1975), Bryson and Ho (1975), and Fleming and Rishel (1975).

Exercise 2.9.1 Consider the two sector model in example 3. However, assume now that we want to minimize the time required to attain the level c in the consumption sector. Thus (63) in example 3 is replaced by $\max \int_0^T (-1) dt$, and "$x_2(T)$ free" is replaced by $x_2(T) \geqslant c$. All the other conditions in example 3 are kept, and we assume that $c > x_2^0$.

(i)   Write down all the information obtained from the Maximum Principle, with p_1 and p_2 as the adjoint functions.
(ii) Prove that $p_2(t) = c_2$ (a constant) and that $p_0 > 0$. (Hint: $(p_1(T^*), p_2(T^*)) \neq 0$.)
(iii) Prove that $u^*(t) = 1$ if $p_1(t) > c_2$, and that $u^*(t)$ is 0 if $p_1(t) < c_2$.
(iv) Prove that $\dot{p}_1(t) < 0$ in $[0, T)$.
(v)   If $(c - x_2^0)/x_1^0 \leqslant 1$, prove that $u^*(t) \equiv 0$ is the optimal control and find the minimal time T^*.
(vi) Find the solution to the problem if $(c - x_2^0)/x_1^0 > 1$.

Exercise 2.9.2 In example 12 let us replace $\dot{K}(t) = y(t)$ with $\dot{K}(t) = y(t) - \delta K(t)$ where $\delta > 0$ is a constant, and thus allow for depreciation of capital. Show that equation (248) in example 12 is then replaced by

$$\dot{x}(t) = -\delta x(t) + \alpha y(t).$$

Except for this replacement, consider the same problem as in example 12.

(i) By using the Maximum Principle prove that an optimal control is again piecewise constant, taking the values u_0 and u_1 only, and with at most one switch.
(ii) Find the equations corresponding to (257) and (258) in example 12.

Exercise 2.9.3   Consider example 12 again. If t_1 is the terminal time, the terminal condition is $x(t_1) = \bar{x}$, $y(t_1) = 0$. Replace this condition with

$$x(t_1) = \bar{x}, \quad y(t_1) \text{ free} \tag{270}$$

(so that we don't care about the value of the flow of imports at the terminal time).

(i) What does theorem 11 tell us about $q(t_1)$?
(ii) Assume that the initial point P_0 is situated as in fig. 2.13 (iii).
     Find the optimal path from P_0 to the terminal point.
     (Hint: $q(t)$ has a constant sign in $[0, t_1]$.)

Exercise 2.9.4   Prove the existence of an optimal solution in example 13 by using theorem 12.

Exercise 2.9.5   In Koopmans (1974) the following problem is studied (we use partly symbols from examples 11 of Chapter 1 and example 11

$$\max \int_0^T U(u(t)) e^{-rt}\, dt, \qquad T \in \left[0, \frac{\bar{x}}{u_0}\right] \tag{271}$$

subject to

$$\int_0^T u(t)\, dt \leqslant \bar{x} \tag{272a}$$

$$u(t) \geqslant u_0, \tag{272b}$$

where T, u_0 and $\bar{x}$ are positive constants and r is a non-negative constant.

We assume that $U'(u)>0$, $U''(u)<0$ for $u \geqslant u_0$, $U(u_0)=0$ and $U'(u)\to\infty$ as $u\to u_0$. An economic interpretation is as follows: At time $t=0$ there is a fixed amount $\bar{x}$ of some resource available. Let $u(t)$ be the rate of consumption of the resource, and let u_0 be the minimum consumption level. In (271) we maximize the total discounted utility arising from the consumption of the resource.

(i) Formulate the problem as a free final time optimal control problem.
(ii) Characterize the unique solution candidate $(x^*(t)$, $u^*(t)$, $T^*)$ by using necessary conditions.
(iii) If $r=0$, $u^*(t)$ is some constant u_0^*. Prove that if $r>0$, $u^*(T)=u_0^*$. Then show that T^* is a decreasing function of r. ("Discounting advances the doomsday".)

Exercise 2.9.6 In example 13 drop condition (262b). Use note 27 to prove that either the candidate in example 13 or the candidate associated with $T=T_2$ is optimal.

Exercise 2.9.7 Consider the problem (cf. exercise 2.3.6)

$$\max \int_0^T \{-2+(1-u)x\}\,dt, \qquad T\in[1,8] \tag{273}$$

$$\dot{x}=u, \qquad x(0)=0, \qquad x(T)\geqslant 1, \qquad u\in[0,1] \tag{274}$$

(i) Prove the existence of an optimal control.
(ii) Solve the problem by using theorem 11 and note 25 (a).
(iii) Show that theorem 13 does not apply. (Check the Arrow concavity condition.)

Exercise 2.9.8 Consider the problem

$$\max \int_0^T \{-3-u+x\}\,dt, \qquad T\in[1,8] \tag{275}$$

$$\dot{x}=u, \qquad x(0)=0, \qquad x(T)\geqslant 1, \qquad u\in[0,1]. \tag{276}$$

(i) Solve the problem by using theorem 13 with note 27.
(ii) Replace $T\in[1,8]$ by $T\in[1,4+2\sqrt{2}]$ in (275), and solve the problem in this case.

Exercise 2.9.9* Consider the problem

$$\min\left(\int_0^1 u\,dt + \int_1^T dt\right), \qquad T \in [1, \infty) \tag{277}$$

$$\dot{x} = u^2 \text{ for } t \leqslant 1, \qquad \dot{x} = 1 - t \text{ for } t > 1. \tag{278}$$

$$u \in (-\infty, \infty), \qquad x(0) = 0, \qquad x(T) \leqslant 0. \tag{279}$$

(i) Prove that $u^*(t) \equiv 0$, $x^*(t) \equiv 0$, $T^* = 1$ is admissible and give the criterion the value 0. Show that the necessary conditions are satisfied only for $p_0 = 0$ for problem $(277) \rightarrow (279)$ with $T = 1$.

(ii) Prove that if $(x_T(t), u_T(t))$ for fixed $T > 1$ satisfies the necessary conditions, then $p_0 = 1$. Prove next that if we put $y = x_T(1)$, then $u_T(t) = -\sqrt{y}$ on $[0, 1]$, with y satisfying $\sqrt{2y} = T - 1$. Verify that the corresponding value of the criterion is > 0. What is then the solution to the original problem?

Exercise 2.9.10* Consider problem $(50) \rightarrow (54)$ with $t_1 \in [T_1, T_2]$. Assuming theorem 2, one can prove the result in note 25 along the following lines, starting with the autonomous case where f_0 and f do not depend explicitly on t.

(a) Consider the differential equation $\dot{y} = vf(y, u)$ with $y(t_0) = x^0$, where $v \in (0, \infty)$ and $u \in U$ are control functions. Let $y(t)$ correspond to the control functions $\bar{u}(t)$ and $\bar{v}(t)$, i.e.

$$y(s) = x^0 + \int_{t_0}^s \bar{v}(\tau) f(y(\tau), \bar{u}(\tau))\,d\tau.$$

By using the rule for substitution in integrals (or otherwise) show that if we define $\alpha(t) = t_0 + \int_{t_0}^t \bar{v}(\tau)\,d\tau$, $u(t) = \bar{u}(\alpha^{-1}(t))$, $x(t) = y(\alpha^{-1}(t))$, then $\dot{x}(t) = f(x(t), u(t))$ on $[t_0, \alpha(t_1)]$, and

$$\int_{t_0}^{\alpha(t_1)} f_0(x(t), u(t))\,dt = \int_{t_0}^{t_1} \bar{v}(t) f_0(y(t), \bar{u}(t))\,dt.$$

(Hint: $y(\tau) = x(\alpha(\tau))$, $\bar{u}(\tau) = u(\alpha(\tau))$.)

(b) Let $(x^*(t), u^*(t), t_1^*)$ be an optimal triple in the problem of note 25 (a).

Assume that f_0 and f are independent of t. Use (a) to show that $\bar{u}(t) \equiv u^*(t)$, $\bar{v}(t) \equiv 1$ are optimal control functions in the fixed final time problem:

$$\max \int_{t_0}^{t_1^*} v f_0(y, u) dt, \quad \dot{y} = f(y, u), \quad y(t_0) = x_0, \quad y(t_1^*) \text{ satisfies (53)},$$

$z_1(t_0) = t_0$, $z_1(t_1^*) \geqslant T_1$, $\dot{z}_1 = \dot{z}_2 = v$, $z_2(t_0) = t_0$, $z_2(t_1^*) \leqslant T_2$.

(c) Applying necessary conditions in the latter problem, prove the conditions stated in note 25(a) and the additional property that the Hamiltonian is constant in this autonomous case.

(d) Let f_0 and f contain t and have continuous partial derivatives also with respect to t. Introduce an auxiliary state variable z by $\dot{z} = 1$, $z(t_0) = t_0$, $z(t_1)$ free.

Writing $f_0(x, u, t)$ and $f(x, u, t)$ as $f_0(x, u, z)$, $f(x, u, z)$ (as can be done since $z = t$), apply the fact that note 25(a) is known to hold in the autonomous case (see (c)) to show that it even holds in the non-autonomous case.

10. General results on linear control problems

In linear control problems it is frequently possible to obtain qualitative results concerning the behaviour of any optimal control.

Consider problem $(50) \to (54)$ and assume that f_0 and f are linear in the state and control variables, with constant coefficients:

$$f_0 = \sum_{j=1}^{n} a_{0j} x_j + \sum_{k=1}^{r} b_{0k} u_k, \tag{280}$$

$$f_i = \sum_{j=1}^{n} a_{ij} x_j + \sum_{k=1}^{r} b_{ik} u_k \quad i = 1, \ldots, n. \tag{281}$$

We also assume that the control region U is defined by:

$$c_j \leqslant u_j(t) \leqslant d_j \quad j = 1, \ldots, r. \tag{282}$$

Thus each control $u_j(t)$ is a separate control in the sense that its range is independent of the other controls.

We need a condition on the coefficients of (280) and (281). Define $a_{i0} = 0$, $i = 0, 1, \ldots, n$, and let A be the $(n+1) \times (n+1)$−matrix $[a_{ij}]$, $i = 0, 1, \ldots, n$,

$j=0,1,\ldots,n$ and let B be the $(n+1)\times r$ — matrix $[b_{ij}]$, $i=0,\ 1,\ \ldots,n$; $j=1,\ldots,r$. For each vector $w\in R^r$, Bw, $ABw,\ldots,A^nBw$ are $n+1$ vectors in R^{n+1}. We introduce the following *general position condition* (G.P.C.):

> For each $w\neq 0$ which is parallel to one of the coordinate axes in R^r(i.e. parallel to one of the edges of U), the vectors
>
> $Bw,\ ABw,\ldots,\ A^nBw$
>
> are linearly independent.   (283)

For problem $(50)\rightarrow(54)$ with f_0, f, and U as defined in $(280)\rightarrow(282)$, one can prove the following results:[17]

(I)   An optimal control $u^*(t)$ is piecewise constant and $u_j^*(t)$ takes the values c_j and d_j only.

(II)   If the matrix A has real characteristic roots only, then the number of switchings (jumps) of each component of $u^*(t)$ is at most n.

(III) If there is at least one control which brings the system from x^0 to a point where (53) is satisfied, then there is an optimal control doing so too.

(IV)   An optimal control is unique if it exists.

Because of property (I) we say that the optimal control is "*bang–bang*": It takes only values at the "corners" (extreme points) of U. Reading (I) as saying just this, then, except for property (II), all the results above also hold if U is any convex polyhedron. Then (283) must be required to hold for any vector w parallel to any two-dimentional edge of U.)

Suppose we consider, in particular, the minimal time problem, replacing (280) by $f_0=-1$, and letting t_1 be free. Then properties $(I)\rightarrow(IV)$ still hold provided we in (283) and in (II) replace n by $n-1$ and A and B by the matrices $[a_{ij}]$, $i,j=1,\ldots,n$, $[b_{ij}]$, $i=1,\ldots,n,j=1,\ldots,r$.

Example 14   Consider the minimal time problem in example 12. In this example the differential equations (248) and (249) are linear in x, y, and u and the matrices A and B are

$$A=\begin{bmatrix} 0 & \alpha \\ 0 & 0 \end{bmatrix}, \qquad B=\begin{bmatrix} 0 \\ 1 \end{bmatrix}.$$

The control region U is defined by $u_0\leqslant u\leqslant u_1$. If w is a real number $\neq 0$,

---

[17]Arguments on pp. 133–134 in Lee and Markus (1967) can be used to show these results. See also Pontryagin (1962b).

then $Bw = [0, w]'$ and $ABw = [\alpha w, 0]'$ are obviously linearly independent. Thus G.P.C. is satisfied and (I)→(IV) in the minimum time case apply.

According to (I), the optimal control is piecewise constant taking the values u_0 and u_1 only. The only characteristic root of A is $\lambda = 0$. Thus (II) tells us that an optimal control can switch at most once. The results are fully in accordance with the results obtained in example 12.

(III) settles the existence problem. We proved that from any initial point there is a unique candidate which brings the system from the initial point to the terminal point. According to (III) these controls are optimal controls.

Some of the above properties of an optimal control can also be obtained in cases in which a_{ij} and b_{ik} are time dependent. For details we refer to Pontryagin (1962b) and Lee and Markus (1967). The latter book contains also a number of results on more general linear problems.

Exercise 2.10.1 Prove that G.P.C. is satisfied in the problem of exercise 2.9.2 and test the relevant version (I)→(IV) in this case.

11. A proof of the Maximum Principle in a special case*

Here we shall give a proof of the Maximum Principle in a special case. Consider the problem (in vector formulation)

$$\max \int_{t_0}^{t_1} f_0(x(t), u(t)) \, dt, \qquad t_0 \text{ fixed}, \qquad t_1 \text{ varies} \tag{284}$$

when

$$\frac{dx(t)}{dt} = f(x(t), u(t)), \qquad x(t_0) = x^0, \qquad x(t_1) = x^1 \tag{285}$$

where $u(t) \in U$. Here U is a fixed set in R^r while x^0 and x^1 are fixed points in the state space R^n.

We have a control problem with a variable time interval of the type studied in section 9, with the simplification that f_0 and f in the present problem are not dependent on time t explicitly. The system is thus *autonomous*. Actually, it turns out that if we solve the above problem with a more general terminal condition, then we can also solve the problems in sections 4 and 9. In both cases this can be accomplished by considering time t as a new state variable, see exercise 2.9.10. (Another approach to the non-autonomous case is outlined in exercise 2.11.1, which is a parallel to the arguments below.)

Let us then consider the problem specified above. We shall assume that there are several points in R^n, from which one can bring the system to the terminal point x^1 by the use of an optimal control. Let S be the set of such points. In particular $x^0 \in S$. For each $y \in S$ there is an optimal path leading to x^1. Let the corresponding value of the integral in (284) be denoted by $I(y)$. Thus $I(y)$ is the greatest "utility" obtainable by bringing the system from y to x^1. We now introduce two important assumptions:

(A) *S is an open set in R^n.*
(B) *The function $y \to I(y)$ is a C^2-function in S.*

With these simplifying assumptions we shall prove the Maximum Principle. Note, however, that the assumptions (A) and (B) are rather restrictive, and in applications they are not satisfied in a number of control problems of interest. Actually, the conditions (A) and (B) are impossible to verify in many problems.

An important property of our problem that will play a crucial role in the proof is the *principle of optimality.* In our case (the autonomous one) it states that *any portion of an optimal path is optimal.* More precisely: Let $u(t)$ be an optimal control with associated optimal path $x(t)$ which steers the system from x^0 to x^1 in the time interval $[t_0, t_1]$, and let $[t_2, t_3] \subset [t_0, t_1]$. Then $u(t)$, restricted to $[t_2, t_3]$, is optimal in the problem of steering $x(t_2)$ to $x(t_3)$ on some interval $[t_2, \alpha)$, α free.

To prove this result, suppose $u_1(t)$ is defined in $[t_2, t_4]$ (with t_4 not necessarily equal to t_3) and assume that $u_1(t)$ brings the system along the path $\tilde{x}(t)$ from $x(t_2)$ to $x(t_3)$. Suppose that a higher value of the criterion functional is obtained by going from $x(t_2)$ to $x(t_3)$ along $\tilde{x}(t)$ than by going along $x(t)$. We shall prove that this assumption leads to a contradiction. Let us define the control $u_2(t)$ by

$$u_2(t) = \begin{cases} u(t) & t \in [t_0, t_2] \\ u_1(t) & t \in (t_2, t_4) \\ u(t - t_4 + t_3) & t \in [t_4, t_4 - t_3 + t_1]. \end{cases}$$

We observe that $u_2(t)$ firstly brings the system along $x(t)$ from x^0 to $x(t_2)$, then moves the system from $x(t_2)$ to $x(t_3)$ along $\tilde{x}(t)$, and, finally, brings the system from $x(t_3)$ to x^1 along $x(t)$. (Note that when t goes from t_4 to $t_4 - t_3 + t_1$, then $t - t_4 + t_3$ goes from t_3 to t_1.) On the first and third section of this path, the value of the criterion functional will be the same as on the first and third section of the original path. On the second portion of the

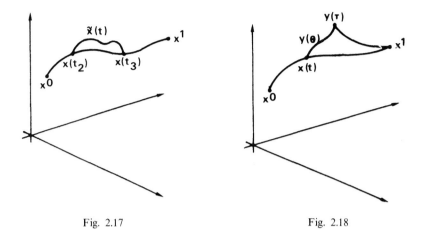

Fig. 2.17   Fig. 2.18

path, by our assumption, the criterion functional attains a lower value along $x(t)$ than along $\tilde{x}(t)$. But this means that $u_2(t)$ brings the system from x^0 to x^1 with a higher total value of the criterion functional than we obtain by using $u(t)$. Since $u(t)$ was assumed to be optimal, we have a contradiction, and the principle of optimality is proved.

Let us go back to the problem in (284) and (285) in which $x^0 \in S$ and $u(t)$ is the optimal control which in the time interval $[t_0, t_1]$ brings the system from x^0 to x^1 along the optimal path $x(t)$. The associated total "utility" is, as above, denoted by $I(x^0)$. For each $t \in [t_0, t_1]$ we know, by the principle of optimality, that

$$I(x^0) = \int_{t_0}^{t} f_0(x(\tau), u(\tau)) \, d\tau + I(x(t)). \tag{286}$$

Differentiating this equation w.r.t. t we obtain for all $t \in [t_0, t_1]$,

$$0 = f_0(x(t), u(t)) + \sum_{i=1}^{n} I'_i(x(t))\dot{x}_i(t). \tag{287}$$

Since $x(t)$ satisfies (285), $\dot{x}_i(t) = f_i(x(t), u(t))$, $i = 1, \ldots, n$. If we define

$$h(x, u) = f_0(x, u) + \sum_{i=1}^{n} I'_i(x)f_i(x, u), \tag{288}$$

then, by (287)

$$h(x(t), u(t)) = 0 \qquad \text{for all } t \in [t_0, t_1]. \tag{289}$$

We now want to prove that if $v \in U$ and $t \in [t_0, t_1]$, then $h(x(t), v) \leqslant 0$. This result together with (289) will tell us that $h(x(t), u)$ has a maximum at $u = u(t)$.

Let t be an arbitrary point in $[t_0, t_1]$ and consider the path $y(\theta) = (y_1(\theta), \ldots, y_n(\theta))$ followed by the system from the point $x(t)$ if we use the control $v \in U$ instead of $u(t)$ in a short time interval (t, τ). By (285), $dy(\theta)/d\theta = f(y(\theta), v)$ for $\theta > t$. Now, if τ is sufficiently close to t, then $y(\tau) \in S$ since S is open and $y(\theta)$ is a continuous function of θ. Note that

$$I(y(t)) \geqslant \int_t^\tau f_0(y(\theta), v)\, d\theta + I(y(\tau)),$$

(the right-hand side is the optimal value in a more restricted control problem: To go in an optimal way from $y(t)$ to x^1 with a prescribed control (the constant v) on $[t, \tau]$). Thus, for $\tau > t$,

$$\frac{I(y(\tau)) - I(y(t))}{\tau - t} \leqslant \frac{-1}{\tau - t} \int_t^\tau f_0(y(\theta), v)\, d\theta. \tag{290}$$

Thus by letting $\tau \to t^+$ in (290) we obtain the right derivative

$$\left[\frac{d^+}{d\tau} I(y(\tau))\right]_{\tau = t} \leqslant -f_0(y(t), v). \tag{291}$$

Here, for $\tau > t$,

$$\frac{d}{d\tau} I(y(\tau)) = \sum_{i=1}^n I_i'(y(\tau)) \dot{y}_i(\tau) = \sum_{i=1}^n I_i'(y(\tau)) f_i(y(\tau), v).$$

Putting $\tau = t$, we obtain

$$\left[\frac{d^+}{d\tau} I(y(\tau))\right]_{\tau = t} = \sum_{i=1}^n I_i'(y(t)) f_i(y(t), v).$$

By inserting this expression into (291), and using the definition (288), we obtain, since $y(t) = x(t)$, $h(x(t), v) \leqslant 0$. This inequality is valid for all $v \in U$ and all $t \in [t_0, t_1]$. Hence, if we also make use of (289), we obtain the following conclusion:

$$h(x(t), v) \leqslant h(x(t), u(t)) = 0 \qquad \text{for all } v \in U. \tag{292}$$

Consequently, we have proved that for $u(t)$ to be optimal, (292) must be valid at each point along the optimal path. This condition is necessary for the optimality of $u(t)$. However, to make use of this result, we must find $I_i'(x(t))$. This appears to be a difficult task since these partial derivatives must be evaluated along the optimal path, which is the unknown in our problem. It turns out that these partial derivatives can be characterized as the solutions of certain differential equations.

To find these differential equations, let $t \in [t_0, t_1]$ and let x be an arbitrary point in S. By the definition of S, there exists an optimal control $u_1(s)$ on some interval $[t', t_1]$ which steers the system from x to x^1. Then by the argument above,

$$h(x, v) \leqslant h(x, u_1(t')) = 0 \qquad \text{for all } v \in U.$$

In particular, by putting $v = u(t)$, $h(x, u(t)) \leqslant h(x, u_1(t')) = 0$. But $h(x(t), u(t)) = 0$. Hence for all $t \in [t_0, t_1]$,

$$h(x, u(t)) \leqslant h(x(t), u(t)) \qquad \text{for all } x \in S.$$

This inequality tells us that $h(x, u(t))$ as a function of $x \in S$ has a maximum at $x = x(t)$. Since the partial derivatives of h w.r.t. $x_1, \ldots, x_n$ exist and S is open, it follows that all the partial derivatives of h w.r.t. $x_1, \ldots, x_n$ are 0 at $x = x(t)$, $\partial h(x(t), u(t))/\partial x_j = 0$, $j = 1, \ldots, n$. By using the definition of h in (288), we obtain

$$\frac{\partial f_0(x(t), u(t))}{\partial x_j} + \sum_{i=1}^{n} I_{ij}''(x(t)) f_i(x(t), u(t))$$

$$+ \sum_{i=1}^{n} I_i'(x(t)) \frac{\partial f_i(x(t), u(t))}{\partial x_j} = 0. \tag{293}$$

Now

$$\frac{d}{dt} I_j'(x(t)) = \sum_{i=1}^{n} I_{ji}''(x(t)) \dot{x}_i(t) = \sum_{i=1}^{n} I_{ij}''(x(t)) f_i(x(t), u(t)). \tag{294}$$

From (293) and (294) we get:

$$\frac{d}{dt} I_j'(x(t)) = -\frac{\partial f_0(x(t), u(t))}{\partial x_j} - \sum_{i=1}^{n} I_i'(x(t)) \frac{\partial f_i(x(t), u(t))}{\partial x_j}. \tag{295}$$

This equation is valid for all $t \in [t_0, t_1]$. Let us define

$$p_j(t) = I_j'(x(t)) \qquad j = 1, \ldots, n \tag{296}$$

and put $p_0 = 1$. Moreover define, as before, the Hamiltonian function by

$H = H(x, u, p) = p_0 f_0(x, u) + \Sigma_{i=1}^n p_i f_i(x, u)$. Then it follows that

$$\frac{\partial H}{\partial x_j} = p_0 \frac{\partial f_0}{\partial x_j} + \sum_{i=1}^{n} p_i \frac{\partial f_i}{\partial x_j}$$

so (295) is equivalent to

$$\dot{p}_j(t) = -\frac{\partial H}{\partial x_j}, j = 1, \ldots, n,$$

where $\partial H/\partial x_j$ is evaluated at $(x(t), u(t))$. By using the Hamiltonian, the condition in (292) becomes

$$H(x(t), v, p(t)) \leqslant H(x(t), u(t), p(t)) = 0 \qquad \text{for all } v \in U.$$

If we compare with theorem 11, we see that we have arrived at the Maximum Principle. With our assumptions we have $p_0 = 1$, while in general one can conclude only that $p_0 = 0$ or 1. Moreover, in this autonomous case, $H(x(t), u(t), p(t))$ is identically equal to 0, not only at $t = t_1$ as in theorem 11.

The proof above is often called "the dynamic programming proof" of the Maximum Principle. Note that the proof was based on the assumptions (A) and (B) above and these are frequently not satisfied in practice.

The adjoint functions $p_1(t), \ldots, p_n(t)$ defined by (296) can be given a price interpretation. $I(x(t))$ was the maximal total "utility" obtainable by bringing the system from $x(t)$ to x^1, and $p_j(t)$ consequently expresses the marginal increase in utility obtained when the jth state variable x_j is slightly increased. This interpretation of $p_j(t)$ is not always valid, since $I(x(t))$ is not, in general, a differentiable function of x. (See Chapter 3, Section 5 for precise results.)

Exercise 2.11.1* Let $V(t, x)$, $x \in R^n$, be the optimal value of the criterion in the problem of reaching the fixed point (t_1, x^1) starting at (t, x), (i.e. reaching x^1 at time t_1 starting at x at time t). Let some starting point (t_0, x^0) be given, and let $(x^*(s), u^*(s))$ be an optimal pair, $x^*(t_0) = x^0$, $x^*(t_1) = x^1$. Assume that $V(t, x)$ is defined and C^2 in a neighbourhood of all points $(t, x^*(t))$, $t \in (t_0, t_1)$. Let $y(t)$ be a solution starting at (t, x) and ending at (t_1, x^1), t some point in (t_0, t_1), with $\dot{y} = f(y(s), v, s)$ for s in $[t, t+\delta]$, $v \in U$, δ a small number > 0, while $y(s)$ restricted to $[t+\delta, t^1]$ is optimal in the problem of reaching (t_1, x^1) from $(t+\delta, y(t+\delta))$.
(a) Show that

$$V(t, x) = \int_t^{t+\delta} f_0(x^*(s), u^*(s), s) \, ds + V(t+\delta, x^*(t+\delta)),$$

(the optimality principle), and that

$$V(t, x) \geqslant \int_t^{t+\delta} f_0(y(s), v, s) \, ds + V(t + \delta, y(t + \delta)),$$

cf. the argument preceding (290). Moving $V(t, x)$ to the right and dividing by δ both in the equality and the inequality, show that $h(x, u^*(t), t) = 0$ and $h(x, v, t) \leqslant 0$ are obtained, with $h(x, u, t) = f_0 + V'_x \cdot f + V_t$.

(b) Using (a), show that along the optimal solution $x^*(t)$, the function $p(t) = V'_x(t, x^*(t))$ satisfies $\dot{p} = -\partial H/\partial x$, cf. (293)→(296).

(c) Using (a) and (b) show that we get a proof of the Maximum Principle in the case $l = n$.

Exercise 2.11.2* Prove the following sufficiency result. Let $(x^*(t), u^*(t))$ be an admissible pair in (50)-(52), (54). Assume that there exists a C^1-function $S(t, x)$ being a solution of the partial differential equation $S'_t = -\hat{H}(x, S'_x, t)$, ($\hat{H}$ the maximized Hamiltonian), with $S(t_1, x) \equiv 0$. Assume that $\hat{H}(x^*(t), S'_x(t, x^*(t)), t) = H(x^*(t), u^*(t), S'_x(t, x^*(t)), t)$. Then $(x^*(t), u^*(t))$ is optimal. [Hint: Let $(x(t), u(t))$ be an arbitrary admissible pair and let S, S'_x, S'_t, f and f_0 be evaluated along $(x(t), u(t))$. Then

$$0 = S(t_1) = S(t_0) + \int_{t_0}^{t_1} [S'_t + S'_x f] \, dt$$

$$\int_{t_0}^{t_1} f_0 \, dt = \int_{t_0}^{t_1} [f_0 + S'_x f + S'_t] dt + S(t_0)$$

$$\leqslant \int_{t_0}^{t_1} [\hat{H}(x(t), S'_x, t) + S'_t] dt + S(t_0) = S(t_0),$$

(equality if $x(t) = x^*(t)$, $u(t) = u^*(t)$).

Exercise 2.11.3* Consider the problem

$$\max \left\{ -\int_{t_0}^{t_1} [x(t)^T A(t)x(t) + u(t)^T B(t)u(t)] \, dt \right\} \tag{297}$$

subject to

$$\dot{x} = F(t)x + G(t)u, \qquad x(t_0) \text{ fixed}, \qquad x(t_1) \text{ free}, \qquad u(t) \in R^r. \tag{298}$$

Here we use matrix notation; $x(t)$ and $u(t)$ are column vectors, $x(t)^T$ and $u(t)^T$ their transposed, the matrices $A(t)$ and $F(t)$ are $n \times n$, $B(t)$ is $r \times r$ and $G(t)$ is $n \times r$. For all t, $A(t)$ is symmetric and non-negative-definite and $B(t)$ is symmetric and positive-definite. The entries in all these matrices are continuous functions of t. Consider the following matrix Riccati equation,

$$\dot{R} = -RF - F^T R + RGB^{-1} G^T R - A. \tag{299}$$

Suppose $R(t)$ is a symmetric $n \times n$ matrix with continuously differentiable entries and suppose $R(t)$ satisfies (299) with $R(t_1) = 0$. Assume that $x^*(t)$ is the solution of (298) with $u = u(t, x) = -B(t)^{-1} G(t)^T R(t)x$, and define $u^*(t) = -B(t)^{-1} G(t)^T R(t)x^*(t)$. Show that the pair $(x^*(t), u^*(t))$ is optimal. (Hint: Use exercise 2.11.2. Show that $H'_u(x, u(t, x), S'_x(t, x), t) = 0$, where $S(t, x) = -x^T R(t)x$. Use the convention that partial derivatives are row vectors, e.g. $(\partial/\partial u)u^T Bu = 2u^T B$.)

Exercise 2.11.4* (Duality: $p(t)$ solves a dual optimization problem) Consider problem (50)→(54) and assume that $\hat{H} = \max_{u \in U} H(x, u, q, t)$ (with $p_0 = 1$) is defined and finite for all (x, t), $q \in A$, A a convex set. Define the function $M(q, s, t) = \inf_x(-sx - \hat{H}(x, q, t))$, $s \in R^n$, and let D be the set of triples (q, s, t) for which this infimum is finite, $q \in A$. Define $Q = \{q(t):(q(t), \dot{q}(t), t) \in D, \; q_i(t_1) \geqslant 0, \; i = l+1, \ldots, m, \; q_i(t_1) = 0, \; i > m\}$, (all $q(t)$'s being continuous and piecewise continuously differentiable). Let $x^1 = (x_1^1, \ldots, x_m^1, 0, \ldots, 0) \in R^n$

(i) By showing (ii)–(v) below, prove that if $(x^*(t), u^*(t), p(t))$, with $p(t) \in A$, satisfies the Arrow sufficient conditions for optimality, then $p(t)$ solves the following dual problem:

$$\max_{q(t) \in Q} \left\{ \int_{t_0}^{t_1} M(q(t), \dot{q}(t), t) \; dt + q(t_1)x^1 - q(t_0)x^0 \right\}. \tag{300}$$

(ii) $M(q, s, t) + \hat{H}(x, q, t) \leqslant -sx$ for all $x, q, s, t, (q, s, t) \in D$.

(iii) If $\bar{x}$ and $(\bar{q}, \bar{s}, t) \in D$ satisfy $\bar{s} = -\hat{H}'_x(\bar{x}, \bar{q}, t)$, and $x \to \hat{H}(x, \bar{q}, t)$ is concave, then (ii) holds with equality for $s = \bar{s}$, $x = \bar{x}$, $q = \bar{q}$. (Hint: $\bar{s} = -\hat{H}'_x$ is a sufficient condition for minimum in the problem defining M.)

(iv) Using $f_0(x, u, t) = f_0 + qf - qf \leqslant \hat{H}(x, q, t) - qf(x, u, t)$ (equality if u maximizes H), show that if $q(t) \in Q$ and $(x(t), u(t))$ is an admissible pair, then (cf. calculations in (121):

$$\int_{t_0}^{t_1} M(q, \dot{q}, t) \, dt + q(t_1)x(t_1) - q(t_0)x^0 + \int_{t_0}^{t_1} f_0(x(t), u(t), t) \, dt \leqslant 0, \quad (301)$$

with equality holding if $q(t) = p(t)$ and $(x(t), u(t)) = (x^*(t), u^*(t))$, the latter pair satisfying the Arrow sufficient conditions with $p(t) \in A$. (Hint: $\dot{p} = -\hat{H}_x$, see note 10.)

(v) Show that $p(t)$ solves (i). (Hint: Insert $x^*(t)$, $u^*(t)$ *in* (301), and use that $q(t_1)x^*(t_1) \geqslant q(t_1)x^1$, with equality if $q(t_1) = p(t_1)$.)

By showing (vi)–(x) below, prove that if $p(t)$ satisfies the sufficient conditions for optimality in problem (300) described in (x) below, with $x^*(t)$ being the adjoint function in these conditions, then $x^*(t)$ is an optimal solution in the original problem.

(vi) $\hat{H}$ is convex in q. Let $\bar{q} \in A$. If $f_0(x, \bar{u}, t) + \bar{q}f(x, \bar{u}, t) = \hat{H}(x, \bar{q}, t)$, then $f(x, \bar{u}, t) \in \partial \hat{H}_q(x, \bar{q}, t)$, where $\partial \hat{H}_q(x, \bar{q}, t)$ is the set of subgradients to $q \to \hat{H}(x, q, t)$ at $\bar{q}$. (Hint: $f_0(x, \bar{u}, t) + qf(x, \bar{u}, t) \leqslant \hat{H}(x, q, t)$, subtract the last equality.)

(vii) With the assumptions in (iii), $-\partial \hat{H}_q(\bar{x}, \bar{q}, t) \subset \partial M_q(\bar{q}, \bar{s}, t)$. (Hint: Use the definition of a subgradient to $\hat{H}$, (ii), and (iii).)

(viii) If $\bar{x}$ and $(\bar{q}, \bar{s}, t) \in D$ satisfy $\bar{x} \in -\partial M_s(\bar{q}, \bar{s}, t)$ and $x \to \hat{H}(x, q, t)$ is concave, then (ii) is satisfied with equality for $s = \bar{s}$, $x = \bar{x}$, $q = \bar{q}$. (Hint: Without proof, use the result from convexity theory that $\hat{H}(x, \bar{q}, t) = \inf_s(-xs - M(\bar{q}, s, t))$ and an argument similar to that in (iii).)

(ix) Let $(x(t), u(t))$ be an admissible pair and let $q(t) \in Q$ with $x \to \hat{H}(x, q(t), t)$ concave. If $x(t) \in -\partial M_s(q(t), \dot{q}(t), t)$ for all t, then equality holds in (301) above, for these functions. (Hint: Use (viii).)

(x) Assume that the partial derivative $\partial M/\partial q$ exists at $(p(t), \dot{p}(t), t)$ for all t. Furthermore, assume that $p(t) \in Q$ satisfies the following sufficient conditions for optimality in problem (300), $x^*(t)$ playing the role of an adjoint function (compare (108), (110)): $x^*(t)$ is continuous and piecewise continuously differentiable, $x^*(t) \in -\partial M_s(p(t), \dot{p}(t), t)$, $\dot{x}^*(t) = -\partial M(p(t), \dot{p}(t), t)/\partial q$,[18] $p(t_1)x^*(t_1) = p(t_1)x^1$. Let $x \to \hat{H}(x, p(t), t)$

---

[18] In some cases, this assumption can be weakened to $\dot{x}^*(t) \in -\partial M_q$ with no assumption of existence of $\partial M/\partial q$. Consider e.g. the case where the control system has the property that the

be concave. $q(t) = p(t)$ solves problem (300). Furthermore, if $u^*(t)$ is chosen such that it maximizes $H(x^*(t), u, p(t), t)$ for each t, then $(x^*(t), u^*(t))$ is an optimal pair in the original problem. (Hint: $H'_x = \hat{H}'_x$. By (vi) and (vii), $\{f(x^*(t), u^*(t), t)\} \subset -\partial M_q = \{\dot{x}^*(t)\}$. Use (ix) with $q(t) = p(t)$.)

(xi) Under the assumptions that $x \to \hat{H}(x, p(t), t)$ is concave and that $\partial M / \partial q$ exists along $p(t)$, observe that $\dot{x}^*(t) \in \partial \hat{H}_q(x^*(t), p(t), t)$, $\dot{p}(t) = -\hat{H}'_x(x^*(t), p(t), t) \Leftrightarrow \dot{x}^*(t) \in -\partial M_q(p(t), \dot{p}(t), t)$, $x^*(t) \in -\partial M_s (p(t), \dot{p}(t), t)$. (Hint: Equality holds in (ii) along the trajectory.)

---

sets N and $\hat{N}$ of (205) and note 19 coincide, and that these sets are convex for each (x, t), with A assumed to be all $R^n_+ = \{q : q \geq 0\}$, and $x \to \hat{H}(x, q, t)$ assumed to be concave for all $q \geq 0$. To see that $\dot{x}^*(t) \in -\partial M_q$ suffices, define $L(x, v, t) = \sup\{f_0(x, u, t) : u \in U, f(x, u, t) = v\}$ $(= -\infty$ if the set is empty). L is always $\leq \hat{H}(x, 0, t)$. Now, $v \to L(x, v, t)$ is concave, (N is convex) and $\hat{H}$ and $-L$ as functions of q and v are conjugate: $\hat{H}(x, q, t) = \sup_v\{vq + L(x, v, t)\}$, $-L(x, v, t) = \sup_q\{vq - \hat{H}(x, q, t)\}$. Thus, $L(x, v, t) = \inf_q\{-vq + \hat{H}(x, q, t)\} = \inf_{q, s}\{-vq - xs - M(q, s, t)\}$. Hence, for each fixed t, $-L$ and $-M$ are conjugate functions. If $\dot{x} \in -\partial M_q(q, s, t)$, we therefore have that $q \in -\partial L_v(x, \dot{x}, t)$, which implies that $L(x, \dot{x}, t)$ is finite, and that $\dot{x}$ maximizes $v \to qv + L(x, s, t)$. Hence, for some $\bar{u} \in U$, $\dot{x} = f(x, \bar{u}, t)$ and $f_0(x, \bar{u}, t) = L(x, \dot{x}, t)$, which implies $\hat{H}(x, q, t) = H(x, \bar{u}, q, t)$.

Note that the equivalence in (xi) still holds in the present case. It enhances the usefulness of the duality results if it is known that the sufficient conditions in (iv) and (x) are also necessary. For the possibility to have such a situation, see theorem 10.

A control system where $N = \hat{N}$ is often a system where $\dot{x} = f$ is replaced by $\dot{x} \leq f(x, u, t)$, so the case considered is related to example 2 in Rockafellar (1970b). This reference contains a general theory of duality for optimal control systems.

CHAPTER 3

EXTENSIONS

In this chapter we shall extend in a number of directions the basic optimal control results of Chapter 2. Additional types of terminal conditions are discussed in section 1, while a more general optimality criterion, which includes a scrap-value function, is presented in section 2.

In sections 3 and 4 we introduce systems where the state variables are not always continuous in time. We look at several alternative models for controlling these "jumps" in the state variable, along with a number of examples.

In section 11 of the previous chapter, we indicated an interesting price interpretation for the adjoint variables. This and other sensitivity results are examined closer in sections 5 and 6, while in sections 7–9 we take up optimal control problems with infinite horizons.

Many optimal control problems in economics are so complicated that the most we can expect to obtain are some qualitative characterizations of the optimal solution, and possibly, some notion of the stationary states. In the last section of this chapter, we present a technique, phase-space analysis, for arriving at such results.

1. Transversality conditions

In this section we shall consider optimal control problems where the terminal point is required to hit a certain curve, a certain surface, etc. Problems of this type are encountered in several areas of economic application.

Suppose we find an optimal solution to such a problem and further that the optimal path terminates at the point x^1. This optimal path must also solve the associated optimal control problem in which all admissible paths terminate at x^1. If follows that the optimal solution must satisfy all the conditions in theorem 2.2 except for the transversality conditions (2.60). To determine the correct transversality conditions for the present problem, we have to compare the optimal path with all admissible paths terminating at other points satisfying the terminal conditions.

Rather than going into analytical details, we shall try to give some intuitive arguments for the correct transversality conditions. We begin by considering a standard optimal control problem with fixed initial point (t_0, x^0) and fixed terminal point (t_1, x^1). Let $V(x^0, x^1, t_0, t_1)$ denote the optimal value of the criterion functional assuming V exists and is well-behaved. A result in exercise 2.11.1 tells us that $p(t_0) = \partial V(x^*(t_0), x^1, t_0, t_1)/\partial x^0$. (See the corresponding result in (2.296).) One can prove the symmetric result $-p(t_1) = \partial V(x^0, x^*(t_1), t_0, t_1)/\partial x^1$. (See (136).) If the system, rather than terminating at a fixed point x^1, is required to terminate at some point on a given surface in R^n, intuitively it is clear that $p(t_1)$ *must be orthogonal to the given surface at the optimal terminal point* (if the vector $- p(t_1)$ has a *nonzero* projection on the tangent plane to the surface, it would increase the value of V if $x^*(t_1)$ were moved slightly in the direction of this projection, contradicting optimality).

Let us see how this terminal condition can be expressed analytically. Observe first the case in which the target surface is $R(x_1, \ldots, x_n) = 0$. Thus we require that $R(x_1(t_1), \ldots, x_n(t_1)) = 0$. Since $p(t_1)$ is orthogonal to the tangent plane of $R(x_1, \ldots, x_n) = 0$ at $x(t_1)$, $p(t_1)$ is parallel to the gradient of R at $x(t_1)$, i.e.

$$p_j(t_1) = \gamma \frac{\partial R(x(t_1))}{\partial x_j}, \qquad j = 1, \ldots, n \qquad \text{for some constant } \gamma. \qquad (1)$$

The corresponding condition for a more general case in which $x(t_1)$ is required to satisfy the following conditions

$$R_j(x_1(t_1), \ldots, x_n(t_1)) = 0, \qquad j = 1, \ldots, r_1'$$
$$R_j(x_1(t_1), \ldots, x_n(t_1)) \geqslant 0, \qquad j = r_1' + 1, \ldots, r_1 \qquad (2)$$

is given in the next theorem. We assume that all R_j are C^1-functions.

Theorem 1   (Necessary conditions. t_1 fixed.)[1]   Let $(x^*(t), u^*(t))$ be an optimal pair in problem $(2.50) \to (2.54)$, replacing (2.53) by (2). Then theorem 2 of Chapter 2 is still valid provided (2.60) is replaced by the

---

[1]See footnote 3.

condition that there exist numbers $\gamma_1, \ldots, \gamma_{r_1}$ such that

$$p_j(t_1) = \sum_{k=1}^{r_1} \gamma_k \frac{\partial R_k(x^*(t_1))}{\partial x_j}, \qquad j = 1, \ldots, n \tag{3}$$

where

$$\gamma_k \geqslant 0 \; (=0 \text{ if } R_k(x^*(t_1)) > 0), \qquad k = r_1' + 1, \ldots, r_1 \tag{4}$$

and provided (2.56) is replaced by

$$(p_0, \gamma_1, \ldots, \gamma_{r_1}) \neq (0, \ldots, 0). \quad \blacksquare \tag{5}$$

Note 1 If the set of gradients $\nabla R_k(x^*(t_1))$ corresponding to all active constraints are linearly independent, then it is easy to see that (5) implies (2.56).

Sufficient conditions for the above problem are contained in theorem 4.

Example 1 Consider the variational problem

$$\max \int_{t_0}^{t_1} F(t, x, \dot{x}) \, dt, \qquad x(t_0) = x_0, \qquad x(t_1) \geqslant x_1.$$

This is a control problem if we put $\dot{x} = u$. If we let $R(x) = x - x_1$, then the terminal condition is $R(x(t_1)) \geqslant 0$. Conditions (3) and (4) tell us that $p(t_1) = \gamma \geqslant 0$, with $\gamma = 0$ if $x^*(t_1) > x_1$. Since $p(t) = -\partial F / \partial \dot{x}$ (see Chapter 2, section 5), we obtain

$$\left(\frac{\partial F}{\partial \dot{x}} \right)_{t=t_1} \leqslant 0 \; (=0 \text{ if } x^*(t_1) > x_1).$$

This result is in accordance with the results of Chapter 1, section 5. (See (1.32b).)

The terminal time is not fixed

Examine next the case in which t_1 is not fixed, but free to vary in some interval $[T_1, T_2]$. We let the terminal constraint functions contain t explicitly and require that

$$R_j(x_1(t_1), \ldots, x_n(t_1), t_1) = 0, \qquad j = 1, \ldots, r_1',$$
$$R_j(x_1(t_1), \ldots, x_n(t_1), t_1) \geqslant 0, \qquad j = r_1' + 1, \ldots, r_1. \tag{6}$$

Again, all R_j are C^1-functions.

Theorem 2   (**Necessary conditions.** $t_1 \in [T_1, T_2].)^2$ Let $(x^*(t), u^*(t), t_1^*)$ be an optimal triple in problem $(2.50) \rightarrow (2.54)$, replacing (2.53) by (6) and letting $t_1 \in [T_1, T_2]$, $t_0 \leqslant T_1 < T_2$. Then theorem 2 of Chapter 2 is valid for $t_1 = t_1^*$ if (2.60) is deleted and if (2.56) is replaced by (5). The transversality condition is: there exist numbers $\gamma_1, \ldots, \gamma_{r_1}$, such that

$$p_j(t_1^*) = \sum_{k=1}^{r_1} \gamma_k \frac{\partial R_k(x^*(t_1^*), t_1^*)}{\partial x_j}, \qquad j = 1, \ldots, n \tag{7}$$

where

$$\gamma_k \geqslant 0 (= 0 \text{ if } R_k(x^*(t_1^*), t_1^*) > 0) \qquad k = r_1' + 1, \ldots, r_1. \tag{8}$$

Moreover,

$$H(x^*(t_1^*), u^*(t_1^*), p(t_1^*), t_1^*) + \sum_{k=1}^{r_1} \gamma_k \frac{\partial R_k(x^*(t_1^*), t_1^*)}{\partial t}$$

$$\begin{aligned} &\leqslant 0 \text{ if } t_1^* = T_1 \\ &= 0 \text{ if } t_1^* \in (T_1, T_2) \\ &\geqslant 0 \text{ if } t_1^* = T_2 \end{aligned} \tag{9}$$

If $t_1^* = T_1 = t_0$, replace $H^*(t_1^*)$ by $\sup_{u \in U} H(x^0, u, p(t_1^*), t_1^*)$ in (9). ■

Sufficient conditions for the present problem is contained in Theorem 6.17.

Exercise 3.1.1   Consider the problem

$$\max \int_0^T (1 - u)(ax + by)dt \tag{10}$$

$$\dot{x} = vu(ax + by), \qquad x(0) = x_0, \qquad \dot{y} = (1 - v)u(ax + by), \qquad y(0) = y_0 \tag{11}$$

$$u \in [0, 1], \quad v \in [0, 1], \quad cx(T) + dy(T) \geqslant A \tag{12}$$

$$a > b, \qquad T > 1/a, \qquad c(ax_0 + by_0)e^{aT} + (ad - bc)y_0 > aA \tag{13}$$

where b, x_0, y_0, c, d, and A are positive.

---

^{2}See footnote 3.

Economic interpretation: We study a two-sector model with production rates ax and by respectively, where x and y are the sector's fixed capital stock at time t. u is the savings rate, v is the fraction of savings allocated to sector no. 1, and thus $1 - v$ the fraction allocated to sector 2. Condition (13) tells us that an index of total production has to attain at least the value A at the terminal time.

(a) Denoting the adjoint variables associated with x and y by p_1 and p_2, respectively, write down the transversality conditions corresponding to (13).
(b) Suppose $c > d$. Put $p_0 = 1$ and find the unique candidate for optimality. (Hint: Prove that $p_1(t) > p_2(t)$ in $[0, T)$.)
(c) Suppose $c < d$. Put $p_0 = 1$ and find the unique candidate for optimality.
(d) Assume that there is some autonomous growth (non-transferable between sectors) in each sector (f and g positive constants):
$$\dot{x} = vu(ax + by) + fx, \qquad \dot{y} = (1 - v)u(ax + by) + gy$$
Put $p_0 = 1$ and find the unique candidate for optimality.

Exercise 3.1.2   Consider a control problem in which the terminal condition is that the first m variables at t_1 have fixed values $\bar{x}_1, \ldots, \bar{x}_m$ while the rest of the variables are free at t_1. Hence we have the special case of (2) with $R_i(x_1, \ldots, x_n) = x_i - \bar{x}_i$ for $i = 1, \ldots, m$, $R_i(x_1, \ldots, x_n) \equiv 0$ for $i = m + 1, \ldots, n$. What does (3) say about $p_1(t_1), \ldots, p_n(t_1)$? Compare with theorem 2 of Chapter 2.

Exercise 3.1.3   Formulate problem (1.30), (1.31c) as a control problem with a terminal condition of the form (6). Prove that (7) and (9) give us the transversality condition (1.32c).

2.   A more general optimality criterion

We consider in this section an optimal control problem in which a function measuring the value (cost or utility) assigned to the terminal size of the state variable is included in the optimality criterion. Specifically we study the problem of maximizing

$$\int_{t_0}^{t_1} f_0(x(t), u(t), t) \, dt + S_1(x(t_1)) \qquad (t_0, t_1 \text{ fixed}) \qquad (14)$$

subject to

$$\dot{x}=f(x,u,t), \qquad x(t_0)=x^0, \qquad u(t)\in U \text{ and terminal condition (2).} \quad (15)$$

S_1 is referred to as a *scrap-value function*. (The corresponding problem in the calculus of variations was discussed in Chapter 1, section 7.) We assume that $R_1,\ldots,R_{r_1}$ and S_1 are all C^1-functions.

Suppose $(x^*(t), u^*(t))$ solves the given problem. Then that pair also solves the associated optimal control problem in which the admissible control functions bring the system from x^0 to $x^*(t_1)$. For each such path the function S_1 has the same value, $S_1(x^*(t_1))$. It follows that $(x^*(t), u^*(t))$ must satisfy all the conditions in theorem 2 in Chapter 2, except the transversality conditions (2.60). To find the correct transversality conditions in the present case, we might use arguments similar to those of a corresponding problem in Chapter 1, section 7 and apply the theorem in the previous section.

Theorem 3 (Necessary conditions. Scrap-value function. t_1 fixed.)[3] Let $(x^*(t), u^*(t))$ be an optimal pair in problem (14), (15). Then theorem 1 is still valid provided (3) is replaced by the condition that for $j=1,\ldots,n$,

$$p_j(t_1)=p_0\frac{\partial S_1(x^*(t_1))}{\partial x_j}+\sum_{k=1}^{r_1}\gamma_k\frac{\partial R_k(x^*(t_1))}{\partial x_j}. \quad \blacksquare \qquad (16)$$

A corresponding sufficiency result is:

Theorem 4 (Sufficient conditions. Scrap-value function. t_1 fixed) Let $(x^*(t), u^*(t))$ be an admissible pair in problem (14), (15). Assume that there exist a function $p(t)=(p_1(t),\ldots,p_n(t))$ and numbers γ_i, $i=1,\ldots,r_1$, such that $(x^*(t), u^*(t))$ satisfies the conditions in theorem 3 for $p_0=1$. Assume, moreover, that

$$\hat{H}(x,p(t),t)(\text{see (2.127)}) \text{ exists and is concave in } x, \qquad (17)$$

$$S_1(x) \text{ is concave, and} \qquad (18)$$

$$\sum_{k=1}^{r_1}\gamma_k R_k(x) \text{ is quasi-concave in } x. \qquad (19)$$

Then $(x^*(t), u^*(t))$ is optimal. $\blacksquare$

---

[3] For a proof, see exercise 3.2.10. The modification needed when the control functions are bounded and measurable are the same as those in footnote 9 in Chapter 2.

Proof. Let $(x, u) = (x(t), u(t))$ be any other admissible pair. Then we must prove that

$$\bar{\Delta} = \int_{t_0}^{t_1} [f_0(x^*, u^*, t) - f_0(x, u, t)] dt + S_1(x^*(t_1)) - S_1(x(t_1)) \geqslant 0.$$

Using the same arguments as in the proof of theorem 5 of Chapter 2 and the concavity of S_1, we find:

$$\bar{\Delta} \geqslant \sum_{j=1}^{n} p_j(t_1)(x_j(t_1) - x_j^*(t_1)) + \sum_{j=1}^{n} \frac{\partial S_1(x^*(t_1))}{\partial x_j} (x_j^*(t_1) - x_j(t_1))$$

$$= \sum_{j=1}^{n} \left(p_j(t_1) - \frac{\partial S_1(x^*(t_1))}{\partial x_j} \right) (x_j(t_1) - x_j^*(t_1))$$

$$= \sum_{j=1}^{n} \left(\sum_{k=1}^{r_1} \gamma_k \frac{\partial R_k(x^*(t_1))}{\partial x_j} \right) (x_j(t_1) - x_j^*(t_1))$$

Thus, if we put $R(x) = \sum_{k=1}^{r_1} \gamma_k R_k(x)$, we see that

$$\bar{\Delta} \geqslant \sum_{j=1}^{n} \frac{\partial R(x^*(t_1))}{\partial x_j} (x_j(t_1) - x_j^*(t_1)). \qquad (20)$$

Now, $R(x(t_1)) - R(x^*(t_1)) = \sum_{k=1}^{r_1} \gamma_k(R_k(x(t_1)) - R_k(x^*(t_1)))$. The first r_1' terms of this sum are 0, since $x(t_1)$ and $x^*(t_1)$ satisfy (2). Take any $k > r_1'$. If $R_k(x^*(t_1)) > 0$, then by (4), $\gamma_k = 0$. If $R_k(x^*(t_1)) = 0$, then $\gamma_k(R_k(x(t_1)) - R_k(x^*(t_1))) = \gamma_k(R_k(x(t_1)) \geqslant 0$ since both factors are $\geqslant 0$. Thus $R(x(t_1)) - R(x^*(t_1)) \geqslant 0$. Since by (19), the function $R(x)$ is quasiconcave, it follows from a well-known characterization of quasiconcavity that the sum in (20) is $\geqslant 0$. (See (B.4).) ∎

Note 2 (Terminal time is not fixed)[4] Suppose we study the problem covered by theorem 2, except that added to the criterion functional is the scrap-value function $S_1(x(t_1), t_1)$. Then the conclusions in theorem 2 are still valid provided (7) is replaced by (16) (with S_1 and R_k now containing t_1^*

---

[4]For a proof in the case where R and S_1 are C^2, see exercise 3.2.11. For the general case, see Neustadt (1976), Chapter V. 2.

explicitly), and (9) is replaced by

$$H^*(t_1^*) + p_0 \frac{\partial S_1(x^*(t_1^*), t_1^*)}{\partial t} + \sum_{k=1}^{r_1} \gamma_k \frac{\partial R_k(x^*(t_1^*), t_1^*)}{\partial t}$$

$$\leqslant 0 \text{ if } t_1^* = T_1$$
$$= 0 \text{ if } t_1^* \in (T_1, T_2)$$
$$\geqslant 0 \text{ if } t_1^* = T_2 \tag{21}$$

where $H^*(t_1^*) = H(x^*(t_1^*), u^*(t_1^*), p(t_1^*), t_1^*)$. If $t_1^* = T_1 = t_0$, replace $H^*(t_1^*)$ by $\sup_{u \in U} H(x^0, u, p(t_1^*), t_1^*)$ in (21)

Sufficient conditions for this problem are contained in theorem 6.17.

Example 2 (**A two-sector model**) Consider a two-sector model for a national economy whose national income Y is the sum of the incomes Y_1 and Y_2 in the two sectors. Let us assume that $Y_1 = b_1 K_1$ and $Y_2 = b_2 K_2$, with K_1 and K_2 the capital stocks in the two sectors and b_1, b_2 positive constants. In each sector a constant fraction of the income is saved, such that the total savings in the economy is $s_1 Y_1 + s_2 Y_2 = s_1 b_1 K_1 + s_2 b_2 K_2, s_1$ and s_2 the savings ratios of the respective sectors. The total savings are pooled in a central agency and the fraction $\beta = \beta(t)$ is allocated to sector 1, while the rest, the fraction $1 - \beta(t)$, is allocated to sector 2. Let the planning period be $[0, T]$ and suppose that the planner's objective is to maximize the national income at time T, $Y_1(T) + Y_2(T)$.

The problem can be formulated as such:

$$\text{maximize } \{b_1 K_1(T) + b_2 K_2(T)\}, \tag{22}$$

subject to

$$\dot{K}_1 = \beta(t)(s_1 b_1 K_1 + s_2 b_2 K_2), \qquad K_1(0) = K_1^0 > 0, \tag{23}$$

$$\dot{K}_2 = (1 - \beta(t))(s_1 b_1 K_1 + s_2 b_2 K_2), \qquad K_2(0) = K_2^0 > 0, \tag{24}$$

$$0 \leqslant \beta(t) \leqslant 1, \ \beta \text{ a control variable.} \tag{25}$$

Here we have a problem precisely of the type discussed above. In particular, $f_0 = 0$ and $S(K_1, K_2) = b_1 K_1 + b_2 K_2$. There are no restrictions of the type (2), so we can put $R_j \equiv 0, r_1' = r_1 = 1$.

Let the adjoint functions associated with (23) and (24) be p_1 and p_2. Then (16) gives us the following transversality conditions for the right endpoint ($p_0 = 1$):

$$p_1(T) = b_1, \qquad p_2(T) = b_2. \tag{26}$$

In exercise 3.2.4 you are asked to study this model more closely.

Initial state subject to choice

Up to now we have always assumed that the admissible paths satisfy $x(t_0) = x^0$, with x^0 as a fixed point in R^n. There are, however, some problems in economics for which it is natural to assume that the initial stock $x(t_0)$ is not necessarily fixed. An example, dealing with the optimal choice of initial capacity for a firm, is given in example 3 below. It is a special case of the following problem:

$$\max\left\{\int_{t_0}^{t_1} f_0(x(t), u(t), t)dt + S_0(x(t_0)) + S_1(x(t_1))\right\}, \qquad t_0, t_1 \text{ fixed} \qquad (27)$$

$$\dot{x}(t) = f(x(t), u(t), t), \qquad u(t) \in U, \qquad x(t_1) \text{ satisfies } (2.53) \qquad (28)$$

with the following conditions on $x(t_0)$: Let N and N' be two disjoint subsets of $\{1, 2, \ldots, n\}$ and let us require that

$$
\begin{aligned}
&x_i(t_0) = x_i^0 \qquad &&\text{for } i \in N \\
&x_i(t_0) \geqslant x_i^0 \qquad &&\text{for } i \in N' \\
&x_i(t_0) \text{ free} \qquad &&\text{for } i \notin N \cup N'.
\end{aligned}
\qquad (29)
$$

As we can see we have imposed initial conditions analogous to the terminal conditions (2.53). For this problem we have the following results:

Theorem 5   (Necessary conditions for problem (27)→(29))[5]   Let $(x^*(t), u^*(t))$ be an optimal pair in problem (27)→(29). Then theorem 2 of Chapter 2 is still valid provided we add the initial transversality conditions

$$\text{no conditions} \qquad\qquad\qquad\qquad\qquad\qquad i \in N \qquad (30a)$$

$$p_i(t_0) + p_0\frac{\partial S_0(x^*(t_0))}{\partial x_i} \leqslant 0 (= 0 \text{ if } x_i^*(t_0) > x_i^0) \qquad i \in N' \qquad (30b)$$

$$p_i(t_0) + p_0\frac{\partial S_0(x^*(t_0))}{\partial x_i} = 0 \qquad\qquad\qquad i \notin N \cup N' \qquad (30c)$$

[5]Neustadt (1976), Chapter V. 2.

and provided (2.60) is replaced by

no conditions $i = 1, \ldots, l$ (31a)

$$p_i(t_1) - p_0 \frac{\partial S_1(x^*(t_1))}{\partial x_i} \geq 0 (=0 \text{ if } x_i^*(t_1) > x_i^1) \qquad i = l+1, \ldots, m \qquad (31b)$$

$$p_i(t_1) - p_0 \frac{\partial S_1(x^*(t_1))}{\partial x_i} = 0 \qquad\qquad i = m+1, \ldots, n. \qquad (31c)$$

Theorem 6 (Sufficient conditions for problem (27)→(29)) If the necessary conditions in theorem 5 are satisfied with $p_0 = 1$, if $\hat{H}(x, p(t), t)$ exists and is concave in x (see(2.127)), and if $S_0(x)$ and $S_1(x)$ are both concave in x, then $(x^*(t), u^*(t))$ solves problem (27)→(29). ∎

A proof of theorem 6 based on the proof of theorem 5 of Chapter 2 (see also the proof of theorem 4) is straightforward.

If for some i's the inequalities in (29) are reversed, the corresponding inequalities in (30) are reversed.

Note 3 Suppose $A(t)$ is a convex set in R^n for each $t \in [t_0, t_1]$, and impose the additional restriction $x(t) \in A$. Then theorem 6 still holds provided $x^*(t)$ is an interior point of $A(t)$. It is then sufficient to assume that $\hat{H}(x, p(t), t)$ is concave only for $x \in A(t)$. (See note 11 of Chapter 2.) The weakened regularity conditions for note 2.11 stated in note 2.12 are sufficient also here.

Note 4* In theorem 4 concavity of $\hat{H}$ can be replaced by the assumption that $-\dot{p}(t)$ is a supergradient of $\hat{H}(x, p(t), t)$ at $x^*(t)$ for all t. See (2.135) in the proof of theorem 2.5. In note 3 it suffices that (2.135) holds for $x \in A(t)$ and we can then drop the assumption that $x^*(t)$ is interior to $A(t)$.

Note 5[6] Suppose we replace (29) by the requirement that $x(t_0)$ satisfies

$$R_j^0(x_1(t_0), \ldots, x_n(t_0)) = 0, \qquad j = 1, \ldots, r_0'$$
$$R_j^0(x_1(t_0), \ldots, x_n(t_0)) \geq 0, \qquad j = r_0' + 1, \ldots, r_0 \qquad (32)$$

with $R_1^0, \ldots, R_{r_0}^0$ C^1-functions. Suppose, moreover, that (2.53) is replaced by (2). Then theorem 5 is still valid provided (30) is replaced by the assumption

---

[6]Neustadt (1976), Chapter V. 2.

that there exist numbers $\delta_1, \ldots, \delta_{r_0}$ such that for $j = 1, \ldots, n$

$$p_j(t_0) + p_0 \frac{\partial S_0(x^*(t_0))}{\partial x_j} + \sum_{k=1}^{r_0} \delta_k \frac{\partial R_k^0(x^*(t_0))}{\partial x_j} = 0 \tag{33}$$

where

$$\delta_k \geqslant 0 (= 0 \text{ if } R_k^0(x^*(t_0)) > 0), \qquad k = r_0' + 1, \ldots, r_0. \tag{34}$$

In addition, (31) must be replaced by the assumption that there exist numbers $\gamma_1, \ldots, \gamma_{r_1}$ such that for $j = 1, \ldots, n$

$$p_j(t_1) - p_0 \frac{\partial S_1(x^*(t_1))}{\partial x_j} - \sum_{k=1}^{r_1} \gamma_k \frac{\partial R_k(x^*(t_1))}{\partial x_j} = 0 \tag{35}$$

where

$$\gamma_k \geqslant 0 \ (= 0 \text{ if } R_k(x^*(t_1)) > 0), \qquad k = r_1' + 1, \ldots, r_1. \tag{36}$$

Finally, (2.56) is replaced by the condition

$$(p_0, \delta_1, \ldots, \delta_{r_0}, \gamma_1, \ldots, \gamma_{r_1}) \neq 0. \tag{37}$$

For existence theorems applying to the above problems, see Chapter 6, section 8.

In the following example, we shall make use of theorem 6 and note 3.

Example 3 (Optimal capacity of a monopolistic firm) Consider the problem

$$\max \left\{ \int_0^T (1 - u(t)) f(x(t)) e^{-rt} dt - cx(0) + dx(T) e^{-rT} \right\} \tag{38}$$

$$\dot{x}(t) = au(t) f(x(t)), \qquad x(0) \text{ free}, \qquad x(T) \text{ free} \tag{39}$$

$$u(t) \in [0, 1] \tag{40}$$

$$d < 1/a < c, \qquad r^{-1}(1 - e^{-rT}) f'(0) > c - de^{-rT} \tag{41}$$

f is a C^2-function on $[0, \infty)$, $f(0) = 0$. At $x' > 0$
f has a maximum, and $f'(x) > 0$, $f''(x) < 0$ in $[0, x')$. $\tag{42}$

$$x(t) \in (0, x'). \tag{43}$$

Here T, r, c, d and a are positive constants.

We offer the following economic interpretation: A monopolist faces a demand structure that gives him a sales revenue $f(x(t))$ when his production (and sales) rate is $x(t)$. He can buy initial production capacity at the unit price c, and can sell it at the end of his planning period, T, at price d. A fraction, $(1-u(t))f(x(t))$, of the sales revenue is consumed; the rest, $u(t)f(x(t))$, is invested in new production capacity bought at a price $1/a$.

The assumption $c > 1/a$ can be explained as follows: The initial capacity is financed by a loan having an interest rate higher than r while the gradual increase in capacity, $uf(x)$, is paid for immediately.

In order to interpret the latter inequality in (41), consider the criterion in (38) when we put $u(t) \equiv 0$. Then by (39), $x(t) \equiv x(0)$ and the latter inequality in (41) ensures that it pays to invest something initially, i.e. to have $x(0) > 0$. (Compute the derivative of the criterion in (38) w.r.t. $x(0)$ at $x(0) = 0$.)

We see that our problem is of the type covered in note 3 with $S_0(x) = -cx$, $S_1(x) = dxe^{-rT}$ and $x(t) \in A(t) = (0, x')$. The sufficient conditions in that note will be used to solve the problem.

The Hamiltonian with $p_0 = 1$ is

$$H = (1-u)f(x)e^{-rt} + pauf(x) = [(1-u)e^{-rt} + apu]f(x), \tag{44}$$

and the candidate for optimality, $(x^*(t), u^*(t))$ satisfies:

$$u^*(t) \text{ maximizes } [(1-u)e^{-rt} + ap(t)u]f(x^*(t)) \text{ for } u \in [0, 1] \tag{45}$$

$$\dot{p}(t) = -\partial H^*/\partial x = -[(1-u^*(t))e^{-rt} + ap(t)u^*(t)]f'(x^*(t)) \tag{46}$$

and from (30) and (31),

$$p(0) = c, \qquad p(T) = de^{-rT}. \tag{47}$$

Since $x^*(t) \in (0, x')$, our assumptions on f imply that $f(x^*(t)) > 0$. Thus from (45), $u^*(t)$ maximizes $e^{-rt} + (ap(t) - e^{-rt})u$ for $u \in [0, 1]$. Hence

$$u^*(t) = 1 \text{ if } p(t) > (1/a)e^{-rt}$$
$$u^*(t) = 0 \text{ if } p(t) < (1/a)e^{-rt}, \tag{48}$$

and $u^*(t) \in [0, 1]$ if $ap(t) = e^{-rt}$. It follows that

$$\dot{p}(t) = -\max[e^{-rt}, ap(t)] f'(x^*(t)), \tag{49}$$

and hence $\dot{p}(t) < 0$ since $x^*(t) \in (0, x')$. Thus $p(t)$ is strictly decreasing from $p(0) = c$ to $p(T) = de^{-rT}$. If we put $\varphi(t) = (1/a)e^{-rt}$, by (41) $\varphi(0) = 1/a < c$ and $\varphi(T) = (1/a)e^{-rT} > de^{-rT} = p(T)$. It follows that the equation $p(t) = \varphi(t)$ has at least one solution in the interval $(0, T)$. It turns out that the solution, T^*, is unique. (See fig. 3.1.)

From (48) we see that $u^*(t) = 1$ in $[0, T^*]$, and $u^*(t) = 0$ in $(T^*, T]$. It

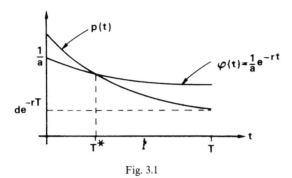

Fig. 3.1

follows from (39) that $x^*(t)\equiv x^*(T)$ in $(T^*, T]$. In turn, $x^*(t)$ becomes uniquely determined in $[0, T^*]$ as the solution to $\dot{x}=af(x)$ and $x^*(T^*)=x^*(T)$. T^* and $x^*(T)$ have to be determined such that the associated function $p(t)$ has the required properties.

Formally the remarks above can now be set aside. Only the arguments to follow are logically necessary in order to solve the problem.

Let T' and y be two parameters, $y\in[0, x']$, $T'\in[0, T]$. Define

$$x(t, T', y)=\begin{cases}\text{a solution of } \dot{x}=af(x) \text{ on } [0, T'], & x(T', T', y)=y \\ y & \text{on } (T', T]\end{cases} \quad (50)$$

$p(t, T', y)$ is a solution of $\dot{p}=-\max(e^{-rt}, ap)f'(x(t, T', y))$,

with $p(T, T', y)=de^{-rT'}$. $\quad (51)$

Moreover, define $T(y)$ as the largest number T' such that

$$p(T', T', y)=\tfrac{1}{a}e^{-rT} \quad (52)$$

if a solution to (52) exists. If not, put $T(y)=0$. Then $p(t, T', y)<\varphi(t)$, $t>T'$.

Let us simplify our notation and write $x(t, T(y), y)=x(t, y)$ and $p(t, T(y), y)=p(t, y)$. We know that for all values of y, $p(T, y)=de^{-rT}$. We want to prove that there exists a $y^*\in(0, x')$ such that $p(0, y^*)=c$.

In the interval $(T(x'), T]$, $0=f'(x')=f'(x(t, x'))$, so from (51), $p(t, x')\equiv de^{-rT}$ in $(T(x'), T]$. Thus for $y=x'$ equation (52) has no solution, so $T(x')=0$, and $p(0, x')=de^{-rT}<c$.

We next prove that for some $x''\in(0, x')$, $p(0, x'')\geqslant c$. By a continuity argument the existence of y^* can be established:

Define $\lambda(y)=(1/r)(1-e^{-rT})f'(y)-(c-de^{-rT})$ for $y\in[0, x']$. Then $\lambda'(y)<0$ in $(0, x')$, $\lambda(0)>0$ by (41) and $\lambda(x')=-(c-de^{-rT})<0$. We conclude that there exists a unique $x'' \in (0, x')$ such that $\lambda(x'') = 0$ or, equivalently,

$$r^{-1}(1 - e^{-rT})f'(x'') = c - de^{-rT}. \tag{53}$$

Define the auxiliary function $\psi(t, y)$ for $t \in [0, T]$, $y \in [0, x']$ by

$$\psi(t, y) = de^{-rT} + \tfrac{1}{r}(e^{-rt} - e^{-rT})f'(y), \tag{54}$$

Note that $p(T, x'') = \psi(T, x'') = de^{-rT}$, and $\dot\psi(t, x'') = -e^{-rt}f'(x'') = \dot p(t, x'')$ for $t \in (T(x''), T]$. Hence $\psi(t, x'') = p(t, x'')$ on $[T(x''), T]$. For $t \leqslant T(x^*)$, by (51), $\dot p(t, x'') \leqslant -e^{-rt} f'(x(t, x'')) \leqslant -e^{-rt} f'(x'') = \dot\psi(t, x'')$. Since $p(T(x''), x'') = \psi(T(x''), x'')$, we see that $p(t, x'') \geqslant \psi(t, x'')$ for $t \leqslant T(x'')$. In particular, $p(0, x'') \geqslant \psi(0, x'') = c$.

We have seen that $p(0, x') < c$ and $p(0, x'') \geqslant c$. We prove below that $p(0, y)$ depends continuously on y. Then for some $y^* \in [x'', x')$, $p(0, y^*) = c$, with $T^* = T(y^*) < T$. Let us show that

$$p(t, y^*) = \begin{cases} > \varphi(t) = (1/a)e^{-rt} & \text{in } [0, T^*) \\ < \varphi(t) & \text{in } (T^*, T] \end{cases} \tag{55}$$

From the definition of T^* and the fact that $p(T, y^*) = de^{-rT}$, the latter inequality in (55) follows. If $T^* > 0$ that inequality implies that for some $t' \in (T^*, T]$, $\dot p(t', y^*) < \dot\varphi(t')$, i.e. $-e^{-rt'} f'(y^*) < -(r/a)e^{-rt'}$, so $-f'(y^*) < -r/a$. Thus for all t, $\dot p(t, y^*) \leqslant -e^{-rt} f'(y^*) < -(r/a)e^{-rt} = \dot\varphi(t)$. Since $p(T^*, y^*) = \varphi(T^*)$. (18) follows.

Using (55) and (48) we obtain the following suggestion for $u^*(t)$, $x^*(t)$ and $p(t)$:

$$u^*(t) = 1 \text{ on } [0, T^*], \qquad u^*(t) = 0 \text{ on } (T^*, T], \tag{56}$$

$x^*(t) = x(t, y^*)$, $p(t) = p(t, y^*)(= p(t, T^*, y^*))$. It follows that (45) is satisfied. Moreover, $p(t)$ satisfies (46) and (47). Finally, the maximized Hamiltonian, $\hat H$, is easily seen to be $[\max(e^{-rt}, ap(t)]f(x)$, which is (strictly) concave in x for $x \in (0, x')$. According to note 3, $(x^*(t), u^*(t))$ is optimal.

The continuity of $p(0, y)$ w.r.t. y can be proved in several ways. For instance, it follows from theorems on approximate solutions (see Hale (1969), p. 37), or by using theorems which guarantee that $x(t, p)$ and $p(t, y)$ depend continuously on terminal conditions. One has to show firstly that $(x(t, y), p(t, y))$, as a solution on $[T(y), T]$, has the property that $(x(T(y), y), p(T(y), y))$ depends continuously on the terminal conditions $x(T, y) = y$, $p(T, y) \equiv de^{-rT}$. Secondly, one must show that $(x(t, y), p(t, y))$ on $[0, T(y)]$ has the property that $(x(0, y), p(0, y))$ depends continuously on y. We shall omit the detailed verification.

We have just seen an interesting economic example where the state variable at the beginning of the planning period was allowed to jump (adjust) to any value. In the next sections we shall study problems in which the state variable is allowed to have jump points also in (t_0, t_1).

Exercise 3.2.1   Consider the problem (T, a, k_0 are positive constants, $T > 1$)

$$\max\left\{\int_0^T (1 - s(t))k(t)dt + ak(T)\right\} \tag{57}$$

$$\dot{k}(t) = s(t)k(t), \qquad k(0) = k_0, \qquad k(T) \text{ free} \tag{58}$$

$$s(t) \in [0, 1] \tag{59}$$

Find the optimal solution and compare with the results in exercise 2.3.3.

Exercise 3.2.2   An economy develops through time according to the system (2.3). Let us assume that one possible (and desirable) path for the economy is the one in which the variables change proportionally:

$$\frac{x_1}{\lambda_1} = \frac{x_2}{\lambda_2} = \cdots = \frac{x_n}{\lambda_n}, \qquad (\lambda_i > 0, \ i = 1, \ldots, n). \tag{60}$$

(In some models such a path is called a *von Neumann path* for the system.) Suppose that at $t = 0$ the system is in some state x^0 outside the path described by (60), and that we wish the system to reach that path. More precisely, let us consider the following problem: Among all admissible control functions $u(t) \in U$ which in the time interval $[0, T]$ bring the system from x^0 to some point x^1 on the path (60), find one which maximizes the distance from the origin to x^1. Explain why the problem is as follows:

$$\max \ [x_1(T)]^2 \qquad \text{when } x_1(T)/\lambda_1 = \ldots = x_n(T)/\lambda_n. \tag{61}$$

What are the transversality conditions for the right endpoint?

Exercise 3.2.3   Formulate the problem in exercise 1.8.4 as a control problem and make use of (2.108) to prove that (16) reduces to the result in exercise 1.8.4(ii).

Exercise 3.2.4   Take a closer look at the problem in example 2. Assume $b_1 > b_2$. What does the Maximum Principle tell us in this case? Prove that $\dot{p}_1 - \dot{p}_2 = (\beta p_1 + (1 - \beta)p_2)(b_2 s_2 - b_1 s_1)$. (For a full analysis of the problem, studied by A. Rahman, M.D. Intriligator and A. Takayama, see Takayama (1974), Chapter 8, B, a.)

Exercise 3.2.5   Show that the conditions in (31) can be derived from those in (16) and (4).
(Hint: Define $R_j(x) = x_j - x_j^1$, $j = 1, \ldots, m$.)

Exercise 3.2.6   Prove that the solution in example 3 is optimal even if we replace (43) by the requirement $x(0) \geqslant 0$.
(Hint: Use note 4 and the fact that $\partial \hat{H}/\partial x = -\dot{p}(t)$.)

Exercise 3.2.7   Prove theorem 4 if (19) is replaced by the assumption that $\gamma_k R_k(x)$ is quasi-concave in x for $k = 1, \ldots, r_1$. (Note that the sum of quasi-concave functions is not necessarily quasi-concave, so the assumption in this exercise does not imply (19).)

Exercise 3.2.8*   Assume in the control problem (14)–(15) that f_0, f and S_1 contain a parameter $w: f_0(x, u, w, t), f(x, u, w, t), S_1(x, w). f_0, f, S_1$ and their partial derivatives w.r.t. x and w are continuous. The vector w is constant over time and is to be chosen from a fixed convex set W in R^s. Assume that $(x^*(t), u^*(t), w^*)$ is an optimal triple and that the rank condition in note 1 is satisfied. Prove that the following set of conditions has to be satisfied: $(2.56) \rightarrow (2.59)$ with H containing w^*. (16) and (4), and for all $w \in W$.

$$\left(\int_{t_0}^{t_1} H'_w(x^*(t), u^*(t), w^*, p(t), t) \, dt + \frac{\partial S_1(x^*(t_1), w^*)}{\partial w} \right) \cdot (w - w^*) \leqslant 0. \qquad (62)$$

(Hint: We make use of a standard trick in control theory. Consider an auxiliary problem on $[t_0 - 1, t_1]$ with an additional state variable y. Let $y(t_0 - 1) = w^*$, $\dot{y}(t) \equiv w$ on $[t_0 - 1, t_0]$, and $\dot{y}(t) \equiv 0$ on $[t_0, t_1]$. Here w is a control variable in $W - \{w^*\}$. Moreover, put $x(t_0 - 1) = x_0$, $\dot{x}(t) \equiv 0$ and $f_0 \equiv 0$ on $[t_0 - 1, t_0)$, while $\dot{x}(t)$ and f_0 are, as usual, on $[t_0, t_1]$. Then $u^*(t)$ and $w(t) \equiv 0$ are optimal controls in the new problem.)
    If $p_0 = 1$ and $\hat{H}$ and S_1 are concave in (x, w), then these conditions are also sufficient. If W is compact and $(2.206) \rightarrow (2.209)$ hold, then there exist optimal triples. ((2.207) must hold for all (x, w, t) and (2.209) must hold for all admissible triples $(x(t), u(t), w(t))$.)

Exercise 3.2.9   An offshore oilfield can be produced by a number w of production platforms that all go into production at time $t = 0$. The full capacity production rate is aw, where a is a constant during the lifetime of the field. The cost of installing this capacity is bw. A control function $u(t) \in [0, 1]$ is that fraction of the installed capacity actually used for production at time t. K is the total amount of recoverable oil, c is the price per unit oil. Consider the problem:

$$\max \left\{ \int_0^T cu(t)awe^{-rt}dt - bw \right\}, \qquad \dot{x}(t)=u(t)aw, \qquad x(0)=0, \quad x(T)\leqslant K$$

assuming that T is a big fixed number, r is the constant discount rate, $br/ca<1$, and $w\in(0,\infty)$. (Disregard the fact that w ought to be a natural number.) Notice that this is a problem of the type considered in exercise 3.2.8. Show that the oilfield is emptied at T', where T' is defined implicitly by the equation $e^{-rT'}(1+rT')=1-br/ca$, and the optimal value of w is $w^*=K/aT'$. (We assume that $T>T'$.)

Exercise 3.2.10* Derive theorem 3 from theorem 2 of Chapter 2.
(Hint: Consider an auxiliary control problem on $[t_0,t_1+1]$ by defining $f_0=S_1$, $f\equiv 0$ on $[t_1, t_1+1]$, and $x(t_1+1)$ free, and by adding a new state variable $y(t)=(y_1(t),\ldots,y_{r_1}(t))$, where $\dot{y}(t)\equiv 0$ on $[t_0, t_1]$, $\dot{y}(t)=(R_1(x),\ldots,R_{r_1}(x))$ on $[t_1,t_1+1]$, $y_i(t_1+1)=0$ for $i=1,\ldots,r'_1$, $y_i(t_1+1)\geqslant 0$, $i=r'_1+1,\ldots,r_1$. If $(x^*(t), u^*(t))$ is optimal in the original problem, then this pair, $x^*(t)$ and $u^*(t)$ constants on $[t_1, t_1+1]$, is optimal in the auxiliary problem. $y^*(t)$ is uniquely determined by $x^*(t)$.)

Exercise 3.2.11* Use the result mentioned in note 25 of Chapter 2 to derive the result in note 2.
(Hint: The hint for exercise 3.2.10 does not work in this case. Instead let $x(t_1)$ be free in an auxiliary problem on $[t_0, t_1]$, with t_1 free to vary in $[T_1, T_2]$. The terminal conditions (6) are now taken care of by introducing an auxiliary state variable $y(t)=(y_1(t),\ldots,y_{r_1}(t))$, where $\dot{y}(t)=(d/dt)R(x(t), t)$, $y(t_0)=R(x^*(t_0), t_0)$, $y_i(t_1)=0$ for $i=1,\ldots,r'_1$, and $y_i(t_1)\geqslant 0$ for $i=r'_1+1,\ldots,r_1$. Here R is a C^2 function, $R(x,t)=(R_1(x,t),\ldots,R_{r_1}(x,t))$. Replace $S_1(x(t),t)$ by an integral as in Chapter 1, section 7.)

Exercise 3.2.12* Consider the problem in Exercise 2.11.3 and add the term $-x(t_1)^T S_1 x(t_1)$ to the criterion with S_1 a symmetric non-negative-definite square matrix. Using theorem 6, show that if $(x^*(t), u(t))$ is an admissible pair such that $u^*(t)=-B(t)^{-1}G(t)^T R(t)x^*(t)$, where $R(t)$ satisfies equation (2.299) in exercise 2.11.3 and $R(t_1)=S_1$, then $(x^*(t), u^*(t))$ is optimal.
(Hint: Let $p(t)=-2x^*(t)^T R(t)$, use the concavity of $H(x,u,p(t),t)$ and that $u^*(t)$ satisfies $H'_u=0$, as shown in exercise 2.11.3. Test $\dot{p}=-H'_x$.)

3. Jumps in the state variables I

In the control problems so far studied, we have assumed that the state variables are continuous functions of time. However, in some economic problems it is reasonable to allow for jumps in the state variables. For instance, a firm might change its capital stock by a sudden purchase of capital goods, or it might suddenly sell a bunch of commodities.

To study the problem of jumps in the state variables, we take as our point of departure the fixed time problem of Chapter 2, section 4 and introduce required modifications.

Suppose some state variable x_i has a jump at $\tau_j \in [t_0, t_1]$. We let $x_i(\tau_j^+)$ denote the right-hand limit of $x_i(t)$ at τ_j, and $x_i(\tau_j^-)$ the corresponding left-hand limit. The magnitude of the jump of x_i at τ_j is then $x_i(\tau_j^+) - x_i(\tau_j^-)$. Of course, the corresponding notation for the jump of the vector $x(t) = (x_1(t), \ldots, x_n(t))$ at τ_j is $x(\tau_j^+) - x(\tau_j^-)$. (If $\tau_1 = t_0$ let $x(\tau_1^-) = x(t_0)$; if $\tau_k = t_1$, let $x(\tau_k^+) = x(t_1)$.)

We shall assume that the number k of jump points as well as the locations of the jump points $\tau_1, \ldots, \tau_k$ in the interval $[t_0, t_1]$ are controlled by the maximizer. The maximizer controls the magnitude of the jumps at τ_j by choosing a *control parameter* $v^j = (v_1^j, \ldots, v_s^j)$. We also assume that the magnitude of the jump at τ_j, in general, depends explicitly on τ_j and the state of the system immediately before the jump. Accordingly we have the following assumption on the magnitude of the jump at τ_j:

$$x_i(\tau_j^+) - x_i(\tau_j^-) = g_i(x(\tau_j^-),\, v^j,\, \tau_j), \qquad j = 1, \ldots, k, \qquad i = 1, \ldots, n, \qquad (63)$$

with $g_i(x, v, t)$ a given function from R^{n+s+1} into R, and $g_i(x, 0, t) = 0$ for all x and all t.

Note 6 The dependence of the jumps' magnitude on the state of the system is rarely considered in the economic literature dealing with optimal control theory. To indicate that (63) can have a useful economic justification, suppose a large firm is able to acquire additional capital at a cost lower than a competing smaller firm. Letting x_i denote the capital stock of a firm (a large firm having a large x_i), the function g_i could represent such a dependency.

Between the jump points we assume that the system develops according to the system of differential equations,

$$\dot{x}_i = f_i(x, u, t) \qquad i = 1, \ldots, n. \qquad (64)$$

Since we allow for jumps at the initial point of time t_0, the initial condition is

$$x(t_0^-) = x^0, \qquad (x^0 \text{ given point in } R^n). \tag{65}$$

As our terminal condition we choose

$$x_i(t_1^+) = x_i^1, \qquad i = 1, \ldots, l, \ x_i^1 \text{ fixed} \tag{66a}$$

$$x_i(t_1^+) \geqslant x_i^1, \qquad i = l+1, \ldots, m, \ x_i^1 \text{ fixed} \tag{66b}$$

$$x_i(t_1^+) \text{ free}, \qquad i = m+1, \ldots, n \tag{66c}$$

As usual, the control vector $u(t)$ is assumed to belong to a fixed control region U in R^r. The jump parameters $v^1, \ldots, v^k$ are assumed to belong to a fixed convex set V in R^s with $0 \in V$. In other words

$$u(t) \in U \text{ and } v^j \in V, \ j = 1, \ldots, k, \ V \text{ convex}, \ 0 \in V. \tag{67}$$

There is a cost or a reward associated with each jump. It is assumed that this cost or reward depends on the state of the system immediately before the jump occurs, on the jump parameter v^j, and explicitly on time τ_j:

$$h(x(\tau_j^-), v^j, \tau_j).$$

Here $h(x, v, t)$ is a given real-valued function of $n + s + 1$ variables with $h(x, 0, t) = 0$ for all x and all t. Our criterion is now the expression

$$\int_{t_0}^{t_1} f_0(x(t), u(t), t) dt + \sum_{j=1}^{k} h(x(\tau_j^-), v^j, \tau_j). \tag{68}$$

Some regularity conditions have to be imposed on the functions involved. Retain condition (2.6) and assume in addition that g_i, $\partial g_i / \partial x_j$, $\partial g_i / \partial v^j$, h, $\partial h / \partial x_j$, and $\partial h / \partial v^j$ are all continuous w.r.t. all variables. Suppose we have a piecewise continuous control vector $u(t)$ satisfying (67), distinct points $\tau_1, \ldots, \tau_k$ in $[t_0, t_1]$, vectors $v^1, \ldots, v^k$ satisfying (67), and a piecewise continuous and piecewise continuously differentiable path $x(t)$ satisfying (63)–(66). We assume that $u(t)$ and $x(t)$ are left-continuous in (t_0, t_1). Then the collection $(x(t), u(t), \tau_1, \ldots, \tau_k, v^1, \ldots, v^k)$ is called *an admissible collection*. (Note that the number and the location of jump points may vary from one admissible collection to another.) For each admissible collection the expression in (68) has a definite value. We search for an admissible collection which maximizes (68).

Necessary conditions

Necessary conditions for this problem are given by the following theorem.

Theorem 7   (Maximum Principle. Jumps in the state variables)[7]   Let $(x^*(t),$ $u^*(t), \tau_1^*, \ldots, \tau_k^*, \bar{v}^1, \ldots, \bar{v}^k)$ be an admissible collection which solves the problem of maximizing (68) subject to (63)–(67). Then there exist a constant p_0 and a piecewise continuous function $p(t) = (p_1(t), \ldots, p_n(t))$ such that,

$$(p_0, p(t_1^+)) \neq (0, 0). \tag{69}$$

If the Hamiltonian H is defined by

$$H(x, u, p, t) = p_0 f_0(x, u, t) + \sum_{i=1}^n p_i f_i(x, u, t),$$

then for all non-jump points of $x^*(t)$,

$$H(x^*(t), u, p(t), t) \leqslant H(x^*(t), u^*(t), p(t), t) \quad \text{for all } u \in U. \tag{70}$$

Except at the points of discontinuities of $x^*(t)$ and $u^*(t)$, $p(t)$ is continuously differentiable and

$$\dot{p}_i(t) = -\frac{\partial H(x^*(t), u^*(t), p(t), t)}{\partial x_i}, \qquad i = 1, \ldots, n. \tag{71}$$

Furthermore,

$$p_0 = 0 \text{ or } p_0 = 1, \tag{72}$$

and the transversality conditions are

$$\begin{aligned}
&p_i(t_1^+) \text{ no condition} & i &= 1, \ldots, l, & \text{(73a)}\\
&p_i(t_1^+) \geqslant 0 \ (=0 \text{ if } x^*(t_1^+) > x_1) & i &= l+1, \ldots, m, & \text{(73b)}\\
&p_i(t_1^+) = 0 & i &= m+1, \ldots, n. & \text{(73c)}
\end{aligned}$$

At the jump points $\tau_1^*, \ldots, \tau_k^*$,

$$p_i(\tau_j^{*+}) - p_i(\tau_j^{*-}) = -p_0 \frac{\partial h(x^*(\tau_j^{*-}), \bar{v}^j, \tau_j^*)}{\partial x_i}$$

$$- \sum_{l=1}^n p_l(\tau_j^{*+}) \frac{\partial g_l(x^*(\tau_j^{*-}), \bar{v}^j, \tau_j^*)}{\partial x_i}, \qquad i = 1, \ldots, n. \tag{74}$$

---

[7]Seierstad (1981b), theorem 3.

and

$$\sum_{l=1}^{s} \left[p_0 \frac{\partial h(x^*(\tau_j^{*-}), \bar{v}^j, \tau_j^*)}{\partial v_l} \right.$$

$$\left. + \sum_{i=1}^{n} p_i(\tau_j^{*+}) \frac{\partial g_i(x^*(\tau_j^{*-}), \bar{v}^j, \tau_j^*)}{\partial v_l} \right] \cdot (\bar{v}_l^j - v_l) \geq 0 \tag{75}$$

for all $v = (v_1, \ldots, v_s) \in V$ and all $j = 1, \ldots, k$.
Moreover, for all $(v_1, \ldots, v_s) \in V$,

$$\sum_{l=1}^{s} \left[p_0 \frac{\partial h(x^*(t), 0, t)}{\partial v_l} + \sum_{i=1}^{n} p_i(t) \frac{\partial g_i(x^*(t), 0, t)}{\partial v_l} \right] \cdot v_l \leq 0 \tag{76}$$

for all t at which there is no jump. ∎

Note 7 The conditions stated in the theorem will sometimes single out one or a few solution candidates. Suppose the partial derivatives of f_0, f, h, and g w.r.t. t also exist and are continuous. Then one can derive other necessary conditions which are most often needed. For instance, condition (2.61) in note 3 of Chapter 2 is valid at all points t of continuity for $u^*(t)$, $t \neq \tau_j^*$, $j = 1, \ldots, k$, Moreover, at all jump points τ_j^*,

$$
\left.
\begin{aligned}
&H(x^*(\tau_j^{*+}), u^*(\tau_j^{*+}), p(\tau_j^{*+}), \tau_j^*) \\
&- H(x^*(\tau_j^{*-}), u^*(\tau_j^{*-}), p(\tau_j^{*-}), \tau_j^*) \\
&- p_0 \frac{\partial h(x^*(\tau_j^{*-}), \bar{v}^j, \tau_j^*)}{\partial \tau} - \sum_{i=1}^{n} p_i(\tau_j^{*+}) \frac{\partial g_i(x^*(\tau_j^{*-}), \bar{v}^j, \tau_j^*)}{\partial \tau}
\end{aligned}
\right\}
\begin{cases}
\geq 0 & \text{if } \tau_j^* = t_0 \\
= 0 & \text{if } \tau_j^* \in (t_0, t_1) \\
\leq 0 & \text{if } \tau_j^* = t_1
\end{cases}
\tag{77}
$$

Note 8* In economics. Arrow and Kurz (1970), II.7 (in addition to Vind (1967)) is a standard reference for control problems with jumps in the state variables. They consider the special case in which

$$g_i(x, v, t) = \begin{cases} v_i \geq 0 & i = 1, \ldots, \bar{s} \\ 0 & i = \bar{s} + 1, \ldots, n \end{cases} \tag{78}$$

$$h(x, v, t) = \sum_{i=1}^{\bar{s}} U_i(t) v_i. \tag{79}$$

Thus they allow upward jumps in the first $\bar{s}$ state variables, but these jumps do not depend on the state of the system before the jumps occur. Moreover, (79) amounts to assuming that a utility (or cost) $U_i(\tau_j)$ is associated with a unit jump in the ith state variable at τ_j.

Let us see what $(74) \rightarrow (77)$ tell us in this case. In the first place, since g_i and h are independent of the state variable, we conclude from (74) that $p(t)$ is also continuous at the jump points. (75) tells us that for all $j = 1, \ldots, k$,

$$\sum_{l=1}^{\bar{s}} [p_0 U_l(\tau_j^*) + p_l(\tau_j^*)](\bar{v}_l^j - v_l) \geqslant 0 \quad \text{for all } (v_1, \ldots, v_{\bar{s}}) \geqslant 0. \tag{80}$$

Putting $v_l = \bar{v}_l^j$, $l \neq i$, then from (80), for all $j = 1, \ldots, k$,

$$[p_0 U_i(\tau_j^*) + p_i(\tau_j^*)](\bar{v}_i^j - v_i) \geqslant 0 \quad \text{for all } (v_1, \ldots, v_{\bar{s}}) \geqslant 0. \tag{81}$$

Suppose $\bar{v}_i^j > 0$. Then we can choose $v_i \geqslant 0$ such that $\bar{v}_i^j - v_i$ is positive as well as negative. Hence from (81)

$$p_0 U_i(\tau_j^*) + p_i(\tau_j^*) = 0 \qquad \text{if } x_i^*(\tau_j^{*+}) > x_i^*(\tau_j^{*-}). \tag{82}$$

Let us turn to condition (76): For all t at which there is no jump,

$$\sum_{l=1}^{\bar{s}} [p_0 U_l(t) + p_l(t)] v_l \leqslant 0 \quad \text{for all } (v_1, \ldots, v_{\bar{s}}) \geqslant 0. \tag{83}$$

(Note that for $l = \bar{s} + 1, \ldots, s$, the corresponding terms in the sum of (76) are all 0.) In particular, putting $v_l = 0$ for $l \neq i$ and letting $v_i > 0$, we obtain from (83),

$$p_0 U_i(t) + p_i(t) \leqslant 0 \quad \text{for } i = 1, \ldots, \bar{s}, \qquad t \text{ not a jump point.} \tag{84}$$

Finally, consider condition (77). Using an obvious notational simplification and remembering that $\bar{v}_i^j = x_i^*(\tau^{*+}) - x_i^*(\tau^{*-})$, we get

$$H^*(\tau_j^{*+}) - H^*(\tau_j^{*-}) = p_0 \sum_{i=1}^{\bar{s}} U_i'(\tau_j^*)(x_i^*(\tau_j^{*+}) - x_i^*(\tau_j^{*-})) \tag{85}$$

at all jump points $\tau_j^* \in (t_0, t_1)$. Except for the fact that they assume that $p_0 = 1$, conditions (82)–(85) are the same as those obtained by Arrow and Kurz (1970), Proposition 11.

Sufficient conditions

Sufficient conditions for the solution of this section's problem are given in the next theorem.

Theorem 8 (Sufficient conditions. Jumps in the state variables)[8]   Let

---

[8]Seierstad (1981b), theorem 1.

$\Gamma = (x^*(t),\ u^*(t),\ \tau_1^*,\ \ldots,\ \tau_k^*,\ \bar{v}^1,\ \ldots,\ \bar{v}^k)$ be an admissible collection with associated adjoint function $p(t)$ which satisfies the conditions in theorem 7 when $p_0 = 1$. Assume that

$$h(x, v, t) + \sum_{i=1}^{n} p_i(t^+) g_i(x, v, t) \tag{86}$$

is concave in (x, v) for each t, and

$$\hat{H}(x, p(t), t) = \max_{u \in U} H(x, u, p(t), t) \tag{87}$$

is concave in x for each t. (We assume that the maximum is attained for each (x, t).) Then Γ solves the problem of maximizing (68) subject to (63)–(67). ∎

Example 4 (**Optimal extraction of a natural resource**) We shall examine a problem concerning the optimal exploitation of a non-renewable natural resource. The model is closely related to example 11 of Chapter 2. At time $t = 0$ there is a fixed amount $\bar{x}$ of the resource which is extractable. We let $x(t)$ be total stock of the resource remaining at time t. Moreover, $q(t)$ is the world market price of the resource (exogeneously given), c is the constant unit cost of production (compared with example 2.11, $C(u, t) = cu$), and r is the discount rate. The planning period $[0,\ T]$ is fixed and it is assumed the planner has two options: Either to produce a more or less steady stream $u(t)$ of the resource, or to produce vast amounts over short periods, the latter modelled by sudden downward jumps in $x(t)$. If we assume that the jump points are $\tau_1, \ldots, \tau_k$ (possibly, $\tau_1 = 0$ and $\tau_k = T$) and that the objective is to maximize the total discounted profit on $[0,\ T]$, then we have the following control problem:

$$\max\left\{ \int_0^T (q(t) - c) u(t) e^{-rt}\, dt + \sum_{j=1}^{k} (q(\tau_j) - c) e^{-r\tau_j} v^j \right\} \tag{88}$$

$$\dot{x}(t) = -u(t),\ x(0^-) = \bar{x},\ x(T^+) \geqslant 0 \tag{89}$$

$$x(\tau_j^+) - x(\tau_j^-) = -v^j \tag{90}$$

$$u(t) \in [0, \infty),\ v^j \in [0, \infty), \qquad j = 1, \ldots, k. \tag{91}$$

Here T, c, r and $\bar{x}$ are positive constants and we assume that $q(t) - c > 0$ for all $t \in [0,\ T]$.

With $p = p(t)$ as the adjoint variable associated with (89), the Hamiltonian is

$$H = p_0(q(t) - c)ue^{-rt} - pu.$$ (92)

Suppose $(x^*(t), u^*(t), \tau_1^*, \ldots, \tau_k^*, \bar{v}^1, \ldots, \bar{v}^k)$ is an optimal collection. Then by theorem 7 there exist a constant p_0 with $p_0 = 0$ or $p_0 = 1$, and a piecewise continuous function $p(t)$ such that

$$(p_0, p(T^+)) \neq (0, 0),$$ (93)

and such that for all non-jump points of $x^*(t)$, $u^*(t)$ maximizes

$$[p_0(q(t) - c)e^{-rt} - p(t)]u$$ (94)

The functions $g(x, v, t) = -v$ and $h(x, v, t) = (q(t) - c)e^{-rt} v$ are independent of x (see (63) and (68)). It follows from (74) that $p(t)$ is continuous even at jump points for $x^*(t)$. From (71), $\dot{p}(t) \equiv 0$, so $p(t) = \bar{p}$ for some constant $\bar{p}$. Hence, applying (73), we obtain

$$p(t) = \bar{p} \geqslant 0 (= 0 \text{ if } x^*(T^+) > 0).$$ (95)

Notice that (75) tells us that for all $j = 1, \ldots, k$,

$$(p_0(q(\tau_j^*) - c)e^{-r\tau_j^*} - \bar{p})(\bar{v}^j - v) \geqslant 0 \text{ for all } v \geqslant 0.$$ (96)

Finally, by (76), at all non-jump points,

$$[p_0(q(t) - c)e^{-rt} - \bar{p}]v \leqslant 0 \text{ for all } v \geqslant 0.$$ (97)

Suppose $\bar{v}^j > 0$ so that there is a jump at τ_j^*. Then by putting $v = 0$ into (96), we see that $A = p_0(q(\tau_j^*) - c)e^{-r\tau_j^*} - \bar{p} \geqslant 0$, while choosing v large positive in (96) leads to $A \leqslant 0$. Hence at all jump points τ_j^*,

$$p_0(q(\tau_j^*) - c)e^{-r\tau_j^*} - \bar{p} = 0, \qquad j = 1, \ldots, k.$$ (98)

From (97) and (98) we can conclude that for all $t \in [0, T]$,

$$p_0(q(t) - c)e^{-rt} - \bar{p} \leqslant 0.$$ (99)

If $p_0 = 1$, then $\bar{p} = 0$ is impossible by (99). If $p_0 = 0$, $\bar{p} = 0$ is impossible from (93). Hence, in all cases, $\bar{p} > 0$, and from (95) we deduce that

$$x^*(T^+) = 0.$$ (100)

If strict inequality holds in (99) for all $t \in [0, T]$, then (98) rules out any jump points, while the Maximum condition (94) requires $u^*(t) \equiv 0$. But then from (89), $x^*(t) \equiv \bar{x}$, contradicting (100). Hence (99) must hold with equality for some $t' \in [0, T]$, and since $\bar{p} > 0$, $p_0 = 0$ is impossible. Therefore $p_0 = 1$.

Thus

$$\bar{p}= \max_{t\in[0,T]} \ [(q(t)-c)e^{-rt}]=(q(t')-c)e^{-rt'}. \tag{101}$$

Let $J=\{t: (q(t)-c)e^{-rt}=\bar{p}\}$. From (101), J is non-empty, and by (98) we know that J contains all the jump points of $x^*(t)$ if there are any.

So far we have only relied upon the necessary conditions of theorem 7. Let us now appeal to theorem 8. Note that the expression (86) for this problem is $(q(t)-c)e^{-rt} v-\bar{p}v$ which is linear and hence concave in (x, v). Moreover, $\hat{H}(x, p(t), t)$ is independent of x, hence concave in x.

Letting $p_0=1$ and $p(t)=\bar{p}$ as defined by (101), pick out any collection $\Gamma_0=(x^*(t), u^*(t), \tau_1^*,\ldots, \tau_k^*, \bar{v}^1,\ldots, \bar{v}^k)$ with the following properties:

(a) $u^*(t)$ is an arbitrary non-negative function satisfying $u^*(t)=0$ for $t\notin J$.
(b) k is an arbitrary finite number and $\tau_1^*,\ldots, \tau_k^*$ are arbitrary numbers in $[0, T]$ with $\tau_j^*\in J$ for $j=1,\ldots, k$.
(c) $\bar{v}^j$ are arbitrary positive numbers, $j=1,\ldots, k$.

If Γ_0 also satisfies the condition

$$\sum_{j=1}^k \bar{v}^j + \int_0^T u^*(t)dt =\bar{x}, \tag{102}$$

then all the conditions in theorem 8 are fulfilled and the collection Γ_0 is optimal.

In general, there is an infinite number of solutions. In order to have a unique solution, J must consist of one point τ_1^* only. Then will $u^*(t)\equiv 0$ (except at τ_1^*), and the single jump in $x^*(t)$ is of the size $\bar{v}^1=\bar{x}$. For instance, if $q(t)$ is a constant $\bar{q}$, with $\bar{q}>c$, then $(\bar{q}-c)e^{-rt}$ has a unique maximum at $t=0$. Then $J=\{0\}$, and $u^*(t)\equiv 0$ in $[0, T]$. In this case the optimal solution amounts to emptying the resource at $t=0$, which obviously makes economic sense.

Note 9   Imagine an economy in which there are many identical producers behaving according to the model above. If the society wants a steady production of the natural resource, then it is necessary that for all producers, $J=[0, T]$. Even if $J=[0, T]$ for all producers, we are far from guaranteed that the total production will be a reasonably even stream. If $J=[0, T]$, then from (101),

$$q(t)-c=e^{rt}\cdot\text{constant}, \qquad t\in [0, T].$$

In the theory concerning competing producers of a natural resource, it is a well-known assertion that at all points of time $q(t) - c$ must grow exponentially with a rate r, if positive production is to be individually rational.

Example 5 (Capital accumulation through savings and purchase of capital equipment) In this example we analyze a problem in optimal growth theory. (The model is a generalization of the one in exercise 2.7.1.)

Let $K(t)$ be the total capital stock of a country or a firm at time t, $u(t)$ the savings rate and $f(K) = K^a$, $0 < a < 1$, the production function. Assume that the price of the output is 1. The distinguishing characteristics of the model is that the capital stock can be increased not only by investments but also by the purchase of capital equipment from abroad at the constant price r per unit. If the planning period is $[0, T]$, we have the following problem:

$$\max \left\{ \int_0^T (1 - u(t))(K(t))^a dt - r \sum_{j=1}^k v^j \right\}, \tag{103}$$

$$\dot{K}(t) = u(t)(K(t))^a, \qquad K(0^-) = K_0 > 0, \qquad K(T^+) \text{ is free} \tag{104}$$

$$K(\tau_j^+) - K(\tau_j^-) = v^j, \qquad j = 1, \ldots, k, \tag{105}$$

$$u(t) \in [0, 1], \qquad v^j \in [0, \infty) \text{ for } j = 1, \ldots, k. \tag{106}$$

The expression we maximize in (103) is total production minus investments minus the outlay for the purchase of capital equipment from abroad. We apply theorem 7. Suppose $\Gamma = (K^*(t), u^*(t), \tau_1^*, \ldots, \tau_k^*, \bar{v}^1, \ldots, \bar{v}^k)$ is an admissible collection which solves problem (103)–(106), and $p(t)$ is the associated adjoint function. Since $K(T^+)$ is free, $p(T^+) = 0$ and thus $p_0 = 1$. In this problem

$$g(K, v, t) = v \qquad \text{and} \qquad h(K, v, t) = -rv, \tag{107}$$

so we conclude from (74) that $p(t)$ is continuous everywhere. The Hamiltonian is

$$H = (1 - u)K^a + puK^a = [1 + (p - 1)u]K^a, \tag{108}$$

so from (71), for v.e. t,

$$\dot{p}(t) = -\frac{\partial H}{\partial K} = -a(K^*(t))^{a-1}[(1 + (p(t) - 1)u^*(t)], \qquad p(T^+) = 0. \tag{109}$$

Since $K_0 > 0$, $u(t) \geqslant 0$, and $\bar{v}^j \geqslant 0$, $K^*(t) > 0$ for all t. Looking at (108) we see that the maximum condition (70) implies

$$u^*(t)=\begin{cases}1 & \text{if } p(t)>1.\\ 0 & \text{if } p(t)<1.\end{cases}$$

Thus $1+(p(t)-1)u^*(t)=\max(1,\ p(t))$ for any t. (Check this equality for $p(t)<1,\ =,\ >1.$) From (109) we see that $\dot{p}(t)=-a(K^*(t))^{a-1}\cdot\max(1,p(t))<0$ for v.e. t. Since $p(t)$ is continuous, we conclude that $p(t)$ is strictly decreasing. According to (75), for all $j=1,\ldots,k$

$$(-r+p(\tau_j^{*+}))(\bar{v}^j-v)\geqslant0 \text{ for all } v\geqslant0, \tag{110}$$

and (76) tells us that at all non-jump points of $K^*(t)$,

$$(-r+p(t))v\leqslant0 \text{ for all } v\geqslant0. \tag{111}$$

From (110), by choosing $v=0$ we see that $p(\tau_j^{*+})\geqslant r$, while choosing v large positive, implies $p(\tau_j^{*+})\leqslant r$. Hence

$$p(\tau_j^*)=r \text{ at all jump points } \tau_j^*,\ j=1,\ldots,k. \tag{112}$$

(Since $p(t)$ is continuous everywhere, we put $p(\tau_j^{*+})=p(\tau_j^*)$.) Moreover, from (111) it is clear that

$$p(t)\leqslant r \text{ at all non-jump points.} \tag{113}$$

Since $p(t)$ is strictly decreasing, we conclude that if there is a jump in $K^*(t)$, that jump must be at $\tau_1^*=0$, with

$$p(0)=r. \tag{114}$$

If there is a jump at 0, we have to determine the size of that jump. Let us put $(K^*(0^-)=K_0)$

$$K^*(0^+)-K_0=v^1, \tag{115}$$

where v^1 is an unknown parameter.

For $v^1\geqslant0$ fixed, our problem on $(0,\ T]$ can be regarded as the following standard control problem

$$\max\left\{\int_0^T(1-u)K^a\,dt-rv^1\right\}$$

$$\dot{K}=uK^a,\qquad K(0)=K_0+v^1,\qquad K(T)\text{ free} \tag{116}$$

$$u\in[0,\ 1].$$

The solution to this problem was given in exercise 2.7.1. (With v^1 fixed, the term $-rv^1$ does not influence the optimal choice of u and K.) We shall make use of the results in that exercise to find suggestions for the optimal solutions to the present problem. If the optimal v^1 is positive, we have to be certain that (114) is satisfied. There are a number of cases to be considered.

(A) $aK_0^{a-1} T \leqslant 1$

Since v^1 is fixed $\geqslant 0$, and $a \in (0, 1)$, $a(K_0 + v^1)^{a-1} \leqslant 1$. From exercise 2.7.1 we know that $u(t) \equiv 0$, $K(t) \equiv K_0 + v^1$, and $p(t) = -a(K_0 + v^1)^{a-1} (t - T)$ solves problem (116). In order to emphasize the dependence of $p(t)$ on v^1, let us put $p_A(t, v^1) = -a(K_0 + v^1)^{a-1} (t - T)$. In particular, $p_A(0, v^1) = a(K_0 + v^1)^{a-1} T$. Note that $p_A(0, v^1)$ is a strictly decreasing function of v^1 and if v^1 goes to ∞, $p_A(0, v^1)$ goes to 0.

How are we to choose v^1? Assume first that

(Aa) $p_A(0, 0) = aK_0^{a-1} T > r$

In view of (113) and (114), $p(t) = p_A(t, 0)$ is an impossible choice for the adjoint function of problem (103)–(106). We therefore increase v^1 so that $p_A(0, v^1)$ decreases. Let $\bar{v}_A^1$ be the unique number for which $p_A(0, \bar{v}_A^1) = r$, i.e. $a(K_0 + \bar{v}_A^1)^{a-1} T = r$, or

$$\bar{v}_A^1 = \left(\frac{r}{aT} \right)^{1/(a-1)} - K_0. \tag{117}$$

Then we have the following suggestion for a solution to problem (103)–(106): $u^*(t) \equiv 0$, $K^*(t) \equiv K_0 + \bar{v}_A^1$, and $p(t) = -a(K_0 + \bar{v}_A^1)^{a-1}(t - T)$, with $\bar{v}_A^1$ given by (117). This case can be referred to as the "one jump, no switch case".

(Ab) $p_A(0, 0) = aK_0^{a-1} T \leqslant r$

This is the "no jump, no switch case". Our suggestion is here $u^*(t) \equiv 0$, $K^*(t) \equiv K_0$, $p(t) = -aK_0^{a-1}(t - T)$.

(B) $aK_0^{a-1}T>1$

If v^1 is a small positive number, then $a(K_0+v^1)^{a-1}\ T$ is also >1 and according to exercise 2.7.1, the optimal solution to problem (116) implies a switch from $u^*(t)=1$ to $u^*(t)=0$ at $t^*=aT-(K_0+v^1)^{1-a}$. The associated a..'int function is given by (204) in that exercise. In particular, the value of the adjoint function at 0 is in this case

$$p_B(0, v^1)=[a(1-a)(K_0+v^1)^{a-1}\ T+a]^{a/(1-a)}. \tag{118}$$

By computing the derivative of $p_B(0, v^1)$ w.r.t. v^1, we find that $p_B\ (0, v^1)$ decreases as v^1 increases.

Again we have to consider two cases.

(Ba) $p_B(0,0)=[a(1-a)K_0^{a-1}\ T+a]^{a/(1-a)}>r$

Increase v^1 from 0. Then $p_B(0, v^1)$ given in (118) starts from a value $>r$ and then decreases. For small values of v^1 we also have

$$a(K_0+v^1)^{a-1}\ T>1. \tag{119}$$

When we increase v^1 further, two things might happen.

(Ba$_1$) $p_B(0,v^1)$ *becomes equal to* r *before* (119) *is violated*

This is clearly the "one jump, one switch case". If $\bar{v}_B^1$ is given by $p_B(0, \bar{v}_B^1)=r$, i.e. if

$$\bar{v}_B^1=\left[\frac{1}{T(1-a)a}(r^{(1-a)/a}-a)\right]^{1/(a-1)}-K_0, \tag{120}$$

then our suggestion for a solution to the problem (103)–(106) is: $u^*(t)=1$ in $[0, t^*]$, and $u^*(t)=0$ in $(t^*, T]$ where $t^*=aT-(K_0+\bar{v}_B^1)^{1-a}$. The associated choices for $K^*(t)$ and $p(t)$ are given by (203) and (204) in exercise 2.7.1 (with K_0 replaced by $K_0+\bar{v}_B^1$).

(Ba$_2$) (119) *is violated before* $p_B(0, v^1)$ *becomes* r

As we increase v^1 from 0, t^* decreases. Stop at that v^1, call it v_C^1, for which $a(K_0+v_C^1)^{a-1}\ T=1$ (still $p_B(0,\ v_C^1)>r$). Then $t^*=0$, and $p_A(0,\ v_C^1)=p_B(0,$

$v_C^1) = 1$. Increase v^1 further so as to obtain $p_A(0, \bar{v}^1) = r$. We are then in the "*one jump, no switch case*" again, and $\bar{v}^1 = \bar{v}_A^1$ given by (117). The solution candidate is as specified in (Aa).

(Bb) $p_B(0, 0) = [a(1 - a)K_0^{a-1} T + a]^{a/(1-a)} \leqslant r$

In this case we suggest putting $\bar{v}^1 = 0$, $u^*(t) = 1$ in $[0, t^*]$, and $u^*(t) = 0$ in $(t^*, T]$ where $t^* = aT - K_0^{1-a}$. The associated choices for $K^*(t)$ and $p(t)$ are given in (203) and (204) in exercise 2.7.1. We have here the "no jump, one switch case".

In all four different cases, we have seen to it that the solution candidates satisfy the necessary conditions in theorem 7 with $p_0 = 1$. Consider the conditions in theorem 8. The expression (86) is here

$$-rv^1 + p(t)v^1 = (p(t) - r)v^1,$$

which is linear, and hence, concave in (K, v^1) for each t. Moreover,

$$\hat{H}(K, p(t), t) = \max_{u \in [0, 1]} [(1 + (p(t) - 1)u]K^a = K^a \max(1, p(t))$$

is concave in K for each t. Theorem 8 therefore tells us that in each of the four cases we have found the optimal solution to problem (103)–(106).

Exercise 3.3.1 (Capital expansion of a firm.) Consider the problem:

$$\max \int_0^T qx(t)dt - \sum_{j=1}^k cv^j e^{-r\tau_j} \tag{121}$$

$$\dot{x}(t) \equiv 0 \text{ between jumps}, \qquad x(0^-) = 0, \qquad x(T^+) \leqslant x_T \tag{122}$$

$$x(\tau_j^+) - x(\tau_j^-) = v^j \geqslant 0, \qquad j = 1, \ldots, k. \tag{123}$$

Assume that T, q, c, r, and x_T are positive numbers, $rc > q$. (*Economic interpretation*: $x(t)$ is the production capacity of a firm. q is the sales price, x_T is the maximum amount that can be sold at any time, and ce^{-rt} is the price of a unit increase in capacity.)

(i)   Prove that there is at most one jump point $\tau^* \in (0, T)$, and compute τ^*.
(ii) Find the adjoint function $p(t)$.
(iii) Find the solution to the problem if $rT > 1 + \ln (cr/q)$

(Hint: Use the results in note 8 and theorem 8. Recall note 4 of Chapter 2.)

4. Jumps in the state variables II

In the previous section we studied optimal control problems with jumps in the state variables where "the maximizer" controlled the number and the location of the jump points as well as the magnitude of the jumps. The jump points could occur anywhere in $[t_0, t_1]$.

Several dynamic optimization problems in economics require modifications of these assumptions. We shall take a look at some of these modifications and briefly indicate the corresponding changes needed in the theorems of the preceeding section.

(A) *Jumps allowed to occur only in* $J' \subset [t_0, t_1]$

In this case we assume that we can single out, a priori, a fixed subset J' of $[t_0, t_1]$ within which all the jumps must occur. For instance, it might be that a jump can occur only at the initial time t_0, in which case $J' = \{t_0\}$.

Theorems 7 and 8 will still be valid in this case provided (76) is restricted so as to hold only for $t \in J'$.[9]

In the rest of the problems studied in this section we can drop the assumptions $0 \in V$, $h(x, 0, t) \equiv g(x, 0, t) \equiv 0$. On the other hand, for problems B, C, and D we need the assumption that $\partial f_0/\partial t$, $\partial f/\partial t$, $\partial h/\partial t$, and $\partial g/\partial t$ exist and are continuous.

(B) *The number k of jump points is fixed, but their location in $[t_0, t_1]$ as well as the jump parameters $v^1, \ldots, v^k$ are subject to choice.*

As an example, consider the problem of developing k oil fields in a prescribed sequence. It is natural to let $x(t)$ be the total production capacity of the developed fields and to let v^j be the size of the investment at field j.

Normally, the same amount of investment yields different increases in the production capacity of different fields. To take account of this feature we might introduce a state variable which jumps up one unit when a new field is developed. This state variable will then be an argument in the functions h and $g_1, \ldots, g_n$.

For this case, theorem 7 is valid provided (76) is deleted and (77) added.[10]

---

[9] Seierstad (1981b), theorem 3.

[10] Seierstad (1981b), remark 1. See also footnote 1, p. 13, and observe that the proof on pp. 10–13 does not make use of $0 \in V$, $h(x, 0, t) \equiv 0$, $g(x, 0, t) \equiv 0$. Getz and Martin (1980) have this result, but the proof is unclear. See Seierstad (1981b), p. 26.

(C) *The number k of jump points and the jump parameters $v^1, \ldots, v^k$ are fixed, but the location of the jump points in $[t_0, t_1]$ is subject to choice.*

An example of such a problem could be deciding the starting points of a prescribed time sequence of investment projects, the sizes of which are fixed.

In this case theorem 7 is valid provided (75) and (76) are replaced by (77), where $\bar{v}^1, \ldots, \bar{v}^k$ are fixed values of the v^j s.[11]

(D) *The number k of jump points and their location in $[t_0, t_1]$ is subject to choice. $v^1 = \ldots = v^k = \bar{v}$, with $\bar{v}$ a fixed number.*

For example: How many schools (of equal size) to build, and when to build them?

In order to solve such problems we use the method indicated in (C) to find the optimal distribution of $\tau_1, \ldots, \tau_k$ for each given value of k. Then for each k we compute the value of the criterion functional, and finally, compare all the values obtained.

(E) *The number k of jump points and their location in $[t_0, t_1]$ are fixed, but the jump parameters $v^1, \ldots, v^k$ are subject to choice.*

This is a very important case. For instance τ_j might denote the beginning of budget year no. j.

Here theorems 7 and 8 are both valid provided (76) is deleted.[12]

Suppose $f_0 \equiv 0$ and $f \equiv 0$, and $\tau_1 = 1, \tau_2 = 2, \ldots$. Then our problem reduces to a discrete time control problem for which the equations of motion are difference equations, and $v^1, v^2, \ldots$ are control variables. Thus we see that our theory covers discrete time control problems as a special case.

Note 10* So far we have assumed that $\tau_1 < \tau_2 < \ldots < \tau_k$, so that the jump points do not coincide. However, it is easy to generalize the results to cover coinciding jump points. If $\tau_j^* = \tau_{j+1}^* = \ldots = \tau_{j''}^*$, then we replace in (74)–(77) the partial derivatives of the h and g_i-functions w.r.t. x_i, v_l and τ by sums over all the coinciding τ's, e.g. $\partial h(x^*(\tau_j^{*-}), \bar{v}^j, \tau_j^*)/\partial x_i$ is replaced by $\Sigma_{j=j'}^{j''} \partial h(x^*(\tau_j^{*-}), \bar{v}^j, \tau_j^*)/\partial x_i$.

[11] See footnote 10.
[12] Seierstad (1981b), remark 7.

Let us make the following digression: Consider two jump points τ_1 and τ_2, and let $\tau_1 \to \tau_2$. Then the criterion as well as the solution, in general, depend discontinuously on τ_1 at τ_2.

Although a number of different types of control problems with jumps in the state variables have been presented, there are still several types we have not yet mentioned. Let us comment briefly on one of them dealing with case (C).

Consider again the problem of developing a number of oil fields. Up until now we have assumed that the fields are developed in a prescribed time sequence. However, there will usually be many different time sequences which are technically feasible. If we were to use the theory discussed so far to solve such a problem, we would have to find the optimal candidate for each choice of ordering and then select the best possible ordering. This method will consequently give us a large number of candidates for optimality. There are necessary conditions available which reduce drastically the number of candidates to consider. The conditions are particularly efficient in reducing the number of solution candidates for problems with co-inciding jump points. See Fourgeaud, Lenclud, Michel (1982) and Michel (1981).[13]

Exercise 3.4.1 **(Capacity expansion for a firm)** Consider the problem

$$\max \int_0^T qx(t)e^{-rt}dt - \sum_{j=1}^k c(a\tau_j - x(\tau_j^-))e^{-r\tau_j} \tag{124}$$

$$\dot{x}(t) \equiv 0 \text{ between jumps}, \qquad x(0^-)=0, \qquad x(T^+)=aT \tag{125}$$

---

[13]Consider again the situation in case (C) when possibly some of the τ_i's coincide. (We omit the stars on the τ_i's here.) In note 10 we indicated necessary conditions for this case. If h and g are independent of x, it seems possible to strengthen these conditions along the following lines: Let L be a subset of $\{1, 2, \ldots, k\}$ with the following two properties: (1) $\{\tau_i : i \in L\}$ consists of non-coinciding points; (2) all τ_i, $i = 1, \ldots, k$ equals $\tau_{i'}$ for some $i' \in L$. Let L'' be the subset of L consisting of indices i' for which $\tau_{i'}$ is different from all other τ_i's and belongs to (t_0, t_1). Put $L' = L - L''$. Choose a partition of L' into sets L^- and L^+ with $i \in L^+$ if $\tau_i = t_0$ and $i \in L^-$ if $\tau_i = t_1$. Define $\check{L}^- = \{\tau_i : \tau_i = \tau_j \text{ for some } j \in L^-\}$ and define $\check{L}^+$ correspondingly.

To each such partition there corresponds an adjoint function such that the necessary conditions of theorem 7 hold, with (75) and (76) deleted. Furthermore, (77) holds with equality for $i \in L''$, (77) with sums as in note 10 holds for the sign $\leqslant$ if $\tau_i \in L^+$, (resp. $\geqslant$ if $\tau_i \in \check{L}^-$). Furthermore, $H(x^*(\tau_j^+), u^*(\tau_j^+), p(\tau_j), \tau_j) - H(x^*(\tau_j^+) - g(\bar{v}^j, \tau_j), u^*(\tau_j^+), p(\tau_j), \tau_j) - p_0(\partial h(\bar{v}^j, \tau_j)/\partial \tau) - p(\tau_j)(\partial g(\bar{v}^j, \tau_j)/\partial \tau) \geqslant 0$ if $\tau_j \in L^+$, and $H(x^*(\tau_j^-), u^*(\tau_j^-), p(\tau_j), \tau_j) - H(x^*(\tau_j^-) + g(\bar{v}^j, \tau_j), u^*(\tau_j^-), p(\tau_j), \tau_j) + p_0(\partial h(\bar{v}^j, \tau_j)/\partial \tau) + p(\tau_j)(\partial g(\bar{v}^j, \tau_j)/\partial \tau) \geqslant 0$ if $\tau_j \in L^-$.

Michel (1981) has given similar conditions in a special case though these conditions are only stated for the partition $L^- = \phi$, $L^+ = L'$.

$$x(\tau_j^+) - x(\tau_j^-) = a\tau_j - x(\tau_j^-), \qquad \tau_1 = 0, \qquad \tau_k = T. \tag{126}$$

Economic interpretation: A firm has a production capacity $x(t)$ which is adjusted in jumps to match the demand, at. More precisely, at a jump point, τ_j, it increases its capacity such that $x(\tau_j^+) = a\tau_j$. Between jumps there is excess demand. While c is the unit cost of investment, q is the price of its goods, and r is the rate of interest.

(a) Fix the number $k \geqslant 3$ and show that the sequence $\tau_2^*, \ldots, \tau_{k-1}^*$ must satisfy the following necessary condition for optimality (apply (77)):

$$p_0[ac + (q - cr)(a\tau_j^* - x^*(\tau_j^{*-}))]e^{-r\tau_j^*} = ap(\tau_j^{*+}),$$
$$j = 2, 3, \ldots, k-1, \tag{127}$$

where $p(t)$ is the adjoint function associated with (127). (Note that problem (124)–(126) is one of case (C) described above. The function g in (63) is here $g(x, v, t) = at - x$ while h in (68) is $h(x, v, t) = -c(at - x)e^{-rt}$; hence there are no jump parameters.)

(b) Show next that $p(\tau_j^{*+}) = (q/r)(e^{-r\tau_j^*} - e^{-r\tau_{j+1}^*}) + ce^{-r\tau_{j+1}^*}$ and that $p_0 = 1$. Since $x^*(\tau_j^{*-}) = a\tau_{j-1}^*$, one has, by (127):

$$1 - r(\tau_j^* - \tau_{j-1}^*) = e^{-r(\tau_{j+1}^* - \tau_j^*)}. \tag{128}$$

A solution of (128) can be obtained in the following manner: Use $\tau_2^* - \tau_1^* = s$ as a parameter. Find, recursively, $\tau_3^* - \tau_2^*$ etc. from (128) and show that the size of these differences increases. Finally, adjust s such that $\tau_k^* = T$. Notice that the criterion in (124) can be expressed solely in terms of $\tau_1, \ldots, \tau_k$. If a number N is given, the criterion functional has a maximum over the set of k-tuples, $k \leqslant N$. This maximum can be found by computing the value of the criterion for the solution obtained from (128) for $k = 1, \ldots, N$.

5. Sensitivity results: Price interpretations

In economics we know the importance of studying the effect of changes in the parameters on the optimal solution of a static optimization problem. It is equally important in control theory to study the sensitivity of the optimal solution to changes in the parameters.

Several of the examples and exercises in this book have considered a special type of sensitivity result concerned with the effect on the optimal value function to changes in the initial (or terminal) conditions. We have obtained in this way several special cases of what is referred to in the literature as *price interpretations* of the adjoint variables. In fact, price interpretations, often based on very intuitive considerations, abound in the economic literature applying optimal control theory. (An early example is Dorfman (1969).)

We begin this section by studying properties of the optimal value function in a general setting.

The optimal value function and price interpretations of the adjoint variables

Consider our standard control problem

$$\max \int_{t_0}^{t_1} f_0(x, u, t) dt, \qquad (t_0 \text{ and } t_1 \text{ fixed}) \tag{129}$$

$$\dot{x} = f(x, u, t), \qquad x(t_0) = x^0 \tag{130}$$

$$u = u(t) \in U, \qquad U \text{ a bounded set in } R^r \tag{131}$$

$$x_i(t_1) = x_i^1, \qquad i = 1, \ldots, l \tag{132a}$$

$$x_i(t_1) \geqslant x_i^1, \qquad i = l+1, \ldots, m \tag{132b}$$

$$x_i(t_1) \text{ free}, \qquad i = m+1, \ldots, n. \tag{132c}$$

Assume that f_0 and f satisfy our standard regularity condition (2.6), and define $x^1 = (x_1^1, \ldots, x_m^1)$.

Suppose we calculate the value of the integral in (129), the criterion functional, for all admissible pairs $(x(t), u(t))$. The supremum (possibly ∞) of the set of real numbers obtained in this way depends, in particular, on x^0, x^1, t_0 and t_1. We denote the supremum by $V = V(x^0, x^1, t_0, t_1)$. (If $m = 0$ in (132), the terminal point is completely free and V does not contain x^1 as an argument.)

Symbolically,

$$V(x^0, x^1, t_0, t_1) = \sup \left\{ \int_{t_0}^{t_1} f_0(x(t), u(t), t) dt : (x(t), u(t)) \text{ admissible.} \right\}$$

The function $(x^0, x^1, t_0, t_1) \to V(x^0, x^1, t_0, t_1)$ is called the *optimal value function*. It is defined only for those quadruples (x^0, x^1, t_0, t_1) for which admissible pairs exist. If an *optimal pair* exists for a given quadruple (x^0, x^1, t_0, t_1), then V is finite and equal to the integral in (129) evaluated at the optimal pair.

We want to study how V changes when x^0, x^1, t_0 or t_1 changes. In the economic literature the most well-known "result" is

$$\frac{\partial V}{\partial x_i^0} = p_i(t_0), \qquad i = 1, \ldots, n, \tag{133}$$

which gives a price interpretation to the adjoint variables: If x_i^0 denotes the amount of some resource, then $p_i(t_0)$ approximately measures the contribution to the optimal value function of a unit increase in x_i^0. Thus, it is reasonable to confer upon a unit of x_i^0 such a value or price. (For special cases, see examples 2, 3 and 6 of Chapter 2.)

The result in (133) assumes, of course, that V is differentiable w.r.t. x_i^0. It turns out that quite restrictive assumptions have to be placed on the control problem in order for V to be differentiable.

Example 6   Consider the extremely simple problem:

$$\max \int_0^1 u(t)x(t)\mathrm{d}t, \qquad \dot{x}(t) = 0,$$

$$x(0) = x^0, \qquad x(1) \text{ free}, \qquad u(t) \in [0, 1]$$

where $u(t)$ and $x(t)$ are scalar functions, and x^0 is a given number. Suppose $(x^*(t), u^*(t))$ is optimal. Then $x^*(t) \equiv x^0$, and since $x(1)$ is free, $p_0 = 1$. Hence $H = ux$, and for $x^0 > 0$, $H(x^*(t), u, p(t), t) = x^0 u$ is maximized by $u = u^*(t) \equiv 1$. For $x^0 < 0$ it is maximized by $u = u^*(t) \equiv 0$. For $x^0 = 0$, $u^*(t)$ is undetermined. Since $V(x^0) = x^0$ for $x^0 \geqslant 0$, and $V(x^0) = 0$ for $x^0 < 0$, it follows that $V'(0^-) = 0$, and $V'(0^+) = 1$. Hence V is not differentiable at $x^0 = 0$.

The next theorem states precise conditions for the differentiability of V. The situation is particularly simple in the case $l = n$ (fixed end case), where it suffices to require uniqueness of the adjoint function and the Arrow sufficient conditions. For $l < n$ the following condition will be needed:

> Either (i) $\hat{H}^*(x, p(t), t)$ is strictly concave in x and (2.208),
> (2.211) hold, or (ii) $x^*(t)$ is a unique optimal solution and
> (2.207), (2.208), (2.211) are satisfied for $t \in [\bar{t}_0, \bar{t}_1]$. $\tag{134}$

In (ii), condition (2.207) can be replaced by the condition that for v.e. t, $u \to H(x, u, p(t), t)$ has a unique maximum in U, for all x such that

$$H^*(t) + \dot{p}(t)x^*(t) = \hat{H}(x, p(t), t) + \dot{p}(t) \cdot x.$$

Theorem 9 (Differentiability of the optimal value function. Fixed time interval)[14] Suppose $(x^*(t), u^*(t))$ solves problem $(129) \rightarrow (132)$ with $x^0 = \bar{x}^0$, $x^1 = \bar{x}^1$, $t_0 = \bar{t}_0$ and $t_1 = \bar{t}_1$, and assume that $(x^*(t), u^*(t))$ satisfies all the conditions in theorem 5 of Chapter 2, the Arrow sufficiency theorem. Suppose further that the adjoint function $p(t)$ associated with $(x^*(t), u^*(t))$ is uniquely determined by (2.57), (2.58), and (2.60) with $p_0 = 1$. Then $V(x^0, x^1, \bar{t}_0, \bar{t}_1)$ is defined and finite near $(\bar{x}^0, \bar{x}^1)$, and $(x^0, x^1) \rightarrow V(x^0, x^1, \bar{t}_0, \bar{t}_1)$ is differentiable at $(\bar{x}^0, \bar{x}^1)$ with partial derivatives given by

$$\frac{\partial V(\bar{x}^0, \bar{x}^1, \bar{t}_0, \bar{t}_1)}{\partial x_i^0} = p_i(\bar{t}_0), \qquad i = 1, \ldots, n \tag{135}$$

$$\frac{\partial V(\bar{x}^0, \bar{x}^1, \bar{t}_0, \bar{t}_1)}{\partial x_i^1} = -p_i(\bar{t}_1), \qquad i = 1, \ldots, m. \tag{136}$$

Furthermore, if either $l = n$, or $l < n$ and (134) holds, then $V(x^0, x^1, t_0, t_1)$ is defined and finite in a neighbourhood of $(\bar{x}^0, \bar{x}^1, \bar{t}_0, \bar{t}_1)$ and is differentiable at this point with partial derivatives given by (135), (136) and

$$\frac{\partial V(\bar{x}^0, \bar{x}^1, \bar{t}_0, \bar{t}_1)}{\partial t_0} = -H(x^*(\bar{t}_0), u^*(\bar{t}_0), p(\bar{t}_0), \bar{t}_0) \tag{137}$$

$$\frac{\partial V(\bar{x}^0, \bar{x}^1, \bar{t}_0, \bar{t}_1)}{\partial t_1} = H(x^*(\bar{t}_1), u^*(\bar{t}_1), p(\bar{t}_1), \bar{t}_1). \quad \blacksquare \tag{138}$$

Example 7 In example 2 of Chapter 2 the conditions in theorem 9 are satisfied and (2.46) agrees with (135) and (136). For that example $H(x^*(T), u^*(T), p(T), T) = U(1 - (x_T - x_0)/T) + U'(1 - (x_T - x_0)/T)(x_T - x_0)/T$, which is easily seen to be equal to the derivative of V in (2.45) w.r.t. T. Hence (138) is confirmed in this case.

In some cases it is possible to verify directly that the optimal value function is differentiable. (See e.g. exercise 3.5.3.) Also the following result is of interest.

Theorem 10 Replace the concavity condition on the maximized Hamiltonian $\hat{H}$ in theorem 9 by the assumption that V is differentiable at $(\bar{x}^0, \bar{x}^1, \bar{t}_0, \bar{t}_1)$. Then V's partial derivatives are still given by $(135) \rightarrow (138)$.[15]

---

[14]See Seierstad (1982) and (1985b, remark 12). The theorem is also valid for bounded, measurable controls $u^*(t)$ provided H^* is replaced by $\hat{H}^*$ in (137) and (138). In (134) (i), instead of the strict concavity, we can assume that $x^*(t)$ is even a unique optimal relaxed solution.

[15]This result follows from note 11. In case $l < n$, (134) is not needed.

Note 11* If we drop the assumption in theorem 9 that the maximized Hamiltonian $\hat{H}(x, p(t), t)$ is concave in x, then $V(x^0, x^1, t_0, t_1)$ is not necessarily differentiable at $\bar{y} = (\bar{x}^0, \bar{x}^1, \bar{t}_0, \bar{t}_1)$. However, one can prove that V is defined in a neighbourhood of $\bar{y}$, and has $(p(\bar{t}_0), -p(\bar{t}_1), -H^*(\bar{t}_0), H^*(\bar{t}_1))$ as a subderivative at $\bar{y}$. (See B.3.) Here $H^*(t)$ is the Hamiltonian evaluated at $(x^*(t), u^*(t), p(t), t)$. For details, see Seierstad (1982, 1985b).

Note 12 Formula (135) informs us that $p_i(t_0)$ approximately measures the change in the optimal value function of a unit increase in x_i^0. Can a similar interpretation be given to $p_i(t)$ for an arbitrary $t \in (t_0, t_1)$? In the following sense the answer is yes.

Consider problem (129)→(132), but assume that all admissible $x(t)$ have a fixed jump $v \in R^n$ at $t \in (t_0, t_1)$, i.e. $x(t^+) = x(t^-) + v$. Assume further that $x(t)$ is continuous elsewhere. The optimal value function for this problem will depend on v, $V = V(v)$. Suppose $(x^*(t), u^*(t))$ is an optimal solution to the problem for $v = 0$, and assume that the conditions in theorem 9 are satisfied. Then $V(v)$ is defined in a neighbourhood of $v = 0$, V is differentiable at $v = 0$ and $V'(0) = p(t)$. (If concavity of $\hat{H}$ is not satisfied, $p(t)$ is a subderivative of V.) Here (134) is not needed.

In theorem 9 we studied the fixed time problem in that t_0 and t_1 were fixed during the optimization. It would be convenient to have a similar result for the differentiability of the optimal value function of free time problems. Actually, in Chapter 2, section 11, where a proof of the Maximum Principle is given for a special case, we did get property (135) for a free time optimal control problem. However, this result was obtained only under the ad hoc assumption that the optimal value function was a C^2-function. (Theorem 11 below tells us that it suffices to assume that V is a C^1-function.) The following trivial example suggests how difficult it can be to obtain general results on the differentiability of the optimal value function of free time problems.

Example 8

$$\max \int_0^T x(t)dt, \qquad \dot{x}(t) \equiv 0, \quad x(0) = x_0, \qquad x(T) \text{ free}, \qquad T \text{ free in } [0, 1].$$

The only admissible function is $x(t) \equiv x_0$. If $x_0 > 0$, we see that $x^*(t) \equiv x_0$, $T^* = 1$ is the optimal solution and the optimal value function is $V(x_0) = x_0 T^* = x_0$. If $x_0 \leqslant 0$, $x^*(t) \equiv x_0$, $T^* = 0$ is optimal, and $V(x_0) = 0$.

Hence $V'(0^+)=1$, $V'(0^-)=0$, so V is not differentiable at 0. In this example $p(t)=T-t$, and $\hat{H}(x, p(t), t)=x$, hence concave in x. Still $V(x_0)$ is not differentiable.

It is possible to prove for free time problems a "subderivative property" of the optimal value function corresponding to the result mentioned in note 11. For details, see Seierstad (1981a).[16] The following theorem is the free time-version of theorem 10.

Theorem 11 Consider problem (129)→(132) with x^0, $x^1 =(x^1_1,\ldots, x^1_m)$ and t_0 fixed and with $t_1\in[T_1, T_2]$, $t_0 \leqslant T_1 < T_2$. Suppose $(x^*(t), u^*(t), t_1^*)$ is an optimal triple which satisfies the necessary conditions for the problem (see theorem 2.11 and note 2.25) with $p_0=1$, $x^0=\bar{x}^0$, $x^1=\bar{x}^1$, $t_0=\bar{t}_0$ and with a unique adjoint function $p(t)$. Then there exist admissible triples for all (x^0, x^1, t_0) in a neighbourhood of $(\bar{x}^0, \bar{x}^1, \bar{t}_0)$ such that the optimal value function $V(x^0, x^1, t_0)$ is defined in this neighbourhood. If $V(x^0, x^1, t_0)$ is differentiable at $(\bar{x}^0, \bar{x}^1, \bar{t}_0)$, then its partial derivatives are given by (135)→(137). ∎

Note 13 If the optimal value function V is differentiable only w.r.t. some of the parameters x^0, x^1, t_0 and t_1 in theorems 10 and 11, then the corresponding formulas in theorem 9 are still valid.

*Changes in the criterion functional when the control is fixed**

Let us briefly mention a sensitivity result of a somewhat different type. Consider problem (129)→(131) with $x^0=\bar{x}^0$, but instead of the terminal conditions (132) we have $x(t_1)$ completely free. Let $\hat{u}(t)$ be an admissible control, a piecewise continuous function on $[t_0, t_1]$ such that $\hat{u}(t)\in U$. Define $\hat{x}(t; \bar{x}^0)$ as the solution through $(t_0, \bar{x}^0)$ of $\dot{x}=f(x, \hat{u}(t), t)$, and let $p(t)$ be the corresponding solution of the adjoint equation with $p_0=1$ and $p(t_1)=0$. If we change $\bar{x}^0$ to x^0, but keep the same control function $\hat{u}(t)$, the solution $\hat{x}(t; x^0)$ will change. Denote $W(x^0)$ as the corresponding value of the criterion functional. Then one can prove that $W(x^0)$ is a C^1-function in

---

[16]The result is as follows (Seierstad (1981a)): In problem (129)–(132) let t_1 be subject to choice in $[T_1, T_2]$, and let $(x^*(t), u^*(t), t_1^*)$ be optimal. If $p(t)$ is uniquely determined by the necessary conditions of this free terminal time problem, then $V(x^0, x^1, t_0)$ is defined in a neighbourhood of $(\bar{x}^0, \bar{x}^1, \bar{t}_0)$ and V has a subderivative at $(\bar{x}^0, \bar{x}^1, \bar{t}_0)$ equal to $(p(\bar{t}_0), -p(t_1^*)$, $-H^*(\bar{t}_0))$. Neither here, nor in Note 11, is (134) needed.

a neighbourhood of $\bar{x}^0$ and that

$$\frac{\partial W(\bar{x}^0)}{\partial x_i^0} = p_i(t_0), \qquad i = 1, \ldots, n. \tag{139}$$

Compare this result with (135) in theorem 9. In the earlier theorem we studied the influence of a small change in the initial stock $\bar{x}^0$ on the optimal value of the criterion functional. In that case a small change in $\bar{x}^0$ implied changes in the optimal control as well as in the optimal path: "The control adjusts optimally". On the other hand, to obtain (139) we start simply with an *admissible* control, and that admissible control is kept fixed while $\bar{x}^0$ changes.

The proof of (135) is quite complicated, whereas the proof of (139) follows from standard properties of the resolvent of the variational equation associated with (130).[17]

Suppose we consider problem (129)→(131) with $x(t_1)$ free, and let $(\hat{x}(t), \hat{u}(t))$ be an optimal pair. Then it follows from (135) and (139) that $\partial V/\partial x_i^0 = \partial W/\partial x_i^0$ for all $i = 1, \ldots, n$. This result is similar to well-known envelope theorems in economics.[18]

If the control function $\hat{u}(t)$ is defined on an interval $(t_0', t_1') \supset [t_0, t_1]$ and if it is continuous at t_0 and t_1, then W as a function of x^0, t_0 and t_1 is C^1, and its derivative is the vector $(p(t_0), -\bar{H}(t_0), \bar{H}(t_1))$, where $\bar{H}(t)$ is the Hamiltonian evaluated at $(\hat{x}(t; \bar{x}^0), \hat{u}(t), p(t), t)$.

Dependence on parameters

We shall in this subsection briefly discuss the influence on the optimal value function of changes in parameters. Consider problem (129)→(132) with $f_0(x, u, t)$ and $f(x, u, t)$ replaced by $f_0(x, u, w, t)$ and $f(x, u, w, t)$, respectively, where $w = (w_1, \ldots, w_s)$ denoting a vector of parameters. Assume that f_0, f, and their partial derivatives w.r.t. x and w are continuous in (x, u, w, t). For all w, define

---

[17] Define $\tilde{f} = (f_0, f)$, $\tilde{x} = (x_0, x)$, and write $\tilde{f}(x, u, t) = \tilde{f}(\tilde{x}, u, t)$. Put $\tilde{x}^*(t) = (x_0^*(t), x^*(t))$ where $x_0^*(t) = \int_{t_0}^t f_0(x^*(s), u^*(s), s) ds$. Generally, the pair $(p_0, p(t)) = \tilde{p}(t)$ considered as a row vector, is a solution of $\dot{\tilde{p}} = -\tilde{p}\tilde{f}_{\tilde{x}}(\tilde{x}^*(t), u^*(t), t)$. Now $\tilde{p}(t) = \tilde{p}_1 \cdot \Phi(t_1, t)$ where $\tilde{p}_1 = \tilde{p}(t_1)$, and where $\Phi(t_1, t)$ is the resolvent of the equation $\dot{y} = \tilde{f}_{\tilde{x}}^* y$. For $\tilde{p}_1 = (1, 0, \ldots, 0)$, $\tilde{p}(t_0) = (1, 0, \ldots, 0) \cdot \Phi(t_1, t_0) = \partial \tilde{x}_0(t_1; \bar{x}^0, t_0)/\partial \tilde{x}^0$, where $\tilde{x}(t; \bar{x}^0, t_0)$ satisfies $\dot{\tilde{x}} = \tilde{f}(\tilde{x}, u^*(t), t)$, $\tilde{x}(t_0) = \bar{x}^0 = (x_0, x^0)$ and $\bar{x}^0 = (0, \bar{x}^0)$. See (A.31) and (A.29).

[18] See e.g. exercise 5.12.4(i) in Sydsæter (1981).

$$V(w) = \sup\left\{\int_{t_0}^{t_1} f_0(x, u, w, t)dt, \quad (x(t), u(t)) \text{ admissible}\right\}. \tag{140}$$

Suppose that the assumptions in theorem 9 (except (134)) are valid in the present case with $w = \bar{w}$, $\bar{w}$ some fixed vector in R^s, and add the following concavity condition on the maximized Hamiltonian ($p_0 = 1$):

$$\hat{H}(x, p(t), w, t) \text{ is concave in } (x, w) \text{ for each } t. \tag{141}$$

Then one can prove that V is defined in a neighbourhood of $\bar{w}$, and is differentiable at $\bar{w}$, with[19]

$$\frac{\partial V(\bar{w})}{\partial w_i} = \int_{t_0}^{t_1} \frac{\partial H(x^*(t), u^*(t), p(t), \bar{w}, t)}{\partial w_i} dt. \tag{142}$$

Note that the derivative under the integral sign is obtained by first differentiating the Hamiltonian w.r.t. w_i, and then evaluating this derivative at the optimal solution. If we know that V is differentiable, then (142) is valid even if condition (141) is not satisfied.

Necessary and sufficient conditions for control problems involving the optimal choice of parameters, are considered in exercise 3.2.8.

Example 9 Let us test (142) on the problem in example 3 of Chapter 2. From (2.79) of that example we know that V is differentiable w.r.t. a for $a \neq 0$, and we can derive

$$\frac{\partial V}{\partial a} = -\frac{2}{a^2} x_1^0 e^{aT-2} + \frac{2T}{a} x_1^0 e^{aT-2} = 2\frac{x_1^0}{a^2} e^{aT-2}(aT-1). \tag{143}$$

On the other hand, from (2.67) and $p_0 = 1$, $H = x_2 + p_1 a u x_1 + p_2 a(1-u)x_1$, so $\partial H/\partial a = p_1 u x_1 + p_2(1-u)x_1$. Applying the results of example 2.3 and exercise 2.4.1, we find that $\partial H^*/\partial a = (2/a)x_1^0 e^{aT-2}$ for $t \in [0, T-2/a]$ and $\partial H^*/\partial a = (T-t)x_1^0 e^{aT-2}$ for $t \in (T-2/a, T]$ where the * indicates evaluation at the optimal solution. Hence,

---

[19]Seierstad (1981a), theorem 6 implies that without imposing (141), the integral in (142) is a subderivative. The trick in exercise 3.2.8 can be used to derive (142) from theorem 9. Note that concavity of $\hat{H}(x, p(t), \bar{w}, t)$ w.r.t. x can replace condition (141) provided $x^*(t)$ is even a unique optimal relaxed solution, and (2.208), (2.211), and the uniqueness property of $u^*(t)$ in C in section 6 are satisfied. See Seierstad (1985b), remark 11. For more general results in more general problems see Malanowski (1984) and references in that paper.

$$\int_0^T \frac{\partial H^*}{\partial a}\, dt = \int_0^{T-2/a} \frac{2}{a} x_1^0 e^{aT-2}\, dt$$

$$+ \int_{T-2/a}^T (T-t)x_1^0 e^{aT-2}\, dt = 2\frac{x_1^0}{a^2} e^{aT-2}(aT-1). \tag{144}$$

Comparing (143) and (144) we see that (142) is confirmed for this problem.

Let us consider further an optimal control problem with w as a vector of parameters, but assume now that $x(t_1)$ is free. If $\hat{u}(t)$ is any piecewise continuous function on $[t_0, t_1]$ with $\hat{u}(t) \in U$, define $\hat{x}(t; w)$ as the solution to $\dot{x} = f(x, \hat{u}(t), w, t)$, $x(t_0) = x^0$, assuming $\hat{x}(t; w)$ exists on $[t_0, t_1]$ for $w = \bar{w}$. Then $\hat{x}(t; w)$ also exists on $[t_0, t_1]$ for w in some neighbourhood of $\bar{w}$. The value $W(w)$ of the criterion functional corresponding to $x = \hat{x}(t; w)$, $u = \hat{u}(t)$ and w, is C^1 and has a derivative at $w = \bar{w}$ which is[20]

$$W'(\bar{w}) = \int_{t_0}^{t_1} \frac{\partial H(\hat{x}(t; \bar{w}), \hat{u}(t), p(t), \bar{w}, t)}{\partial w}\, dt. \tag{145}$$

Exercise 3.5.1 Consider the problem

$$\max \int_{t_0}^{t_1} (1 - u(t))x(t)\,dt, \qquad \dot{x}(t) = u(t)x(t),$$

$$x(t_0) = x^0, \qquad x(t_1) \geqslant x^1 \tag{146}$$

with $u(t) \in [0, 1]$. Assume that t_0, t_1, x^0 and x^1 are fixed positive constants with $t_1 > t_0 + 1$ and $e^{t_1 - t_0} > x^1/x^0 > e^{t_1 - t_0 - 1}$

(i) Prove that the optimal control $u^*(t)$ is given by

$$u^*(t) = 1 \text{ on } [t_0, t^*], \qquad u^*(t) = 0 \text{ on } (t^*, t_1] \tag{147}$$

where $t^* = t_0 + \ln(x^1/x^0)$, and the associated adjoint function (with $p_0 = 1$), is

$$p(t) = e^{-(t-t^*)} \text{ on } [t_0, t^*], \qquad p(t) = -(t-t^*)+1 \text{ on } (t^*, t_1]. \tag{148}$$

---

[20] (145) follows from (139) by using the trick in exercise 3.2.8. $p(t)$ satisfies
$\dot{p} = -\partial H(\hat{x}(t; \bar{w}), \hat{u}(t), \bar{w}, t)/\partial x$, $p(t_1) = 0$.

(ii) Prove that the optimal value function in this case is

$$V(x^0, x^1, t_0, t_1) = x^1(t_1 - t^*) = x^1[t_1 - t_0 - \ln(x^1/x^0)] \tag{149}$$

and verify $(135) \rightarrow (138)$ in theorem 9 for this problem.

Exercise 3.5.2 Examine the problem in exercise 2.8.3.

(i) Explain why the Arrow sufficient conditions (theorem 2.5) fail to be met.
(ii) Compute the optimal value function $V(x_0)$ and verify that $V'(x_0) = p(0) = ex_0$. (This confirms theorem 10.)

Exercise 3.5.3 Return to example 2.13, and assume that $x_1/a > 1$. In this case the optimal control switches from 1 to 0 at $\hat{T} = \ln(q/a)$.

(i) Prove that the optimal value function $V(x_1)$ is a C^1-function for x_1's satisfying (262), and $x_1/a > 1$. (It is not necessary to calculate $V(x_1)$.)
(ii) Use theorem 11 and note 13 to prove that $V'(x_1) = (x_1 - q)/a$.
(iii) Verify the result in (ii) by calculating $V(x_1)$ and $V'(x_1)$.

Exercise 3.5.4* Using theorem 9 – more specifically (138) – prove theorem 2.13 in the case $l = n$, assuming that the function $p_T(t)$ in theorem 2.13 is unique for each T.

6. Some further sensitivity results

In the previous section we were concerned with the effect on the criterion functional to changes in parameters like t_0, t_1, x^0, x^1 and w. Of course, changes in these parameters will change the optimal control as well as the optimal path. In applications it is often important to study these changes, but general results of any practical value turn out to be very complicated, and hard to apply. The reason is obvious. A change in one or more of the parameters might change the optimal solution in a variety of ways. For instance, a bang-bang control might change to a continuous control, switch-points might change drastically, etc.

However, it is possible to give some results of a qualitative nature. Consider problem $(129) \rightarrow (132)$, assuming that f_0 and f contain a vector w of parameters, $w \in W \subset R^s$. Let $f_0, f, \partial f_0/\partial x$ and $\partial f/\partial x$ be continuous in (x, u, w, t). Define V^* as a set of parameter values $v = (t_0, t_1, x^0, x^1, w)$, and assume that for each $v \in V^*$ there exist admissible solutions. The following properties then hold:

(A) Assume that U is compact, that $N(x, U, t)$ is convex for all (x, t), $w \in W$ (see (2.205)), and that for some $b > 0$, for all t, $\|x(t)\| \leqslant b$ for any admissible solution $x(t)$ corresponding to any $v \in V^*$. Then the optimal value function $V(t_0, t_1, x^0, x^1, w) = V(v)$ is upper semicontinuous at all $\bar{v} \in V^*$, i.e. $V(\bar{v}) \geqslant \lim \sup V(v')$ when $v' \to \bar{v}$, $v' \in V^*$.[21]

(B) Let $\bar{v} \in V$. Assume that there exists an optimal pair $(x^*(t), u^*(t))$ corresponding to $\bar{v}$, and that the necessary conditions are satisfied by this pair only for $p_0 = 1$, not for $p_0 = 0$. Then there exist admissible pairs corresponding to all v in some neighbourhood of $\bar{v}$ in $R^{n+m+s+2}$. Moreover, if the assumptions of (A) also hold, $V(v)$ is continuous at $\bar{v}$.[22]

(C) Assume that the conditions in (A) and (B) are satisfied. Assume further that $x^*(t)$ is the only optimal solution corresponding to $\bar{v} = (\bar{t}_0, \bar{t}_1, \bar{x}^0, \bar{x}^1, \bar{w}^1)$. ($u^*(t)$ may or may not be unique). Let $v^n \in V^*$ and $(x^n(t), u^n(t))$ be an optimal pair corresponding to v^n, $n = 1, 2, \ldots$. If $v^n \to \bar{v}$, then $x^n(t) \to x^*(t)$ for any $t \in (t_0, t_1)$.

If $u^*(t)$ for v.e. t is the only vector in U which maximizes $H(x^*(t), u, p(t), t)$, where $p(t)$ is some adjoint function corresponding to $(x^*(t), u^*(t))$, then $\{u^n(t)\}$ converges to $u^*(t)$ in the sense that

$$\int_a^b \|u^n(s) - u^*(s)\| ds \to 0, \qquad \text{for any interval } [a, b] \subset (\bar{t}_0, \bar{t}_1)$$

If in addition there exists only one adjoint function for which $(x^*(t), u^*(t))$ satisfies the necessary conditions when $p_0 = 1$, then $p^n(t) \to p(t)$, for any $t \in (\bar{t}_0, \bar{t}_1)$. Here $p^n(t)$ is some (perhaps non-unique) adjoint function corresponding to $(x^n(t), u^n(t))$.

Finally, if $\hat{H}(x, p(t), t)$, with $p_0 = 1$, is strictly concave in x, where $p(t)$ is some adjoint function corresponding to $(x^*(t), u^*(t))$, then $x^*(t)$ is automatically unique, and in this case the conclusions in (B) and (C) hold without the assumption that $N(x, U, t)$ is convex.[23]

These continuity results give meaning to the following argument: Suppose $U = \{u: h(u) \geqslant 0\}$, with h a given vector function. Then the maximi-

---

[21] Rockafellar (1973), Cesari (1983), theorem 10.8(ii). See also Seierstad (1985b), theorem 4. Convexity of $N(x, U, t)$ can be replaced by the condition that there exists an admissible pair $(x^*(t), u^*(t))$ corresponding to $\bar{v}$ that satisfies the Arrow sufficient conditions.

[22] Seierstad (1985b), theorem 6. The remark in footnote 21 applies also in the present case. To obtain B a well-known property of normal fixed terminal point control problems ($p_0 \neq 0$) can be used: For suitable small perturbations of $u^*(t)$, all points x^1 near $\bar{x}^1$ can be reached at any time near $\bar{t}_1$.

[23] Seierstad (1985b), theorems 8 and 9. The conclusions in (C) hold even if N is non-convex, provided $x^*(t)$ is a unique optimal relaxed solution.

zation of the Hamiltonian results in a nonlinear programming problem. The set of "junction points" for this problem, i.e. the time points where the set of active constraints change, include the discontinuity points for the optimal control function. Let $x(t, v)$, $u(t, v)$ and $p(t, v)$ be the solution to the Maximum Principle corresponding to the choice of parameters v. These functions are then determined by the relations $\dot{x} = f$, $\dot{p} = -H'_x$ and the first-order nonlinear programming conditions. The continuity results above imply that $x(t, v)$, $u(t, v)$, and $p(t, v)$ are close to $x(t, \bar{v})$, $u(t, \bar{v})$, and $p(t, \bar{v})$ respectively, when v is close to $\bar{v}$. In order to find out how $x(t, v)$, $u(t, v)$ and $p(t, v)$ in the first order depend on v, one can therefore linearize the relations around $x(t, \bar{v})$, $u(t, \bar{v})$, $p(t, \bar{v})$, and $\bar{v}$. (A separate treatment is needed between each pair of discontinuity points for $u(t, \bar{v})$.) Additionally one will have to derive the linearized conditions for the junction points. These first-order approximations are discussed in some detail by Oniki (1973). The general formulas are rather unwieldy.

Perturbation of controls

We study next the effect on the criterion functional of a special type of perturbation of the control function: Consider problem (129)→(131) and replace (132) by the assumption that $x(t_1)$ is free. Let $(\hat{x}(t), \hat{u}(t))$ be an admissible pair. If $\tau \in [t_0, t_1]$ we say that $\hat{u}(t)$ undergoes a perturbation u near τ, provided $\hat{u}(t)$ is replaced by the constant $u \in R^r$ on some interval $A(s)$, where $A(s) = [\tau, \tau + s]$ if $s \geqslant 0$, $A(s) = (\tau + s, \tau]$ if $s < 0$. The value of the criterion functional for a control perturbed to the left or to the right of τ in this manner is denoted by $W(s)$ (u is kept constant). The left and right derivatives dW^-/ds and dW^+/ds exist and are given by the formulas:

$$\frac{d^{\pm} W(0)}{ds} = \pm [H(\hat{x}(\tau), u, p(\tau), \tau) - H(\hat{x}(\tau), \hat{u}(\tau^{\pm}), p(\tau), \tau)] \tag{150}$$

where the $+$ and $-$ signs correspond to each other, and where $p(t)$ is determined by $\dot{p} = -H'_x(\hat{x}(t), \hat{u}(t), p(t), t)$, $p(t_1) = 0$.[24]

---

[24] If $\hat{u}(t)$ is optimal in problem (129)→(131) with $x(t_1)$ free, then W has a maximum for $s = 0$ so, $d^+ W(0)/ds \leqslant 0$ and $d^- W(0)/ds \geqslant 0$ for $u \in U$. Applying (150), the maximum condition (2.57) follows from either of these inequalities, and thus we have a proof of theorem 2.2 for this case.

Let us sketch a proof of (150): Let $s > 0$, and let $\tilde{x}(t,s)$ be the solution of $\dot{\tilde{x}} = \tilde{f}(\tilde{x}, u, t) \chi_{A(s)} + \tilde{f}(\tilde{x}, u^*(t), t)\chi_{\mathcal{Q}4(s)}$, $\tilde{x}(t_0) = (0, x^0)$, using the notation in footnote 17. In the first order, $\tilde{x}(\tau + s, s) - \hat{x}^*(\tau + s) \simeq (\tilde{x}(\tau^+, s) - \dot{x}^*(\tau^+))s = [\tilde{f}(\hat{x}^*(\tau), u, s) - \dot{\hat{x}}^*(\tau^+)]s = \delta$. Using $\tilde{p}(t_0) = \partial \tilde{x}_0(t_1; \bar{x}^0, t_0)/\partial \tilde{x}^0$ with $t_0 = \tau + s$, we get $W(s) - W(0) \approx \tilde{p}(\tau + s)\delta$. Dividing by s and letting $s \rightarrow 0$, we derive (150). A similar proof goes for $s < 0$. See e.g. Pallu de la Barrière (1980).

Consider again problem (129)→(132). Assume now that $(x^*(t), u^*(t))$ is an optimal pair for this problem satisfying the Arrow sufficient conditions (theorem 5 of Chapter 2) for a unique adjoint function $p(t)$. Assume, moreover, that (134) holds. Let τ be any point in (t_0, t_1), and $V(s)$ the supremum of the criterion in problem (129)→(132) when the restriction $u(t) \in U$ is replaced by $u(t) \in U$ for $t \notin A(s)$, $u(t) \in \{u'\} \cup U$ for $t \in A(s)$, where u' is some fixed vector in R^r. Then $V(s)$ has right and left derivatives at $s=0$ given by[25]

$$\frac{d^{\pm}V(0)}{ds} = \pm \max[0, H(x^*(\tau), u', p(\tau), \tau) - H(x^*(\tau), u^*(\tau^{\pm}), p(\tau), \tau)]. \tag{151}$$

Results (150) and (151) can be used to calculate the effect on the criterion functional caused by allowing the control functions to take on an alternative value on short time intervals. In example 10 we shall see, in particular, how (150) can be used in a differential game problem, and in exercise 3.6.3 we look at an application of (151).

Note 14   Almost all the results presented so far in this chapter hold for the weaker regularity conditions on f_0 and f stated in note 2.6. We list the few exceptions: theorem 2, note 2, note 7 (and thus (85) of note 8), and points (B), (C) and (D) in section 4. In theorems 9, and 10 we obtain differentiability only w.r.t. x^0 and x^1 provided $\bar{t}_0$ and $\bar{t}_1$ belong to the set C (note 2.6(b)), while the results are valid as stated if $\bar{t}_0$ and $\bar{t}_1$ do not belong to C. This qualification pertains also to other results on the differentiability w.r.t. t_0 and t_1. Finally, (150) and (151) fail if $\tau \in C$.

Example 10 (Workers versus capitalists. Lancaster (1973), Hoel (1978))   Consider the following model:

$$W = \int_0^T u(t)K(t)dt, \qquad u(t) \in [a, b], \qquad 0 < a < b < 1 \tag{152}$$

$$C = \int_0^T (1 - v(t))(1 - u(t))K(t)dt, \qquad v(t) \in [0, 1] \tag{153}$$

---

[25]See Seierstad (1985b), theorem 12. If $x^*(t)$ is a unique *relaxed* solution, convexity of $N(x, U, t)$ can be dropped. The derivative in (151) is at least a subderivative even if the convexity of $N(x, U, t)$, the uniqueness of $x^*(t)$ as a relaxed or an ordinary solution, and the concavity of $\hat{H}$ all fail.

$$\dot{K}(t) = v(t)(1 - u(t))K(t), \qquad K(0) = K_0 > 0, \qquad K(T) \text{ free} \tag{154}$$

where T, a, b, and K_0 are constants, $T > 1/b$ and $T > 1/(1-b)$. $K = K(t)$ denotes the capital stock of the economy, and the rate of production is proportional to K. We choose the time unit such that the constant of proportionality is 1. Within limits a and b, workers decide their share $u = u(t)$ of production. The remaining production, $(1-u)K$, is controlled by the capitalists, who invest a fraction $v = v(t)$ (see (154)) and consume the remaining portion, $(1-v)(1-u)K$. Both workers and capitalists want to maximize their own total consumption, (152) and (153), respectively. Although workers usually do gain future benefits from investment, their willingness to sacrifice consumption can be exploited by the capitalists. On the other hand, a willingness to invest will be less effective if the workers too soon press their share towards the limit b.

Nash non-cooperative equilibrium

The two "players" workers and capitalists, are placed in a game situation. To solve this game we borrow a solution concept, *Nash non-cooperative equilibrium, from game theory*. (We refrain from discussing the merits and demerits of this solution concept for the present problem. All the solution concepts usually suggested are indeed questionable.) A *Nash equilibrium* for our game is a pair of control functions $(u^*(t), v^*(t))$ such that $u^*(t)$ solves the problem

$$\max_{u(t)} W \text{ for } v(t) = v^*(t) \tag{155}$$

while $v^*(t)$ solves the problem

$$\max_{v(t)} C \text{ for } u(t) = u^*(t). \tag{156}$$

Hence $u^*(t)$ maximizes the workers' utility (total consumption) when the capitalists control function is kept fixed equal to $v^*(t)$. Symmetrically, $v^*(t)$ maximizes the capitalists' utility when the workers' control function is kept fixed equal to $u^*(t)$.

We apply the Arrow sufficiency theorem (theorem 2.5) to find this models' Nash equilibrium. The Hamiltonians for the two problems (with $p_0 = 1$) are

$$H_W = uK + pv^*(t)(1-u)K, \quad H_C = (1-v)(1-u^*(t))K + qv(1-u^*(t))K \tag{157}$$

where $p = p(t)$ and $q = q(t)$ are the corresponding adjoint functions.

For problem (155) we know that $u^*(t)$ maximizes $uK^*(t) + p(t)v^*(t)(1 - u)K^*(t)$ and also $(1 - p(t)v^*(t))uK^*$ for $u \in [a,b]$. Since $K^*(t) > 0$, we see that

$$u^*(t) = a \text{ if } p(t)v^*(t) > 1, \qquad u^*(t) = b \text{ if } p(t)v^*(t) < 1, \tag{158}$$

while $u^*(t) \in [a, b]$ if $p(t)v^*(t) = 1$. The adjoint function $p(t)$ satisfies

$$\dot{p}(t) = -u^*(t) - p(t)v^*(t)(1 - u^*(t)), \qquad p(T) = 0. \tag{159}$$

For problem (156) we find in the same manner:

$$v^*(t) = 0 \text{ if } q(t) < 1, \qquad v^*(t) = 1 \text{ if } q(t) > 1 \tag{160}$$

while $v^*(t) \in [0, 1]$ if $q(t) = 1$, and

$$\dot{q}(t) = [-(1 - v^*(t)) - q(t)v^*(t)](1 - u^*(t)), \qquad q(T) = 0. \tag{161}$$

From (160) and (161) we see that $\dot{q}(t) = -\max(1, q(t))(1 - u^*(t))$, so $\dot{q}(t) < 0$ everywhere, and thus $q(t)$ is strictly decreasing. Since $q(T) = 0$, $q(t) < 1$ in an interval $(t', T]$. From (160), $v^*(t) \equiv 0$ in $(t', T]$, regardless of the form of $u^*(t)$ (although t' itself may depend on $u^*(t)$). It follows from (158) that $u^*(t) = b$ in $(t', T]$, and thus from (161), $\dot{q}(t) = 1 - b$, and $q(t) = (1 - b)(T - t)$. By letting t' be determined by $q(t') = (1 - b)(T - t') = 1$, $t' = T - 1/(1 - b)$. From (159), $\dot{p}(t) = -b$ and $p(t) = b(T - t)$. Hence we have found a proposal for $u^*(t)$, $v^*(t)$, $p(t)$ and $q(t)$ on $(t', T]$, and (158)→(161) are all satisfied on $(t', T]$. Observe that from our assumptions, $0 < t' < T$.

Since $q(t') = 1$ and $q(t)$ is strictly decreasing, $q(t) > 1$ in $[0, t')$. Therefore from (160), $v^*(t) = 1$ in $[0, t')$.

Note that we have been able to determine $v^*(t)$ uniquely, without knowing much about $u^*(t)$. This is a rather exceptional situation in differential game models; it is due to the extreme simplicity of the model.

Next let us determine $u^*(t)$ on $[0, t']$. We have to consider two cases:
Suppose $b \geqslant 1/2$. Then $t' = T - 1/(1 - b) \leqslant T - 1/b$. Since $p(t) = b(T - t)$ in $[t', T]$, $p(t) = 1$ for $t = T - 1/b$. By using the bouncing off results (A.14), we see that $p(t) \geqslant 0$ for all $t \in (0, T]$. It follows from (159) that $\dot{p}(t) < 0$ in $(0, T]$, so $p(t)$ is strictly decreasing. Hence $p(t) > 1$ in $[0, t')$, so $p(t)v^*(t) > 1$ in that interval. Hence from (158), $u^*(t) = a$ in $[0, t']$. Thus we have the following proposal for optimal controls in the two problems when $b \geqslant 1/2$:

$$\begin{aligned} &u^*(t) = a \text{ and } v^*(t) = 1 \text{ in } [0, t'] \\ &u^*(t) = b \text{ and } v^*(t) = 0 \text{ in } (t', T] \end{aligned} \qquad t' = T - \frac{1}{1 - b} \tag{162}$$

From (154) it is easy to find $K^*(t)$ while $p(t)$, $q(t)$ on $[0, t']$ are determined by (159) and (161). Conditions (158)→(161) are all satisfied on $[0, T]$.

We turn next to the case $b < 1/2$. In this case $p(t') = b(T - t') = b/(1 - b) < 1$. Thus $p(t) < 1$ in some interval $(t'', t']$. Choose t'' as the smallest t'' with this property. It turns out that $t'' > 0$. Then $p(t'') = 1$, and in $(t'', t']$, $p(t)v^*(t) = p(t) < 1$, so $u^*(t) = b$. By (159), the differential equation for p in $[t'', t']$ is $\dot{p} = -b - p(1 - b)$ with $p(t') = b/(1 - b)$. Therefore $p = De^{-(1-b)t} - b/(1 - b)$, with $b/(1 - b) = De^{-(1-b)t'} - b/(1 - b)$, i.e.

$$p(t) = \frac{2b}{1 - b} e^{-(1-b)(t - t')} - \frac{b}{1 - b}. \tag{163}$$

The number t'' is determined by the equation $1 = p(t'')$, Solving for t'', we find

$$t'' = t' - \frac{1}{1 - b} \ln \frac{1}{2b}. \tag{164}$$

It is an easy exercise in calculus to prove that since $T > 1/b$, $t'' > 0$.

As in the case $b \geq 1/2$, $p(t)$ is strictly decreasing and $p(t) > 1$ in $[0, t'')$ and thus $p(t)v^*(t) = p(t) > 1$ in this interval. It follows from (158) that $u^*(t) = a$ in $[0, t'']$. Thus we have the following proposal for optimal controls in the two problems when $b < 1/2$

$$
\begin{aligned}
&u^*(t) = a \text{ in } [0, t''], \ u^*(t) = b \text{ in } (t'', T] \qquad t' \text{ given in (162)}\\
&v^*(t) = 1 \text{ in } [0, t'], \ v^*(t) = 0 \text{ in } (t', T] \qquad t'' \text{ given in (164)}
\end{aligned} \tag{165}
$$

Again it is easy to find $K^*(t)$, $p(t)$ and $q(t)$ such that all the conditions (158)→(161) are satisfied on $[0, T]$.

Note that the maximized Hamiltonian for problem (155) is $\hat{H}_W(K, p(t), t) = K \cdot \max_{u \in [a, b]}(u + p(t)v^*(t)(1 - u))$, with the maximum independent of K. Hence $\hat{H}_W$ is linear in K and thus concave. By a similar argument, the maximized Hamiltonian for problem (156) is also concave in K. By applying theorem 2.5 and note 2.11, we conclude that $u^*(t)$ and $v^*(t)$ in (162) for $b \geq 1/2$, and in (165) for $b < 1/2$, solve problems (155) and (156), respectively.

In this model we found an equilibrium pair. Note, however, that in non-cooperative differential games, equilibrium pairs do not necessarily exist.

Cooperation versus non-cooperation

Suppose the capitalists and the workers cooperate in the maximization of a common criterion, the sum of the workers and the capitalists consumption. Then one can prove that the optimal solution leads to a higher total consumption than in the non-cooperative case ("the dynamic inefficiency of capitalism" in the terminology of Lancaster (1973)). See exercise 3.6.2.

We next discuss a way in which the workers and the capitalists can jointly improve the result in the Nash solution by cooperating in their choice of controls. To this end we use the result on pertubation of controls in (150) and rely on a general result in game theory.

A game theory result

Consider a game with two players. Let $F(r, s)$ and $G(r, s)$ denote the payoffs to player 1 and player 2, respectively, when player 1 chooses $r \in R$ and player 2 chooses $s \in S$. Here R and S are two prescribed sets of real numbers. The pair (r^*, s^*) is *a Nash equilibrium* if r^* maximizes $F(r, s^*)$ for $r \in R$, while s^* maximizes $G(r^*, s)$ for $s \in S$.

Let us make the following assumptions:

(i) F and G are differentiable at an equilibrium point (r^*, s^*)
(ii) r^* and s^* are interior points in R and S respectively.
(iii) $F'_2(r^*, s^*) \neq 0$, $G'_1(r^*, s^*) \neq 0$.

From the definition of an equilibrium pair, and (i) and (ii), it follows that $F'_1(r^*, s^*) = G'_2(r^*, s^*) = 0$. Note that the assumptions in (iii) seem reasonable. (There is no reason why s^* should be chosen so as to maximize $F(r^*, s)$, in which case we would get $F'_s(r^*, s^*) = 0$.) We now claim:

(iv) Provided (i), (ii) and (iii) are satisfied, the two players can jointly increase their payoffs by properly modifying r^* and s^*.

In order to prove (iv), let δ and γ have the same signs as $G'_1(r^*, s^*)$ and $F'_2(r^*, s^*)$, respectively. If $|\delta|$ and $|\gamma|$ are small,

$$F(r^* + \delta, s^* + \gamma) \approx F(r^*, s^*) + F'_1(r^*, s^*)\delta + F'_2(r^*, s^*)\gamma$$
$$= F(r^*, s^*) + F'_2(r^*, s^*)\gamma > F(r^*, s^*)$$

since $F'_1(r^*, s^*) = 0$ and $F'_2(r^*, s^*)\gamma > 0$. Similarly we prove the corresponding inequality for player 2's payoff function. Therefore we have the conclusion in (iv).

Back to the original model

Assume that $b < 1/2$. (The case $b \geq 1/2$ is treated in exercise 3.6.2).

Let $u(t) = a$ on $[0, t'' + r]$, $u(t) = b$ on $(t'' + r, T]$. $v(t) = 1$ on $[0, t' + s]$, $v(t) = 0$ on $(t' + s, T]$. If these functions and the corresponding solution for $K(t)$ are inserted into (152) and (153), we obtain payoff functions $F(r, s)$ and $G(r, s)$, defined in neighbourhoods R of $r = 0$ and S of $s = 0$, respectively. Note that for $r > 0$, $u(t)$ is the function obtained by giving $u^*(t)$ the perturbation $u = a$ on $[t'', t'' + r)$. For $r < 0$, $u(t)$ is the function obtained by giving $u^*(t)$ the perturbation $u = b$ on $(t'' + r, t'']$. An analogous result holds for $v(t)$.

Note that the game with payoff functions F and G differs from the original game as the new classes of $u(t)$ and $v(t)$ functions are more restricted. However, $(r, s) = (0, 0)$ is obviously a Nash equilibrium pair for the new game as well.

It is an easy, but tedious job to find explicit expressions for $F(r, s)$ and $G(r, s)$. To find one for $F(r, s)$ we proceed as follows: Put $u = a$, $v = 1$ on $[0, t'' + r]$, $u = b$, $v = 1$ on $(t'' + r, t' + s]$, and $u = b$, $v = 0$ on $(t' + s, T]$ into (154) and determine the development of $K(t)$ on $[0, T]$. Then from (152) we find for $|r|$ and $|s|$ small,

$$F(r, s) = \int_0^{t'' + r} a K_0 e^{(1 - a)t} dt + \int_{t'' + r}^{t' + r} b K_0 e^{(t'' + r)(b - a) + (1 - b)t} dt$$

$$+ \int_{t' + r}^{T} b K_0 e^{(t'' + r)(b - a) + (1 - b)(t' + s)} dt.$$

It follows that $F(r, s)$ is differentiable in a neighbourhood of $(0, 0)$ and by differentiating w.r.t. s and then putting $r = s = 0$, we obtain $F'_2(0,0) = K_0 b e^{t''(b - a) + (1 - b)t'} > 0$. In the same way we find $G(r,s)$ it being differentiable with $G'_1(0, 0) > 0$. We conclude from (iv) above that the workers and the capitalists can jointly increase their payoffs by modifying their control functions slightly.

We can also reach the same conclusions by relying on the result in (150). Since it is usually impossible to obtain explicit expressions for the payoff functions, it is instructive to see how (150) can be put to use.

Notice that $F(0, s)$ and $G(0, s)$ are the values of the integrals in (152) and (153), respectively obtained by giving $u^*(t)$ and $v^*(t)$ the perturbed values

$u=b$ and $v=1$ on $(t', t'+s]$ for $s>0$, $u=b$, $v=0$ on $(t'+s, t']$ for $s<0$. (Thus $u^*(t)$ is actually not perturbed in this case.) We can now use (150) to calculate $F_2'^+(0,0)$. Putting $H_W=H_W(K, u, v, p(t), t)=uK+p(t)v(1-u)K$, we obtain

$$F_2'^+(0,0)=H_W(K^*(t'), b, 1, p(t'), t')-H_W(K^*(t), b, 0, p(t'), t')$$

$$=bK^*(t')+p(t')\cdot 1\cdot(1-b)K^*(t')-bK^*(t')-p(t')\cdot 0\cdot(1-b)K^*(t')$$

$$=(1-b)p(t')K^*(t')=bK^*(t')>0.$$

By similar calculation we find $F_2'^-(0,0)=bK^*(t')$, and also $G_2'^+(0,0)=G_2'^-(0,0)=0$.

Notice next that $F(r, 0)$ and $G(r, 0)$ are the values of the integrals in (152) and (153) obtained by giving $u^*(t)$ and $v^*(t)$ the perturbations $u=a$, $v=1$ on $[t'', t''+r]$ if $r>0$, $u=b$, $v=1$ on $(t''+r, t'']$ if $r<0$. Again applying (150) we find

$$F_1'^+(0,0)=F_1'^-(0,0)=0,$$

$$G_1'^+(0,0)=G_1'^-(0,0)=(b-a)q(t'')K^*(t'')>0.$$

By this procedure we have again established that F and G have partial derivatives at $(0, 0)$. The differentiability at $(0, 0)$ can be established by using the result in the note below. (A slight reediting of the arguments leading to (iv) would make it possible to restrict our attention here to right derivatives of F and G at $(0, 0)$.)

Note 15   In connection with the result in (150) note the following fact: Give $\hat{u}(t)$ perturbations u_i on distinct intervals $A_i(s_i)$, $i=1,\ldots,i^*$, where $A_i(s_i)=[\tau_i, \tau_i+s_i)$ if $s_i>0$, $A_i(s_i)=(\tau_i+s_i, \tau_i]$ if $s_i<0$. The value of the criterion functional for these perturbations of $\hat{u}(t)$ will depend on $s_1,\ldots,s_{i^*}$. Denoting this value $W(s_1,\ldots,s_{i^*})$, the partial derivatives $\partial W^\pm(0,\ldots,0)/\partial s_i$ are given by (150) for $u=u_i$, $\tau=\tau_i$. Moreover,

$$W(s_1,\ldots,s_{i^*})-W(0,\ldots,0)-\sum_{s_i\geqslant 0} s_i\frac{\partial W_i^+(0,\ldots,0)}{\partial s_i}$$

$$-\sum_{s_i<0} s_i\frac{\partial W_i^-(0,\ldots,0)}{\partial s_i}$$

is of the second order in $s_1,\ldots,s_{i^*}$.

Exercise 3.6.1   (Illustration of (150).). Examine the problem in exercise 2.3.3. Let $\tau\in(0, T)$ and put $s_r(t)=1$ in $[0, \tau+r]$, $s_r(t)=0$ in $(\tau+r, T]$. Find the corresponding state variable $k_r(t)$, and adjoint variable $p(t)$ corresponding to $\hat{s}(t) = s_0(t)$. Compute the corresponding value of the criterion functional, $W(r)$, and prove that $W'(r) = k_0 e^{\tau + r}(T - \tau - r - 1)$. Verify (150) for this case.

Exercise 3.6.2 (Lancaster 1973). With reference to example 10, suppose we maximize $W + C$ subject to (154). Show that by introducing $s(t) = v(t)(1 - u(t))$ as a new control function, the problem reduces to the following problem (see exercise 2.3.3)

$$\max \int_0^T (1 - s(t))K(t)dt \qquad (T > 1) \tag{166}$$

$$\dot{K}(t) = s(t)K(t), \qquad K(0) = K_0 > 0, \qquad K(T) \text{ free} \tag{167}$$

$$s(t) \in [0, 1 - a] \tag{168}$$

(i) Prove that the optimal control is $s^*(t) = 1 - a$ on $[0, T - 1]$, $s^*(t) = 0$ on $(T - 1, T]$. Show that the corresponding value of the integral in (166) is $A = K_0[e^{(1-a)(T-1)} - a]/(1 - a)$.

(ii) Show that when $b \geqslant 1/2$ the game solution in example 10 gives $W + C$ the value

$$B = \left(\frac{a}{1-a} + \frac{1}{1-b}\right)K_0 e^{(1-a)t'} - \frac{a}{1-a}K_0.$$

(iii) Prove that $A > B$.
   (Hint: $e^x > 1 + x$ for $x > 0$.) Thus cooperation is "better" than non-cooperation in this case.

Exercise 3.6.3 Consider again example 10 assuming $b < 1/2$, so that $t'' < t'$. Show that if one uses $u(t) = a$ even on an interval $(t'', t'' + s)$, the effect is the same as assuming that $v = 1$ is replaced by $v' = (1 - a)/(1 - b)$ on $(t'', t'' + s)$ in (154). $v' > 1$, so v' is outside of $[0, 1]$. Consider problem (156) and calculate $dV^+(0)/ds$ by applying (151).

7. Infinite horizons

One of the main areas in economics for the application of variational methods is growth theory. When constructing optimal growth models, the problem of choosing the length of the planning period arises. If a finite horizon model is specified, the problem of deciding the capital stock one wants to leave at the end of the planning period is crucial. It is especially important for those living after the end of the period since the terminal capital stock limits those consumption paths possible in the post-planning period. Because there is this problem of stipulating the terminal capital stock, and because the process of capital accumulation for the economy as a whole has no natural stopping date in the foreseeable future, it has become customary in optimal growth models to introduce the fiction that the planning period is infinite.

Since the world is predicted to be of a finite duration, this choice of an infinite horizon is also somewhat objectionable. In fact, in economic models it makes sense only provided what happens in the very distant future has almost no influence on the optimal solution for the period in which one is interested.

The choice of an infinite horizon usually simplifies formulas and conclusions. On the other hand some new mathematical problems arise and must be sorted out.

We take as our starting point a system whose evolution for $t \geq t_0$ is determined by

$$\dot{x}(t) = f(x(t), u(t), t), \qquad x(t_0) = x^0 \tag{169}$$

and

$$u(t) \in U, \ U \text{ a fixed subset of } R^r. \tag{170}$$

The control function $u(t)$ is piecewise continuous on $[t_0, \infty)$. Moreover, there might be some terminal conditions on $x(t)$ as $t \to \infty$. A simple one is the following

$$\lim_{t \to \infty} x(t) = x^1 \qquad (x^1 \text{ fixed vector in } R^n). \tag{171}$$

A pair $(x(t), u(t))$ is *admissible* if it satisfies (169) and (170) for all $t \geq t_0$, and in addition, if the terminal conditions (if any) are satisfied. When (171) is imposed, the requirement for admissibility is that the limit exists and is equal to x^1.

One of the problems in connection with infinite horizon models is the choice of optimality criterion. Let us consider several possibilities.

As a direct generalization of the finite horizon case, we might search for an admissible pair which maximizes the improper integral

$$\int_{t_0}^{\infty} f_0(x(t), u(t), t)\mathrm{d}t. \tag{172}$$

For this criterion to make sense the integral in (172) must converge for all admissible pairs. For example, if $f_0(x, u, t) = U(x, u)\mathrm{e}^{-at}$, with a as a positive constant, and if for some constant M, $|U(x, u)| \leqslant M$ for all (x, u), then the integral in (172) converges for all admissible pairs.

The simplest infinite horizon problem to discuss is $(169) \rightarrow (171)$ with maximization of (172) as the optimality criterion. Such problems can be solved in a manner similar to the corresponding finite horizon problems. In fact, the Maximum Principle is fully valid for this particular infinite horizon problem (see theorem 12), and the Mangasarian and Arrow sufficiency theorems (theorems 2.4 and 2.5) carry over in an obvious sense.

On the basis of these results it is quite natural to expect that all results for finite horizon problems can be carried over to the infinite horizon case.

For instance, if (171) is replaced by the assumption that there is no restrictions on $x(t)$ as t tends to infinity, we might expect that the correct transversality condition would be that $p(t) \rightarrow 0$ as $t \rightarrow \infty$. A well-known counter example due to Halkin shows that this conclusion is not generally valid. (See example 14 below.) In general, without imposing restrictions on the functions involved, we cannot expect that finite horizon transversality conditions carry over to the infinite horizon case.

We are, in fact, faced with the following two problems:

(A) What are reasonable optimality criteria when the integral in (172) does not converge for all admissible pairs?
(B) What are the correct transversality conditions?

Optimality criteria

Suppose the integral in (172) does not converge for some admissible path. Then maximization of (172) does not make sense. We want to discuss four alternative optimality criteria presented in an increasing order of strictness.

(I) An admissible pair $(x^*(t), u^*(t))$ is *piecewise optimal* (PW-optimal) if for every $T \geqslant t_0$, the restriction of $(x^*(t), u^*(t))$ to $[t_0, T]$ is optimal in the corresponding fixed horizon problem with $x(T) = x^*(T)$ as the terminal condition and with maximization at $\int_0^T f_0(x, u, t)\mathrm{d}t$ as the optimality criterion.

Let $(x(t), u(t))$ be any admissible pair and $(x^*(t), u^*(t))$ the pair we want to test for optimality. Define

$$D(t) = \int_{t_0}^{t} f_0(x^*(\tau), u^*(\tau), \tau) d\tau - \int_{t_0}^{t} f_0(x(\tau), u(\tau), \tau) d\tau. \tag{173}$$

Then:

(II) $(x^*(t), u^*(t))$ is *optimal according to the sporadically catching up criterion* (SCU-optimal) if

$$\overline{\lim_{t \to \infty}} \, D(t) \geq 0 \text{ for all admissible pairs } (x(t), u(t)). \tag{174}$$

According to the definition of $\overline{\lim}$ (see (B.10)), the inequality in (174) is satisfied provided for every $\varepsilon > 0$ and every t' there exist some $t \geq t'$ such that $D(t) \geq -\varepsilon$.

(III) $(x^*(t), u^*(t))$ is *optimal according to the catching up criterion* (CU-optimal) if

$$\underline{\lim_{t \to \infty}} \, D(t) \geq 0 \text{ for all admissible pairs } (x(t), u(t)). \tag{175}$$

According to the definition of $\underline{\lim}$ (see (B.9)), the inequality in (175) is satisfied provided for every $\varepsilon > 0$ there exists a t' such that if $t \geq t'$, then $D(t) \geq -\varepsilon$.

(IV) $(x^*(t), u^*(t))$ is *optimal according to the overtaking criterion* (OT-optimal) if

there exists a number t' such that $D(t) \geq 0$ for all $t \geq t'$. $\qquad$ (176)

We have the following chain of implications:

$$\text{OT-optimality} \Rightarrow^{(1)} \quad \text{CU-optimality} \Rightarrow^{(2)} \quad \text{SCU-optimality} \Rightarrow^{(3)}$$
$$\text{PW-optimality} \tag{177}$$

(1) is obvious from the definitions, (2) derives from the general property $\overline{\lim} \geq \underline{\lim}$, while (3) is proved as follows:

Suppose (x^*, u^*) is SCU-optimal, but not PW-optimal. By the latter assumption there exists a $T \geq t_0$ and an admissible pair $(\hat{x}, \hat{u})$ defined on $[t_0, T]$ with $\hat{x}(T) = x^*(T)$ such that

$$\varepsilon = \int_{t_0}^{T} f_0(\hat{x}(t), \hat{u}(t), t)\mathrm{d}t - \int_{t_0}^{T} f_0(x^*(t), u^*(t), t)\mathrm{d}t > 0.$$

Define the admissible pair $(\tilde{x}(t), \tilde{u}(t))$ as $(\hat{x}(t), \hat{u}(t))$ for $t \in [t_0, T]$ and as $(x^*(t), u^*(t))$ for $t \in (T, \infty)$. Since (x^*, u^*) is SCU-optimal and $(\tilde{x}(t), \tilde{u}(t))$ is admissible, associated with the positive number $\frac{1}{2}\varepsilon$, there exists a $t' \geq T$ such that

$$D(t') = \int_{t_0}^{t'} f_0(x^*(t), u^*(t), t)\mathrm{d}t - \int_{t_0}^{t'} f_0(\tilde{x}(t), \tilde{u}(t), t)\mathrm{d}t \geq -\tfrac{1}{2}\varepsilon. \tag{178}$$

Notice that by the definition of $(\tilde{x}(t), \tilde{u}(t))$, $D(t') = D(T) = -\varepsilon$, contradicting (178). Hence the assumption that (x^*, u^*) is not PW-optimal must be rejected.

We have now established the logical relationships between the four different optimality criteria. Note that the simplest criteria in which we required the integral in (172) to converge for every admissible pair, is a special case of the CU and of the SCU criteria, since in case of convergence, $\lim D(t) = \overline{\lim} D(t) = \underline{\lim} D(t)$ as $t \to \infty$. The choice of optimality criterion depends, of course, on the model we study and on the preferences we have with regard to optimality.

Terminal conditions

The simple terminal condition $x(t) \to x^1$ as $t \to \infty$ was already mentioned. An alternative terminal condition is the requirement that $x(t)$ tends to a limit $\geq x^1$ as t tends to ∞. With this terminal condition, however, paths with some components tending to ∞ or with some components exhibiting undamped cyclical movements will automatically be inadmissible since the appropriate limits do not exist. Among the several possible alternative terminal conditions we shall consider, in particular, the following one:

$$\lim_{t \to \infty} x_i(t) \quad \text{exists and is equal to } x_i^1, \qquad i = 1, \ldots, l \tag{179a}$$

$$\varliminf_{t \to \infty} x_i(t) \geq x_i^1, \qquad i = l+1, \ldots, m \tag{179b}$$

$$\text{no conditions on } x_i(t) \text{ when } t \to \infty, \qquad i = m+1, \ldots, n \tag{179c}$$

Necessary conditions

Suppose we find *necessary conditions* for PW-optimality of an admissible pair $(x^*(t), u^*(t))$. If $(x^*(t), u^*(t))$ is PW-optimal, then these conditions must be satisfied. It follows immediately from (177) that the same necessary conditions must also be satisfied for $(x^*(t), u^*(t))$ to be SCU-, CU- or OT-optimal. Hence the following theorem due to Halkin (1974) applies to any of the four types of optimality criteria. (Halkin's proof covers PW-optimality, although his theorem deals with SCU-optimality.)

Theorem 12 (**Necessary conditions. Infinite horizon**) An admissible pair $(x^*(t), u^*(t))$ for problem (169) (170), (179) is piecewise (PW-) optimal only provided all the conditions in the Maximum Principle (theorem 2.2) with the exception of the transversality conditions (2.60), are satisfied. ■

Actually, the theorem is valid for any type of terminal conditions. If the terminal condition (171) is imposed, then theorem 12 contains enough information to single out one or a few candidates for optimality.

If (179) is the terminal condition, theorem 12 does not contain enough information. As we have indicated previously, the "natural" transversality conditions $\lim p_i(t) \geq 0$, $i = l+1, \ldots, m$, $\lim p_i(t) = 0$, $i = m+1, \ldots, n$ are *not* necessary conditions for optimality in the general case. In section 9 we shall see that the conditions $\lim p_i(t) = 0$, $i = m+1, \ldots, n$ are necessary provided certain growth conditions are imposed on the system. (See note 22.) A full set of transversality conditions is given in theorem 16.

Sufficient conditions

The following sufficiency result is a direct generalization of the corresponding theorem for finite horizon problems. We restrict our attention to CU-optimality since it seems to be the most frequently used optimality criterion in the literature. (See note 18 for alternatives.)

Theorem   13 (The   Mangasarian   sufficiency   theorem.   Infinite horizon) Consider the system (169), (170), (179) with U convex, and the CU-optimality criterion (175), (173). Assume (in addition to (2.6)) that the partial derivatives of $f_0, f_1, \ldots, f_n$ w.r.t. u are continuous. Suppose $(x^*(t), u^*(t))$ is an admissible pair and that there exists a continuous function $p(t)$

such that for $p_0 = 1$, and for all $t \geqslant t_0$,

$u^*(t)$ maximizes $H(x^*(t), u, p(t), t)$ for $u \in U$. $\qquad$ (180)

$$\dot{p}(t) = -\frac{\partial H^*}{\partial x}, \text{ v.e.} \qquad (181)$$

$H(x, u, p(t), t)$ is concave in (x, u) for each $t \geqslant t_0$. $\qquad$ (182)

$\lim\limits_{t \to \infty} p(t) \cdot (x(t) - x^*(t)) \geqslant 0$ for all admissible $x(t)$. $\qquad$ (183)

Then $(x^*(t), u^*(t))$ is CU-optimal. With strict concavity imposed in (182), $(x^*(t), u^*(t))$ is unique. ∎

Proof Using (2.121) with $t_1 = t$ and $\varDelta = D(t)$, where $D(t)$ is given in (173), we obtain

$$D(t) \geqslant p(t) \cdot (x(t) - x^*(t)). \qquad (184)$$

By taking the $\underline{\lim}$ on both sides of (184), the theorem follows. Uniqueness is proved in a similar way as in the proof of theorem 2.4. ∎

Note 16 Condition (183) is implied if the following conditions are valid for all admissible $x(t)$:

$\underline{\lim}\limits_{t \to \infty} p_i(t)(x_i^1 - x_i^*(t)) \geqslant 0 \qquad\qquad i = 1, \ldots, m \quad$ (185(i))

There exists a constant M such that
$|p_i(t)| \leqslant M$ for all t $\qquad\qquad\qquad i = 1, \ldots, m \quad$ (185(ii))

Either there exists a number t' such that
for all $t \geqslant t'$, $0 \leqslant p_i(t)$, or there exists a
number P such that $|x_i(t)| < P$ for all t,
and $\lim\limits_{t \to \infty} p_i(t) \geqslant 0$ $\qquad\qquad\qquad i = l+1, \ldots, m \quad$ (185(iii))

There exists a number Q such that
$|x_i(t)| < Q$ for all t and $\lim\limits_{t \to \infty} p_i(t) = 0$ $\qquad\qquad i = m+1, \ldots, n \quad$ (185(iv))

The proof that (185) ⇒ (183) is based on the splitting of the scalar product in (183) as in (97) of Seierstad and Sydsæter (1977). None of the conditions in (185) can be dropped, contrary to claims implicitly or explicitly made in other economic literature, see e.g. Arrow and Kurz (1970), p. 46 and exercise 3.8.2. Note that (183) does not imply (185), as is clear from example 14.

Note 17 Let us apply theorem 13 to the calculus of variation problem

$$\text{CU-opt} \int_{t_0}^{\infty} F(t, x, \dot{x}) dt, \qquad x(t_0) = x^0 \text{ and (179)} \tag{186}$$

with $x = (x_1, \ldots, x_n)$. Assume that $F(t, x, \dot{x})$ is a C^2-function and concave in $(x, \dot{x})$. If $x^*(t)$ is a solution of the associated Euler equation (1.61) on $[t_0, \infty)$, and if (183) or (185) is satisfied with $p(t) = -\partial F(t, x^*(t), \dot{x}^*(t))/\partial \dot{x}$, then $x^*(t)$ is CU-optimal for problem (186).

If we add to the requirements for admissibility in (186) that $h(t, x, \dot{x}) > 0$ for all t, and h is quasiconcave in $(x, \dot{x})$ for each t, then F needs to be concave only for those $(x, \dot{x})$ where $h(t, x, \dot{x}) > 0$. (See (1.23).) For an application, see example 11.

The next result generalizes the Arrow sufficiency theorem 2.5 (using also the result in note 2.10) to infinite horizon problems.

Theorem 14 (**The Arrow sufficiency theorem. Infinite horizon**) Consider the system: (169), (170), and (179) with the CU-optimality criterion (175), (173), and with the additional admissibility requirement that $x(t)$ belongs to a given convex set $A(t)$ for each $t \geq t_0$. Suppose $(x^*(t), u^*(t))$ is an admissible pair for this problem such that $x^*(t)$ belongs to the interior of $A(t)$ for each $t \geq t_0$. Assume that there exists a continuous function $p(t)$ such that for $p_0 = 1$ and all $t \geq t_0$, (180) and (181) are satisfied. Moreover, assume that

$\hat{H}(x, p(t), t) = \max_{u \in U} H(x, u, p(t), t)$ exists and is

concave in x for $x \in A(t)$ and each $t \geq t_0$ $\tag{187}$

and that for all admissible $x(t)$,

$$\lim_{t \to \infty} p(t) \cdot (x(t) - x^*(t)) \geq 0. \tag{188}$$

Then $(x^*(t), u^*(t))$ is CU-optimal. ∎

For a proof of this result notice that in Chapter 2, section 6 the inequality (2.121) was proved also in the Arrow case. Hence (184) holds in the present case. Of course, condition (185) again implies (188).

Note 18 Theorems 13 and 14 provide sufficient conditions for CU-optimality. In Seierstad and Sydsæter (1977), section 9, similar results are obtained for OT- and SCU-optimality.

Note 19 Theorems 12–14 are valid also for the weaker regularity conditions of note 2.6. If admissibility sets $A(t)$ are introduced as in notes 2.8 and 2.11, the regularity conditions for theorems 13 and 14 can be as described in note 2.12.

An existence theorem

We end this section by stating an existence result for infinite horizon problems.

Theorem 15 (Existence. Infinite horizon)[26]* Consider the problem (169), (170), (172) and (179) assuming merely that $f_0(x, u, t)$ and $f(x, u, t)$ are continuous. Suppose that U is closed and bounded and that there exist piecewise continuous functions $\phi_i(t) \geqslant 0$, $i = 0, \ldots, m$ with $\int_{t_0}^{\infty} \phi_i(t) \mathrm{d}t < \infty$ such that

$$f_i(x(t), u(t), t) \leqslant \phi_i(t), \qquad i = 0 \text{ and } i = l+1, \ldots, m,$$
$$|f_i(x(t), u(t), t)| \leqslant \phi_i(t), \qquad i = 1, \ldots, l \tag{189}$$

for all admissible pairs $(x(t), u(t))$ and all $t \geqslant t_0$. Suppose further that there exist piecewise continuous, non-negative functions $a(t)$ and $b(t)$ such that

$$\|f(x, u, t)\| \leqslant a(t) \cdot \|x\| + b(t) \qquad \text{for all } (x, u, t). \tag{190}$$

Assume finally that the set

$$N(x, U, t) = \{(f_0(x, u, t) + \gamma, \ f(x, u, t)) : u \in U, \gamma \leqslant 0\} \tag{191}$$

is convex for all (x, t). Then the existence of an admissible pair in the problem implies the existence of an optimal pair $(x^*(t), u^*(t))$. (Here $u^*(t)$ might only be measurable.) ∎

In theorem 15 the criterion takes values in $[-\infty, \infty)$. Either all admissible pairs have criterion values equal to $-\infty$, or for some admissible pair the value is $> -\infty$, implying that the optimal value is also $> -\infty$.

8. Examples of infinite horizon problems

In this section we shall study several examples illustrating the theory in section 7. We examine first a calculus of variation problem.

---

[26]See Baum (1976) for essentially this result.

Example 11   Consider the problem (using the catching up criterion)

$$\text{CU-opt} \int_0^\infty \frac{1}{1-v}(bK - \dot K)^{1-v} e^{-\rho t}\, dt,$$

$$K(0)=K_0, \ \lim_{t\to\infty} K(t)\geqslant 0, \quad bK(t)-\dot K(t)>0, \tag{192}$$

assuming that $v\in(0,1)$, $b>0$, $K_0>0$ and $b-\rho<bv$.

We shall solve this problem by relying on note 17.

The function $h(t,K,\dot K)=bK-\dot K$ is linear and hence quasiconcave. The integrand $F(t,K,\dot K)$ is concave in $(K,\dot K)$. Moreover, $F'_{\dot K}= -e^{-\rho t}(bK-\dot K)^{-v}$. Our candidate for optimality, $K^*(t)$, satisfies the Euler equation and $K(0)=K_0$. Hence (see exercise 1.2.5):

$$K^*(t)=(K_0 - A)e^{at} + Ae^{bt} \qquad \text{with } a=\frac{b-\rho}{v}. \tag{193}$$

Admissibility requires $bK^* - \dot K^* = (b-a)(K_0 - A)e^{at}>0$, i.e. $K_0>A$ and $\lim_{t\to\infty} K^*(t)\geqslant 0$, so $A\geqslant 0$. Carrying out some algebra, we find that

$$p(t)= -\frac{\partial F^*}{\partial \dot K} = e^{-bt}[(K_0 - A)(b-a)]^{-v}. \tag{194}$$

Condition (185) is satisfied provided

$$\lim_{t\to\infty} (-p(t)K^*(t))$$

$$= \lim_{t\to\infty} -[(K_0 - A)(b-a)]^{-v}[A+(K_0 - A)e^{-(b-a)t}]\geqslant 0 \tag{195}$$

and provided there exist numbers t' and N such that for $t>t'$,

$$0\leqslant p(t)\leqslant N, \quad \text{that is } 0\leqslant e^{-bt}[(K_0 - A)(b-a)]^{-v}\leqslant N. \tag{196}$$

Since $b>a$, the limit in (195) exists and is 0 if $A=0$. For $A=0$, (196) is also satisfied, with $N=[K_0(b-a)]^{-v}$. We conclude from the result in note 17 that

$$K^*(t)=K_0\, e^{at}, \qquad a=(b-\rho)/v, \tag{197}$$

solves problem (192).

Example 12 (**Resource extraction with waste**) Consider the problem

$$\text{CU-opt} \int_0^\infty [f(u(t)) - ay(t) - qu(t)]e^{-rt}\,dt, \qquad u(t) \in [0,b] \tag{198}$$

$$\dot{x}(t) = -u(t), x(0) = x_0 > 0, \ \lim_{t \to \infty} x(t) \geqslant 0, \tag{199}$$

$$\dot{y}(t) = cf(u(t)), \ y(0) = y_0 \geqslant 0, \ y(\infty) \text{ free}. \tag{200}$$

Assume that f is a C^2-function, $f \geqslant 0$, $f' > 0$, $f'' < 0$, $d = 1 - ac/r > 0$, $df'(b) < q$ and a, b, c, r positive constants. (*Suggested economic interpretation:* $x(t)$ is the amount of a natural resource left in the ground, $u(t)$ is the rate of extraction. By means of the resource, a good is produced at the rate $f(u(t))$; q is the unit cost of production. $y(t)$ is the stock of a non-decaying waste which accumulates at the rate $cf(u(t))$. The produced good has price 1. The unit cost of producing the resource is q. The cost of eliminating the negative effects of the accumulated waste is ay.)

We solve the problem by finding a triple $(x^*(t), y^*(t), u^*(t))$ which satisfies all conditions in theorem 13. With $p_0 = 1$, the Hamiltonian is

$$H = (f(u) - ay - qu)e^{-rt} + p_1(-u) + p_2\,cf(u)$$

$$= (e^{-rt} + p_2 c)f(u) - (qe^{-rt} + p_1)u - aye^{-rt} \tag{201}$$

and

$$\dot{p}_1 = -\frac{\partial H^*}{\partial x} = 0, \qquad p_1(t) = \bar{p}_1 \text{ (constant)} \tag{202}$$

$$\dot{p}_2 = -\frac{\partial H^*}{\partial y} = ae^{-rt}, \qquad p_2(t) = -\frac{a}{r}e^{-rt} + K \qquad (K \text{ constant}) \tag{203}$$

We guess that $\lim_{t \to \infty} p_2(t) = K = 0$ and that $\bar{p}_1 \geqslant 0$. With this guess $H(x, y, u, p_1(t), p_2(t), t) = e^{-rt}\,(1 - ac/r)f(u) - (qe^{-rt} + \bar{p}_1)u - aye^{-rt}$ is concave in (x, y, u), so condition (182) would be satisfied. For $u \in [0, b]$, the function $u^*(t)$ must maximize

$$g(u) = e^{-rt}(1 - ac/r)f(u) - (qe^{-rt} + \bar{p}_1)u = (df(u) - qu)e^{-rt} - \bar{p}_1 u. \tag{204}$$

Here $g'(u) = (df'(u) - q)e^{-rt} - \bar{p}_1$, $g''(u) = df''(u)e^{-rt}$. Since $g''(u) < 0$, g is strictly concave and therefore $u^*(t) > 0$ can maximize $g(u)$ for $u \in [0, b]$ only provided $g'(0) > 0$, i.e. $(df'(0) - q)e^{-rt} > \bar{p}_1 \geqslant 0$, which cannot be satisfied unless $df'(0) > q$. We consider two cases:

Case 1. df'(0)≤q.
Then $g'(0)\leq 0$ so $u^*(t)\equiv 0$ is the only suggestion for an optimal control. We find $x^*(t)\equiv x_0$, and $y^*(t)=cf(0)t+y_0\geq 0$. If we put $p_1(t)=0$ and $p_2(t)=-(a/r)e^{-rt}$, all the conditions in theorem 13 are satisfied. In particular (183) reduces to

$$\lim_{t\to\infty}\left(-\frac{a}{r}\right)e^{-rt}(y(t)-y^*(t))\geq 0 \text{ for all admissible } y(t). \tag{205}$$

For an admissible $y(t)$, we obtain from (200),

$$y(t)-y_0=\int_0^t cf(u(\tau))d\tau\leq\int_0^t cf(b)d\tau=cf(b)t,$$

so $y(t)\leq y_0+cf(b)t$, and thus

$$(-a/r)e^{-rt}\cdot y(t)\geq(-a/r)e^{-rt}(y_0+cf(b)t)\to 0 \text{ as } t\to\infty.$$

Since $(a/r)e^{-rt}y^*(t)\geq 0$, (205) follows. We see that in case 1 the unit cost of production is so high that the optimal policy is to extract nothing at all.

Case 2. df'(0)>q.
By our initial assumptions, $g'(b)=(df'(b)-q)e^{-rt}-\bar{p}_1<0$. Hence $u^*(t)=b$ is not possible at any point. For $u^*(t)>0$ to be possible at some point, $u^*(t)$ must solve the equation $g'(u)=0$, i.e.

$$f'(u)=\frac{1}{d}(q+\bar{p}_1 e^{rt}). \tag{206}$$

If $\bar{p}_1=0$, then the solution of (206) is some constant $u=\bar{u}$ which is nonzero since $df'(0)>q$. However, if $u(t)\equiv\bar{u}>0$, then from (199) $x(t)=-\bar{u}t+x_0$, violating the terminal condition in (199). Thus we suggest $\bar{p}_1>0$ which indicates that $\lim_{t\to\infty}x(t)=0$. This implies that $u^*(t)\neq 0$ for some t. Since $f''(u)<0$, we deduce from (206) that $u^*(t)$ is a decreasing function of t on intervals where $u^*(t)\in(0,b)$. Notice that $g'(0)=(df'(0)-q)e^{-rt}-\bar{p}_1$ eventually becomes <0 if t is large enough, whatever the value of $\bar{p}_1>0$. Hence for t sufficiently large, $u^*(t)=0$. We thus suggest that $u^*(t)>0$ is a solution of (206) on some interval $[0,t']$, while $u^*(t)=0$ for $t>t'$. Let $t'(\bar{p}_1)$ be the solution of $(q+\bar{p}_1 e^{rt})/d=f'(0)$, and let $u(t;\bar{p}_1)$ be the solution of (206) on $[0,t'(\bar{p}_1)]$, while $u(t;\bar{p}_1)=0$ for $t>t'(\bar{p}_1)$. If $\bar{p}_1$ is close to 0, we can get $u(t;\bar{p}_1)$ close to a nonzero constant $\bar{u}$ over a large interval $[0,t'(\bar{p}_1)]$, and we more than exhaust the resource. On the other hand, for $\bar{p}_1$ very large, $t'(\bar{p}_1)\simeq 0$.

Consequently, for a value $\bar{p}_1^*$ of $\bar{p}_1$ in between the two extremes, $x(t'(\bar{p}_1^*))=0$. (Formally, we could appeal to the continuity of $x(t'(\bar{p}_1))=x_0-\int_0^{t'(\bar{p}_1)}u(s;\bar{p}_1)ds$ w.r.t. $\bar{p}_1$, and the intermediate value theorem.) The value $\bar{p}_1^*$ fixes the value of $t'(\bar{p}_1^*)$ and defines our candidate $u^*(t)=u(t;\bar{p}_1^*)$ along with the associated $x^*(t)$ and $y^*(t)$. Put $t^*=t'(\bar{p}_1^*)$.

We have now obtained a proposal for a solution with associated adjoint functions. Applying theorem 13 we shall prove optimality.

Using the concavity of H w.r.t. u and the fact $H'_u=0$ for $t\leqslant t^*$ and $H'_u<0$ for $t>t^*$, we see that (180) is satisfied. In this case condition (183) requires that for all admissible $x(t)$ and $y(t)$,

$$\lim_{t\to\infty}\left[\bar{p}_1^*(x(t)-x^*(t))+\left(-\frac{a}{r}e^{-rt}\right)(y(t)-y^*(t))\right]\geqslant 0. \tag{207}$$

Clearly (205) holds for this case, and thus (207) follows if we can prove that $\underline{\lim}_{t\to\infty}\bar{p}_1^*(x(t)-x^*(t))\geqslant 0$ for all admissible $x(t)$. Since $x(t)\geqslant 0$ for all t and $x^*(t)=0$ for $t>t^*$, the required inequality is obvious.

Example 13 **(A turnpike result)** Reconsider the problem in example 2.6. We want to find the solution to the problem if we put $T=\infty$ and if catching up is used as the optimality criterion.

Assuming that $0<K_0<1$, let us see what happens to the solutions in (2.165) and (2.166) of example 2.6 when $T\to\infty$. Notice that $t^*\to\infty$ and $t_*=T-1-\ln\ (K_0e^{T-1}+1)=T-1-\ln\ (K_0e^{T-1}(1+e^{1-T}/K_0))=T-1-\ln K_0-\ln e^{T-1}-\ln (1+e^{1-T}/K_0)=-\ln K_0-\ln (1+e^{1-T}/K_0)\to -\ln K_0$ as $T\to\infty$. For $0\leqslant t\leqslant -\ln K_0$, (2.165) suggests that $u^*(t)=1$, while (2.165) and (2.164) suggest that for $t>-\ln K_0$, $u^*(t)=1/(t+1+\ln K_0)$. From (2.166) we find the associated suggestion for $K^*(t)$: For $0\leqslant t\leqslant -\ln K_0$, $K^*(t)=K_0e^t$, for $t>-\ln K_0$, $K^*(t)=t+1+\ln K_0$. The associated adjoint function is found by passing to the limit as $T\to\infty$ in (2.167): $p(t)=e^{-t}/K_0$ for all $t\geqslant 0$.

For the infinite horizon problem we now have candidates for $(K^*(t), u^*(t))$, and $p(t)$. Let us use theorem 14 to prove optimality. To this end put $A(t)=[0,\ \infty)$ for all t, $p_0=1$, and note that $(K^*(t),u^*(t))$ is admissible. The function $p(t)$ is easily seen to satisfy the differential equation in (2.153). Moreover, the Hamiltonian $\varphi(u)=H(K^*(t),u,p(t),t)=1-\exp(-(1-u)K_0e^t)+u$ for $t\in[0,\ -\ln K_0]$, and $\varphi(u)=1-\exp(-(1-u)(t+1+\ln K_0))+(1/K_0)e^{-t}(t+1+\ln K_0)u$ for $t>-\ln K_0$. The function φ is strictly concave (see comments after equation (2.154)). Since $\varphi'(1)=1-K_0e^t\geqslant 0$ in $[0,\ -\ln K_0]$, $\varphi(u)$ is maximized by $u^*=1$ in this interval. Moreover, for $t>-\ln K_0$, $u^*=1/(t+1+\ln K_0)$ maximizes $\varphi(u)$ since $\varphi'(u^*)=0$. The concavity condition on $\hat{H}$ was verified in example 2.6. It remains to verify the limit condition (188). By integrating the separable equation $\dot{K}=uK$ with $K(0)=K_0$, we derive $K(t)=K_0 \exp \int_0^t u(s)ds$. Hence for any admissible $K(t)$,

$$\lim_{t\to\infty} p(t)K(t)=\lim_{t\to\infty} e^{-t}\exp\left(\int_0^t u(s)ds\right)\geqslant 0$$

and $p(t)K^*(t)=(1/K_0)e^{-t}(t+1+\ln K_0)\to 0$ as $t\to\infty$. It follows that $\lim_{t\to\infty} p(t)(K(t)-K^*(t))\geqslant 0$ for all admissible $K(t)$, therefore theorem 14 secures optimality of $(K^*(t), u^*(t))$.

Let us denote the solution for $K_0<1$ in example 2.6 by $K^*(t,T)$. Now, $t^*\to\infty$ and $t_*\to -\ln K_0$ when $T\to\infty$, and on $(-\ln K_0,\infty)\cap(t_*,T-1)$, $K^*(t, T)$ is close to $K^*(t)=t+1+\ln K_0$. Thus when T is large, $K^*(t, T)$ is close to $t+1+\ln K_0$, most of the time. This is often referred to as a *turnpike property*.

Example 14   (Halkin (1974)) Consider the problem

$$\text{CU-opt} \int_0^\infty u(t)(1-x(t))dt, \qquad u(t)\in[0,1] \tag{208}$$

$$\dot{x}(t)=u(t)(1-x(t)), \qquad x(0)=0, \qquad x(\infty) \text{ free.} \tag{209}$$

The optimal solution can be found directly. Integrating the separable differential equation in (209) with $x(0)=0$, we obtain, $x(t)=1-e^{-F(t)}$, with $F(t)=\int_0^t u(s)ds$. Moreover,

$$\int_0^T u(t)(1-x(t))dt = \int_0^T \dot{x}(t)dt = x(T)=1-e^{-F(T)}.$$

It follows that *any* choice of control $u(t)$ such that $F(t)$ approaches infinity as t approaches infinity, is optimal.

In particular, $u^*(t)=1/2$, $x^*(t)=1-e^{-t/2}$ is optimal. We use theorem 12 to find the associated adjoint function. The Hamiltonian is $H=p_0u(1-x)+pu(1-x)=(p_0+p)(1-x)u$, and the differential equation for p is $\dot{p}=-\partial H/\partial x=(p_0+p)u$. Observe that $u^*(t)=1/2$ can only maximize $(p_0+p(t))(1-x^*(t))u$ for $u\in[0,1]$ provided $p_0+p(t)=0$ for all t. Consequently $p_0=0$ is impossible, since $(p_0, p(t))\neq(0,0)$ for all t. Hence $p_0=1$

and $p(t) \equiv -1$. We conclude, therefore that $\lim_{t\to\infty} p(t) = -1$. Thus the adjoint function does not satisfy the "natural" transversality condition $\lim_{t\to\infty} p(t) = 0$, corresponding to the terminal condition $x(\infty)$ free.

Notice that condition (183) *is* satisfied in the present case:

$$p(t)(x(t) - x^*(t)) = -1(1 - e^{-F(t)} - (1 - e^{-t/2})) = e^{-F(t)} - e^{-t/2},$$

with $F(t) = \int_0^t u(s)ds$, so $\underline{\lim}_{t\to\infty}\ p(t)(x(t) - x^*(t)) \geqslant 0$. (We have here an example showing that (185) is not equivalent to (183).)

In exercise 3.8.3 you are asked to test theorem 14 on this example.

Exercise 3.8.1 Consider the problem

$$\text{CU-opt} \int_0^\infty \frac{1}{1-\delta}(rA(t) + w - \dot{A}(t))^{1-\delta} e^{-\rho t}dt,$$

$$A(0) = A_0, \ \lim_{t\to\infty} A(t) \geqslant -\frac{w}{r},$$

and $rA(t) + w - \dot{A}(t) > 0$. Assume that $0 < r < \rho$. Put $a = (\rho - r)/\delta$ and prove that $A(t) = (A_0 + w/r)e^{-at} - w/r$ solves the problem. (See exercise 1.5.3 for an economic interpretation.)

Exercise 3.8.2 Prove that the following implication is incorrect:
$\lim_{t\to\infty}\ p(t) \geqslant 0$, $\lim_{t\to\infty}\ p(t)x^*(t) = 0$ and $x(t) \geqslant 0$ for all $t \Rightarrow$
$\lim_{t\to\infty} p(t) \cdot (x(t) - x^*(t)) \geqslant 0$.
(Hint: $p(t) = -e^{-t}$, $x(t) = e^t$, $x^*(t) = 1$.)

Exercise 3.8.3 Use theorem 14 to prove that $x^*(t) = 1 - e^{-t/2}$, $u^*(t) = 1/2$ is optimal in the problem in example 14.

9. Further results on infinite horizon problems

We discussed in section 7 the difficulties in obtaining a full set of necessary conditions for infinite horizon optimal control problems. Mechanically adapting transversality conditions from corresponding conditions for the finite horizon case often turns out to be wrong.

The following result however does give a full set of necessary conditions.[27]

Theorem 16 (Necessary transversality conditions. Infinite horizon) Consider the system (169), (170), (179), and assume that $(x^*(t), u^*(t))$ is optimal according to the catching up criterion. Suppose (in addition to the usual regularity condition (2.6)) that $\int_{t_0}^{\infty} |f_i(x^*(t), u^*(t), t)| dt < \infty$, $i = 0, 1, \ldots, m$, and that there exist non-negative numbers A, B, C, a, b, k with $a > 0$ and $b > k$ such that the following "growth conditions" are satisfied for all $t \geqslant t_0$ and all x:

Defining $G_{ij} = \partial f_i(x, u^*(t), t) / \partial x_j$,
for $i = 0, 1, \ldots, m$:

$$|G_{ij}| \leqslant A e^{-at}, \qquad j = 1, \ldots, m, \tag{210(i)}$$

$$|G_{ij}| \leqslant B e^{-bt}, \qquad j = m+1, \ldots, n \tag{210(ii)}$$

for $i = m+1, \ldots, n$:

$$|G_{ij}| \leqslant C e^{kt}, \qquad j = 1, \ldots, m, \tag{211(i)}$$

$$|G_{ij}| \leqslant k, \qquad j = m+1, \ldots, n \tag{211(ii)}$$

Then there exist a number p_0, $p_0 = 0$ or $p_0 = 1$, and a continuous and piecewise continuously differentiable adjoint function $p(t)$, such that for all $t \geqslant t_0$

$$H(x^*(t), u^*(t), p(t), t) \geqslant H(x^*(t), u, p(t), t) \qquad \text{for all } u \in U. \tag{212}$$

Furthermore, consider the differential equation:

$$\dot{p} = -\frac{\partial H(x^*(t), u^*(t), p, t)}{\partial x} \qquad \text{v.e. } t. \tag{213}$$

The function $p(t)$ satisfies the following conditions: There exists a vector $p^* \in R^n$ such that, if $p(t, T)$ is the solution of equation (213) in $[t_0, T]$ with $p(T, T) = p^*$, then

$$p(t) = \lim_{T \to \infty} p(t, T) \qquad \text{(the limit does exist).} \tag{214}$$

---

[27]Seierstad (1977b). Sometimes the following generalization of the theorem is needed. Instead of (210) and (211), assume that $|G_{ij}| \leqslant a_{ij}(t)$ for all (x, t), $i \neq j$, $i = 0, \ldots, n$, $j = 1, \ldots, n$, $a'_{ii}(t) \leqslant G_{ii} \leqslant a_{ii}(t)$ for all (x, t), $i = 1, \ldots, n$, $a'_{ii}(t)$, $a_{ij}(t)$ locally integrable. Assume also that $a_{ij}(t) b^j(t)$ is integrable for $i = 0, \ldots, m$, $j = 1, \ldots, n$, where $b(t) = (b^1(t), \ldots, b^n(t))$ is the solution of $\dot{b}(t) = A(t) b(t)$, $b^j(t_0) = 1$, $j = 1, \ldots, n$, $A(t)$ an $n \times n$ matrix with elements $a_{ij}(t)$.

If we assume that all admissible solutions lie in a subset G of R^{n+1}, then the above conditions need only hold for $(x, t) \in G$.

These results also hold if $u^*(t)$ is merely measurable and not necessarily bounded, (with (212) and (213) holding a.e.).

Moreover, $(p_0, p(t)) \neq (0, 0)$ for all t and p^* satisfies the conditions $(p_0, p^*) \neq (0, 0)$ and

no condition on p_i^*	$i = 1, \ldots, l$	(215a)
$p_i^* \geqslant 0$ ($=0$ if $\overline{\lim_{t \to \infty}} x_i^*(t) > x_i^l$)	$i = l+1, \ldots, m$	(215b)
$p_i^* = 0$	$i = m+1, \ldots, n$ ■	(215c)

Note 20 (a) Conditions (214) and (215) imply that $p^* = \lim_{t \to \infty} p(t)$ exists and satisfies (215). Moreover, $p(t)$ satisfies (213).
(b). For $i = j$ the absolute value sign in (211(ii)) can be dropped, if $G_{ii} \geqslant K$ for some constant K for all (x, t).

Note 21 The conditions (210) and (211) can be replaced by the following condition: There exists a subset D of R^n such that all solutions $x(t)$ of (169) take values in D, and such that (210) and (211) (or its modification in the preceding Note) are valid for all $x \in D$.

Sometimes it can be useful to derive some knowledge of $u^*(t)$ from theorem 12 before trying to show that the growth conditions of theorem 16 are satisfied.

Note 22 For $j \geqslant m+1$ ("$x_j(\infty)$ free"), it follows from the conditions in the theorem that $p_j(t) \to 0$ as $t \to \infty$.

For the infinite horizon, free end problem, where f is not a function of t, $f^0 = e^{-rt} g(x, u)$ (r, a real number), and where admissibility requires the convergence of the criterion functional, Michel (1982) proves that $\bar{H}(t) = H(x^*(t), u^*(t), p(t), t) \to 0$ as $t \to \infty$. This additional necessary condition is useful especially for problems in which conditions (210) and (211) are not satisfied. Michel further shows that the property $\bar{H}(\infty) = 0$ implies $p(\infty) = 0$ when the following two conditions are satisfied: For some numbers $b > 0$ and $c \leqslant 0$, $B(0, b) \subset \text{co} \{ f(x^*(t), u) : u \in U \}$ and $g(x^*(t), u) \geqslant cr$, $u \in U$, for all large t. Within the context of theorem 16, if $\partial f_i(x, u, t) / \partial t, i = 0, 1, \ldots, n$, is continuous and $\partial f_i(x, u^*(t), t) / \partial t$ satisfies (210(ii)) and (211(ii)), then by letting t be a new state variable, one can easily see that $\bar{H}(t) \to 0$ as $t \to \infty$. and $d\bar{H}(t) / dt = \bar{H}_t'$. Normally, however, neither of these two conditions contains essential new information *in this case*.

A sufficiency result closely related to theorem 16 is given in the next theorem.[28]

Theorem 17 (**Sufficient conditions. Infinite horizon**) Consider the system (169), (170) and (179), with the CU-optimality criterion (175), (173). Assume that $\partial f_0/\partial u$ and $\partial f/\partial u$ exist and are continuous, and that

> f_0 and f are non-decreasing in x for each (u, t), and concave in
> (x, u) for each t. $\hspace{6cm}$ (216)

Let $(x^*(t), u^*(t))$ be an admissible pair for which there exists a vector p^* satisfying (215), with the condition $p_i^* \geqslant 0$ extended to $i = 1, \ldots, l$ and with $p^* = p(T, T)$ where $p(t, T)$ is a solution to (213) such that

$$\lim_{T \to \infty} \int_{t_0}^{T} \frac{\partial H(x^*(t), u^*(t), p(t, T), t)}{\partial u} (u^*(t) - u(t)) dt \geqslant 0 \qquad (217)$$

for all admissible controls $u(t)$. Then $(x^*(t), u^*(t))$ is optimal. ∎

Note 23 Condition (217) can be replaced by the following conditions: (1) U is bounded and convex, (2) $\lim_{T \to \infty} p(t, T)$ exists and is equal to $p(t)$ for all t, (3) with $p(t)$ given in (2), $u^*(t)$ satisfies (180) and, finally, (4)

$$\lim_{T \to \infty} \int_{t_0}^{T} \left\| \frac{\partial f(x^*(t), u^*(t), t)}{\partial u} (p(t, T) - p(t)) \right\| dt = 0 \qquad (218)$$

Note 24 For $i = 1, \ldots, n$ condition (216) can be replaced by the condition that $f_i(x, u, t) = g_i(x, u, t) + a_i(t)x_i$, with $a_i(t)$ continuous and g_i satisfying (216). An even weaker condition can replace the monotonicity: $\partial f_i^*/\partial x_j \geqslant 0$ for $i \neq j$, $i = 0, \ldots, n, j = 1, \ldots, n$.

Note 25* Let L be the set of piecewise continuous control functions taking values in U. If $u(t) \in L$, assume that there exists a unique solution $x^u(t)$ of (169). Then condition (216) can be replaced by the condition that the

---

[28]Seierstad (1977a).

mapping from L into $\mathbf{R}^{n+1}$ defined by

$$u(t) \rightarrow \left(\int_0^T f_0(x^u(t), u(t), t) dt, \int_0^T f(x^u(t), u(t), t) dt \right) \tag{219}$$

is concave for all T.

In exercises 3.9.1 and 3.9.2, problems are given for which theorem 17, but not theorem 14 applies.

Example 15 Consider the problem

$$\text{CU-opt} \int_0^\infty (x-u) e^{-t} dt, \qquad \dot{x} = (u-1)x,$$

$$x(0) = x_0 > 0, \ x(\infty) \text{ free}, \ u \in [0,1]. \tag{220}$$

To reduce the number of cases to be discussed we assume $x_0 \neq 1$ and $x_0 \neq 2$.

We want to use theorem 16. Notice that $m=0$, and that $|f_0(x(t), u(t), t)| = |x(t) - u(t)| e^{-t}$ is integrable on $[0, \infty)$ in the sense that $\int_0' |f_0(x(t), u(t), t)| dt$ exists for all admissible pairs $(x(t), u(t))$ since $u(t) \in [0,1]$ and $0 < x(t) \leqslant x_0$ (See (A.20).) Moreover, condition (210) is satisfied (only (ii) applies since $m=0$):

$$\frac{\partial}{\partial x} ((x-u) e^{-t}) = e^{-t}.$$

and (211) as modified in note 20(b) is satisfied since

$$\frac{\partial}{\partial x} ((u-1)x) = u - 1 \leqslant 0 = k \text{ for all } u \in [0, 1].$$

Suppose $(x^*(t), u^*(t))$ solves problem (220). Then, from the properties of p_0 and p^*, we see that $p^* = 0$ by (215) and $p_0 = 1$. Hence the Hamiltonian is

$$H = (x-u) e^{-t} + p(u-1)x = xe^{-t} - px + (px - e^{-t})u, \tag{221}$$

and the adjoint function $p(t)$ satisfies the equation

$$\dot{p} = -\frac{\partial H^*}{\partial x} = -e^{-t} + p(1 - u^*(t)). \tag{222}$$

According to the maximum condition, $u^*(t)$ maximizes $H(x^*(t), u, p(t), t)$, i.e.

for each $t \geqslant 0$,

$$u^*(t) \text{ maximizes } (p(t)x^*(t) - e^{-t})u, \qquad u \in [0, 1]. \tag{223}$$

In addition, $p(t, T)$ satisfies (222) on $[0, T]$ with $p(T, T) = p^* = 0$. Hence, applying (A.20), we find for $t \in [0, T]$,

$$p(t, T) = \int_t^T e^{-\tau} \exp\left(-\int_t^\tau (1 - u^*(s))ds\right)d\tau. \tag{224}$$

Since $\tau \in [t, T]$ and $u^*(s) \in [0, 1]$, the integrand in the outer integral is $\leqslant e^{-\tau}$. Thus the outer integral converges when $T \to \infty$, and from (214),

$$p(t) = \lim_{T \to \infty} p(t, T) = \int_t^\infty e^{-\tau} \exp\left(-\int_t^\tau (1 - u^*(s))ds\right)d\tau$$

$$\leqslant \int_t^\infty e^{-\tau}d\tau = e^{-t}. \tag{225}$$

It is easy to see that if $u^*(t) \equiv 0$ on $[t, \infty)$, then $p(t) = \frac{1}{2}e^{-t}$, while $u(t) \equiv 1$ on $[t, \infty)$ implies $p(t) = e^{-t}$.

Let us define for $t \geqslant 0$ (see 223)

$$d(t) = p(t)x^*(t) - e^{-t} \tag{226}$$

From (223) it follows that if $d(t) < 0$, then $u^*(t) = 0$, while $d(t) > 0$ implies $u^*(t) = 1$. Note also (for later use) that from the differential equation for $x^*(t)$ and from (222), we find

$$\dot{d}(t) = \dot{p}(t)x^*(t) + p(t)\dot{x}^*(t) + e^{-t} = e^{-t}(1 - x^*(t)). \tag{227}$$

Suppose $x_0 < 1$. Then $0 < x^*(t) \leqslant x_0 < 1$ for all $t \geqslant 0$ and therefore, since $p(t) \leqslant e^{-t}$, $d(t) < e^{-t} - e^{-t} = 0$. Hence, if $x_0 < 1$, $u^*(t) \equiv 0$ is the only possible optimal control.

Suppose $x_0 > 1$. Assume that $d(0) = p(0)x_0 - 1 \leqslant 0$. Let $t' = \sup\{t : x^*(s) > 1$ for $s \in [0, t]\}$. For $t \in [0, t')$ we see from (227) that $\dot{d}(t) < 0$, and hence $d(t) < 0$ on $(0, t']$. By continuity of $d(t)$, $d(t) < 0$ on $(0, t' + \varepsilon]$ for some $\varepsilon > 0$. Therefore $u^*(t) \equiv 0$ on $(0, t' + \varepsilon]$, and $x^*(t' + \varepsilon) < 1$. Again, for $t \geqslant t' + \varepsilon$, $x^*(t) \leqslant x^*(t' + \varepsilon) < 1$, so by (226), $d(t) < 0$, and $u^*(t) = 0$ for all t. The associated adjoint function is $p(t) = \frac{1}{2}e^{-t}$ so $p(0) = 1/2$. The condition $p(0)x_0 - 1 \leqslant 0$ therefore reduces to

$x_0 \leqslant 2$. We conclude that for $1 < x_0 \leqslant 2$, $u^*(t) \equiv 0$ is a possible optimal control.

Consider next the case $x_0 > 1$ and $d(0) = p(0)x_0 - 1 > 0$. Let $t^* = \sup\{t: d(s) > 0 \text{ on } [0, t]\}$. Then $t^* \leqslant \infty$, while $u^*(t) = 1$ and $x^*(t) = x_0$ on $[0, t^*)$. We claim that $t^* < \infty$ is impossible. Suppose $t^* < \infty$. Then $d(t^*) = 0$ and $x^*(t^*) = x_0 > 1$, so by the same argument as above, $u^*(t) = 0$ on (t^*, ∞). But then $p(t^*) = \frac{1}{2}e^{-t^*}$, so $0 = d(t^*) = \frac{1}{2}e^{-t^*}x_0 - e^{-t^*}$ which implies $x_0 = 2$, a contradiction. We therefore conclude that $t^* = \infty$, so $u^*(t) = 1$ for all t. The associated adjoint function is $p(t) = e^{-t}$ with $p(0) = 1$, so the condition $p(0)x_0 - 1 > 0$ reduces to $x_0 > 1$.

We have now the following suggestions for optimal controls: For $x_0 < 1$, $u^*(t) \equiv 0$; for $1 < x_0 < 2$, either $u^*(t) \equiv 0$ or $u^*(t) \equiv 1$; for $x_0 > 2$, $u^*(t) \equiv 1$. Corresponding to $u^*(t) \equiv 0$ we get $x^*(t) \equiv x_0 e^{-t}$ and

$$\int_0^\infty (x^*(t) - u^*(t))e^{-t}dt = \int_0^\infty x_0 e^{-2t}dt = \frac{1}{2}x_0,$$

while $u^*(t) \equiv 1$ gives $x^*(t) \equiv x_0$ and

$$\int_0^\infty (x^*(t) - u^*(t))e^{-t}dt = \int_0^\infty (x_0 - 1)e^{-t}dt = x_0 - 1.$$

For $x_0 < 2$, then $\frac{1}{2}x_0 > x_0 - 1$. Hence we get the following unique suggestion for an optimal control: For $0 < x_0 < 2$, $x_0 \neq 1$, $u^*(t) \equiv 0$. For $x_0 > 2$, $u^*(t) \equiv 1$.

In this problem none of our sufficiency theorems apply. Notice in particular, that condition (187) is not satisfied since $\hat{H}$ is convex, rather than concave in x. Theorem 15, however, secures the existence of an optimal control. Conditions (189) and (190) are easily verified, and the set $N(x, U, t) = \{((x - u)e^{-t} + \gamma, (u - 1)x): u \in [0, 1], \gamma \leqslant 0\}$ is "an infinite trapezium" for each (x, t), and hence convex. All the conditions in the theorem are satisfied, hence there exists an optimal solution whatever is the value of x_0.

Exercise 3.9.1 Consider the problem

$$\text{CU-opt} \int_0^\infty x dt, \qquad \dot{x} = u,$$

$$x(0) = 0, \qquad x(\infty) \text{ free}, \qquad u \in [0, 1].$$

(i) The solution is clearly $u^*(t)=1$. Show that theorem 17 confirms this result, while theorem 14 fails.

(ii) Show that theorem 12 gives $p_0=0$, $p(t)=c$ (c arbitrary constant $\geqslant 0$) for $u(t)=u^*(t)=1$. There are an infinite number of other candidates.

Exercise 3.9.2 Consider the problem

$$\text{CU-opt} \int_0^\infty -x \, dt, \qquad \dot{x}=-x+u,$$

$$x(0)=0, \qquad x(\infty) \text{ free}, \qquad u \in [0,1].$$

The solution is clearly $u^*(t)\equiv 0$. Show that theorem 17 confirms this result. (Use note 24) Show that (188) in theorem 14 is not satisfied.

Exercise 3.9.3 Consider example 14 with $u^*(t)=1/2$, $x^*(t)=1-e^{-t/2}$. Show that the growth conditions in theorem 16 fail. Find the solution $p(t, T)$ of (213) with $p(T, T)=p^*=0$, and show that $\lim_{T\to\infty} p(t, T)= -1 = p(t)$. (In this problem we get the correct adjoint function using (214) and (215), even if the growth conditions fail.)

10. Phase-space analysis

In economic applications of optimal control theory we often study problems in which the differential equations obtained from the Maximum Principle are not explicitly solvable. All the same, in a number of these cases we are able to obtain qualitative characterizations of the optimal solution.

There is a geometric method, called *phase-space analysis*, which is frequently used to shed light on the structure of the solutions to autonomous systems of differential equations. The recent literature on optimal growth theory abounds with examples illustrating this technique.

Usually, phase-space analysis is complicated for systems of more than two differential equations. We shall therefore restrict our attention to autonomous systems of differential equations in the plane.

Autonomous systems in the plane

Let F and G be C^1-functions and consider the autonomous system of differential equations,

$$\dot{x} = F(x, p), \qquad \dot{p} = G(x, p). \tag{228}$$

We call the xp-plane the *phase-space* for (228). A solution $(x(t), p(t))$ of (228) will trace out a solution curve in the phase-space, with a velocity vector $(\dot{x}(t), \dot{p}(t))$ which is tangent to the solution curve. Note that the direction of this tangent vector at any point depends only on that point. (For non-autonomous systems such as $\dot{x} = F(x, p, t)$, $\dot{p} = G(x, p, t)$, the tangent vector depends also on the time at which the solution curve passes through the point.)

Phase space analysis of the system (228) is concerned with studying the behaviour of solution paths on the phase-space, based on qualitative properties of the functions F and G.

In order to investigate (228), we first draw the two curves (see fig. 3.2).

$$\dot{x} = F(x, p) = 0, \qquad \dot{p} = G(x, p) = 0. \tag{229}$$

The curve $F(x, p) = 0$ graphs the condition $\dot{x} = 0$, which says that x is not changing at any point on $F(x, p) = 0$. Likewise, $G(x, p) = 0$ graphs the condition $\dot{p} = 0$, so p is not changing at any point on $G(x, p) = 0$. An intersection $E = (\bar{x}, \bar{p})$ of these curves, at which neither x nor p is changing, is *an equilibrium point*, and $x(t) \equiv \bar{x}, p(t) \equiv \bar{p}$ is a solution of (228).

The curves $F(x, p) = 0$ and $G(x, p) = 0$ might divide the xp-plane in quite complicated ways. Let us consider, in particular, the case illustrated in fig. 3.2. Here $F(x, p) > 0$ above the curve $\dot{x} = 0$, $F(x, p) < 0$ below that curve. With respect to the curve $\dot{p} = 0$, we assume that $G(x, p)$ is < 0 to the left of the curve, > 0 to the right. Then the signs of the derivatives of a solution $x = x(t)$, $p = p(t)$ to (228) will be as indicated in the diagram. For instance, for an integral curve passing through the point P in region (II) $\dot{x} > 0$, while $\dot{p} < 0$. The fact that x is increasing and p is decreasing at P is indicated by arrows in the diagram. Corresponding movements in the other three regions are also shown in fig. 3.2.

In fig. 3.3. we have indicated some integral curves which move in accordance with the sign distributions for $\dot{x}$ and $\dot{p}$. Observe how some of the curves in regions (II) and (IV) show convergence towards E, while the

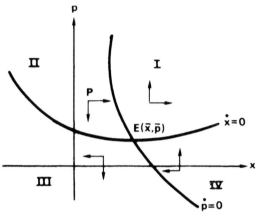

Fig. 3.2

integral curves indicated in regions (I) and (III) diverge (move away) from E.

The phase diagram in fig. 3.2 indicates that the equilibrium point E is of a special type called a *saddle point*.

Actually, the precise nature of an equilibrium point for (228) cannot be conclusively decided on the basis of a phase diagram, but such a diagram strongly suggests the character of the equilibrium point. Precise characterizations can be attained by using results available in the differential equation literature. The behaviour of the integral curves near an equilibrium point

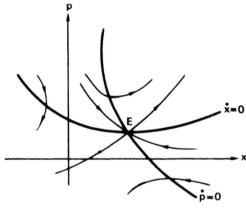

Fig. 3.3

depends on the characteristic roots (the eigenvalues) of the Jacobian matrix.

$$\begin{bmatrix} F'_1(\bar{x}, \bar{p}), & F'_2(\bar{x}, \bar{p}) \\ G'_1(\bar{x}, \bar{p}), & G'_2(\bar{x}, \bar{p}) \end{bmatrix} \tag{230}$$

We shall consider only one case in detail, the saddle point case.

Theorem 18 (Local Saddle Point Theorem.)[29] Consider the system

$$\dot{x} = F(x, p), \qquad \dot{p} = G(x, p) \tag{231}$$

and assume that F and G are C^1 in a neighbourhood of an equilibrium point $E = (\bar{x}, \bar{p})$. Assume further that the two characteristic roots of the matrix (230) are real and of the opposite signs, or equivalently,

$$(F'_1 - G'_2)^2 + 4F'_2 G'_1 > 0 \tag{232(i)}$$

and

$$F'_1 G'_2 - F'_2 G'_1 < 0 \tag{232(ii)}$$

where all the partial derivatives are evaluated at $E = (\bar{x}, \bar{p})$.

Then there exist exactly two solutions $(x_1(t), p_1(t))$ and $(x_2(t), p_2(t))$ on some interval $[t^0, \infty)$ which converge to $E = (\bar{x}, \bar{p})$. These solutions converge to E from opposite directions and both are tangent to a line through E with the same direction as the characteristic vector corresponding to the negative characteristic root. Such an equilibrium point is called a saddle point. ∎

Note that the trajectories of the solutions guaranteed by theorem 18 could have their starting point very close to $(\bar{x}, \bar{p})$, although their solutions are defined on an infinite time interval. In optimal control problems in which we appeal to theorem 18, we have to establish the existence of a solution starting from some fixed initial point. Thus we need a global version of the theorem.

A global saddle point theorem

In order to obtain a global version of theorem 18 we begin by analyzing the geometric implications of the inequalities in (232).

We shall assume, for simplicity, that

$$F'_2 G'_1 > 0 \qquad \text{at } (\bar{x}, \bar{p}). \tag{233}$$

[29]See e.g. Pontryagin (1962a). Actually, (232(i)) implies that the roots are real; (232(ii)) implies (232(i)).

In economics, optimal control problems with one state variable x and one control variable frequently lead to the study of system (228), where p is an adjoint variable. In a number of such problems, condition (233) *is* satisfied. (Examples are studied below.) In fact, we could assume that (233) is satisfied with $F_2' > 0$, since an easy calculation shows that the case $F_2' < 0$ is taken care of by replacing x with a new state variable $y = -x$.

If (233) is valid, then (232(i)) is necessarily satisfied. (232(ii)) is related to the slopes of the curves $F(x, p) = 0$ and $G(x, p) = 0$ at $(\bar{x}, \bar{p})$. (Recall that $-H_1'(x, p)/H_2'(x, p)$ is the slope of the curve $H(x, p) = 0$.) There are four main cases to be examined, assuming that $F_2' > 0$, $G_1' > 0$. (Note: all partial derivatives are evaluated at $(\bar{x}, \bar{p})$.)

(A) $F_1' < 0$, $G_2' > 0$ $(F_2' > 0$, $G_1' > 0)$.
In this case the curve $\dot{x} = 0$ has a positive slope at $(\bar{x}, \bar{p})$, while $\dot{p} = 0$ has a negative slope. (232) is always satisfied. The situation is graphed in fig. 3.4.

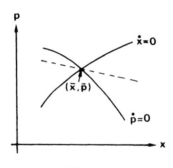

Fig. 3.4

(B) $F_1' < 0$, $G_2' < 0$ $(F_2' > 0$, $G_1' > 0)$.
The curves $\dot{x} = 0$ and $\dot{p} = 0$ both have positives slopes at $(\bar{x}, \bar{p})$. (232(ii)) is satisfied if and only if $-F_1'/F_2' < -G_1'/G_2'$, which means that the slope of the $\dot{p} = 0$ curve is steeper than that of $\dot{x} = 0$. See fig. 3.5.

(C) $F_1' > 0$, $G_2' > 0$ $(F_2' > 0$, $G_1' > 0)$.
The curves $\dot{x} = 0$ and $\dot{p} = 0$ both have negative slopes at $(\bar{x}, \bar{p})$. (232(ii)) is satisfied if and only if $-F_1'/F_2' > -G_1'/G_2'$, which means that the slope of the $\dot{x} = 0$ curve is less steep than the slope of the $\dot{p} = 0$ curve at $(\bar{x}, \bar{p})$. See fig. 3.6.

(D) $F_1' > 0$, $G_2' < 0$ $(F_2' > 0$, $G_1' > 0)$.

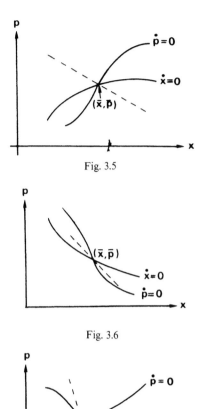

Fig. 3.5

Fig. 3.6

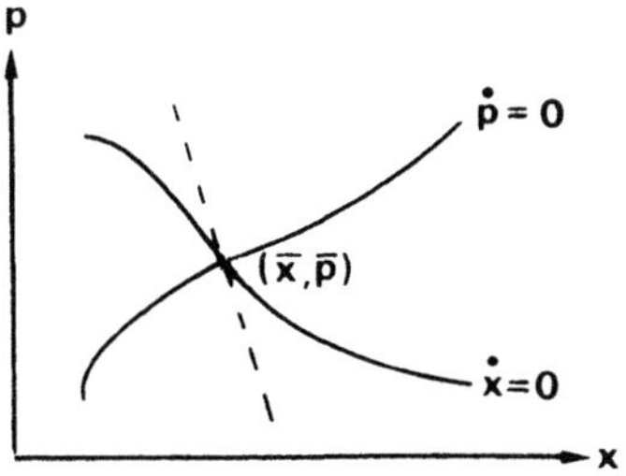

Fig. 3.7

The curve $\dot{x}=0$ has a negative slope, while $\dot{p}=0$ has a positive slope at $(\bar{x}, \bar{p})$. (232) is always satisfied. See fig. 3.7.

Each of the four figures describes the situation in a neighbourhood of the equilibrium point. As we shall see, the dotted lines indicate the direction of the characteristic vector corresponding to the negative characteristic root,

$$\lambda = \tfrac{1}{2}\{F'_1 + G'_2 - [(F'_1 - G'_2)^2 + 4F'_2 G'_1]^{1/2}\}. \tag{234}$$

A characteristic vector $(\tilde{x}, \tilde{p})$ is a vector $\neq (0, 0)$ which satisfies

$$(F_1' - \lambda)\tilde{x} + F_2'\tilde{p} = 0 \tag{235}$$

$$G_1'\tilde{x} + (G_2' - \lambda)\tilde{p} = 0$$

Using (233) we see that $\tilde{x} = 0$ or $\tilde{p} = 0$ is impossible, so we can put $\tilde{x} = 1$. Then from (235),

$$\tilde{p} = -\frac{F_1'}{F_2'} + \lambda, \tag{236a}$$

$$\tilde{p}\left(1 - \frac{\lambda}{G_2'}\right) = -\frac{G_1'}{G_2'}. \tag{236b}$$

Since $\lambda < 0$, we see from (236a) that $\tilde{p} < -F_1'/F_2'$ which shows that the slope of the characteristic vector corresponding to λ is always less than the slope of $F(x, p) = 0$ at $(\tilde{x}, \tilde{p})$.

Consider next the relationship between the slope of $G(x, p) = 0$ at $(\tilde{x}, \tilde{p})$ and the number $\tilde{p}$. If $G_2' > 0$ as in figs. 3.4 and 3.6, it follows from (236) that since $\lambda < 0$, $1 - \lambda/G_2' > 1$, and hence $0 > \tilde{p} > -G_1'/G_2'$. Thus the dotted lines in figs. 3.4 and 3.6 are correctly indicated.

If $G_2' < 0$, suppose first that $F_1' > 0$ which is case D. Then, since $\tilde{p} < -F_1'/F_2'$, we obtain $\tilde{p} < 0$ and $\tilde{p} < -G_1'/G_2'$, so the dotted line in fig. 3.7 is correctly located.

Finally, with $G_2' < 0$ suppose $F_1' < 0$, which is case B. We claim that $\tilde{p} < 0$. Looking at (236b), observe that it suffices to prove $1 - \lambda/G_2' < 0$. Now, $\lambda/G_2' > 0$, so $1 - \lambda/G_2' < 1$. By contradiction, if $1 - \lambda/G_2' > 0$, then by (236b), $\tilde{p} > -G_1'/G_2' > -F_1'/F_2'$, a contradiction. Hence, $1 - \lambda/G_2' \leqslant 0$ with equality not possible since $\tilde{p} \neq 0$ (Use (236b) again.) Thus, $\tilde{p} < 0$ in case B. Since the slope of $\dot{p} = 0$ is positive for this case, the dotted line in fig. 3.5 is also correctly indicated.

For the borderline case $G_2' = 0$, the curve $\dot{p} = 0$ has a vertical tangent at $(\bar{x}, \bar{p})$, while in the case $F_1' = 0$, the curve $\dot{x} = 0$ has a horizontal tangent at $(\bar{x}, \bar{p})$. We may subsume these cases under the cases in figs. 3.6 and 3.4.

We are now ready to prove the following result. It deals with the case $x_0 < \bar{x}$, while the case $x_0 > \bar{x}$ is taken care of in note 27.

Theorem 19 (Global Saddle Point Theorem) Consider the system

$$\dot{x} = F(x, p), \qquad \dot{p} = G(x, p) \tag{237}$$

where F and G are continuous functions being C^1 in a neighbourhood of an equilibrium point $E = (\bar{x}, \bar{p})$, and let $x_0 < \bar{x}$. Assume that $F'_2(\bar{x}, \bar{p}) \cdot G'_1(\bar{x}, \bar{p}) > 0$ and that

$F(x, p) = 0$ has a unique C^1-solution $p = p_1(x)$ for all $x \in [x_0, \bar{x}]$.   (238)

If $S^+ = \{(x, p): x \in [x_0, \bar{x}], p > p_1(x)\}$ and $S^- = \{(x, p): x \in [x_0, \bar{x}], p < p_1(x)\}$, let $S = S^+$ if $F(x, p) > 0$ in S^+ and $S = S^-$ if $F(x, p) > 0$ in S^-. Assume that $(\bar{x}, \bar{p})$ is a unique equilibrium point in $\bar{S}$. Assume, moreover, that (232(ii)) holds at $(\bar{x}, \bar{p})$ and that there exist positive constants D and B such that

$$\left| \frac{G(x, p)}{F(x, p)} \right| \leqslant D|p| \text{ for all } |p| \geqslant B, \qquad (x, p) \in S. \qquad (239)$$

Then for any t_0 there exists a pair of solutions $(x(t), p(t))$ of (237) on $[t_0, \infty)$, such that $(x(t), p(t)) \in S$ for all $t \in [t_0, \infty)$, $x(t_0) = x_0$, and such that $x(t) \to \bar{x}$, and $p(t) \to \bar{p}$ as $t \to \infty$.

Note 26(a)   The solution $(x(t), p(t))$ has the same tangency property as the solutions in theorem 18.
(b)   Assume that $G(x, p) = 0$ has a C^1-solution $p_2(x)$, with $(x, p_2(x)) \in S$ for each $x \in [x_0, \bar{x}]$ and such that $p'_2 > 0$ if $S = S^-$, $p'_2 < 0$ if $S = S^+$. Then condition (239) can be dropped.[30]

Note 27   Theorem 19 and note 26 are also valid if $x_0 > \bar{x}$, provided the interval $[x_0, \bar{x}]$ is replaced by $[\bar{x}, x_0]$ and the inequalities in the definition of S are reversed.

Proof of theorem 19*   As we remarked above, we can assume that $F'_2(\bar{x}, \bar{p}) > 0$, so $S = \{x, p): x \in [x_0, \bar{x}], p > p_1(x)\}$. We argue in terms of the situation in fig. 3.6 (case C) and therefore assume that F'_1, G'_2, F'_2 and G'_1 are all > 0. With obvious modifications the proof applies equally well to cases A, B and D.

   Consider fig. 3.8 which corresponds to fig. 3.6. The dotted line has the same slope as the characteristic vector corresponding to the negative characteristic root. According to theorem 18, there exists a solution $(x(t), p(t))$ to (237) whose graph is tangent to the dotted line at $(\bar{x}, \bar{p})$. Since the dotted line has a slope which is strictly between the slopes of $\dot{x} = 0$ and $\dot{p} = 0$ at $(\bar{x}, \bar{p})$, we can assume that $(x(t), p(t))$ is defined on some interval $[t^0, \infty)$

---

[30]R. Hartl made this observation and provided the reference to Hartman (1964), Chapter VIII.

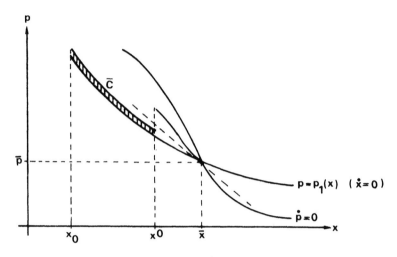

Fig. 3.8

with $x^0 = x(t^0) < \bar{x}$, and $(x(t), p(t)) \in S$ for all $t \geq t^0$. If for some $T \in [t^0, \infty)$, $x(T) = x_0$, the proof is already complete: Since the system is autonomous, we can replace the solution $(x(t), p(t))$ defined on $[T, \infty)$, by $(x(t+T-t_0),$ $p(t+T-t_0))$ which is defined on $[t_0, \infty)$, hence the required extension is found.

If no $T \in [t^0, \infty)$ satisfies $x(T) = x_0$, then $x(t^0) > x_0$, and we proceed as follows:

Define $A = \{(t, x, p): -\infty < t < \infty, x \in (x_0, \bar{x}), p > p_1(x)\}$. Then $(t^0, x(t^0), p(t^0)) \in A$. Let $(x(t), p(t))$ defined on (T, ∞) be some maximal extension of the solution through $(x(t^0), p(t^0))$, subject to the condition that $(t, x(t), p(t)) \in A$, (See theorem A.11.) We want to prove: (1) T is finite, (2) $\lim_{t \to T^+} x(t)$ exists and is equal to x_0, and (3) $\lim_{t \to T^+} p(t)$ exists. With these properties established, the solutions $x(t)$ and $p(t)$ asserted in theorem 19 are obtained by the same translation of time as used above.

Since $\dot{x}(t) > 0$ in A, $x(t)$ is strictly increasing and $x(t) < \bar{x}$ for $t \in (T, \infty)$. Hence $x(t)$ has a continuously differentiable inverse $r(x)$. The function $p(t)$ is clearly bounded from below by the bounded function $p_1(x)$ on $[x_0, \bar{x}]$. We claim that $p(t)$ is bounded from above as well: Define $\check{p}(x) = p(r(x))$ for $x \in (x', x^0]$, where $x' = \lim_{x \to T^+} x(t)$. Then $\check{p}'(x) = (\mathrm{d}p/\mathrm{d}t)/(\mathrm{d}x/\mathrm{d}t) = G(x, \check{p})/F(x, \check{p})$, and $\check{p}(x^0) = p(t^0)$. Assume that $B > |p(t^0)| = |\check{p}(x^0)|$. Let $s(x)$ be the solution of the differential equation $s' = -1 - Ds$, $s(x^0) = B$. We claim that $\check{p}(x) < s(x)$ for all $x \in (x', x^0)$:

Assume by contradiction that $\check{p}(\tilde{x}) \geq s(\tilde{x})$ for some $\tilde{x} \in (x', x^0)$. Then for some $\hat{x} \in (x', x^0)$, $\check{p}(\hat{x}) = s(\hat{x})$. Let $\hat{x}$ be the largest $\hat{x}$ with this property. Define

$\check{x} = \inf \{x : \check{p}(x) \leqslant B\}$. Since $s(x)$ is strictly decreasing, $s(\hat{x}) > B$, and thus $\check{x} > \hat{x}$. On $[\hat{x}, \check{x}]$, $s(x) \geqslant \check{p}(x) > B$, and hence by (239), $\check{p}' \geqslant -D|\check{p}| \geqslant -Ds > -1 - Ds = s'$. Since $s(\hat{x}) = \check{p}(\hat{x})$, and s decreases more rapidly than $\check{p}$, we have $s(\check{x}) < \check{p}(\check{x})$, which contradicts the fact that $s(\hat{x}) \geqslant s(x^0) = B = \check{p}(\check{x})$. We have shown that $\check{p}(x) < s(x)$ in (x', x^0), and $s(x)$ is obviously bounded, so we conclude that $\check{p}(x)$ is bounded above on (x', x^0), say $\check{p}(x) \leqslant B'$.

We prove next that $T > -\infty$. Since $(d/dx)G(\bar{x}, p_1(\bar{x})) > 0$ by (232(ii)) and $G(\bar{x}, p_1(\bar{x})) = 0$, then $G(x, p_1(x)) < 0$ for $x < \bar{x}$ and x close to $\bar{x}$.

Due to the uniqueness of $(\bar{x}, \bar{p})$, $G(x, p_1(x))$ cannot change sign and is thus < 0 in $[x_0, \bar{x})$. By the compactness of $[x_0, x^0]$, for some number a, $0 > a > G(x, p_1(x))$ for all $x \in [x_0, x^0]$. Hence for some number $b > 0$, $a/2 > G(x, p)$ for $x \in [x_0, x^0]$, $p_1(x) \leqslant p \leqslant p_1(x) + b$. Let $c = |a|/2K$, where $K = \max \{|p_1'(|x|)| : x \in [x_0, \bar{x}]\}$. Define moreover $C = \{(x, p) : x \in (x_0, x^0), p_1(x) < p < p_1(x) + b'\}$, where $b' > 0$ is chosen so small that $b' < b$, $|F(x, p)| < c$ for all $(x, p) \in C$ and $(x^0, \check{p}(x^0)) \notin \bar{C}$, the closure of C. We claim that $(x(t), p(t))$ cannot belong to C for any $t \in (T, t^0)$. Assume, by contradiction, that $(x(t'), p(t')) \in C$ for some $t' \in (T, t^0)$. Since C is open, there exists an interval $[t', t'')$, $t'' > t'$ such that $(x(s), p(s)) \in C$ for all $s \in [t', t'')$. Let $[t', t'')$ be a maximal interval with this property. Then $t'' < t^0$ since $(x(t^0), p(t^0)) \notin \bar{C}$. Furthermore, $x_0 < x(t'') < x^0$ since $\dot{x} > 0$ for all $t \in (T, t^0)$. Observe next that $(x(t''), p(t'')) = (x(t''), p_1(x(t'')) + b')$ is impossible: Firstly, $p(t') < p_1(x(t')) + b'$, and secondly, $\dot{p} < (d/dt)(p_1(x(t)) + b') = (d/dt)p_1(x(t)) = p_1'(x(t))\dot{x}(t)$, where the inequality is due to the fact that $|p_1' \cdot \dot{x}| = |p_1'| \cdot |\dot{x}| \leqslant K \cdot c = K \cdot |a|/2K = |a|/2$, and thus $p_1' \cdot \dot{x} \geqslant -|a|/2 = a/2 > \dot{p}$. Hence $p(t)$ does not meet the curve $p_1 = p_1(x(t)) + b'$ for $t = t''$. Finally, by the definition of T, for no $t \in (T, t^0)$ is $p(t) = p_1(x(t))$. In particular, $p(t'') \neq p_1(x(t''))$, and thus $(x(t''), p(t''))$ belongs to the open set C, which contradicts the maximality of $[t, t'')$. We have therefore shown that for no $t \in (T, t^0)$ does $(x(t), p(t))$ belong to C. Hence, for $t \in (T, t^0)$, $(x(t), p(t))$ belongs to the compact set $\{(x, p) : x \in [x_0, x^0], p \in [A', B'], p \geqslant p_1(x) + b'\}$, where A' and B' satisfy $A' \leqslant p_1(x(t)) \leqslant p(t) = \check{p}(x(t)) \leqslant B'$. On this set $F(x, p)$ has a minimum value $d > 0$, so $\dot{x} > d$ for all $t \in (T, t^0)$. Since $x_0 \leqslant x' = \lim_{t \to T^+} x(t)$ and $x(t^0) = x^0$, $T = -\infty$ and $\dot{x}(t) > d > 0$ on (T, t^0), give a contradiction. Hence $T > -\infty$.

Since T is finite, we obtain from (A.49) that

$$\lim_{t \to T^+} d((t, x(t), p(t)), \complement A) = 0 \tag{240}$$

By note A.7, $\lim_{t \to T^+} x(t)$ and $\lim_{t \to T^+} p(t)$ both exist and we denote them by $x(T)$ and $p(T)$. From (240),

$$x(T) = x_0 \tag{241a}$$

or

$$p(T) = p_1(x(T)) \tag{241b}$$

But it is evident that the latter equality is impossible: Since $\dot{p}(T) < [(\mathrm{d}/\mathrm{d}t)$ $(p_1(x(t))]_{t=T} = p_1'(x(T)) \cdot \dot{x}(T) = 0$, it would imply $p(t) < p_1(x(t))$ for $t > T$ and t close to T. This inequality contradicts the fact that $(x(t), p(t)) \in A$. By (241) we conclude that $x(T) = x_0$. ∎

The next example illustrate the usefulness of the global saddle point theorem when solving an optimal growth problem.

Example 16 (**Optimal growth**) Consider the following problem (for related problems see examples 1.1, 2.7, 11):

$$\text{CU-opt} \int_0^\infty \frac{1}{1-v} [C(t)]^{1-v} \mathrm{d}t \tag{242}$$

$$\dot{K}(t) = aK(t) - b(K(t))^2 - C(t), \qquad K(0) = K_0, \lim_{t \to \infty} K(t) \geqslant 0 \tag{243}$$

$$C(t) > 0 \qquad \text{for all } t \geqslant 0, \tag{244}$$

where a, b, v and K_0 are positive constants, $v < 1$, and $K_0 < a/2b$. The function C is the control variable.

We shall ultimately appeal to theorem 13. Thus we assume $p_0 = 1$ and the Hamiltonian is $H = C^{1-v}/(1-v) + p(aK - bK^2 - C)$. From the maximum condition, $H_C' = 0$, i.e. $C^{-v} - p = 0$ and hence $\dot{p} = -vC^{-v-1}\dot{C}$. On the other hand, the differential equation for p is $\dot{p} = -H_K' = -p(a - 2bK)$. Eliminating $\dot{p}$ and substituting $p = C^{-v}$, we obtain $\dot{C} = w(a - 2bK)C$, where $w = 1/v$.

Consider the pair of differential equations:

$$\dot{K} = aK - bK^2 - C = F(K, C)$$
$$\dot{C} = w(a - 2bK)C = G(K, C), \tag{245}$$

We study the phase diagram of the system disregarding (for a while) the condition $C > 0$: The equilibrium point is $(\bar{K}, \bar{C}) = (a/2b, a^2/4b)$. $F_2'(\bar{K}, \bar{C}) \cdot G_1'(\bar{K}, \bar{C}) = (-1)(-2wb\bar{C}) = 2wb\bar{C} > 0$, and the equation $F(K, C) = 0$ clearly has a unique C^1-solution $C = aK - bK^2$. Since $F(K, C) > 0$ in S^-, $S = S^- = \{(K, C): K \in [K_0, a/2b], C \leqslant aK - bK^2\}$. The equilibrium point $(\bar{K}, \bar{C})$ is obviously unique in $[K_0, \bar{K}] \times \bar{S}$. Moreover, $F_1' G_2' - F_2' G_1' = -2wb\bar{C} < 0$ at the equilibrium point.

Since C is bounded from above in S, we need to verify (239) only for

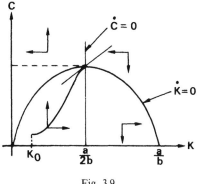

Fig. 3.9

$C < 0$, and $|C|$ large. For $C < 0$ and $K \in [K_0, a/2b]$, we have

$$\left| \frac{G(K,C)}{F(K,C)} \right| = \frac{w(a-2bK)|C|}{K(a-bK)+|C|} \leqslant \frac{w(a-2bK)|C|}{|C|} \leqslant w(a-2bK_0),$$

hence (239) is clearly satisfied.

According to theorem 19 we conclude that there exists a pair of solutions to (245), $(K^*(t), C^*(t))$, defined on $[0, \infty)$, with $K^*(0) = K_0$, which converges to the equilibrium point $(a/2b, a^2/4b)$ as $t \to \infty$, and such that $(K^*(t), C^*(t)) \in S$ for all $t \geqslant 0$, i.e. $\dot{K}^*(t) > 0$ and $\dot{C}^*(t) > 0$. Notice, since $C^*(t)$ satisfies the separable differential equation $\dot{C} = w(a - 2bK^*(t))C$, if $C^*(t) = 0$ for some t, then $C^*(t) \equiv 0$. Hence, $C^*(t) > 0$ for all $t \geqslant 0$.

For the proof of sufficiency, it turns out that we need two extra conditions for admissibility: $K(t)$ and $\dot{K}(t)$ must both approach limits as $t \to \infty$. We refer to exercise 3.10.2 (iv).

Our next example is somewhat more complicated.

Example 17 (Optimal exploitation of a fish population) Consider the problem

$$\text{CU-opt} \int_0^\infty e^{-rt}[ag(x(t), u(t)) - u(t)]dt \tag{246}$$

$$\dot{x}(t) = f(x(t)) - g(x(t), u(t)), \qquad x(0) = x_0 > 0 \tag{247}$$

$$u(t) \geqslant 0, \qquad x(t) > 0. \tag{248}$$

Here r is a (positive) discount rate, a is the (positive) price of fish, $x(t)$ is the fish population at time t, $u(t)$ is the yearly fishing effort measured in money spent, $g(x(t), u(t))$ is the yearly catch, and $f(x(t))$ is the natural rate of increase in the fish population.

We shall assume that f and g are defined and continuous for $x \geqslant 0$ and $u \geqslant 0$, and that f and g are C^2 for $x > 0$, $u > 0$. Moreover, we assume that g'_x exists even for $x > 0$, $u = 0$ and is continuous at such points. Finally we assume that

$$f(x) > 0 \text{ in } (\bar{x}, \bar{\bar{x}}), \quad f(x) < 0 \text{ for } x \notin [\bar{x}, \bar{\bar{x}}], \quad f(\bar{x}) = f(\bar{\bar{x}}) = 0, \qquad (249(\text{i}))$$
$$0 < \bar{x} < x_0 < \bar{\bar{x}}, \quad f''(x) < 0 \text{ for } x > 0, \quad f'(x_0) > r$$

$$g(x, u) \geqslant 0, \quad g(x, 0) = 0, \quad g(x, \infty) = \infty \text{ for } x > 0, \quad g'_x(x, u) > 0,$$
$$g'_u(x, u) > 0, \quad g'_u(x, \infty) = 0 \text{ for } x > 0, g''_{uu}(x, u) < 0, \quad g''_{xx}(x, u)g''_{uu}(x, u)$$
$$- (g''_{xu}(x, u))^2 > 0, \quad g''_{xu}(x, u) > 0 \text{ for } x > 0, \quad u > 0. \qquad (249(\text{ii}))$$

The Hamiltonian for this problem, with $p_0 = 1$, is

$$H = e^{-rt}[ag(x, u) - u] + q(t)(f(x) - g(x, u)), \qquad (250)$$

with $q(t)$ as the adjoint variable associated with (247).

It is convenient to introduce a "current value adjoint variable" $p(t)$ defined by $p(t)e^{-rt} = q(t)$. Then $\dot{q} = -H'_x$ reduces to

$$\dot{p} = (p - a)g'_x(x, u) + (r - f'(x))p. \qquad (251)$$

From (250),

$$\frac{\partial H}{\partial u} = e^{-rt}[(a - p)g'_u(x, u) - 1].$$

An optimal control $u = u(t)$ must maximize the Hamiltonian for $u \in [0, \infty)$. In particular, for $x > 0$,

$$u > 0 \Rightarrow (a - p)g'_u(x, u) = 1 \qquad (252(\text{i}))$$

$$u = 0 \Rightarrow (a - p)g'_u(x, 0^+) \leqslant 1 \qquad (252(\text{ii}))$$

(If $g'_u(x, 0^+) = \infty$, replace the inequality in (252(ii)) by $p \geqslant a$.) Observe that if (252(i)) is valid, then $a \neq p$ and

$$g'_u(x, u) = \frac{1}{a - p} \qquad (253)$$

Let $p^0(x)$ be the value of p for which we have equality in (252(ii)). Then

$$
p^0(x) = \begin{cases} a - \dfrac{1}{g'_u(x,0^+)} & \text{if } g'_u(x,0^+) < \infty \\ a & \text{if } g'_u(x,0^+) = \infty \end{cases}
\tag{254}
$$

Since g''_{ux} is >0 for $x>0$, $u>0$, $g'_u(x,0^+)$ is non-decreasing in x, which implies that $p^0(x)$ is non-decreasing in x.

Let $x>0$ and $p \in (-\infty, \infty)$. We claim that (252) defines u as a function of x and p, $u = u(x, p)$.

Suppose $x>0$. If $p \geqslant a$, then from (252), $u(x, p) = 0$. If $p < a$ and $(a-p)g'_u(x, 0^+) \leqslant 1$, again $u(x, p) = 0$. Since $g'''_{uu}(x, u) < 0$, $g'_u(x, 0^+) > g'_u(x, u)$ for all $u>0$, it follows that (252(i)) cannot be satisfied. If $(a-p)g'_u(x, 0^+) > 1$, u is uniquely determined by (252(i)) due to the monotonicity of $g'_u(x, u)$ w.r.t. u, and the fact that $g'_u(x, \infty) = 0$.

We claim next that $p < p^0(x) \Leftrightarrow u(x, p) > 0$: Suppose $p < p^0(x)$. If $g'_u(x, 0^+) = \infty$, then $a = p^0(x) > p$, so $(a-p)g'_u(x, 0^+) = \infty$ and (252) implies $u(x, p) > 0$. If $g'_u(x, 0^+) < \infty$ then $(a-p)g'_u(x, 0^+) > (a-p(x^0))g'_u(x, 0^+) = 1$ by (254), so again from (252), $u(x, p) > 0$. If $p \geqslant p^0(x)$ and $g'_u(x, 0^+) = \infty$, $p \geqslant a$, so $u(x, p) = 0$ from (252). If $g'_u(x, 0^+) < \infty$, from (254), $(a-p^0(x))g'_u(x, 0^+) = 1$, so $(a-p)g'_u(x, 0^+) \leqslant 1$, and thus $u(x, p) = 0$. Hence, the equivalence asserted above is proved.

For pairs (x, p) satisfying $p < p^0(x)$, we have that $u(x, p)$ satisfies (253) since $u(x, p) > 0$. According to the implicit function theorem, $u(x, p)$ is C^1, and by implicit differentiation we obtain, using (249),

$$
u'_x(x, p) = \frac{-g''_{ux}}{g''_{uu}} > 0, \qquad u'_p(x, p) = \frac{1}{(a-p)^2 g''_{uu}} < 0,
\tag{255}
$$

valid for all (x, p) where $x>0$ and $p < p^0(x) \leqslant a$. Thus $u(x, p)$ is strictly increasing w.r.t. x and strictly decreasing w.r.t. p for $x>0$ and $p < p^0(x)$. If $p \to p^0(x)^-$, then it is easy to see that $u(x, p) \to 0$, and $u(x, p)$ is continuous at $(x, p^0(x))$, $x > 0$.

Substituting $u = u(x, p)$ into (247) and (251) we obtain the system

$$
\dot{x} = f(x) - g(x, u(x, p)) = F(x, p)
$$
$$
\dot{p} = (p-a)g'_x(x, u(x, p)) + (r - f'(x))p = G(x, p).
\tag{256}
$$

Let us agree that a reasonable conjecture is that an optimal policy leads to a convergence to a steady state. Hence, we shall look for an equilibrium point. Consequently, we shall apply theorem 19.

Consider the equation $F(x, p) = 0$. We claim that for each $x \in (\bar{x}, \bar{\bar{x}})$ it has a unique solution $p = p_1(x)$. (The arguments are presented in a somewhat informal manner.) Notice first that if p increases, $u(x, p)$ decreases so by (249(ii)), $g(x, u(x, p))$ decreases. Hence we see that $F(x, p)$ increases as p increases. If $p = -\infty$, then by (252), $g'_u(x, u) = 0$, so $u(x, -\infty) = \infty$ from (249). Then $F(x, -\infty) = f(x) - g(x, \infty) = -\infty$. If $p = p^0(x)$ we have shown that $u = 0$. For $x \in (\bar{x}, \tilde{x})$ we then have $F(x, p^0(x)) = f(x) - g(x, 0) = f(x) > 0$. Since $F(x, p)$ is strictly increasing in p, for each $x \in (\bar{x}, \tilde{x})$ there is a unique value of $p \in (-\infty, p^0(x))$ such that $F(x, p) = 0$. This value of p depends on x, say $p = p_1(x)$. Then $F(x, p_1(x)) = 0$ for all $x \in (\bar{x}, \tilde{x})$, and by the implicit function theorem $p_1(x)$ is C^1, with $p'_1(x) = -F'_1/F'_2$ where $F'_2 = -g'_u u'_p > 0$.

Consider next the equation $G(x, p) = 0$. Applying (249(i)), we see that there exists a unique $\check{x} \in (x_0, \tilde{x})$ such that $f'(\check{x}) = r$. Then for $x > \check{x}$, $r - f'(x) > 0$. We assert that if $x \in (\check{x}, \tilde{x})$, there is a unique value of $p \in (0, p^0(x))$ such that $G(x, p) = 0$. Note first that if $\varphi(p) = (p - a)g'_x(x, u(x, p))$, then $\varphi'(p) = g'_x + (p - a)g''_{xu} \cdot u'_p > 0$ for $p \in (0, p^0(x))$, so $G(x, p)$ is increasing w.r.t. p. $G(x, 0) = -ag'_x < 0$. $G(x, p^0(x)) = (p^0(x) - a)g'_x(x, 0) + (r - f'(x))p^0(x) = (r - f'(x))p^0(x) > 0$. Hence there exists a unique p such that $G(x, p) = 0$. Denote this value of p by $p_2(x)$. The function $p_2(x)$ is defined implicitly by the equation $G(x, p_2(x)) = 0$, i.e.

$$(p_2(x) - a)g'_x(x, u(x, p_2(x))) + (r - f'(x))p_2(x) = 0. \tag{257}$$

Using the implicit function theorem, it follows that $p_2(x)$ is C^1, and by implicit differentiation w.r.t. x, we obtain (simplifying the notation),

$$p'_2 g'_x + (p_2 - a)(g''_{xx} + g''_{xu}(u'_x + u'_p p'_2)) - f'' p_2 + (r - f')p'_2 = 0, p_2(x) > 0.$$

Inserting (255),

$$[g'_x + (p_2 - a)g''_{xu} u'_p + (r - f'(x))]p'_2$$

$$= f'' p_2 + (a - p_2)\left(\frac{g''_{xx} g''_{uu} - (g''_{xu})^2}{g''_{uu}}\right). \tag{258}$$

By the assumptions in (249), for $x \in (\check{x}, \tilde{x})$ and for $p_2(x) \in (0, p^0(x)) \subset (0, a)$, and by the inequalities in (255), we see from (258) that $p'_2 < 0$, so $p_2(x)$ is strictly decreasing.

We claim next that $p_2(\check{x}^+) = p^0(\check{x})$. Suppose $p_2(\check{x}^+) < p^0(\check{x})$. For $x \in (\check{x}, \tilde{x})$, (257) is valid, and by passing to the limit we obtain, using $r = f'(\check{x})$,

$$(p_2(\check{x}^+) - a)g'_x(\check{x}^+, u(\check{x}^+, p_2(\check{x}^+))) = 0.$$

Here $p_2(\check{x}^+) - a < 0$, so $g'_x(\check{x}^+, u(\check{x}^+, p_2(\check{x}^+))) = 0$; thus by (249(ii)), $u(\check{x}^+, p_2(\check{x}^+)) = 0$, so $p_2(\check{x}^+) = p^0(\check{x})$. We have obtained a contradiction. Based on

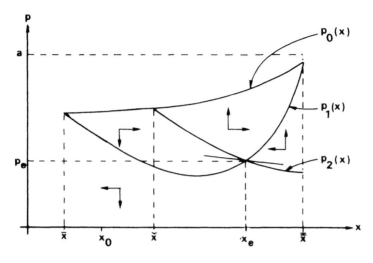

Fig. 3.10

our results concerning $p_0(x), p_1(x)$ and $p_2(x)$, we can draw the phase diagram for (256).

Note, in particular, that for $x = \bar{x}$ and $x = \bar{\bar{x}}, f(x) = 0$, which implies that $F(x, p) = 0$ reduces to $g(x, u(x, p)) = 0$ and then $u(x, p) = 0$. It follows that $p_0(x) = p_1(x)$ for $x = \bar{x}$ and $x = \bar{\bar{x}}$.

Since $p_0(x)$ is non-decreasing and $p_2(x)$ is decreasing, there is at least one equilibrium point (x_e, p_e) for (256), i.e. $p_1(x_e) = p_2(x_e)$, with $x_e \in (\check{x}, \bar{\bar{x}})$, $p_e < p^0(x_e)$. We now make the following assumptions

The point (x_e, p_e) is unique   (259(i))

$$p'_2(x_e) < p'_1(x_e)$$ (259(ii))

Using (249) and (255) we see that

$$F'_2 \cdot G'_1 = (-g'_u \cdot u'_p)[(p-a)(g''_{xx} + g''_{xu} u'_x) - f'' \cdot p]$$

$$= (-g'_u u'_p)\left[(p-a)\frac{(g''_{xx} g''_{uu} - (g''_{xu})^2)}{g''_{uu}} - f'' \cdot p\right] > 0.$$ (260)

Moreover, since $F'_2 > 0$ and $G'_2 > 0$,

$$F'_1 G'_2 - F'_2 G'_1 < 0 \Leftrightarrow \frac{F'_1}{F'_2} < \frac{G'_1}{G'_2} \Leftrightarrow -\frac{F'_1}{F'_2} > -\frac{G'_1}{G'_2}$$ (261)

Since we have assumed that $p'_1(x_e) > p'_2(x_e)$, $-F'_1/F'_2 > -G'_1/G'_2$ at (x_e, p_e), which is exactly (261). Using (260) and (261) we see that condition (232) is satisfied.

The set S of theorem 19 equals $S^+ = \{(x, p): x \in [x_0, x_e], p > p_1(x)\}$. It remains to check (239). We get

$$\rho = \left| \frac{G(x, p)}{F(x, p)} \right| = \left| \frac{(p-a)g'_x(x, u(x, p)) + (r - f'(x))p}{f(x) - g(x, u(x, p))} \right|.$$

Choose the number B such that $B > |p_1(x)|$ for all $x \in [x_0, x_e]$ and $B > a$. Then for $|p| > B$, $p > a$ and $u(x, p) = 0$, so $\rho = |(r - f'(x))p/f(x)| \leqslant D|p|$ for some constant D for all $x \in [x_0, x_e]$. It follows that (239) is satisfied. We conclude from theorem 19 that there exists a pair of solutions $(x^*(t), p(t))$ to (256) on $[0, \infty)$ such that $x^*(0) = x_0$, $x^*(t) \to x_e$ and $p(t) \to p_e$ as $t \to \infty$. Moreover, $(x^*(t), p(t)) \in S$, so $\dot{x}^*(t) > 0$. Finally, $p(t) > 0$, $(G(x, 0) < 0)$.

In order to completely solve our problem, we appeal to theorem 14, putting $A(t) = (0, \infty)$. The optimal pair we propose is $(x^*(t), u^*(t))$, where $u^*(t) = u(x^*(t), p(t))$. It satisfies the maximum condition, $((252)$, with $x = x^*(t)$, $p = p(t)$, and $u = u^*(t)$ is satisfied). Clearly, $q(t) = p(t)e^{-rt}$ satisfies $\dot{q} = -H_x^*$. Consider next condition (188). Take any admissible $x(t)$. Then

$$q(t)(x(t) - x^*(t)) = e^{-rt}p(t)(x(t) - x^*(t)). \tag{262}$$

Since $e^{-rt}p(t) \to 0 \cdot p_e = 0$ and $x^*(t) \to x_e$ as $t \to \infty$, and since $0 < x(t) \leqslant \bar{x}$, the $\underline{\lim}$ of the expression in (262) is equal to 0; and thus (188) is satisfied.

The Hamiltonian is

$$H = e^{-rt}(a - p(t))g(x, u) + p(t)e^{-rt}f(x).$$

If $a \geqslant p(t)$, H is concave in (x, u) due to (249). If $a < p(t)$, H is a decreasing function of u and has a maximum at $u = 0$, which is the value prescribed in (252) for this case. For $u = 0$, $\hat{H} = pe^{-rt}f(x)$, which is concave in x. We therefore conclude from theorem 14 that $(x^*(t), u^*(t))$ solves our problem.

Exercise 3.10.1 In example 17, let $f(x) = x(\bar{x} - x)$ (thus $\bar{x} = 0$), $g(x, u) = 3b^{2/3}x^{2/3}u^{1/3}$, $b > 0$. Assume that $2x_0 < \bar{x} - r$. (Note that g is not strictly concave as required in (249).)

(i) Show (using (252)) that for $p \geqslant a$, $u(x, p) = 0$, $F(x, p) = x(\bar{x} - x)$ and $G(x, p) = (r - \bar{x} + 2x)p$. For $p < a$, show that $u(x, p) = bx(a - p)^{3/2}$, $F(x, p) = x(\bar{x} - x) - 3bx(a - p)^{1/2}$, and $G(x, p) = -2b(a - p)^{3/2} + (r - \bar{x} + 2x)p$.

(ii) Put $b = 1/3$, $\bar{x} = 3/2$, $r = 2/3$, $a = 5$. Verify that $(x_e, p_e) = (1/2, 4)$ is an equilibrium point.

(iii) Apply theorem 19 to the problem with parameter values in (ii), $x_0 = 1/4$.

Exercise 3.10.2   Consider the variational problem:

$$\text{CU-opt} \int_0^\infty U(f(K) - \dot{K}) e^{-\rho t} \, dt,$$

$$K(0) = K_0 > 0, \qquad \varliminf_{t \to \infty} K(t) \geqslant 0, \tag{263}$$

with $f(K) - \dot{K} > 0$ everywhere. We assume that U and f are C^2-functions and

$$U' > 0, \ U'' < 0, \ \lim_{C \to 0} U'(C)/U''(C) = 0, \tag{264}$$

$$f'' < 0, f'(0) > \rho \geqslant 0. \text{ For some } \bar{K} > K_0, f'(\bar{K}) = \rho, f(0) \geqslant 0. \tag{265}$$

(i)   Consider the phase diagram for the system

$$\dot{K} = f(K) - C, \ \dot{C} = (\rho - f'(K)) U'(C)/U''(C) \tag{266}$$

(The second equation in (266) is the Euler equation for problem (263) according to (1.21) in example 1.3.) The equilibrium point is $(\bar{K}, \bar{C})$ where $\bar{C} = f(\bar{K})$. Extend the definition of the function $U'(C)/U''(C)$ by letting it have the value 0 in $(-\infty, 0]$. Use theorem 19 to show that there exists a solution $(K^*(t), C^*(t))$ to (266) such that $K^*(0) = K_0$ and $(K^*(t), C^*(t)) \to (\bar{K}, \bar{C})$ as $t \to \infty$ with $\dot{K}^*(t) > 0$, and $\dot{C}^*(t) \geqslant 0$ for all t.

(ii)   Suppose from now on that U'/U'' is locally Lipschitzian in a neighbourhood of 0 (see (A.12)). Show that $C^*(t) > 0$.
(Hint: Consider the equation $\dot{C} = (\rho - f'(K^*(t))(U'(C)/U''(C))$.)

(iii)   Use the sufficient conditions in note 17 to prove the CU-optimality of $K^*(t)$ when $\rho > 0$.

(iv)   Put $\rho = 0$. Add to the requirements for admissibility that the limits $K(\infty) = \lim_{t \to \infty} K(t)$ and $\dot{K}(\infty) = \lim_{t \to \infty} \dot{K}(t)$ exist or are $+\infty$. Observe that $K(\infty) \geqslant 0$ implies $\dot{K}(\infty) \geqslant 0$ and observe next that if $K(\infty) < \bar{K}$, then $U(f(K(\infty)) - \dot{K}(\infty)) < U(f(K^*(\infty)) - \dot{K}^*(\infty))$. By considering the criterion functional directly, show that $K^*(t)$ is CU-better than any admissible $K(t)$ with $K(\infty) < \bar{K}$. Use note 17 to prove that $K^*(t)$ is CU-optimal among all admissible $K(t)$ for which $K(\infty) \geqslant \bar{K}$.

CHAPTER 4

MIXED CONSTRAINTS

In this chapter we begin a discussion of control problems having inequality constraints on their state variables. These constraints also involve the control variables and are therefore called *mixed* constraints. We restrict our attention to problems which can be briefly formulated as follows:

$$\max_{u} \int_{t_0}^{t_1} f_0(x, u, t)\mathrm{d}t, \qquad \dot{x} = f(x, u, t), \qquad h(x, u, t) \geqslant 0, \tag{1}$$

with initial condition $x(t_0) = x^0$ and with the standard terminal conditions (2.53). As usual, $x = x(t)$ is an n-dimensional state vector, $u = u(t)$ is an r-dimensional control vector, and f is an n-dimensional vector function. The function h is an s-dimensional vector function so that the inequality in (1) represents the inequalities

$$h_k(x(t), u(t), t) \geqslant 0, \qquad k = 1, \dots, s. \tag{2}$$

All restrictions on the control vector $u(t)$ itself have to be incorporated into (2), as we – to begin with – do not allow additional restrictions of the type $u(t) \in U$. (This results in only a slight loss of generality, since in practice U is nearly always specified in terms of inequalities.)

Whether or not restrictions of the form (2) are present in a given control problem is partly a question of the form in which the problem is stated. For instance, in example 2.7 we studied an optimal growth model in which investment $\dot{K}$ was a fraction $u \in [0, 1]$ of production $f(K)$, u the control variable. We could, however, have taken the amount of investment $\dot{K}$ as the control variable. With $\dot{K} = u$, the constraints on the control variable would be $u \geqslant 0$, $f(K) - u \geqslant 0$, the latter being "truly" mixed. We avoided mixed constraints by choosing the former formulation.

There are, however, a number of important control problems in economics for which it is unnatural or impossible to redefine the control variable so as to prevent the occurrence of mixed constraints. A case in point is the growth model of example 2.7 with the additional condition $f(K) - \dot{K} \geqslant C_0, C_0$ a positive constant denoting a subsistence level of consumption.

In this chapter we shall usually impose on the h_k-functions a constraint qualification guaranteeing that the control variable $u(t)$ enters the functions h_k in an essential way. Optimal control problems with pure state variable constraints (i.e. with $u(t)$ lacking in the h_k-functions) are more difficult to handle. We shall consider such problems in the next chapter.

1. Preliminary considerations

Let us consider the special case of problem (1) above in which $h(x, u, t)$ is independent of x and t, such that (2) reduces to

$$h_k(u(t)) \geqslant 0, \qquad k = 1, \dots, s. \tag{3}$$

Then (1) becomes an ordinary control problem of the type studied in Chapters 2 and 3. To solve the problem, we would again introduce the Hamiltonian

$$H(x, u, p, t) = p_0 f_0(x, u, t) + \sum_{i=1}^{n} p_i f_i(x, u, t), \tag{4}$$

where $p_1, \dots, p_n$ are the adjoint variables associated with the differential equations for $x_1, \dots, x_n$. From theorem 2.2 we know that an essential part of the necessary conditions for the solution of the problem is the maximization of H subject to the constraints in (3). Generally, this is a non-linear programming problem, and in order to solve it we usually use the Kuhn–Tucker theorem. We then introduce a multiplier associated with each of the s constraints in (3) and form the resulting Lagrangian function. It turns out that a similar procedure is an essential part of the necessary conditions for problem (1).

The correct theorem will be formulated in the next section. However, let us here indicate the main content of the result. Associate multipliers $q_1(t), \dots, q_s(t)$ with each of the constraints in (2) and form the *Lagrangian* or *generalized Hamiltonian L* defined by

$$L(x, u, p, q, t) = p_0 f_0(x, u, t) + \sum_{i=1}^{n} p_i f_i(x, u, t) + \sum_{k=1}^{s} q_k h_k(x, u, t), \tag{5}$$

where $q = (q_1, \ldots, q_s)$. Thus $L = H + q \cdot h(x, u, t)$ where H is the Hamiltonian defined by (4). Briefly formulated, the main necessary conditions for an admissible pair $(x^*(t), u^*(t))$ to solve the problem are:

$$u^*(t) \text{ maximizes } H(x^*(t), u, p(t), t) \text{ subject to } h(x^*(t), u, t) \geqslant 0, \tag{6}$$

$$\partial L^*/\partial u_j = 0, \qquad j = 1, \ldots, r, \tag{7}$$

$$q_k(t) \geqslant 0 (= 0 \text{ if } h_k(x^*(t), u^*(t), t) > 0), \qquad k = 1, \ldots, s, \tag{8}$$

$$\dot{p}_i(t) = -\partial L^*/\partial x_i, \qquad i = 1, \ldots, n, \tag{9}$$

Terminal conditions (2.60). $\tag{10}$

In (7) and (9) the derivatives of L are evaluated along the optimal solution.

For (6)–(10) to hold, a constraint qualification must be presupposed. Imagine that we write down the first-order necessary conditions for the maximization problem in (6), using $q_k(t)$ as names of the multipliers associated with the constraints. We then get (7) and (8). The $q_k(t)$'s occuring in (6)–(10) have, however, further properties: They also occur in (9).

Let us also make the observation that if the Hamiltonian is concave in (x, u) and the h_k-functions are quasi-concave in (x, u), then (7) and (8) imply (6). (See e.g. Sydsæter (1981) p. 317.)

Note carefully that the differential equations for the p_i-functions are different from the ones in the ordinary maximum principle since $\partial H^*/\partial x_i$ is replaced by $\partial L^*/\partial x_i$. Thus $\dot{p}$ will now also depend on the q_k-functions.

Just as in the ordinary maximum principle, $p(t)$ is a continuous function and (9) is valid at all points of continuity of $u^*(t)$. As for the multipliers $q_k(t)$ associated with the constraints $h_k(x, u, t) \geqslant 0$, it turns out that they are only piecewise continuous, with possible jumps at the points of discontinuity of $u^*(t)$.

Suppose the Hamiltonian (4) is concave in (x, u) and the h_k functions are all quasi-concave in (x, u). Then we shall see (in theorem 5) that an admissible pair $(x^*(t), u^*(t))$, with associated functions $p(t)$ and $q(t)$ satisfying all the conditions in (6)–(10), with $p_0 = 1$, is indeed optimal. (A constraint qualification is unnecessary in this case.)

Before proceeding, let us look at a simple example.

Example 1 Reconsider the problem in example 2.1, assuming in addition $x(t) + u(t) \leqslant 2$ for all $t \in [0, 1]$. The problem can then be formulated in the following way:

$$\max_{u} \int_{0}^{1} x(t)dt \tag{11}$$

$$\dot{x}(t) = x(t) + u(t), \qquad x(0) = 0, x(1) \text{ free}, \tag{12}$$

$$h_1(x(t), u(t), t) = 1 - u(t) \geqslant 0, \tag{13}$$

$$h_2(x(t), u(t), t) = 1 + u(t) \geqslant 0, \tag{14}$$

$$h_3(x(t), u(t), t) = 2 - x(t) - u(t) \geqslant 0. \tag{15}$$

In example 2.1 we proved that if constraint (15) is disregarded, the optimal solution is $u_1^*(t) \equiv 1$, $x_1^*(t) \equiv e^t - 1$; and the value of the integral in (11) equal to $e - 2$. Of course, if $(x_1^*(t), u_1^*(t))$ also satisfies (15), then this pair would solve our new problem. However, $x_1^*(t) + u_1^*(t) = e^t$ which varies between 1 and $e = 2.718 \ldots$ in the interval $[0, 1]$, so (15) *is* violated.

Let us consider the conditions in (6)–(10) in this case. The Lagrangian is (with $p_0 = 1$)

$$L = x + p(x + u) + q_1(1 - u) + q_2(1 + u) + q_3(2 - x - u) \tag{16}$$

and the conditions necessary for $(x^*(t), u^*(t))$ to solve the problem are:

$u^*(t)$ maximizes $x^*(t) + p(t)x^*(t) + p(t)u$ subject to

$$u \in [-1, 1], \quad u \leqslant 2 - x^*(t) \tag{17}$$

$$\partial L^*/\partial u = p(t) - q_1(t) + q_2(t) - q_3(t) = 0 \tag{18}$$

$$q_1(t) \geqslant 0 (= 0 \text{ if } u^*(t) < 1) \tag{19}$$

$$q_2(t) \geqslant 0 (= 0 \text{ if } u^*(t) > -1) \tag{20}$$

$$q_3(t) \geqslant 0 (= 0 \text{ if } u^*(t) < 2 - x^*(t)) \tag{21}$$

$$\dot{p}(t) = -\partial L^*/\partial x = -1 - p(t) + q_3(t), \qquad p(1) = 0. \tag{22}$$

Since the Hamiltonian is concave (in fact linear) in (x, u) and h_1, h_2, and h_3 are quasi-concave (in fact linear) in (x, u), we know that if we find an admissible pair $(x^*(t), u^*(t))$ and functions $p(t)$ and $q_1(t)$, $q_2(t)$, $q_3(t)$ which satisfy (18)–(22), then $(x^*(t), u^*(t))$ is optimal. (As we noted above, the constraint qualification is unnecessary in this case, and (17) is a consequence of (18)–(22).) Therefore let us try to guess the optimal solution, and then verify its optimality by checking that all conditions are satisfied.

We want to maximize the integral of $x(t)$. Since $u \in [-1, 1]$, it seems from

(12) that it must be optimal to start out with $u^*(t)=1$, $x^*(t)=e^t-1$ as in example 2.1. With $x^*(t)+u^*(t)=e^t$, constraint (15) becomes binding when $e^t=2$, i.e. when $t=\ln 2$. At $t=\ln 2$, $x^*(\ln 2)=e^{\ln 2}-1=1$ and for $t>\ln 2$ it must be optimal to increase $x(t)$ as fast as constraint (15) allows i.e. to put $\dot{x}^*(t)=x^*(t)+u^*(t)\equiv 2$, provided (13) and (14) are not violated. Now, $\dot{x}^*(t)\equiv 2$, $t\geqslant\ln 2$ with $x^*(\ln 2)=1$ implies $x^*(t)=2t+1-2\ln 2$. Then $u^*(t)=2-x^*(t)=1+2\ln 2-2t$, and we easily verify that for $t\in[\ln 2, 1]$, $u^*(t)\in[-1, 1]$. Thus we arrive at the following suggestion for an optimal solution to problem (11)–(15): In $[0, \ln 2]$ let $u^*(t)=1$ and $x^*(t)=e^t-1$. In $(\ln 2, 1]$ let $u^*(t)=1+2\ln 2-2t$ and $x^*(t)=2-u^*(t)$.

We know that $(x^*(t), u^*(t))$ defined as above is admissible. It remains for us to find the appropriate multipliers $p(t)$, $q_1(t)$, $q_2(t)$, and $q_3(t)$.

During the interval $[\ln 2, 1]$, $-1<u^*(t)<1$, so according to (19) and (20) we have to put $q_1(t)=q_2(t)=0$ in this interval. Then from (18), $p(t)=q_3(t)$ and from (22), $\dot{p}(t)=-1$, $p(1)=0$, i.e. $p(t)=1-t$. In particular, $p(\ln 2)=1-\ln 2$. During the interval $[0, \ln 2]$, $u^*(t)>-1$ and $x^*(t)+u^*(t)=e^t<2$, so according to (20) and (21), $q_2(t)=q_3(t)=0$. From (22) it follows that $\dot{p}(t)=-1-p(t)$. Solving this linear differential equation on $[0, \ln 2]$ with $p(\ln 2)=1-\ln 2$, gives $p(t)=(4-2\ln 2)e^{-t}-1$. In turn, from (18) we get $q_1(t)=p(t)$. All in all, the complete suggestion for a solution and its associated multipliers is:

	$x^*(t)$	$u^*(t)$	$p(t)$	$q_1(t)$	$q_2(t)$	$q_3(t)$
$[0, \ln 2]$	e^t-1	1	$(4-2\ln 2)e^{-t}-1$	$(4-2\ln 2)e^{-t}-1$	0	0
$(\ln 2, 1]$	$2t+1-2\ln 2$	$1+2\ln 2-2t$	$1-t$	0	0	$1-t$

Having checked that $(x^*(t), u^*(t))$ is admissible and that conditions (18)–(22) are all satisfied, we conclude $(x^*(t), u^*(t))$ is optimal. An easy computation reveals that $\int_0^1 x^*(t)\mathrm{d}t=3-\ln 2(4-\ln 2)\approx 0{,}708$, not far from $e-2\approx 0{,}718$, the corresponding value for the problem when condition (15) is deleted.

It is logically unnecessary to check condition (17) for this problem, but let us do so just the same. Notice that the only term in the Hamiltonian which depends on u is $p(t)u$. If $t\in[0, \ln 2]$, constraint (15) is not binding and $p(t)\geqslant 1-\ln 2>0$, so $p(t)u$ is maximized at $u^*(t)=1$. If $t\in(\ln 2, 1]$, we must seek the solution to the problem of maximizing $p(t)u=(1-t)u$, subject to $u\in[-1, 1]$ and $u\leqslant 2-x^*(t)=2-2t-1+2\ln 2=1+2\ln 2-2t$. We have indicated the constraint set of this problem, A, in fig. 4.1. Since $1-t$ is positive in the interval $[\ln 2, 1)$, it is obvious that for each $t\in[\ln 2, 1)$, $(1-t)u$

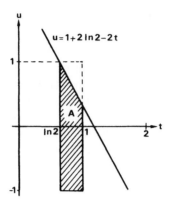

Fig. 4.1

is maximized if u is chosen as the corresponding point on the line $u = 1 + 2 \ln 2 - 2t$. Therefore, the above suggested optimal choice of $u^*(t)$ is confirmed.

The graphs of $x^*(t)$, $u^*(t)$, and the multipliers are as follows:

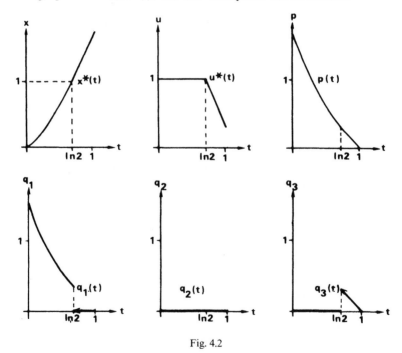

Fig. 4.2

Observe, in particular, that $p(t)$ is continuous everywhere (but has a kink at $t = \ln 2$), whereas $q_1(t)$ and $q_3(t)$ are both discontinuous at $t = \ln 2$.

Exercise 4.1.1 Consider the model in example 1.
Let $\tilde{u}(t) = 1$ if $t \in [0, \ln 2]$, $\tilde{u}(t) = u_0$ (constant) if $t \in (\ln 2, 1]$.

(i)   Find the associated solution $\tilde{x}(t)$ of (12).
(ii) Determine the restriction which has to be imposed on u_0 in order that $(\tilde{x}(t), \tilde{u}(t))$ is admissible.
(iii) Compute the associated value in (11) and maximize it w.r.t. u_0.
(iv) Test how (17)–(22) reveal the non-optimality of $(\tilde{x}(t), \tilde{u}(t))$.

2. Necessary conditions. Existence

In this and the following section, we examine more closely the theory indicated in section 1. First, we shall write down a precise version of the main results used in solving this type of constrained optimization problems. Later, in section 4, we shall offer some extensions and generalizations.
    Consider the following problem,

$$\max \int_{t_0}^{t_1} f_0(x(t), u(t), t)\,\mathrm{d}t, \qquad (t_0, t_1 \text{ fixed}) \tag{23}$$

subject to the vector differential equation and the initial condition

$$\dot{x}(t) = f(x(t), u(t), t), \qquad x(t_0) = x^0 \ (x^0 \text{ fixed}), \tag{24}$$

the terminal conditions

$$
\begin{array}{llll}
x_i(t_1) = x_i^1 & i = 1, \ldots, l & (x_i^1 \text{ all fixed}) & \text{(25a)} \\
x_i(t_1) \geqslant x_i^1 & i = l+1, \ldots, m, & (x_i^1 \text{ all fixed}) & \text{(25b)} \\
x_i(t_1) \text{ free} & i = m+1, \ldots, n, & & \text{(25c)}
\end{array}
$$

and for all $t \in [t_0, t_1]$, the constraints

$$h_k(x(t), u(t), t) \geqslant 0, \qquad k = 1, 2, \ldots, s. \tag{26}$$

In addition to our standard assumption (2.6) on f_0 and f, we shall assume throughout this chapter that $\partial f_0/\partial u$ and $\partial f/\partial u$ exist and are continuous. We assume, moreover, that $h = (h_1, \ldots, h_s)$ is a continuous function of (x, u, t) and that $\partial h_k/\partial x_i$ and $\partial h_k/\partial u_j$ exist and are continuous functions of (x, u, t).

We call $(x(t), u(t))$ *an admissible pair* if $u(t)$ is piecewise continuous, $x(t)$ is continuous and piecewise continuously differentiable, and (24)–(26) are satisfied.

In order to formulate the Maximum Principle for this problem, a constraint qualification has to be imposed on the h_k-functions. The qualification is a condition on the rank of the matrix of partial derivatives w.r.t. $u_1, \ldots, u_r$ of those h_k-functions which are binding along the optimal pair.

Let $(x^*(t), u^*(t))$ be an optimal pair and define the two sets I_t^- and I_t^+ by

$$I_t^- = \{k: h_k(x^*(t), u^*(t^-), t) = 0, k = 1, \ldots, s\}$$
$$I_t^+ = \{k: h_k(x^*(t), u^*(t^+), t) = 0, k = 1, \ldots, s\}. \tag{27}$$

Then the constraint qualification we shall use is as follows:

For every $t \in [t_0, t_1]$,

If $I_t^- \neq \phi$, the matrix $\{\partial h_k(x^*(t), u^*(t^-), t)/\partial u_i\}$, $k \in I_t^-, i = 1, \ldots,$
r, has a rank equal to the number of elements in I_t^-. $\tag{28(i)}$

If $I_t^+ \neq \phi$, the matrix $\{\partial h_k(x^*(t), u^*(t^+), t)/\partial u_i\}$, $k \in I_t^+, i = 1, \ldots,$
r, has a rank equal to the number of elements in I_t^+. $\tag{28(ii)}$

(If $t = t_0$, drop (i), if $t = t_1$, drop (ii).)

Theorem 1 (Necessary conditions. Fixed time interval. Mixed constraints.)[1] Let $(x^*(t), u^*(t))$ be an admissible pair which solves problem (23)–(26). Assume that the constraint qualification (28) is satisfied. Then there exist a number p_0, vector functions $p(t) = (p_1(t), \ldots, p_n(t))$ and $q(t) = (q_1(t), \ldots, q_s(t))$, where $p(t)$ is continuous and piecewise continuously differentiable and $q(t)$ piecewise continuous, such that for all $t \in [t_0, t_1]$:

$$(p_0, p_1(t), \ldots, p_n(t)) \neq (0, 0, \ldots, 0), \tag{29}$$

$H(x^*(t), u^*(t), p(t), t) \geqslant H(x^*(t), u, p(t), t)$ for all u such that
$h_k(x^*(t), u, t) > 0$, $k = 1, \ldots, s$. (H defined by (4).) $\tag{30}$

$$\partial L^*/\partial u_j = 0, \qquad j = 1, \ldots, r. \tag{31}$$

Here $\partial L^*/\partial u_j$ means $\partial L/\partial u_j$ evaluated at $(x^*(t), u^*(t), p(t), q(t), t)$ (L defined by (5).)

---

[1] Neustadt (1976) VI 3. The theorem also holds for a bounded and measurable $u^*(t)$ with the following modifications: $q(t)$ is bounded and measurable, (30), (31) and (33) hold a.e. and (28) is changed as follows: Define Ω to be the closure of the set $\{(u^*(t), t): t \in [t_0, t_1]\}$ in R^{r+1}. The matrix with elements $\partial h_k(x^*(t), u, t)/\partial u_i$, $k \in I(u, t) = \{k: h_k(x^*(t), u, t) = 0\}$, $i = 1, \ldots, r$, has a rank equal to the number of elements in $I(u, t)$ for all $(u, t) \in \Omega$.

$$q_k(t) \geq 0(=0 \text{ if } h_k(x^*(t), u^*(t), t) > 0), \qquad k = 1, \ldots, s. \tag{32}$$

$$\dot{p}_i(t) = -\partial L^*/\partial x_i, \ i = 1, \ldots, n \text{ for v.e.t.} \tag{33}$$

Here $\partial L^*/\partial x_i$ means $\partial L/\partial x_i$ evaluated at $(x^*(t),\ u^*(t),\ p(t),\ q(t),\ t)$.

$$p_0 = 0 \text{ or } p_0 = 1. \tag{34}$$

$p_i(t_1)$ no conditions	for $i = 1, \ldots, l,$	(35a)
$p_i(t_1) \geq 0(=0 \text{ if } x_i^*(t_1) > x_i^1)$	for $i = l+1, \ldots, m,$	(35b)
$p_i(t_1) = 0$	for $i = m+1, \ldots, n.$	(35c)

Note 1 In some examples we shall need the following trivial consequences of (31): $\partial L^*(t^-)/\partial u = 0$ for all $t \in (t_0, t_1]$ and $\partial L^*(t^+)/\partial u = 0$ for all $t \in [t_0, t_1)$.

Note 2 One implication of the conditions stated in the theorem is that $L^* = L(x^*(t), u^*(t), p(t), q(t), t)$ is continuous everywhere. Moreover, if $f_0, f,$ and h are C^1-functions of (x, u, t), it can be proved that at all continuity points for $u^*(t)$, the total time derivative of L^*, $\dot{L}^*$, is equal to the partial derivative $\partial L^*/\partial t$, i.e. $\partial L/\partial t$ evaluated at $(x^*(t), u^*(t), p(t), \dot{q}(t), t)$.

$$\dot{L}^* = \frac{\partial L^*}{\partial t} \text{ at all continuity points for } u^*(t). \tag{36}$$

(See exercise 4.2.1 for a rough proof.)

In particular, if the system is autonomous, such that f_0, f and h do not depend explicitly on t, then $\partial L^*/\partial t = 0$ at all continuity points of $u^*(t)$. Since L^* is continuous everywhere, we conclude that L^* is a constant. An immediate consequence of (32) is that $P^* = \Sigma_{k=1}^s q_k(t)h_k(x^*(t), u^*(t), t) = 0$ for all t. (In fact, each term in the sum is 0.) Since $L^* = H^* + P^*$, where $H^* = H(x^*(t), u^*(t), p(t), t)$, we have the following conclusion:

If problem (23)–(26) is autonomous, the Hamiltonian is constant along the optimal path.

Note 3 (**Equality constraints**)[2] Suppose that the inequality sign $\geq$ in (26) is replaced by the equality sign $=$. It turns out that theorem 1 is still valid provided we require that the matrix with elements $\partial h_k(x^*(t), u, t)/\partial u_i$, $k = 1, \ldots, s$, $i = 1, \ldots, r$, has rank equal to s for all u such that $h(x^*(t), u, t) = 0$, and drop (32) and change the strict inequality sign in (30) to an equality sign.

---

[2] Makowski and Neustadt (1974).

In Chapter 2, section 4 we indicated why the ordinary Maximum Principle contains enough conditions to single out one or a few candidates for optimality. Let us study the conditions in theorem 1 in the same spirit.

Just as in the case of theorem 2.2, the terminal conditions (25) and the transversality conditions (35) amount, essentially, to n conditions on the pair $(x^*(t_1), p(t_1))$.

Consider the first-order conditions: $0 = \partial L/\partial u = \partial H(x, u, p, t)/\partial u + q\partial h(x, u, t)/\partial u = 0$, $qh(x,u,t)=0$. It is well known that these conditions determine $u = u(t)$ and the "Kuhn–Tucker multipliers" $(q_1,\ldots,q_s)$ quite well. Assume that u and q are uniquely determined in this way, $u = u(p, x)$, $q = q(p, x)$. Imagine inserting these functions into the differential equations $\dot{x} = f(x, u, t)$, $\dot{p} = -\partial L/\partial x$ and solving these equations with $x(t_0) = x^0$ and $p(t_0) = p^0$ as initial conditions, where $p^0 \in R^n$ is an unknown parameter. One could then determine p^0 by using the n conditions obtained from (25) and (35). (Note again that this argument should not necessarily be regarded as a procedure for solving actual control problems.)

Note 4 (**Constraint qualifications**) The constraint qualification (28) rather severely qualifies the constraint functions. In particular, at least one of the control variables must be an argument in each active constraint, since in the contrary case at least one of the rows in the matrices of (28) consists only of zeros, therefore violating the rank condition. Then theorem 1 helps not at all in singling out any candidates.

If constraint qualifications have been introduced into the formulation of necessary conditions, one correct way of obtaining *all* the candidates for optimality is the following:

(i) First find all admissible pairs for which the constraint qualification (here (28)) fails for some $t \in [t_0, t_1]$.

(ii) Then find all admissible pairs that satisfy the necessary conditions (here (29)→(35)) *and* the constraint qualification.

In most cases this is the best procedure. However, there is an alternative way of finding all the candidates for optimality which is sometimes used:

(i)′ First, find all admissible pairs which satisfy the necessary conditions (disregarding the constraint qualification).

(ii)′ Then find all admissible pairs which violate the constraint qualification, *other than those obtained in* (i)′.

(If we let A denote the set of admissible pairs satisfying the necessary conditions and B denote the set of all admissible pairs which fail to satisfy the constraint qualification, then the candidates for optimality are those belonging to $A \cup B$. In (i)–(ii) and (i)′–(ii)′ we make use of the identities $A \cup B = B \cup (A - B) = A \cup (B - A)$.

We offer this somewhat pedantic note in order to warn against a procedure which makes no logical sense, but which is nevertheless frequently encountered in the literature: A unique candidate is obtained from the necessary conditions and only that candidate is checked against the constraint qualifications. Yet it is the only candidate that need not be tested against the constraint qualifications!

When constraint qualifications fail, we usually get a large number of candidates from (i), or (ii)′. In principle, the value of the criterion should be calculated for all candidates and the "best ones" picked out. This often becomes impractical when there is a large number of candidates.

Example 2 Let us briefly consider the application of necessary conditions (theorem 1) to the problem in example 1. We shall follow the procedure (i) and (ii) in note 4.

The matrix $[\partial h_1/\partial u, \ \partial h_2/\partial u, \ \partial h_3/\partial u]' = [-1, \ 1, \ -1]'$ and all its submatrices have rank 1. If we look at the three constraints (13), (14), and (15) in example 1, clearly not all can be active, i.e. neither I_t^- nor I_t^+ can have three elements. The pair (13), (14) cannot be active either. Consider the pair (14), (15). When these are both active, $x(t) + u(t) = 2$ and $u(t) = -1$, so $x(t) = 3$. Moreover, $\dot{x}(t) = x(t) + u(t) \leqslant x(t) + 1$, so $x(t)$ is $\leqslant y(t)$ where $\dot{y}(t) = y(t) + 1$, $y(0) = 0$. Here $y(t) = e^t - 1 \in [0, e - 1]$ for $t \in [0, 1]$. Hence an admissible $x(t)$ is always $\leqslant e - 1 < 3$. The only *pair* which can be active for any admissible solution is thus (13), (15), and *if* this pair is active, the rank condition in (28) fails. If only one of the constraints (13)–(15) is active, the c.q. *is* satisfied. We conclude that the c.q. only fails for admissible pairs $(x(t), u(t))$ for which $x(t) = 1$ and $u(t^+)$ or $u(t^-)$ is 1 for some $t \in [0, 1]$.

We have now carried out (i) in note 4. As for (ii), it is easy to see that the necessary conditions imply $p(t) > 0$ in $[0, 1)$. The maximum condition (30) (which implies (17) in example 1) gives $u^*(t) = \min (1, 2 - x^*(t))$. Thus we get the pair $(x^*(t), u^*(t))$ found in example 1. But this pair does *not* satisfy the c.q., so (ii) gives no candidate.

From (i) we obtained, unfortunately a large number of candidates so theorem 1 is not of much help in this case.

The problem is that the c.q. (28) is too demanding. Fortunately the c.q. can be weakened as explained in the next note.

Note 5* Theorem 1 is still valid if (28) holds for all t, except at a finite

number of points, provided the following weaker c.q. holds at these exceptional points: ·
For some $u \in R^r$,

$$\frac{\partial h_k(x^*(t), u^*(t^-), t)}{\partial u} \cdot (u - u^*(t^-)) > 0 \qquad \text{for all } k \in I_k^- \tag{37a}$$

and, for some $u \in R^r$,

$$\frac{\partial h_k(x^*(t), u^*(t^+), t)}{\partial u} \cdot (u - u^*(t^+)) > 0 \qquad \text{for all } k \in I_k^+. \tag{37b}$$

In fact, if we require (37) to hold for all t, instead of requiring (28), theorem 1 is still valid with the modification that $q(t)$ and $p(t)$ are less well-behaved ($q(t)$ is bounded and measurable, $p(t)$ is absolutely continuous, and (33) holds a.e.).

These modified c.q.'s turn out to be satisfied for all admissible pairs in example 2, and $(x^*(t), u^*(t))$ becomes the unique candidate.

Note 6   Suppose we add the extra condition $u(t) \in U$ for some convex set U to problem (23)–(26). For a suitable constraint qualification the conclusions in theorem 1 hold provided we introduce the following modifications: (i), $p(t)$ is absolutely continuous and (33) holds a.e., (ii), $q(t)$ is bounded and measurable, (iii) for all t

$$(30) \text{ holds for } u \in U, \qquad h_j(x^*(t), u, t) > 0, \qquad j = 1, \ldots, s. \tag{38a}$$

$$\frac{\partial L^*}{\partial u} \cdot (u - u^*(t)) \leqslant 0 \text{ for all } u \in U. \tag{38b}$$

((38b) replaces (31).)

Finally, a constraint qualification that is sufficient for these conclusions to hold, can be obtained by replacing the expression $u \in R^r$ in (37) by $u \in U$, and assuming (37) to hold for all t.

Note 7   (**"Almost necessary conditions"**)   For the problem discussed in the previous note, Neustadt (1976) has proved more general necessary conditions, not using any constraint qualifications (c.q.). The most essential change in the necessary conditions is that we must allow for jumps in $p(t)$ at points where the c.q. fails.

The precise statement of the Neustadt result involves concepts which are beyond the level of this book (e.g. finitely additive measures). We shall now formulate conditions that may be called *almost necessary conditions* (ANC)[3]. which allow for a finite number of jump points in $p(t)$. These conditions are almost necessary in the sense that, very rarely, problems and optimal solutions are encountered which require multipliers and functions $p(t)$ with a more complicated structure. Thus the risk of erring is very slight if one assumes that the true necessary conditions take on the more convenient form of ANC. (True necessary conditions are given in footnote 3).)

[3]ANC is a "simplification" of necessary conditions stated in Neustadt (1976). For measurable $u^*(t)$'s, the multipliers connected with the h_k's have to be represented by finitely additive measures.

In the case where $u^*(t)$ is piecewise continuous, the multipliers can be represented by continuous, non-decreasing functions $Q_k(t)$ and "left and right" jump factors β_j^-, β_j^+ at each of a possibly countable number of discontinuity points of $p(t)$. Thus, true necessary conditions, when no constraints qualifications are assumed, read as follows:

There exist a number p_0, a function $p(t)$ of bounded variation, continuous, non-decreasing functions $Q_k(t)$, vectors $\beta_j^- = (\beta_{j1}^-, \ldots, \bar{\beta}_{js}^-)$, $\beta_j^+ = (\beta_{j1}^+, \ldots, \beta_{js}^+)$ associated with the jump points $\tau_j \in [t_0, t_1]$ of $p(t)$, $j = 1, 2, \ldots$ such that (30) holds both for $u^*(t^+)$ and $u^*(t^-)$ for all $t \in [t_0, t_1]$ with $u \in U$, $h_j(x^*(t), u, t) > 0$. Moreover

$$\int_{t_0}^{t_1} H_u'^*(t)(u(t) - u^*(t))\,dt + \sum_k \int_{[t_0, t_1]} h_{ku}'^*(t)(u(t) - u^*(t))\,dQ_k(t) \leqslant 0 \qquad \text{(a)}$$

for all piecewise continuous functions $u(t)$, $u(t) \in U$. Furthermore, for $j = 1, 2, \ldots$

$$\sum_k \beta_{jk}^- h_{ku}'^*(\tau_j^-)(u - u^*(\tau_j^-)) \leqslant 0 \qquad \text{for all } u \in U \qquad \text{(b)}$$

$$\sum_k \beta_{jk}^+ h_{ku}'^*(\tau_j^+)(u - u^*(\tau_j^+)) \leqslant 0 \qquad \text{for all } u \in U \qquad \text{(c)}$$

$Q_k(t)$ is constant on any interval on which
$h_k(x^*(t), u^*(t), t) > 0, k = 1, \ldots, s$

$$p(t) = p(t_1) + \int_t^{t_1} H_x'^*(t)\,dt + \sum_k \int_{[t, t_1]} h_{kx}'^*(t)\,dQ_k(t) + \sum_{\tau_j \geqslant t} \sum_k \beta_{jk}^+ h_{kx}'^*(\tau_j^+)$$

$$+ \sum_{\tau_j > t} \sum_k \beta_{jk}^- h_{kx}'^*(\tau_j^-) \qquad \text{for } t \in [t_0, t_1). \qquad \text{(d)}$$

Condition (41) is satisfied and $\Sigma_k \beta_{jk}^- < \infty$, $\Sigma_k \beta_{jk}^+ < \infty$. (If $\tau_j = t_0$, $\beta_j^- = 0$; if $\tau_j = t_1$, $\beta_j^+ = 0$.) $\qquad$ (e)

Not all the numbers p_0, $p_i(t_1)$, $Q_k(t_1) - Q_k(t_0)$, β_{jk}^-, and β_{jk}^+ are zero. $\qquad$ (f)

Conditions (34) and (35) are satisfied. $\qquad$ (g)

Almost necessary conditions (ANC)

Let $(x^*(t), u^*(t))$ be an admissible pair which solves problem (23)–(26) with $u(t) \in U$, where U is a convex subset of R^r. Then there exist a number p_0, piecewise continuous functions $p(t)$ and $q(t)$, and vectors $\beta_k^- = (\beta_{k1}^-, \ldots, \beta_{ks}^-)$, $\beta_k^+ = (\beta_{k1}^+, \ldots, \beta_{ks}^+)$, associated with each jump point $\tau_k \in [t_0, t_1]$, $k = 1, \ldots, N$ of $p(t)$ such that the following conditions hold: (38) for v.e.t. (32), (33), ($p(t)$ being piecewise continuously differentiable), (34), and (35). Furthermore, at the jump points τ_k:

$$p_i(\tau_k^-) - p_i(\tau_k^+) = \sum_{j=1}^{s} \beta_{kj}^+ \frac{\partial h_j(x^*(\tau_k), u^*(\tau_k^+), \tau_k)}{\partial x_i}$$

$$+ \sum_{j=1}^{s} \beta_{kj}^- \frac{\partial h_j(x^*(\tau_k), u^*(\tau_k^-), \tau_k)}{\partial x_i} \tag{39}$$

(If $\tau_k = t_0$, $\beta_k^- = 0$ and we let $p_i(\tau_k^-)$ be $p_i(t_0)$. If $\tau_k = t_1$, $\beta_k^+ = 0$ and we put $p_i(\tau_k^+) = p_i(t_1)$.)
For all $u \in U$:

$$\sum_{j=1}^{s} \beta_{kj}^+ \frac{\partial h_j(x^*(\tau_k), u^*(\tau_k^+), \tau_k)}{\partial u} [u - u^*(\tau_k^+)] \leq 0 \tag{40a}$$

$$\sum_{j=1}^{s} \beta_{kj}^- \frac{\partial h_j(x^*(\tau_k), u^*(\tau_k^-), \tau_k)}{\partial u} [u - u^*(\tau_k^-)] \leq 0 \tag{40b}$$

Moreover,

$$\beta_{kj}^+ \geq 0 (= 0 \text{ if } h_j(x^*(\tau_k), u^*(\tau_k^+), \tau_k) > 0) \tag{41a}$$

$$\beta_{kj}^- \geq 0 (= 0 \text{ if } h_j(x^*(\tau_k), u^*(\tau_k^-), \tau_k) > 0) \tag{41b}$$

Finally, for some $t \in [t_0, t_1]$,

$$(p_0, p(t), q(t), \beta_1^-, \ldots, \beta_N^-, \beta_1^+, \ldots, \beta_N^+) \neq (0, \ldots, 0). \quad \blacksquare \tag{42}$$

To show that $p_0 \neq 0$, sometimes even the following necessary condition is useful: Defining $p(\tau_k) = p(\tau_k^+) + \beta_k^+ \partial h^*(\tau_k^+)/\partial x$ if $\tau_k \in (t_0, t_1)$, then (38a) holds for *all* $t \in [t_0, t_1]$ and both for $u^*(t)$ equal to $u^*(t^+)$ and $u^*(t^-)$ if $t \in (t_0, t_1)$.

The possible jump points of $p(t)$ are not a-priori known. If we want to examine if $t' \in [t_0, t_1]$ is a possible jump point, we put $t' = \tau_k$ and look for possible values of β_{kj}^+ and β_{kj}^- that are consistent with (40) and (41). Using (39) we can then often see if t' is a jump point or not. Note, in particular, that if $h_j^*(t'^+) > 0$ and $h_j^*(t'^-) > 0$ for all j, then by (41) $\beta_{kj}^+ = \beta_{kj}^- = 0$ for all j, so t' is not a jump point.

If $U = R^r$ and if the c.q. (28(i)) holds for $t'^- = \tau_k^-$, then $\beta_{kj}^- = 0$ and if (28(ii)) holds for $t'^+ = \tau_k^+$, then $\beta_{kj}^+ = 0$. For these conclusions to hold, (28) can be replaced by condition (37). Of course, when $U = R^r$, (40a, b) reduce to $\Sigma \beta_{kj}^+ (\partial h_j^*(\tau_k^+)/\partial u) = 0$ and $\Sigma \beta_{kj}^- (\partial h_j^*(\tau_k^-)/\partial u) = 0$.

(For $p_0 = 1$, the conditions in the theorem are sufficient if the concavity conditions (59) and (60) in the next section hold. The condition (59) can be replaced by (63) if $U = R^r$ and if the c.q. (28) holds v.e. along the candidate proposed.)

Example 3 Consider again example 1 but change the time interval to $[0, \frac{1}{2} + 2 \ln 2]$. Thus the criterion is now

$$\max \int_0^T x(t)dt, \qquad T = \tfrac{1}{2} + 2 \ln 2. \tag{43}$$

For this problem all the constraint qualifications mentioned above fail at $t = T$ for any admissible pair $x(t)$ where $x(T) = 3$. If we carry out the procedure (i), (ii) in note 4 using theorem 1 together with note 5, we run into the difficulty of getting a large number of candidates.

A way out in this problem is to use the ANC result, where no constraint qualification is needed. In fact, one can prove that there is a unique pair $(x^*(t), u^*(t))$ with $x^*(T) = 3$ which, together with appropriate multipliers, satisfy all the ANC conditions. Here we shall only argue heuristically in order to find (x^*, u^*), and then see that all the ANC conditions are satisfied for this pair.

As in example 1 it is reasonable to suggest that $u^* \equiv 1$ and $x^* = e^t - 1$ on $[0, \ln 2]$, while $u^* = 1 + 2 \ln 2 - 2t$ and $x^* = 2 - u^*$ on some interval $[\ln 2, t']$. In the present case $t' = T$ is impossible, since $u^* = 1 + 2 \ln 2 - 2t$ is less than -1 in $(1 + \ln 2, \frac{1}{2} + 2 \ln 2]$.

Looking at (43) we see that it pays to have $x(t)$ as high as possible. Since $x(t) \leqslant 2 - u(t)$ and $-u(t) \leqslant 1$, $x(t) \leqslant 3$. Note, however, that if $x(t) = 3$ for some $t' < T$, then $\dot{x}(t) = x(t) + u(t)$ becomes positive for t larger than t', by (4), contradicting $x(t) \leqslant 3$. We guess that $x^*(T)=3$, and that $u^*(t)=-1$ in $(t', T]$. From $\dot{x}^*(t)=-1+x^*(t)$, $x^*(T)=3$ we find $x^*(t)=e^t/2\sqrt{e}+1$ on $(t', T]$. We found above another expression for $x^*(t)$ on $[\ln 2, t']$. By the continuity of $x^*(t)$ at t', it follows that t' is defined by the equation

$$\frac{1}{2\sqrt{e}}e^{t'}+1=2t'-2 \ln 2+1$$

that is

$$\frac{1}{4\sqrt{e}}e^{t'}-t'+\ln 2=0. \tag{44}$$

This equation is satisfied for $t'=\frac{1}{2}+\ln 2$, and it is easy to see that equation (44) has no other root $t' \in [\ln 2, \frac{1}{2}+2 \ln 2]$.

We continue by trying to find an appropriate adjoint function $p(t)$. We claim first that $p(t')=0$. In fact, since $U=R$, (38) reduces to (30) and (31), and condition (30) implies (17). At $t=t'=\frac{1}{2}+\ln 2$, $u^*(t)$ switches from $2-x^*(t)>-1$ to $u^*(t)=-1$, so from (17), $p(t')=0$. Since (15) is not binding on $[t', T)$, $q_3(t)=0$, so from (22), $\dot{p}(t)=-1-p(t)$, with $p(t')=0$, and hence $p(t)=2e^{1/2}e^{-t}-1$. Moreover, $q_1(t)=0$ and from (18), $q_2(t)=-p(t)$. On $(\ln 2, t')$, $q_1(t)=q_2(t)=0$, so from (18), $p(t)=q_3(t)$ and from (22), $\dot{p}(t)=-1$, i.e. $p(t)=\frac{1}{2}+\ln 2-t$. In $[0, \ln 2]$, $q_2(t)=q_3(t)=0$, $p(t)=q_1(t)$ and $p(t)=3e^{-t}-1$.

The only possible jump point is $\tau_N=T$. Since $p(t)=2\sqrt{e}\,e^{-t}-1$ on (t', T), $p(T^-)=2\sqrt{e}\,e^{-T}-1=-\frac{1}{2}$, and since $x(T)$ is free, $p(T^+)=p(T)=0$. Hence using (39),

$$-\frac{1}{2}=p(T^-)-p(T)=\beta_{N1}^- \cdot 0+\beta_{N2}^- \cdot 0+\beta_{N3}^-(-1). \tag{45}$$

(By the remark below (39), $\beta_{N1}^+=\beta_{N2}^+=\beta_{N3}^+=0$.) From (45) it follows that $\beta_{N3}^-=\frac{1}{2}$, and by (41b), $\beta_{N1}^-=0$, since $1-u^*(T)>0$. As for β_{N2}^-, since $u \in (-\infty, \infty)$, we obtain from (40b) that

$$\beta_{N1}^- \cdot (-1)+\beta_{N2}^- \cdot 1+\beta_{N3}^-(-1)=0 \tag{46}$$

which implies $\beta_{N2}^-=\beta_{N3}^-=\frac{1}{2}$.

We have now specified a pair $(x^*(t), u^*(t))$, functions $p(t)$, $q_1(t)$, $q_2(t)$, and $q_3(t)$, all defined on $[0, T]$, with $p(t)$ having a single jump at T, with $\beta_{N1}^+=\beta_{N2}^+=\beta_{N3}^+=0$, $\beta_{N1}^-=0$, $\beta_{N2}^-=\frac{1}{2}$, and $\beta_{N3}^-=\frac{1}{2}$, such that all the conditions in ANC are satisfied. (It is true, but not proven, that this is the only solution to ANC.) Using the remark at the end of note 7 we see that we have indeed found the solution to our problem.

Existence

The Filippov–Cesari existence theorem (theorem 2.8) can be generalized to cover this section's problem. The result is as follows:[4]

Theorem 2 (Filippov–Cesari. Existence) Consider problem (23)–(26). Assume that:

There exists an admissible pair $(x(t), u(t))$. (47)

The set

$$N(x, t) = \{(f_0(x, u, t) + \gamma, \ f(x, u, t)): \ \gamma \leqslant 0, \ h(x, u, t) \geqslant 0\}$$
is convex for all x and all $t \in [t_0, t_1]$. (48)

There exists a number b such that $\|x(t)\| \leqslant b$ for all admissible pairs $(x(t), u(t))$, and all $t \in [t_0, t_1]$. (49)

There exists a ball $B(0, b_1)$ in R^r which, for all x with $\|x\| \leqslant b$ and all $t \in [t_0, t_1]$, contains the set
$U(x, t) = \{u: h(x, u, t) \geqslant 0\}$. (50)

Then there exists an optimal (measurable) control. ■

Again, we can ensure only the existence of a measurable control. (See the comments following the statement of theorem 2.8.) The functions f_0, f, and h need only be continuous for theorem 2 to be valid.

Free final time problems

Consider next the case in which t_1 is not fixed, but is free to vary in an interval $[T_1, T_2]$, $t_0 \leqslant T_1 < T_2$, T_1, T_2 fixed. Then an admissible triple $(x(t), u(t), t_1)$ consists of a number t_1 in $[T_1, T_2]$, and a pair of functions $(x(t), u(t))$ defined on $[t_0, t_1]$ satisfying (24)–(26). In this case we have the following theorem:

Theorem 3 (Necessary conditions. Free final time)[5] Let $(x^*(t), u^*(t), t_1^*)$ be

[4]Cesari (1983), sections 9.3, 9.5.
[5]If $u^*(t)$ is measurable, the modifications needed are the same as those needed in theorem 1 (see footnote 1). In addition, the left-hand side of (51) must be replaced by sup $\{H(x^*(t_1^*), u, p(t_1^*), t_1^*): h_k(x^*(t), u, t) > 0$ for all $k\}$ (which is finite). The theorem is derived from theorem 1 by treating t as a state variable. At least when $u^*(t)$ is left continuous (or left regular) at t_1, the theorem apparently holds without assuming that $\partial f_i / \partial t$ and $\partial h_k / \partial t$ exist.

an admissible triple solving problem (23)–(26), with $t_1 \in [T_1, T_2]$, and t_1 free. Assume that $\partial f_i/\partial t$, $i = 0, 1, \ldots, n$ and $\partial h_k/\partial t$, $k = 1, \ldots, s$ exist and are continuous. Suppose also that the constraint qualification (28) is satisfied for all $t \in [t_0, t_1^*]$. Then all the conclusions (29)–(35) are valid for $t_1 = t_1^*$. In addition

$$H(x^*(t_1^*), u^*(t_1^{*-}), p(t_1^*), t_1^*) \begin{cases} \leqslant 0 \text{ if } t_1^* = T_1 \\ = 0 \text{ if } T_1 < t_1^* < T_2 \\ \geqslant 0 \text{ if } t_1^* = T_2. \end{cases} \tag{51}$$

Sufficient conditions for a considerably more general problem are given in theorem 6.17.

Theorem 2.12 which dealt with existence for free final time problems can also be generalized:

Theorem 4 (**Filippov–Cesari. Existence**)[6] Consider problem (23)–(26) but assume that t_1 is free to vary in an interval $[T_1, T_2]$. Suppose that the conditions in theorem 2 are satisfied on $[t_0, T_2]$. Then there exists an optimal (measurable) control. ■

Exercise 4.2.1 Referring to note 2, give a rough "proof" of (36). (Hint:

$$\dot{L}^* = \frac{\partial L^*}{\partial x} \cdot \dot{x}^* + \frac{\partial L^*}{\partial u} \cdot \dot{u}^* + \frac{\partial L^*}{\partial p} \cdot \dot{p} + \frac{\partial L^*}{\partial q} \cdot \dot{q} + \frac{\partial L}{\partial t}. \tag{52}$$

Using (33), (31), $\partial L^*/\partial p = f^* = \dot{x}^*$, $\partial L^*/\partial q = h^*$, show that

$$\dot{L}^* = h^* \cdot \dot{q} + \partial L/\partial t. \tag{53}$$

By (32), $q \cdot h^* \equiv 0$, i.e. $h^* \cdot \dot{q} = -\dot{h}^* \cdot q \equiv 0$ (if $q(t) > 0$, $h^* \equiv 0$ near t, so $\dot{h}(t) \equiv 0, \ldots$).)

Exercise 4.2.2 Solve the problem

$$\max \int_0^T u(t)dt, \quad \dot{x}(t) = ax(t) - u(t),$$

$$x(0) = x_0 > 0, \quad x(T) \geqslant x_T \tag{54}$$

---

[6]See footnote 4.

$$c \leqslant u(t) \leqslant ax(t), \tag{55}$$

$$a > 0, \qquad c > 0, \qquad T > 1/a, \qquad ax_0 > c,$$
$$x_0 \leqslant x_T < (x_0 - c/a)e^{aT} + c/a, \qquad (T \text{ fixed}). \tag{56}$$

(Interpretation: A simple growth model with subsistence level c.)

Exercise 4.2.3 Consider the problem

$$\max \int_0^T (-u(t) - x(t)) dt, \qquad \dot{x}(t) = -u(t),$$

$$x(0) = 1, \qquad x(T) \text{ free}, \qquad T \text{ fixed} > 1. \tag{57}$$

$$0 \leqslant u(t) \leqslant x(t). \tag{58}$$

(i) Show that all admissible solutions satisfy $x(t) > 0$, and that the constraint qualification is satisfied for all admissible pairs.
(ii) Solve the problem.

3. Sufficient conditions

Earlier, in section 1 we indicated a sufficiency theorem for problem (23)–(26), and with example 1 saw an application of the result. We now want to formulate the precise theorem.

Theorem 5 (**Sufficient conditions. Mangasarian type**) Let $(x^*(t), u^*(t))$ be an admissible pair for problem (23)–(26). Assume that this pair together with a continuous and piecewise continuously differentiable function $p(t)$ and a piecewise continuous function $q(t)$, satisfy (31)–(33), and (35) with $p_0 = 1$. Moreover, assume that for each $t \in [t_0, t_1]$,

$$H(x, u, p(t), t) \text{ is concave in } (x, u), \tag{59}$$

and

$$h_k(x, u, t), \ k = 1, \dots, s \text{ are quasi-concave in } (x, u). \tag{60}$$

Then $(x^*(t), u^*(t))$ solves the problem. If $H(x, u, p(t), t)$ is strictly concave in (x, u), then the pair $(x^*(t), u^*(t))$ is unique. ∎

Proof The proof runs along the same lines as the proof of Theorem 2.4.

We let $(x, u) = (x(t), u(t))$ be an admissible pair for our problem, and we let * denote evaluation along the optimal pair (x^*, u^*). A compact scalar-product and matrix notation is used. (You might prefer to read the proof as if $n = r = s = 1$.)

$$\Delta = \int_{t_0}^{t_1} (f_0^* - f_0) dt = \int_{t_0}^{t_1} (H^* - H) dt + \int_{t_0}^{t_1} p \cdot (\dot{x} - \dot{x}^*) dt$$

$$\overset{(1)}{\geqslant} \int_{t_0}^{t_1} \left[\left(-\frac{\partial H^*}{\partial x} \right) \cdot (x - x^*) + \left(-\frac{\partial H^*}{\partial u} \right) \cdot (u - u^*) \right] dt + \int_{t_0}^{t_1} p \cdot (\dot{x} - \dot{x}^*) dt$$

$$\overset{(2)}{=} \int_{t_0}^{t_1} \left[\left(\dot{p} + q\frac{\partial h^*}{\partial x} \right) \cdot (x - x^*) + \left(q\frac{\partial h^*}{\partial u} \right) \cdot (u - u^*) \right] dt + \int_{t_0}^{t_1} p \cdot (\dot{x} - \dot{x}^*) dt$$

$$= \int_{t_0}^{t_1} [\dot{p} \cdot (x - x^*) + p \cdot (\dot{x} - \dot{x}^*)] dt$$

$$+ \int_{t_0}^{t_1} q \left[\frac{\partial h^*}{\partial x} \cdot (x - x^*) + \frac{\partial h^*}{\partial u} \cdot (u - u^*) \right] dt \overset{(3)}{\geqslant} 0.$$

Explanation:
(1): The inequality follows from the concavity of H (see (2.117)).
(2): Use the fact that $H = L - q \cdot h$, (33) and (31) to obtain, $-\partial H^*/\partial x = -\partial L^*/\partial x + q\partial h^*/\partial x = \dot{p} + q\partial h^*/\partial x$, $\quad -\partial H^*/\partial u = -\partial L^*/\partial u + q\partial h^*/\partial u = q\partial h^*/\partial u$.
(3): Follows from the fact that both integrals to the left of (3) are $\geqslant 0$. In fact, the first integral is $\geqslant 0$ by the same arguments as we used in connection with (2.121). As to the second integral, note that it is equal to

$$\sum_{k=1}^{s} \int_{t_0}^{t_1} q_k [(\partial h_k^*/\partial x) \cdot (x - x^*) + (\partial h_k^*/\partial u) \cdot (u - u^*)] dt.$$

Here $(x, u) \to q_k h_k(x, u, t)$ is quasiconcave, and by the admissibility of (x, u) and (32), we see that $q_k h_k(x, u, t) \geqslant q_k h_k(x^*, u^*, t) = 0$, so $q_k(\partial h^*/\partial x) \cdot (x - x^*) + q_k(\partial h^*/\partial u) \cdot (u - u^*) \geqslant 0$. (See (B.4).) Hence this second integral is also $\geqslant 0$, and $(x^*(t), u^*(t))$ is optimal. Uniqueness follows in the same way as in the proof of theorem 2.4. ∎

Arrow-type sufficient conditions

The concavity of the Hamiltonian w.r.t. (x, u) was a crucial condition in the previous theorem. Unfortunately, a number of economic models lead to control problems which fail to satisfy this concavity condition. For this reason we want to generalize the Arrow sufficiency theorem (theorem 2.5) to the present control problem. In the mixed constraint case it is natural to define the maximized Hamiltonian $\hat{H}$ by

$$\hat{H}(x, p, t) = \max \{H(x, u, p, t): u \in U(x, t)\}$$
$$\text{where } U(x, t) = \{u \in R^r: h_k(x, u, t) \geqslant 0, \ k = 1, \ldots, s\} \tag{61}$$

Examining the results in Chapter 2, section 6, it is tempting to conjecture that theorem 5 remains valid if the concavity condition on H is replaced by concavity of $\hat{H}(x, p(t), t)$ w.r.t. all "relevant values" of x. This conjecture is wrong, however. In fact, it turns out that we have to impose on $h_1, \ldots, h_s$ a constraint qualification. On the other hand, quasi-concavity of $h_1, \ldots, h_s$ is no longer required. The precise result is as follows:

Theorem 6 (**Sufficient conditions. Arrow-type.**) Theorem 5 remains valid if we assume that the c.q. (28) is satisfied for v.e. $t \in [t_0, t_1]$, if

$$\hat{H}(x^*(t), p(t), t) = H(x^*(t), u^*(t), p(t), t) \qquad \text{for v.e. } t \tag{62}$$

and if, in addition, we drop (60), and replace (59) by the following assumption: For v.e. t,

if we define $A_1(t) = \{x: \text{for some } u, \ h_k(x, u, t) \geqslant 0 \text{ for } k = 1, \ldots, s\}$,
then $\hat{H}(x, p(t), t)$ is concave on $A_1(t)$ if $A_1(t)$ is convex.
If $A_1(t)$ is not convex, we assume that $\hat{H}$ has a concave
extension to $\mathrm{co}(A_1(t))$, the convex hull of $A_1(t)$. ∎ (63)

A proof of this result is given in Seierstad and Sydsæter (1977), section 7. While a constraint qualification (c.q.) is a standard requirement in the formulation of necessary conditions for problems with mixed constraints, one might really wonder if a c.q. is "necessary for sufficiency". In exercise

4.3.1 we exhibit an example which shows that condition (28) cannot be dropped from theorem 6. (In fact, this example will also indicate that condition (28) cannot be replaced by any of the other weaker standard constraint qualifications occurring in necessary conditions.)

No constraint qualification is needed however in theorem 5. Nevertheless, if a constraint qualification is not satisfied, the conditions in the theorem will frequently be satisfied only if we allow $p(t)$ to be discontinuous. (See note 7 and theorem 6.2).

Note 8   Assume we impose on the pair $(x(t), u(t))$ as an additional condition for admissibility, that $x(t)$ has to belong to a fixed convex set $A(t)$ for each t. Provided $x^*(t)$ is an interior point in $A(t)$ for each t, theorem 6 still holds, and $\hat{H}(x, p(t), t)$ need to be concave (or have a concave extension) only on $A(t) \cap \mathrm{co}A_1(t)$.

For sufficient conditions for free final time problems we refer to theorem 6.7.

Note 9* (Weaker regularity conditions)   In the whole of this chapter, except for theorem 3, it suffices to assume that the functions f_i and h_k and their partial derivatives w.r.t. x_j and u_j are "regular" in the sense of note 2.6(b). In fact, the partial derivatives need only exist and be regular in an open set D containing the points $(x^*(t), u^*(t^+), t)$ and $(x^*(t), u^*(t^-), t)$ for all $t \in [t_0, t_1]$.

In the necessary conditions the constraint qualification has to hold for both left, and right limits of h'_u at any $t \in C$, (i.e. for $h'_u(x^*(t), u^*(t^-), t^-)$ and $h'_u(x^*(t), u^*(t^+), t^+)$.

In note 8 (the Arrow case), f_0 and f need only be defined and be regular for $x \in A(t)$.

In theorem 5 (the Mangasarian case), if $D \supset A = \{(x, u, t): h(x, u, t) \geqslant 0, t \in [t_0, t_1]\}$, f_0 and f need only be defined and regular in D.

This note is e.g. of use in the following situation: If some of the restrictions $h_k \geqslant 0$ are required to hold only in a subinterval I_k of $[t_0, t_1]$, then we can redefine h_k to be identically equal to 1 outside I_k, and formally require that for this redefined h_k, $h_k(x(t), u(t), t) \geqslant 0$ for all t. Clearly, this inequality is inactive for $t \notin I_k$.

Exercise 4.3.1   (Seierstad and Sydsæter (1977).) Consider the problem

$$\max \int_0^1 [(u-1)^2 + x]dt, \qquad \dot{x} = u, \qquad x(0) = 0, \qquad x(1) \text{ free}, \qquad (64)$$

$$h_1(x, u, t) = u - \tfrac{1}{2}x \geqslant 0, \qquad h_2(x, u, t)$$

$$= 2 - u \geqslant 0, \qquad h_3(x, u, t) = u \geqslant 0. \tag{65}$$

(i) Show that $(\bar{x}, \bar{u}) = (0, 0)$ and $(x^*, u^*) = (2t, 2)$ are both admissible pairs, and compute the value of the criterion function in each case.

(ii) Show that with $p(t) \equiv 0$, $q_1(t) \equiv 2$, $q_2(t) \equiv q_3(t) \equiv 0$, all the conditions in theorem 6 are satisfied by $(\bar{x}, \bar{u})$ except the c.q. (28).

(iii) Prove the optimality of (x^*, u^*) by applying theorem 6.

Exercise 4.3.2 Solve the problem

$$\max \int_a^{1/2} [-(u(t))^2 - x(t)] dt,$$

$$\dot{x}(t) = -u(t), \qquad x(a) = 7/4, \qquad x(1/2) \text{ free} \tag{66}$$

$$x(t) \geqslant u(t) \tag{67}$$

where $a = -1 - \ln 3$.

4. Some extensions and some examples

Let us study some variants and extensions of the results in the previous section, and also investigate some interesting examples.

A minor but useful variant of the results in sections 2 and 3

Consider an optimal control problem with mixed constraints and assume that there are additional constraints on the control functions which do not involve the state variables. In such cases it is often inconvenient, and indeed unnecessary, to associate multipliers to each of the pure control function constraints.

Consider problem (23)–(26) and require, in addition, that

$$h_k(x(t), u(t), t) = \hat{h}_k(u(t), t) \geqslant 0, \qquad k = s + 1, \ldots, \hat{s}. \tag{68}$$

With I_t^- and I_t^+ defined as in (27), with $k \in \{1, \ldots, \hat{s}\}$, we can easily derive from theorem 1 the following result:

Theorem 7 (Necessary conditions) Consider problem (23)–(26), (68). Theorem 1 remains valid if $k = 1, \ldots, s$ in (27), (28) and (30) are replaced by

$k = 1, \ldots, \hat{s}$, and if (31) is replaced by the condition that for all t,

$$\sum_{j=1}^{r} \frac{\partial L^*}{\partial u_j} (u_j - u_j^*(t)) \leqslant 0 \qquad \text{for all } u = (u_1, \ldots, u_r) \in \hat{U}(t), \tag{69}$$

where

$$\hat{U}(t) = \{u: \sum_{j=1}^{r} (\partial \hat{h}_k^* / \partial u_j)(u_j - u_j^*(t)) \geqslant 0 \text{ for all } k > s, \, \hat{h}_k(u^*(t), t) = 0\}. \quad \blacksquare$$

Note 10 Let $\varphi(t) = \Sigma_{j=1}^{r}(\partial L^*/\partial u_j)(u_j - u_j^*(t))$. From (69) it follows that for all $u \in \hat{U}(t^-)$, $\varphi(t^-) \leqslant 0$ for all $t \in (t_0, t_1]$ and for all $u \in \hat{U}(t^+)$, $\varphi(t^+) \leqslant 0$ for all $t \in [t_0, t_1)$, where the definitions of $\hat{U}(t^-)$ and $\hat{U}(t^+)$ should be obvious. The result is used in examples below.

The sufficient conditions in theorem 5 are easily adapted to the present problem.

Theorem 8 (Sufficient conditions) Consider problem (23)–(26), (68). Theorem 5 remains valid provided (31) is replaced by

$$\sum_{j=1}^{r} \frac{\partial L^*}{\partial u_j} (u_j - u_j^*(t)) \leqslant 0 \qquad \text{for all } (u_1, \ldots, u_r) \in U(t)$$

where $U(t) = \{u: \hat{h}_k(u, t) \geqslant 0, \, k = s+1, \ldots, \hat{s}\}. \quad \blacksquare$ (70)

(The proof of this result is similar to the proof of theorem 5: In the present case we replace $-(\partial H^*/\partial u) \cdot (u - u^*)$ in (1) of the proof by $(-\partial L^*/\partial u) \cdot (u - u^*) + q(\partial h^*/\partial u) \cdot (u - u^*)$. Because of (70), the equality in (2) becomes $\geqslant$.)

Note 11 If $\hat{h}_k(u, t)$, $k = s+1, \ldots, \hat{s}$ are quasi-concave in u, then (70) implies (69).

The regularity conditions on f_0, f and h in theorems 7 and 8 can be those described for the necessary conditions and the Mangasarian case in note 9.

In the next example we see an application of theorems 7 and 8.

Example 4 (Resource extraction in an open economy) Consider a country with a stock of an exhaustible natural resource which is exported, but whose volume of export is too small to influence world prices of the commodity. We shall develop a model by which optimal strategies can be determined for the extraction of the resource. (A more general model is studied by Aarrestad (1979). See also example 5.)

Let $x(t)$ be the total stock of the resource (say oil) remaining at time t and let $u(t)$ denote the rate of extraction. Assume that $u(t)$ cannot exceed $\bar{u}$, the maximal rate of extraction (determined by the production equipment). Suppose the country can lend or borrow abroad, $y(t)$ being the net foreign credit balance at time t. Letting $v(t) = \dot{y}(t)$, assume that $v(t)$ cannot fall below $-\bar{v}$. The absorption rate v can be interpreted as the maximum rate of reduction the international capital market will allow in the country's net credit position. The loan interest is a. The price of the resource equals e^{rt}. Income from other economic activities equals the constant $\bar{c}$. All together, $\bar{c} + u(t)e^{rt} + ay(t) - v(t)$ is available for consumption. Suppose additionally that the country demands a non-negative contribution to consumption from the resource extraction and from its international financial activities i.e. $u(t) + ay(t) - v(t) \geq 0$ for all t. Assuming total consumption is maximized during the planning period $[0, T]$, we confront the following control problem:

$$\max_{u,v} \int_0^T [\bar{c} + u(t)e^{rt} + ay(t) - v(t)]dt, \tag{71}$$

$$\dot{x}(t) = -u(t), \qquad x(0) = x_0 > 0, \qquad x(T) \geq 0, \tag{72}$$

$$\dot{y}(t) = v(t), \qquad y(0) = y_0 > 0, \qquad y(T) \geq 0, \tag{73}$$

$$u(t)e^{rt} + ay(t) - v(t) \geq 0, \tag{74}$$

$$u(t) \in [0, \bar{u}], \qquad v(t) \in [-\bar{v}, \infty), \qquad \bar{u} \text{ and } \bar{v} \text{ positive constants.} \tag{75}$$

We shall further assume that:

$$r > a, \tag{76(i)}$$

$$\bar{u}T > x_0, \tag{76(ii)}$$

$$\bar{v}T > y_0, \tag{76(iii)}$$

$$\bar{v} > a\bar{u}Te^{rT}. \tag{76(iv)}$$

By assuming (76(i)), we insist that the rate of growth of the resource price is higher than the rate of interest on foreign bonds. Condition (76(ii)) amounts to assuming that the economy has the productive capacity to more than exhaust the resource stock in the planning period. Assumption (76(iii)) implies that borrowing at the maximal rate for all t results in a negative credit position at the terminal time. Condition (76(iv)) has an ad-hoc nature. It reduces the number of cases needing consideration. The restrictions (74) and (75) can be represented in the form (26) and (68) if we define

$$h_1(x, y, u, v, t) = ue^{rt} + ay - v, \qquad \hat{h}_2(u, v, t) = u,$$

$$\hat{h}_3(u, v, t) = \bar{u} - u, \qquad \hat{h}_4(u, v, t) = v + \bar{v}.$$

In the present case it turns out that *all* admissible pairs satisfy the constraint qualification. In fact, consider the matrix

$$\begin{bmatrix} \partial h_1/\partial u & \partial h_1/\partial v \\ \partial \hat{h}_2/\partial u & \partial \hat{h}_2/\partial v \\ \partial \hat{h}_3/\partial u & \partial \hat{h}_3/\partial v \\ \partial \hat{h}_4/\partial u & \partial \hat{h}_4/\partial v \end{bmatrix} = \begin{bmatrix} e^{rt} & -1 \\ 1 & 0 \\ -1 & 0 \\ 0 & 1 \end{bmatrix}.$$

If only one of the constraints $h_1 \geq 0$, $\hat{h}_2 \geq 0$, $\hat{h}_3 \geq 0$, $\hat{h}_4 \geq 0$ is active at some point t for an admissible pair $(x(t), y(t), u(t), v(t))$, then the rank condition is obviously satisfied. We observe that the constraints $\hat{h}_2 \geq 0$ and $\hat{h}_3 \geq 0$ cannot be active at the same time. Moreover, if any other pair of constraints are active at some t, then again the rank condition is clearly satisfied. The rank condition will fail only if the constraints $h_1 \geq 0$ and $\hat{h}_4 \geq 0$ are active together with either $\hat{h}_2 \geq 0$ or $\hat{h}_3 \geq 0$ at some point $t \in [0, T]$. In fact, we claim that an admissible quadruple $(x^*(t), y^*(t), u^*(t), v^*(t))$ cannot satisfy both $h_1 = 0$ and $\hat{h}_4 = 0$ for the same value of t. Suppose, on the contrary, that

$$u^*(t')e^{rt'} + ay^*(t') - v^*(t') = 0 \quad \text{and} \quad v^*(t') = -\bar{v}. \tag{77}$$

Then $ay^*(t') = -\bar{v} - u^*(t')e^{rt'} \leq -\bar{v}$, so $y^*(t') \leq -\bar{v}/a$. Let $T' = \sup \{t: y^*(s) < 0 \text{ on } [t', t)\}$. For $t \in [t', T')$, $y^*(t) < 0$ and, using (74), (75), $\dot{y}^*(t) = v^*(t) \leq u^*(t)e^{rt} + ay^*(t) \leq \bar{u}e^{rT'}$. Hence,

$$y^*(T') - y^*(t') = \int_{t'}^{T'} \dot{y}^*(t)dt \leq \bar{u}e^{rT'}(T' - t') \leq \bar{u}T'e^{rT'}.$$

Thus $y^*(T') \leq y^*(t') + \bar{u}T'e^{rT'} \leq -\bar{v}/a + \bar{u}Te^{rT} < 0$, because of (76(iv)). This means that $T' = T$ and $y^*(T) < 0$, which contradicts admissibility. In this way we have taken care of the constraint qualification.

We proceed by examining the necessary conditions. If we let $p_1 = p_1(t)$

and $p_2 = p_2(t)$ be the adjoint variables associated with (72) and (73), and $q = q(t)$ be the multiplier associated with (74), the Lagrangian is

$$L = p_0(\bar{c} + ue^{rt} + ay - v) + p_1(-u) + p_2 v + q(ue^{rt} + ay - v). \tag{78}$$

Suppose $(x^*(t), y^*(t), u^*(t), v^*(t))$ solves problem (71)–(75). Then, there exist a number p_0, either $p_0 = 0$ or $p_0 = 1$, continuous functions $p_1(t)$, $p_2(t)$, and a piecewise continuous function $q(t)$, such that for v.e. $t \in [0, T]$: $(p_0, p_1(t), p_2(t)) \neq (0, 0, 0)$,

$$(p_0 e^{rt} - p_1(t))(u^*(t) - u) + (-p_0 + p_2(t))(v^*(t) - v) \geq 0$$

for all (u, v) where

$$ue^{rt} + ay^*(t) - v > 0, \quad u \in (0, \bar{u}), \quad \text{and} \quad v > -\bar{v} \tag{79}$$

(a slight rearranging of (30)), and according to (69), for v.e. t,

$$\frac{\partial L^*}{\partial u}(u - u^*(t)) + \frac{\partial L^*}{\partial v}(v - v^*(t))$$

$$= (p_0 e^{rt} - p_1(t) + q(t)e^{rt})(u - u^*(t))$$

$$+ (-p_0 + p_2(t) - q(t))(v - v^*(t)) \leq 0 \tag{80}$$

for all (u, v) such that

$$\frac{\partial \hat{h}_2^*}{\partial u}(u - u^*(t)) + \frac{\partial \hat{h}_2^*}{\partial v}(v - v^*(t)) \geq 0 \qquad \text{if } u^*(t) = 0$$

$$\frac{\partial \hat{h}_3^*}{\partial u}(u - u^*(t)) + \frac{\partial \hat{h}_3^*}{\partial v}(v - v^*(t)) \geq 0 \qquad \text{if } u^*(t) = \bar{u} \tag{81}$$

$$\frac{\partial \hat{h}_4^*}{\partial u}(u - u^*(t)) + \frac{\partial \hat{h}_4^*}{\partial v}(v - v^*(t)) \geq 0 \qquad \text{if } v^*(t) = -\bar{v}$$

(81) reduces to

$$u \geq 0 \text{ if } u^*(t) = 0, \qquad u \leq \bar{u} \text{ if } u^*(t) = \bar{u},$$

$$v \geq -\bar{v} \text{ if } v^*(t) = -\bar{v}. \tag{82}$$

The set $\hat{U}(t)$ in (69) can therefore be described as follows:

if $u^*(t) = 0$	and	$v^*(t) = -\bar{v}$,	$\hat{U}(t) = [0, \infty) \times [-\bar{v}, \infty)$	(83)
if $u^*(t) = 0$	and	$v^*(t) > -\bar{v}$,	$\hat{U}(t) = [0, \infty) \times (-\infty, \infty)$	(84)
if $u^*(t) = \bar{u}$	and	$v^*(t) = -\bar{v}$,	$\hat{U}(t) = (-\infty, \bar{u}] \times [-\bar{v}, \infty)$	(85)

$$\text{if } u^*(t)=\bar{u} \qquad \text{and} \qquad v^*(t)>-\bar{v}, \qquad \hat{U}(t)=(-\infty,\bar{u}]\times(-\infty,\infty) \quad (86)$$

$$\text{if } u^*(t)\in(0,\bar{u}) \quad \text{and} \quad v^*(t)=-\bar{v}, \qquad \hat{U}(t)=(-\infty,\infty)\times[-\bar{v},\infty) \quad (87)$$

$$\text{if } u^*(t)\in(0,\bar{u}) \quad \text{and} \quad v^*(t)>-\bar{v}, \qquad \hat{U}(t)=(-\infty,\infty)\times(-\infty,\infty). \quad (88)$$

Suppose $u^*(t)=0$ and $v^*(t)=-\bar{v}$. Then by (83), the inequality in (80) is valid for all $(u,v)\in[0,\infty)\times[-\bar{v},\infty)$. If we put $v=v^*(t)=-\bar{v}$ and (say) $u=1$, it follows that $p_0 e^{rt}-p_1(t)+q(t)e^{rt}\leqslant 0$. Putting $u=u^*(t)=0$ and (say) $v=v^*(t)+1=-\bar{v}+1$, $-p_0+p_2(t)-q(t)\leqslant 0$. With analogous arguments we can obtain the following implications of (80), and (83)–(88):

$$\frac{\partial L^*}{\partial u}=p_0 e^{rt}-p_1(t)+q(t)e^{rt}\begin{cases}\leqslant 0 \text{ if } u^*(t)=0 \\ =0 \text{ if } u^*(t)\in(0,\bar{u}) \\ \geqslant 0 \text{ if } u^*(t)=\bar{u}\end{cases} \qquad (89)$$

$$\frac{\partial L^*}{\partial v}=-p_0+p_2(t)-q(t)\begin{cases}\leqslant 0 \text{ if } v^*(t)=-\bar{v} \\ =0 \text{ if } v^*(t)>-\bar{v}\end{cases} \qquad (90)$$

Moreover, according to (32), (33), and (35), we have

$$q(t)\geqslant 0(=0 \text{ if } u^*(t)e^{rt}+ay^*(t)-v^*(t)>0) \qquad (91)$$

$$\dot{p}_1(t)=-\frac{\partial L^*}{\partial x}\equiv 0, \qquad p_1(T)\geqslant 0(=0 \text{ if } x^*(T)>0) \qquad (92)$$

$$\dot{p}_2(t)=-\frac{\partial L^*}{\partial y}=-a(p_0+q(t)),$$

$$p_2(T)\geqslant 0 \;(=0 \text{ if } y^*(T)>0). \qquad (93)$$

We next show that for $p_0=0$ the necessary conditions cannot be satisfied by any admissible quadruple $(x^*(t),y^*(t),u^*(t),v^*(t))$. Suppose $p_0=0$. From (92) and (93), $p_1(t)\equiv\bar{p}_1\geqslant 0$ and $p_2(t)\geqslant 0$ for all t, and for any t, either $\bar{p}_1>0$ or $p_2(t)>0$, since $(p_0,p_1(t),p_2(t))\neq(0,0,0)$ for all t. Assume that $p_2(t')>0$ for some t'. If $t'=T$, then $p_2(t)>0$ for $t<T$, t close to T, so we assume $t'<T$. Then by the continuity of $p_2(t)$, $p_2(t)>0$ in some interval $[t',\alpha)$. By (90), $p_2(t)\leqslant q(t)$, thus $q(t)>0$ in $[t',\alpha)$. Hence from (91), for $t\in[t',\alpha)$, $v^*(t)=u^*(t)e^{rt}+ay^*(t)$. But then according to the arguments below (77), $v^*(t)>-\bar{v}$, so from (90), $p_2(t)=q(t)$ and from (93), $\dot{p}_2(t)=-ap_2(t)$, so $p_2(t)=p_2(t')e^{-a(t-t')}$ in $[t',\alpha)$. It follows that $p_2(\alpha)>0$ and by a maximal interval argument, $\alpha=T$. Thus $h_1\geqslant 0$ is active in $(t',T]$. In fact, $h_1\geqslant 0$ is active on $[0,t']$ as well, since if $t\in[0,t']$, $0<p_2(t')\leqslant p_2(t)\leqslant q(t)$, since $p_2(t)$ is

decreasing. It follows that $\dot{y}^*(t) = v^*(t) = u^*(t)e^{rt} + ay^*(t)$ on $[0, T]$ and since $y_0 > 0$, $y(T) > 0$ (use formula (A.20)). But $p_2(T) > 0$, which contradicts (93).

If $p_2(t) \equiv 0$, then $\bar{p}_1 > 0$. From (93) $q(t) \equiv 0$, and by (89), $u^*(t) \equiv 0$, so $x^*(T) = x^0 > 0$, which contradicts $\bar{p}_1 > 0$.

Thus we can proceed by assuming that $p_0 = 1$. Note first that from (92),

$$p_1(t) \equiv \bar{p}_1 \geqslant 0 \qquad \text{for some constant } \bar{p}_1. \tag{94}$$

From (89) we see that if $\partial L^*/\partial u > 0$ for all $t \in [0, T]$, then $u^*(t) \equiv -\bar{u}$. It follows from (72) that $x^*(t) = -\bar{u}t + x_0$, so $x^*(T) = -\bar{u}T + x_0 \geqslant 0$, which contradicts (76(ii)). Hence, from (89) and (94), for some $t' \in [0, T]$,

$$e^{rt'} + q(t')e^{rt'} \leqslant \bar{p}_1. \tag{95}$$

It follows that $\bar{p}_1 > 0$ so from (92), since $x^*(T) \geqslant 0$,

$$x^*(T) = 0. \tag{96}$$

Hence the resource stock is exhausted at time T.

Suppose $v^*(t) = -\bar{v}$ for all t. Then from (73), $y^*(t) = -\bar{v}t + y_0$ and therefore $y^*(T) = -\bar{v}T + y_0 \geqslant 0$, which contradicts (76(iii)). Hence $v^*(t) > -\bar{v}$ for at least one $t \in [0, T]$. Define

$$t^* = \inf\{t : v^*(s) = -\bar{v} \text{ for all } s \in (t, T]\}, \tag{97}$$

If the set in (97) is empty, put $t^* = T$. In any case, $t^* > 0$. From the definition of t^* it follows that there exists t arbitrarily close to, and less than t^* such that $v^*(t) > -\bar{v}$. For such values of t, $\partial L^*/\partial v = 0$, i.e. $p_2(t) = 1 + q(t)$. Take a sequence of such t values which converges to t^*. Since $p_2(t)$ is continuous and $q(t)$ has left (and right) hand limits everywhere, it follows that

$$p_2(t^*) = 1 + q(t^{*-}). \tag{98}$$

Since $p_0 = 1$ and $q(t) \geqslant 0$, (93) implies that $p_2(t)$ is strictly decreasing. Take any $t < t^*$. Then, using (90), (91), and (98),

$$q(t) \geqslant p_2(t) - 1 > p_2(t^*) - 1 = q(t^{*-}) \geqslant 0.$$

We therefore conclude that

$$q(t) > 0 \qquad \text{in} \qquad [0, t^*). \tag{99}$$

From (91) and (74), we see that

$$v^*(t) = u^*(t)e^{rt} + ay^*(t) \qquad \text{in} \qquad [0, t^*), \tag{100}$$

and by (73), for all $t < t^*$,

$$\dot{y}^*(t) = u^*(t)e^{rt} + ay^*(t), \qquad y^*(0) = y_0 \geqslant 0, \tag{101}$$

from which it follows that $y^*(t) \geqslant 0$ in $[0, t^*]$. (Use e.g. (A.20).)
Hence from (100), $v^*(t) \geqslant 0$ in $[0, t^*)$, so from (90), $q(t) = -1 + p_2(t)$. Inserting this result into (93) yields $\dot{p}_2(t) = -ap_2(t)$, and thus

$$p_2(t) = p_2(0)e^{-at}, \qquad t \in [0, t^*). \tag{102}$$

Note that since $p_2(t)$ is strictly decreasing, $p_2(0) > 0$.

We prove next that $t^* < T$. Suppose on the contrary that $t^* = T$. Then from (102), $p_2(T) > 0$, so by (93), $y^*(T) = 0$. On the other hand, (101) is now valid on all of $[0, T]$. Since $u^*(t)$ is piecewise continuous, non-negative and not identically 0 ($u^*(t) \equiv 0$ contradicts (72) and (96)), (101) implies that $y^*(T) > 0$, a contradiction. Hence $t^* \in (0, T)$.

Let us examine the behaviour of $u^*(t)$. According to (89), we have to study the behaviour of $\partial L^*/\partial u$. Applying $q(t) = -1 + p_2(t)$, (94), (102) and (89), we obtain

$$\frac{\partial L^*}{\partial u} = p_2(0)e^{(r-a)t} - \bar{p}_1, \qquad t \in [0, t^*). \tag{103}$$

Since $r > a$ by (76(i)), and since $p_2(0) > 0$, $\partial L^*/\partial u$ is strictly increasing in $[0, t^*)$. In fact, we maintain it is strictly increasing in all of $[0, T]$. In $[t^*, T]$, $v^*(t) = -\bar{v}$. Since $\dot{y}^*(t) = -\bar{v}$ and $y^*(T) \geqslant 0, y^*(t) > 0$ in $[t^*, T)$. But then we also see that in $[t^*, T]$,

$$u^*(t)e^{rt} + ay^*(t) - v^*(t) = u^*(t)e^{rt} + ay^*(t) + \bar{v} > 0,$$

so from (91)

$$q(t) = 0 \qquad \text{in} \qquad (t^*, T]. \tag{104}$$

It follows from (89) that

$$\frac{\partial L^*}{\partial u} = e^{rt} - \bar{p}_1, \qquad t \in (t^*, T]. \tag{105}$$

Clearly (105) implies that $\partial L^*/\partial u$ is strictly increasing in $(t^*, T]$. Although we have also proved that $\partial L^*/\partial u$ is strictly increasing in $[0, t^*)$, $q(t)$ might be discontinuous at t^*, so that $\partial L^*/\partial u$ might behave as in fig. 4.3. In order to prove that $\partial L^*/\partial u$ is, in fact, strictly increasing on all of $[0, T]$, it suffices to prove that $q(t)$ is continuous at t^* for then it follows from (89) that $\partial L^*/\partial u$ is also continuous at t^*.

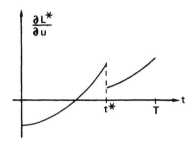

Fig. 4.3.

We already proved that $q(t)=0$ in $(t^*, T]$. Hence from (90),

$$\frac{\partial L^*}{\partial v} = -1 + p_2(t) \leqslant 0 \qquad \text{for} \qquad t \in (t^*, T].$$

Hence $p_2(t^*) \leqslant 1$, and from (98) it follows that $q(t^{*-}) \leqslant 0$. We conclude that $q(t)$ is continuous at t^*, and $q(t^*)=0$, so that $p_2(t^*)=1$.

Since $u^*(t) \equiv 0$ contradicts (72) and (96), from (89) we know that $\partial L^*/\partial u < 0$ for all $t \in (0, T)$ is impossible. Earlier we demonstrated that $\partial L^*/\partial u > 0$ for all $t \in (0, T)$ is impossible. Hence there exists a unique $t'' \in (0, T)$ such that $\partial L^*/\partial u < 0$ in $[0, t'')$ and $\partial L^*/\partial u > 0$ in $(t'', T]$. From (89) it follows that

$$u^*(t) = \begin{cases} 0 & \text{in } [0, t''] \\ \bar{u} & \text{in } (t'', T] \end{cases} \tag{106}$$

Applying (72) and (96), se see that

$$x^*(t) = \begin{cases} x_0 & \text{in } [0, t''] \\ \bar{u}(T-t) & \text{in } (t'', T] \end{cases} \tag{107}$$

Since $x^*(t)$ is continuous at t'', $x_0 = \bar{u}(T - t'')$, i.e.

$$t'' = T - (x_0/\bar{u}). \tag{108}$$

Notice that $t'' \in (0, T)$ is confirmed by (76(ii)) and the positivity of x_0 and $\bar{u}$. We see from (106) that the optimal rate of extraction is bang-bang: first an initial period with no extraction, then a final period in which the rate of extraction is maximal. Examining (108) we see that the length of the initial period is greater the greater is T or $\bar{u}$, and lower the higher is x_0. This accords with our intuition.

We already proved that $p_2(t^*)=1$. Since $p_2(t)$ is a continuous function, substituting into (102) gives $p_2(0)e^{-at^*} = 1$, i.e.

$$p_2(0) = e^{at^*}. \tag{109}$$

In $(t^*, T]$, $q(t) = 0$ so from (93), $\dot{p}_2(t) = -a$ and $p_2(t) = -a(t - T) + p_2(T.)$ In particular, $1 = p_2(t^*) = -a(t^* - T) + p_2(T)$, hence,

$$p_2(T) = a[t^* - (T - 1/a)]. \tag{110}$$

The expression for $p_2(t)$ becomes

$$p_2(t) = -a(t - t^*) + 1, \qquad t \in (t^*, T]. \tag{111}$$

Since $p_2(T) \geqslant 0$, $t^* \geqslant T - 1/a$.

How is t^* determined? We previously established that t^* is the unique point at which $v^*(t)$ switches from being determined by (100) to being equal to $-\bar{v}$. During the first interval, $y^*(t)$ is the solution to (101), in the second interval, $y^*(t) = -\bar{v}(t - T) + y^*(T)$.

Let us define $y_1(t)$ as the solution to (101) on all of $[0, T]$ and define $y_2(t) = -\bar{v}(t - T)$. The functions y_1 and y_2 are graphed in fig. 4.4.

Let t_* be the unique solution to $y_1(t) = y_2(t)$. \hfill (112)

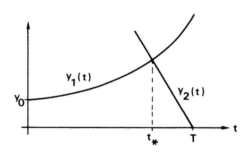

Fig. 4.4.

We see that $t^* \geqslant t_*$, since in the contrary case, $y^*(T) < 0$. Note also that if $t^* > t_*$, $y^*(T) > 0$.

Suppose first that $T - 1/a < 0$. Then from (110), $p_2(T) > 0$, so that $y^*(T) = 0$, hence we put $t^* = t_*$. On the other hand, if $T - 1/a \geqslant 0$, we consider two cases.

If (a) $t_* \geqslant T - 1/a$, $t^* \geqslant t_* \geqslant T - 1/a$. Then $t^* > t_*$ is impossible, since it implies $y^*(T) > 0$ and from (110), $p_2(T) > 0$, contradicting (93). Hence $t^* = t_*$.

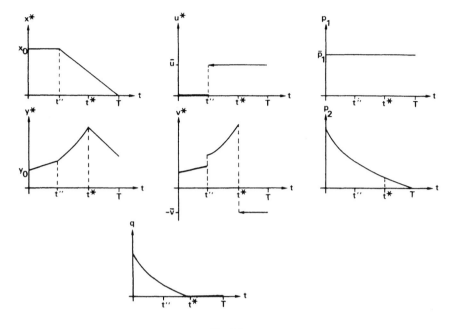

Fig. 4.5.

If (b) $t_* < T - 1/a$, then $t^* = t_*$ is impossible, since from (110) $p_2(T) < 0$. Hence $t^* > t_*$, and then $y^*(T) > 0$ so $p_2(T) = 0$, and hence from (110), $t^* = T - 1/a$.

Summarizing, for $T \geqslant 1/a$,

$$t^* = t_* \text{ if } t_* \geqslant T - 1/a, \qquad t^* = T - 1/a \text{ if } t_* < T - 1/a. \tag{113}$$

We can now conclude that the necessary conditions are satisfied by a unique candidate $(u^*(t), v^*(t))$, the form of which is described above. Since the question of constraint qualification is already resolved, it remains only to prove the existence of an optimal control. (Even though the necessary conditions produce a unique candidate for optimality, we have no guarantee that an optimal solution exists.) We apply theorem 2.

Note first that $(x(t), y(t), u(t), v(t)) \equiv (x_0, y_0, 0, 0)$ is admissible. The set $N(x, y, t) = \{(\bar{c} + ue^{rt} + ay - v) + \gamma, -u, v): \gamma \leqslant 0, ue^{rt} + ay - v \geqslant 0, u \in [0, \bar{u}], v \in [-\bar{v}, \infty)\}$ is easily proved to be convex by direct verification. Clearly, if $(x(t), y(t), u(t), v(t))$ is admissible, $|x(t)| \leqslant x_0$ and $\dot{y}(t) = v(t) \geqslant -\bar{v}$ so $y(t) \geqslant y(t_0) - \bar{v}(t - t_0) \geqslant y(t_0) - \bar{v}(T - t_0)$ for all $t \in [0, T]$. If $\dot{z}(t) = u(t)e^{rt} + az(t)$, $z(t_0) = y_0$, then using (A.20) we see that $z(t)$ is bounded above on $[0, T]$, say $\dot{z}(t) \leqslant M$ for some constant M. Since $\dot{y}(t) = v(t) \leqslant u(t)e^{rt} + ay(t)$, $y(t_0) = y_0$, it follows that $y(t) \leqslant M$ on $[0, T]$. Thus $|x(t)|$ and $|y(t)|$ are both bounded on $[0, T]$, and thus (49) is satisfied. Finally, the set $U(x, y, t)$ in (50) is here: $\{(u, v): ue^{rt} + ay - v \geqslant 0, \ u \in [0, \bar{u}], \ v \in [-\bar{v}, \infty)\}$. Thus, if $(u, v) \in U(x, y, t)$, then $u \in [0, \bar{u}]$ and $-\bar{v} \leqslant v \leqslant ue^{rt} + ay$. It follows that if $|y|$ is bounded, $U(x, y, t)$ is bounded for all $t \in [0, T]$. We therefore conclude from theorem 2 that there exists an optimal control. Since the necessary conditions produced a unique candidate it is optimal. For the case in which $t^* > t''$, fig. 4.5 illustrates the solution.

Note 12   The concavity/quasi-concavity conditions in theorem 8 are trivially satisfied for the problem in example 4, and the optimality of the solution candidate could therefore have been proved in a much easier way. In particular, there would have been no need to prove that $p_0 \neq 0$, and no need to examine the constraint qualification.

Note 13 **(Sensitivity)**   In Chapter 3, sections 5 and 6 we discussed a number of sensitivity results for control problems without constraints of the form $h(x, u, t) \geqslant 0$. For sensitivity results in more general settings, see Malanowski (1984) and references therein. For the problem in example 4 you are asked to verify some sensitivity results in exercise 4.4.1.

Example 5 **(Resource extraction in an open economy)**   We want to consider a variant of the problem in example 4:

$$\max_{u, v} \int_0^T U(\bar{c} + u(t)e^{rt} + ay(t) - v(t))dt, \tag{114}$$

$$\dot{x}(t) = -u(t), \qquad x(0) = x_0 > 0, \qquad x(T) \geqslant 0, \tag{115}$$

$$\dot{y}(t) = v(t), \qquad y(0) = y_0 \geqslant 0, \qquad y(T) \geqslant 0, \tag{116}$$

$$u(t)e^{rt} + ay(t) - v(t) \geqslant 0, \tag{117}$$

$$u(t) \in [0, \bar{u}], \ \bar{u} \text{ positive constant}, \qquad v(t) \in (-\infty, \infty) \tag{118}$$

$$\text{(i) } r > a, \qquad \text{(ii) } \bar{u}T > x_0. \tag{119}$$

In contrast to the situation in example 4 we now have a nonlinear utility function in (114). On the other hand in the present case we assume that there are no limitations on the rate of reduction of the country's net credit position, i.e. the control variable $v(t)$ is completely free.

In addition to the above assumptions we impose on the utility function the following condition:

U, U' and U'' are continuous on $(0, \infty)$, $U' > 0$, $U'' < 0$ and
$-U'(z)/U''(z) \leqslant kz$ for some constant k. (120)

We shall use the sufficient conditions of theorem 8, so we put $p_0 = 1$. Then the Lagrangian is,

$$L = U[\bar{c} + ue^{rt} + ay - v] - p_1 u + p_2 v + q(ue^{rt} + ay - v). \tag{121}$$

Suppose $(x^*(t), y^*(t), u^*(t), v^*(t))$ solves (114)–(119). Then using theorem 8 and putting $c^*(t) = \bar{c} + u^*(t)e^{rt} + ay^*(t) - v^*(t)$, we obtain the following conditions:

$$\frac{\partial L^*}{\partial u} = U'[c^*(t)]e^{rt} - p_1(t) + q(t)e^{rt} \begin{cases} \leqslant 0 & \text{if } u^*(t) = 0. \\ = 0 & \text{if } u^*(t) \in (0, \bar{u}) \\ \geqslant 0 & \text{if } u^*(t) = \bar{u} \end{cases} \tag{122}$$

$$\frac{\partial L^*}{\partial v} = -U'[c^*(t)] + p_2(t) - q(t) = 0 \tag{123}$$

$$q(t) \geqslant 0 (= 0 \text{ if } u^*(t)e^{rt} + ay^*(t) - v^*(t) > 0) \tag{124}$$

$$\dot{p}_1(t) = -\frac{\partial L^*}{\partial x} = 0, \qquad p_1(T) \geqslant 0 (= 0 \text{ if } x^*(T) > 0) \tag{125}$$

$$\dot{p}_2(t) = -\frac{\partial L^*}{\partial y} = -a(U'[c^*(t)] + q(t)),$$

$$p_2(T) \geqslant 0 (= 0 \text{ if } y^*(T) > 0) \tag{126}$$

From (125) and the continuity of $p_1(t)$ we see that $p_1(t) \equiv \bar{p}_1 \geqslant 0$ for some constant $\bar{p}_1$. Moreover, from (123) and (126), $\dot{p}_2 = -ap_2$ for v.e. t, so by continuity of $p_2(t)$,

$$p_2(t) = p_2(0)e^{-at}, \qquad \text{with } p_2(0) > 0. \tag{127}$$

$(p_2(0) = 0$ implies $p_2(t) \equiv 0$, which contradicts $U' > 0$ and $q \geqslant 0$.) In particular, (127) implies $p_2(T) > 0$, so from (126),

$$y^*(T) = 0. \tag{128}$$

Solving (123) for U' and using (127) we obtain in (122) for v.e. t,

$$\frac{\partial L^*}{\partial u} = p_2(0)e^{(r-a)t} - \bar{p}_1. \tag{129}$$

Here the right-hand side is strictly increasing in t since $r > a$ and $p_2(0) > 0$. From (122) it follows that $u^*(t)$ has at most one switch from 0 to $\bar{u}$. The extreme policy $u^*(t) \equiv 0$ is impossible: In this case from (115), $x^*(t) \equiv x_0 > 0$, so by (125), $\bar{p}_1 = 0$. Hence, from (122) and (129), $p_2(0)e^{(r-a)t} \leqslant 0$, a contradiction. Moreover, $u^*(t) \equiv \bar{u}$ is also impossible due to (115) and (119(ii)). We therefore conclude from (122) that for some $t'' \in (0, T)$, $\partial L^*/\partial u < 0$ in $[0, t'')$ and $\partial L^*/\partial u > 0$ in $(t'', T]$. Consequently, observing (129), it is obvious that $\bar{p}_1 = 0$ is impossible. Hence $\bar{p}_1 > 0$, and from (125), $x^*(T) = 0$; the resource stock is exhausted at T. From (122) we deduce the behaviour of $u^*(t)$ and, in turn, we obtain $x^*(t)$ from (115):

$$u^*(t) = \begin{cases} 0 & \text{in } [0, t''] \\ \bar{u} & \text{in } (t'', T] \end{cases} \tag{130}$$

$$x^*(t) = \begin{cases} x_0 & \text{in } [0, t''] \\ -\bar{u}(t - T) & \text{in } (t'', T]. \end{cases} \tag{131}$$

Since $x^*(t)$ is continuous at t'', $x_0 = -\bar{u}(t'' - T)$, so

$$t'' = T - (x_0/\bar{u}). \tag{132}$$

Next note that the term $ay(t)$ in the integrand of (114) along with the assumptions on U indicates that it pays to save in the beginning and consume later. We guess therefore that if consumption is at the minimal level $\bar{c}$ in some interval (i.e. (117) is active), then that interval occurs at the beginning, say $[0, t']$. Hence we assume that $v^*(t) = u^*(t)e^{rt} + ay^*(t)$ and thus $c^*(t) = \bar{c}$ on $[0, t']$. Then $y^*(t)$ must be a solution of

$$\dot{y} = u^*(t)e^{rt} + ay, \qquad y(0) = y_0, \qquad t \in [0, t']. \tag{133}$$

On $(t', T]$ we assume that (117) is inactive (i.e. $c^*(t) > \bar{c}$) and thus $q(t) = 0$. Combining (123) and (126), it is seen that $c^*(t) = \bar{c} + u^*(t)e^{rt} + ay^*(t) - \dot{y}^*(t)$ must be a solution of $(d/dt)U'(c) = -aU'(c)$. Moreover, using the properties of U we can prove quite easily, (see Seierstad and Sydsæter, 1981), that $c^*(t)$ is continuous also at t'. Hence, on $[t', T]$, $c^*(t)$ must satisfy $c^*(t') = \bar{c}$ and be a solution of

$$\dot{c} = -\frac{aU'[c]}{U''[c]}. \tag{134}$$

Once $c^*(t)$ is given as a solution of (134), $y^*(t)$ is a continuous solution on $[t', T]$ of the linear differential equation

$$\dot{y} - ay = \bar{c} + u^*(t)e^{rt} - c \qquad \text{with} \qquad c = c^*(t), \qquad t \neq t''. \tag{135}$$

Corresponding to the two possibilities $t' = 0$ and $t' > 0$, we define two types of solutions:

A pair $(c(t), y(t))$ of continuous solutions to (134) and (135) on $[0, T]$, is called *a type-A solution* if $c(0) \geqslant \bar{c}$ and $y(0) = y_0$. Notice from (134) and (120) that $\dot{c}(t) > 0$ for all t, so for type-A solutions $c(t) > \bar{c}$ for all $t > 0$, i.e. (117) is inactive on $(0, T]$.

Let $y_1(t)$ be a continuous solution to (133) on $[0, T]$. A pair of continuous functions $(c(t), y(t))$ on $[0, T]$ is called *a type-B solution* if on some interval $[0, t']$, $c(t) = \bar{c}$ and $y(t) = y_1(t)$, while on $(t', T]$, $(c(t), y(t))$ satisfies (134) and (135). (On $(t', T]$, $c(t) > \bar{c}$, by (134) and (120).)

Note that a type-A, or a type-B solution represents a real solution candidate only provided $y(T) = 0$. To find a real solution candidate we carry out the following program.

(I)   Prove that there exists a unique type-A solution $(c(t), y(t))$ for any value of $c(0)$, $c(0) \geqslant \bar{c}$.

(II)   Prove that for all $t' \in [0, T)$ there exists a unique type-B solution.

(III) Prove that there exists a type-A solution with $y(T) < 0$.

(IV)   Prove that there exists a type-B solution with $y(T) > 0$.

(V)   Prove that in (III) $y(T)$ varies continuously as $c(0)$ varies.

(VI)   Prove that in (IV) $y(T)$ varies continuously as t' varies.

(VII) From (III)–(VI) conclude that there always exists (for all admissible values of the problem data) either a type-A solution or a type-B solution where $y(T) = 0$.

Let us consider the proofs of these propositions in order:

(I)   It suffices to prove that equation (134) with $c(0) \geqslant \bar{c}$ has a unique solution on $[0, T]$, because in that case the linear differential equation (135) will obviously have a solution on $[0, T]$ with $y(0) = y_0$.

By the ordinary existence theorem for differential equations, (134) does have a solution $c(t) > 0$ on some interval $[0, \alpha)$. From (120), $0 < \dot{c}(t) \leqslant a \cdot k \cdot c(t)$, so that $c(0) < c(t) \leqslant c(0) e^{akt}$ on $(0, \alpha)$. By a standard extension theorem the solution $c(t)$ can be extended to all of $[0, T]$. (Use theorem A.4 with $A = (0, \infty)$, $A'' = [\bar{c}, \infty)$.) Uniqueness follows from theorem A.5.

(II)   The existence of a solution $y_1(t)$ to (133) on $[0, t']$ is clear since the differential equation is linear. On $(t', T]$ the existence proof is the same as in the proof of (I).

(III) If $(c(t),\ y(t))$ is a type-A solution on $[0, T]$, then by integrating the linear differential equation in (135), we find

$$y(T) = y_0 e^{aT} + \int_0^T [(\bar{c} + u^*(t)e^{rt})e^{(T-t)a} - c(t)e^{(T-t)a}]dt$$

$$\leqslant y_0 e^{aT} + \int_0^T [(\bar{c} + \bar{u}e^{rT})e^{Ta} - c(0)]dt$$

$$\leqslant y_0 e^{aT} + (\bar{c} + \bar{u}e^{rT})e^{Ta} T - c(0) T,$$

using the inequalities $u^*(t) \leqslant \bar{u}$, $c(t) \geqslant c(0)$. By choosing $c(0)$ sufficiently large, it follows that $y(T) < 0$.

(IV) From the definition of a type-B solution we see that if t' is close to T, then $y(T) > 0$. In fact, if $t' \approx T$, then $y(T) \approx y(t') = y_1(t') \approx y_1(T) > 0$.

(V) A type-A solution $(c(t),\ y(t))$ satisfies the system (134)–(135). Using theorem A.8 on $[0, t'']$ and on $(t'', T]$, we see that $y(T)$ depends continuously on $c(0)$.

(VI) A type-B solution $(c(t),\ y(t))$ satisfies on $(t', T]$ (134)–(135) with $c(t') = \bar{c}$, $y(t') = y_1(t')$. The continuous dependence of $y(T)$ on t' again follows from theorem A.8, in a way similar to the proof of (V).

(VII) Let the type-B solution corresponding to $t' = 0$ be denoted by $(\tilde{c}(t), \tilde{y}(t))$. It coincides with the type-A solution with $c(0) = \bar{c}$. If $\tilde{y}(T) \leqslant 0$, then the intermediate value theorem gives the existence of a $t' \in [0, T)$ such that the corresponding type-B solution $(c(t),\ y(t))$ satisfies $y(T) = 0$. If $\tilde{y}(T) > 0$, then the intermediate value theorem gives the existence of a type-A solution $(c(t),\ y(t))$ with $c(0) > \bar{c}$ for which $y(T) = 0$. Thus we conclude that there exists always (for all admissible values of the problem data) either a type-A solution or a type-B solution with $y(T) = 0$. Such a solution is denoted by $(c^*(t), y^*(t))$, and the associated control function is $v^*(t)$.

The final step is to verify that $P^* = (x^*(t), y^*(t), u^*(t), v^*(t))$ (whether $(c^*(t), y^*(t))$) is a type-A solution or a type-B solution) is admissible, and satisfies all the conditions (122)–(126). In addition, we must verify that

$$H = U[\bar{c} + ue^{rt} + ay - v] - p_1(t)u + p_2(t)v \text{ is concave as a}$$

function of (x, y, u, v) in the set $A(t) = \{(x, y, u, v): ue^{rt} + ay - v \geqslant 0\}$.   (136)

and

$h(x, y, u, v, t) = ue^{rt} + ay - v$ is quasi-concave as a function of
(x, y, u, v) (137)

From the fact that U is increasing and concave, (136) follows. Since h is linear, and hence quasi-concave in (x, y, u, v), (137) is also satisfied. It remains to verify that P^* is admissible and satisfies (122)–(126).

We have a unique suggestion for $x^*(t)$ and $u^*(t)$ given in (130) and (131) with the switch point $t'' = T - x_0/\bar{u}$. Obviously, (115) and (118) are satisfied (with $x^*(T) = 0$). Depending on the values of the problem data, we have either a type-A or a type-B solution for $y^*(t)$, with an associated control function $v^*(t)$. We must verify that in either case (116) and (117) are satisfied, so that P^* is admissible. Moreover, we must specify completely, continuous functions $p_1(t)$, $p_2(t)$ and a piecewise continuous function $q(t)$ such that along the admissible P^*, (122)–(126) are satisfied.

Let $(c^*(t), y^*(t))$ be a type-A solution with $y^*(T) = 0$ and $c^*(0) > \bar{c}$. If we define $v^*(t) = \dot{y}^*(t)$ we see that P^* *is* admissible. Since (117) is satisfied with strict inequality, we put $q(t) \equiv 0$. Then (124) is satisfied. The differential equation in (125) is satisfied if we put $p_1(t) \equiv \bar{p}_1$. Concerning $p_2(t)$, let us for the moment forget our suggestion in (127), and define $p_2(t) = U'[c^*(t)]$. Then (123) is satisfied and $\dot{p}_2(t) = U''[c^*(t)]\dot{c}^*(t) = -aU'[c^*(t)] = -ap_2(t)$, so (126) is also satisfied, and indeed $p_2(t)$ is as given in (127). We have left only to determine a non-negative $\bar{p}_1$ and to verify that (122) is satisfied. Since $u^*(t)$ switches from 0 to $\bar{u}$ at t'', using (123) and (127), (122) reduces to

$$p_2(0)e^{(r-a)t} - \bar{p}_1 \begin{cases} \leqslant 0 & \text{if } t < t'' \\ \geqslant 0 & \text{if } t > t''. \end{cases} \tag{138}$$

By putting $\bar{p}_1 = p_2(0)e^{(r-a)t''}$, (138) *is* satisfied. This takes care of the case in which $(c^*(t), y^*(t))$ is a type-A solution.

Suppose $(c^*(t), y^*(t))$ is a type-B solution with $t' \geqslant 0$ and $y^*(T) = 0$. Then on $[0, t']$, if we put $v^*(t) = ay^*(t) + u^*(t)e^{rt}$, then (116) and (117) are satisfied. On $(t', T]$, $c^*(t) > \bar{c}$, so (117) is satisfied, and with $v^*(t) = \dot{y}^*(t)$, then (116) is again satisfied. Hence P^* is admissible.

As in the previous case, put $p_1(t) \equiv \bar{p}_1$ on $[0, T]$. Then the differential equation in (125) is satisfied. As to $p_2(t)$ and $q(t)$, define on $[0, t']$, $p_2(t) = U'[\bar{c}]e^{at'}e^{-at}$, and $q(t) = p_2(t) - U'[\bar{c}] = U'[\bar{c}](e^{a(t'-t)} - 1)$. We then observe that (123) and (126) are satisfied and $q(t) \geqslant 0$. On $(t', T]$ we define $p_2(t) = U'[c^*(t)]$, $q(t) = 0$. Then (123) and (126) are also satisfied and again $p_2(t)$ is as given in (127) with $p_2(0) = U'[\bar{c}]e^{at'}$.

It remains to determine a non-negative $\bar{p}_1$ and to verify (122) for the type-B solution. As in the case above, (122) reduces to (138), and if we choose

$\bar{p}_1 = p_2(0)e^{(r-a)t''}$ again, we have satisfied (122). This completes the verification of all conditions (122)–(126).

In conclusion, the optimal control $u^*(t)$ is as given in (130). As for $v^*(t)$, there are two cases: Take the type-B solution $\tilde{y}(t)$ defined above. If $\tilde{y}(T) > 0$, then a type-A solution is optimal. If $\tilde{y}(T) \leqslant 0$, then a type-B solution is optimal. The numerical values of the problem data determine which of the cases prevails. (It *is* easy to show in the present model that the solution is unique.)

The analysis above has demonstrated that, depending on the numerical values of the problem data, two qualitatively quite different types of solutions are possible. The two types of solutions for the amount lent (borrowed), $y^*(t)$, are indicated in fig. 4.6. The inspiration for example 4 as well as for the present model was Aarrestad (1979). Aarrestad, however, considers only type-A solutions, due to his ad-hoc assumption "$c_v > 0$ for all $t \in [0, T]$", on p. 555.

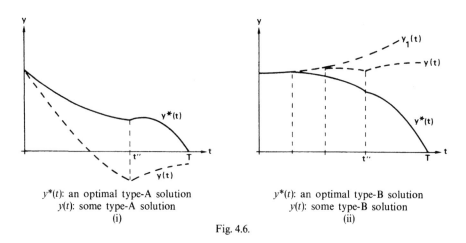

$y^*(t)$: an optimal type-A solution
$y(t)$: some type-A solution
(i)

$y^*(t)$: an optimal type-B solution
$y(t)$: some type-B solution
(ii)

Fig. 4.6.

Aarrestads' paper had as its background the problems faced by a small country like Norway, planning the exploitation of an exhaustible natural resource (North Sea oil) with ample possibilities of borrowing abroad.

Actually, Norway did follow a kind of type-B development in the seventies. In the middle of that decade it was decided to borrow abroad to keep consumption at roughly the previous level, waiting for oil-money to pour in around 1980. Notice that in the above model one can expect the possibility of a type-B solution increasing as y_0, x_0 or r decreases, or T

increases. We could prove (using the formula for $y(T)$ in the proof of III) that if $-U'/U'' \geqslant dc$ for some constant $d > 1$, where $r < ad$, then for T sufficiently large, only type-B solutions are possible.

Exercise 4.4.1 Re-examine the problem in example 4, and consider the solution for the case $t_* < T - 1/a$, $t^* = T - 1/a$, $t^* \geqslant t'' = T - x_0/\bar{u}$. Define

$$V(x_0, y_0, T) = \int_0^T [\bar{c} + u^*(t)e^{rt} + ay^*(t) - v^*(t)]dt.$$

(i) Prove that

$$y^*(t) = \begin{cases} y_0 e^{at} & [0, t''] \\ y_0 e^{at} + \dfrac{\bar{u}}{r-a}(e^{rt} - e^{at+(r-a)t''}) & (t'', t^*] \\ -\bar{v}(t - t^*) + y^*(t^*) & (t^*, T] \end{cases}$$

and

$$V(x_0, y_0, T) = A + y_0 e^{at^*} + \frac{\bar{u}}{r-a}[e^{rt^*} - e^{at^*+(r-a)t''}]$$

where $A = \bar{c}T + \bar{v}/2a + \bar{u}(e^{rl} - e^{rt^*})/r$ (which is independent of x_0 and y_0).
(ii) By (103) and $\partial L^*(t'')/\partial u = 0$, $\bar{p}_1 = e^{at^*+(r-a)t''}$. Verify that

$$\frac{\partial V}{\partial x_0} = p_1(0) = \bar{p}_1, \qquad \frac{\partial V}{\partial y_0} = p_2(0).$$

Exercise 4.4.2 Reconsider problem (114)–(119) in example 5 and assume that $U[c] = c^{1/2}$, $T = 50$, $r = 2a$, $\bar{u} = 1$, $x_0 = 30$, $y_0 = 0$, a and $\bar{c}$ are positive constants.

(i) Find suggestions for $u^*(t)$ and $x^*(t)$ and determine t''.
(ii) Find the solution $y_1(t)$ of (133) with $y_1(0) = 0$ on all of $[0, 50]$.
(iii) Assume $t' > 0$ and find the solution of (134) on $(t', 50]$.
(iv) Suppose $t' < 20$. Find for the present case the suggestion for $y^*(t)$ given in example 5.
   Hint: On $[0, t']$, $y^*(t) \equiv 0$. On $(t', 20]$ $y^*(t)$ must satisfy (135) with $y^*(t') = 0$, and $u^*(t) = 0$. On $(20, 50]$ $y^*(t)$ satisfies (135) with $u^*(t) = 1$.

Prove that on $(20, 50]$,

$$y^*(t) = \left(\frac{2\bar{c}}{a} e^{-at'} - \frac{1}{a}e^{20a}\right)e^{at}$$

$$-\frac{\bar{c}}{a} + \frac{1}{a}e^{2at} - \frac{\bar{c}}{a}e^{-2at'}e^{2at} \tag{139}$$

(v) Put $y^*(50) = 0$ in (139) and determine an admissible value for $z = e^{-at'}$. (You will get a quadratic equation in z.) Put $a = 1/10$, $\bar{c} = 10$ and compute the associated value of t'. (In this way you have found a type-B solution for $y^*(t)$.)

(vi) Prove that if $a = 1/50$, $\bar{c} = 1$, then there is a type-A solution. (Solve the problem using $c(0) = \alpha$ as a parameter and determine α such that $y^*(50) = 0$.)

Exercise 4.4.3   Prove in the general situation of example 5 with $t' < t'' < T$, that $\dot{y}^*(t)$ has an upward jump at t'' and find $\dot{y}^*(t''^+) - \dot{y}^*(t''^-)$.
(Hint: Make use of the facts that $c^*(t)$ and $y^*(t)$ are continuous at t'', while $u^*(t)$ jumps from 0 to $\bar{u}$ at t''.)

Exercise 4.4.4   Solve the problem

$$\max \int_0^1 (v(t) - x(t))dt, \qquad \dot{x}(t) = u(t), \qquad x(0) = 1/8, \qquad x(1) \text{ free} \tag{140}$$

$$u(t) \in [0, 1], \qquad (v(t))^2 \leqslant x(t), \qquad v(t) \in (-\infty, \infty). \tag{141}$$

(u and v are the controls.)

Exercise 4.4.5   (**Growth with foreign lending**)   Consider the problem

$$\max \int_0^T (1 - s)(bK + rB - u)\,dt \tag{142}$$

$$\dot{K} = s(bK + rB - u), \qquad K(0) = K_0, \qquad K(T) \geqslant K_T \tag{143}$$

$$\dot{B} = u, \qquad B(0) = 0, \qquad B(T) \geqslant 0 \tag{144}$$

$$bK + rB - u \geqslant d \tag{145}$$

$$u \geqslant \underline{u}, \qquad s \in [0, 1] \qquad (u \text{ and } s \text{ are controls}) \tag{146}$$

$$b > r, \quad T > 1/b, \qquad d/b < K_0 < K_T < K_0\, e^{bT}. \tag{147}$$

Here T, b, r, K_0, K_T, and d are positive constants, $\underline{u}$ is a negative constant. ($K(t)$ is real capital, $B(t)$ is the stock of foreign bonds, $s(t)$ is the savings rate, $bK(t)$ is production, $(1 - s(t))\,(bK(t) + rB(t) - u(t))$ is consumption, and r is the interest rate.)

(i) Suppose $(K^*(t), B^*(t), u^*(t), s^*(t))$ satisfies the necessary conditions in theorem 7 with $p_0 = 1$. Let $p_1(t)$ and $p_2(t)$ be the adjoint variables associated with (143) and (144) respectively. Prove that

$$p_2(t) - p_2(T) = \frac{r}{b}(p_1(t) - p_1(T)). \tag{148}$$

(ii) Use (69) to prove that

$$p_1(t) > 1 \Rightarrow s^*(t) = 1, \qquad p_1(t) < 1 \Rightarrow s^*(t) = 0. \tag{149}$$

$$p_2(t) < \max(1, p_1(t)) \Rightarrow u^*(t) = \underline{u} \tag{150}$$

$$p_2(t) > \max(1, p_1(t)) \Rightarrow bK^*(t) + rB^*(t) - u^*(t) = d. \tag{151}$$

(iii) Show that there exist numbers $t_1 < t' \leqslant t_2$ where $[t_1, t_2] \neq [0, T]$, such that

$$s^*(t) = 1 \text{ in } [0, t'], \qquad s^*(t) = 0 \text{ in } (t', T] \tag{152}$$

$$u^*(t) = \underline{u} \text{ in } [0, t_1], \qquad u^*(t) = bK^*(t) + rB^*(t) - d \text{ in } (t_1, t_2]$$

$$u^*(t) = \underline{u} \text{ in } (t_2, T]. \tag{153}$$

Here $p_1(t') = 1$. If $t_1 > 0$, then $p_2(t_1) = p_1(t_1)$. If $t_2 < T$, then $p_2(t_2) = 1$.

(iv) Show that for any solution of the above type, $b\dot{K}^* + r\dot{B}^* \geqslant b(bK^* + rB^*)$ in $(0, t_1)$ and $b\dot{K}^* + r\dot{B}^* \geqslant r(bK^* + rB^*)$ in (t_1, t'). Use (A.13) to show $bK^* + rB^* \geqslant 0$ in $[0, t']$. Thus $b\dot{K}^* + r\dot{B}^* \geqslant 0$, hence $bK^*(t) + rB^*(t) \geqslant bk^*(0) + rB^*(0) > d$ in $[0, t']$. Show that $\dot{B}^* \geqslant 0$ in (t', t_2) and that $bK^*(t) \geqslant bK^*(0) > d$ and $rB^*(t) \geqslant 0$ in $(t_2, T]$. Use this to show that for such solutions, (145) is never active with $u = \underline{u}$. Then show that (28) holds for such solutions. Finally, show that (63) is satisfied.

(v) Put $p_1(T) = 0$. Use $p_2(T)$ as a parameter and show that if $p_2(T)$ is so small that $t_1 = t_2 = t'$, then $B^*(T) < 0$, while $B^*(T)$ is > 0 if $p_2(T)$ is large. Conclude that for some t_1, t_2, $B^*(T) = 0$. If also $K^*(T) \geqslant K_T$, we have an ⌣ptimal solution.

(vi) Assume that we instead got $K^*(T) < K_T$ in (v). Let also $p_1(T)$ be a parameter. Show that the equations $B^*(T) = 0$, $K^*(T) = K_T$ have a solution for certain values of $p_1(T)$, $p_2(T)$, which entails that we have found an optimal solution. (Hint: Similar to (v), for each fixed $p_1(T)$, $B^*(T) = 0$ has a solution for some $t_1 = t_1(p_1(T))$, $t_2 = t_2(p_1(T))$. For $p_1(T) = 1$, $s^*(t) \equiv 1$, and $u^*(t)$ with switch points t_1^1, t_2^1 entail a $K^*(t)$ for which $K^*(T) \geqslant K_0 e^{bT}$, because in this case $u^*(t)$ is optimal in the problem of maximizing $\int_0^T (bK + rB - u) \, dt = K(T) - K_0$, ($s$ fixed equal to 1). Sufficient conditions can be used to show this last fact. Hence, the inequality $K^*(T) \geqslant K_0 e^{bT} > K_T$ then follows from the fact that $K(t) = K_0 e^{bt}$ is an admissible solution. For $p_1(T) = 0$, by assumption $K^*(T) < K_T$. For some value of $p_1(T)$ between 0 and 1 ....)

CHAPTER 5

PURE STATE CONSTRAINTS

In this chapter we shall study optimal control problems with pure state variable constraints. Briefly formulated, the problems we shall consider are of the following type:

$$\max_u \int_{t_0}^{t_1} f_0(x, u, t)\mathrm{d}t, \qquad \dot{x} = f(x, u, t), \qquad g(x, t) \geq 0, \qquad u \in U, \tag{1}$$

with initial condition $x(t_0) = x^0$ and with the standard terminal conditions (2.53). As before, $x = x(t)$ is an n-dimensional state vector, $u = u(t)$ is an r-dimensional control vector, U is a fixed set in R^r, and f is an n-dimensional vector function. The function g is an s-dimensional vector function so that the inequality $g(x, t) \geq 0$ represents the s inequalities

$$g_j(x(t), t) \geq 0, \qquad j = 1, \ldots, s, \qquad \text{for all } t \in [t_0, t_1]. \tag{2}$$

Note that in contrast to the situation in Chapter 4 the control function $u(t)$ does not appear in the constraints (2).

Due to the complexity of the present problem we begin this chapter with some preliminary considerations which, although not quite precise, do convey the main idea. We then will proceed to present a precise version of a sufficiency result for problem (1), and study an example. We depart from our previous routine of presenting necessary conditions first. This is done because sufficient conditions are easier to grasp than necessary conditions for the present problem.

In the literature there are two seemingly quite different ways of handling pure state constraints. Probably the most natural one is to associate a multiplier to each constraint in (2) and then try to proceed in much the same way as in the mixed constraint case. This is the approach we follow in sections 1 and 2.

An alternative approach is to associate multipliers to the time derivatives of the constraints. This approach is adopted in section 3.

1. Preliminary considerations

In order to solve the problem indicated above it seems reasonable to suggest the introduction of multipliers $\lambda_1(t), \ldots, \lambda_s(t)$ associated with the constraints (2) and the forming of the Lagrangian

$$L(x, u, p, \lambda, t) = H(x, u, p, t) + \sum_{j=1}^{s} \lambda_j(t) g_j(x, t) \tag{3}$$

where

$$H(x, u, p, t) = p_0 f_0(x, u, t) + \sum_{i=1}^{n} p_i f_i(x, u, t)$$

is the usual Hamiltonian.

From our previous experience, let us try to anticipate conditions on $(x^*(t), u^*(t))$ that are sufficient to solve problem (1). Our suggestion is as follows: With $p_0 = 1$,

$u^*(t)$ maximizes $H(x^*(t), u, p(t), t)$ for $u \in U$ (4)

$\lambda_j(t) \geqslant 0 (=0$ if $g_j(x^*(t), t) > 0),$ $j = 1, \ldots, s$ (5)

$\dot{p}_i(t) = -\partial L^*/\partial x_i,$ $i = 1, \ldots, n$ (6)

terminal conditions (2.60) (7)

$H(x, u, p(t), t)$ concave in (x, u) (8)

$g_j(x, t)$ quasi-concave in $x, j = 1, \ldots, s.$ (9)

Roughly speaking, the conditions in (4)–(9) are sufficient for $(x^*(t), u^*(t))$ to solve our problem. However, it turns out that these conditions are often not satisfied if we keep the condition that $p(t)$ is continuous. At least we have to allow $p(t)$ to jump at $t = t_1$. Let us consider an example.

Example 1 (Inventory policy)

$$\min_{u} \int_{0}^{T} (cu(t) + hx(t)) dt \tag{10}$$

$$\dot{x}(t) = u(t) - at, \qquad x(0) = 1, \qquad x(T) \text{ free} \tag{11}$$

$$g(x(t), t) = x(t) \geqslant 0 \tag{12}$$

$$u(t) \geqslant 0, \qquad (\text{i.e. } u \in U = [0, \infty)). \tag{13}$$

Assume that c, h, T and a are positive constants and that

$$T > (2/a)^{1/2}. \tag{14}$$

This problem, which has a completely trivial solution, is given the following interpretation: A shop has at time t a stock $x(t)$ of some commodity. It fulfils a demand given by at. The rate $u(t)$ at which the shop supplements its inventory is its control variable. The shop's goal is to minimize its total cost over the planning period, $[0, T]$. The cost per unit of time is the cost of acquiring the commodity, $cu(t)$, plus the storing costs, $hx(t)$.

Let us see what the conditions (4)–(9) tell us in the present cases. We have one state variable $x(t)$, one control variable $u(t)$ and one state constraint (12). Let the multipliers associated with (11) and (12) be $p(t)$ and $\lambda(t)$ respectively, and define the Hamiltonian and the Lagrangian by ($p_0 = 1$)

$$H = -(cu + hx) + p(u - at) = (p - c)u - (hx + pat) \tag{15}$$

$$L = H + \lambda x. \tag{16}$$

(We write $-(cu + hx)$ in the expression for H since (10) is being minimized rather than maximized.) Now, condition (4) reduces to

$$u^*(t) \text{ maximizes } (p(t) - c)u \qquad \text{for } u \geqslant 0. \tag{17}$$

We see that (17) requires

$$u^*(t) \equiv 0 \qquad \text{if } p(t) < c. \tag{18}$$

If $p(t) = c$, (17) tells us nothing, while $p(t) > c$ makes it impossible for any $u^*(t)$ to maximize (17).

The conditions (5)–(7) are:

$$\lambda(t) \geqslant 0 \quad (= 0 \text{ if } x^*(t) > 0) \tag{19}$$

$$\dot{p}(t) = -\frac{\partial L^*}{\partial x} = h - \lambda(t), \qquad p(T) = 0 \tag{20}$$

We see that (8) and (9) are satisfied since H is linear in (x, u) and g is linear in x.

Intuition suggests that it must pay to keep inventories as low as possible, i.e. to run the inventories down to zero as fast as possible. We propose therefore, that the optimal solution is "bang-bang": In the period $[0, t']$ let

$u^*(t) \equiv 0$ (buy nothing) until the inventory is driven to 0. In this periode, by (11), $\dot{x}^*(t) = -at$, $x^*(0) = 1$, so $x^*(t) = 1 - \frac{1}{2}at^2$. The requirement $x^*(t') = 0$ implies $t' = (2/a)^{1/2}$. In the interval $(t', T]$ let $u^*(t) \equiv at$, such that $x^*(t) \equiv 0$ in $[t', T]$. Hence our suggestions for $u^*(t)$ and $x^*(t)$ are

$$u^*(t) \equiv 0 \text{ and } x^*(t) = 1 - \tfrac{1}{2}at^2 \qquad \text{in } [0, (2/a)^{1/2}]$$

$$u^*(t) \equiv at \text{ and } x^*(t) = 0 \qquad \text{in } ((2/a)^{1/2}, T] \tag{21}$$

Since $x^*(t) > 0$ in $[0, t')$, we must, according to (19), put $\lambda(t) = 0$ in $[0, t')$. From (20), $\dot{p}(t) = h$ in $[0, t')$ so $p(t) = ht + K$ for some constant K. On $(t', T]$ we suggested $u^*(t) = at$. It follows from (17) that $p(t) = c$ for $t > t'$. Hence $\dot{p}(t) = 0$, so by (20), $\lambda(t) = h$ for $t > t'$. If we let $ht' + K = c$, i.e. $K = c - ht'$, $p(t)$ becomes continuous at t'. Then all the conditions (17)–(20) are satisfied except $p(T) = 0$. Since we strongly believe $u^*(t)$ to be optimal, it seems to be unavoidable that $p(t)$ jumps at $t = T$ from $p(T^-) = c$ to $p(T) = 0$, so that

$$p(T^-) - p(T) = c. \tag{22}$$

Hence in the present case $p(t)$ is discontinuous at T. Theorem 1 in the next section will confirm the optimality of $u^*(t)$. See example 2. (For a generalization of this model, see exercise 5.2.6)

It turns out to be the normal situation for optimal control problems with pure state space constraints that the adjoint variables associated with the differential equations have jump discontinuities at the final time point t_1 when the constraints are active at t_1. Sometimes we also have jump discontinuities in (t_0, t_1). Theorem 1 and note 1 will give us information about the direction and magnitudes of these jumps.

2. Sufficient conditions

We now offer a precise version of the sufficiency result mentioned in section 1. Consider the problem

$$\max \int_{t_0}^{t_1} f_0(x(t), u(t), t)\mathrm{d}t \tag{23}$$

subject to the vector differential equation and the initial condition,

$$\dot{x}(t) = f(x(t), u(t), t), \qquad x(t_0) = x^0 \qquad (x^0 \text{ fixed}) \tag{24}$$

the terminal conditions

$$x_i(t_1) = x_i^1 \qquad i = 1, \ldots, l, \qquad (x_i^1 \text{ all fixed}) \qquad (25a)$$

$$x_i(t_1) \geqslant x_i^1 \qquad i = l+1, \ldots, m, \qquad (x_i^1 \text{ all fixed}) \qquad (25b)$$

$$x_i(t_1) \text{ free} \qquad i = m+1, \ldots, n \qquad (25c)$$

the control variable restriction,

$$u(t) \in U, \ U \text{ a fixed set in } R^r, \qquad (26)$$

and the constraints

$$g_j(x(t), t) \geqslant 0, \qquad j = 1, \ldots, s. \qquad (27)$$

In addition to our standard continuity assumptions (2.6) on f_0 and f, we assume

$$g = (g_1, \ldots, g_s) \text{ and the matrix } g_x' \text{ are continuous w.r.t. } (x, t) \qquad (28)$$

If $u(t)$ is a piecewise continuous control which in the time interval $[t_0, t_1]$ brings $x(t)$, via (24) from x^0 to a point $x(t_1)$ satisfying (25) such that (26) and (27) always hold, then we call $(x(t), u(t))$ *an admissible pair*.

As in section 1 we introduce multipliers $\lambda_1(t), \ldots, \lambda_s(t)$ associated with the constraints (27) and we make use of the Lagrangian L defined by (3). The following result can then be proved:

Theorem 1   (Sufficient conditions. Pure state constraints)   Let $(x^*(t), u^*(t))$ be an admissible pair for problem (23)–(28). Assume that there exists a vector function $p(t) = (p_1(t), \ldots, p_n(t))$, which is continuous and piecewise continuously differentiable in $[t_0, t_1)$. Assume further that there exist a piecewise continuous vector function $\lambda(t) = (\lambda_1(t), \ldots, \lambda_s(t))$ and numbers β_j, $j = 1, \ldots, s$ such that the following conditions are satisfied with $p_0 = 1$:

$$u^*(t) \text{ maximizes } H(x^*(t), u, p(t), t) \text{ for } u \in U, \text{ for v.e. } t \qquad (29)$$

$$\lambda_j(t) \geqslant 0 (= 0 \text{ if } g_j(x^*(t), t) > 0), \qquad j = 1, \ldots, s, \text{ for all } t \qquad (30)$$

$$\dot{p}_i(t) = -\frac{\partial L(x^*(t), u^*(t), p(t), \lambda(t), t)}{\partial x_i}, \ i = 1, \ldots, n \text{ for v.e. } t. \qquad (31)$$

At t_1, $p_i(t)$ can have a jump discontinuity in which case:

$$p_i(t_1^-) - p_i(t_1) = \sum_{j=1}^{s} \beta_j \frac{\partial g_j(x^*(t_1), t_1)}{\partial x_i}; \qquad i = 1, \ldots, n \qquad (32)$$

$$\beta_j \geqslant 0 \; (=0 \text{ if } g_j(x^*(t_1), t_1) > 0), \qquad j = 1, \ldots, s \tag{33}$$

$p_i(t_1)$ no conditions	for $i = 1, \ldots, l$ (34a)
$p_i(t_1) \geqslant 0 (=0 \text{ if } x_i^*(t_1) > x_i^1)$	for $i = l+1, \ldots, m$ (34b)
$p_i(t_1) = 0$	for $i = m+1, \ldots, n.$ (34c)

$\hat{H}(x, p(t), t) = \max_{u \in U} H(x, u, p(t), t)$ is concave in $x \in R^n$

for v.e. t. (assuming that the maximum value is attained). $\tag{35}$

$g(x, t)$ is quasi-concave in x. $\tag{36}$

Then $(x^*(t), u^*(t))$ solves problem (23)–(27). ∎

When $l = n$ (fixed terminal point) we can always assume that all $\beta_j = 0$. A proof of the theorem is presented below.

Note 1 The conditions in the theorem are in some cases too restrictive. In particular, we must sometimes allow $p(t)$ to have discontinuities also at interior points of $[t_0, t_1]$.

Consider the situation where x and g are scalars, $(n = 1, s = 1)$. Whether or not $p(t)$ is continuous at some point $\tau_k \in (t_0, t_1)$, depends on the behaviour of $g(x^*(t), t)$ at τ_k. If τ_k is a point where $g(x^*(t), t)$ becomes or ceases to be active, and if $g(x^*(t), t)$ has a kink at τ_k, in the sense that $(d/dt)g(x^*(t), t)$ is discontinuous at τ_k, then it turns out that $p(t)$ is continuous at τ_k. (See fig. 5.1.). (Notice that $g(x^*(t), t)$ can only have a kink at τ_k if $u^*(t)$ is discontinuous at τ_k.)

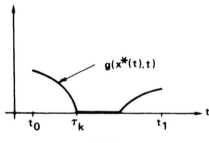

$g(x^*(t), t)$

$t_0 \qquad \tau_k \qquad t_1$

Fig. 5.1.

On the other hand, if $(d/dt)g(x^*(t), t)$ is coninuous at such points τ_k, as in fig. 5.2, then we may have to allow $p(t)$ to be discontinuous at τ_k. However, such a discontinuity is rather unusual. (See note 3 (e).) Similar remarks can also be made for the case $n > 1, s > 1$.

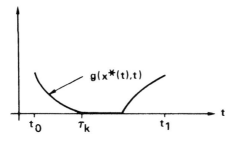

Fig. 5.2.

Since discontinuities in $p(t)$ at interior points cannot be ruled out, the following modification of theorem 1 is sometimes needed.

Assume in the setting of theorem 1 that $p(t)$ is piecewise continuous and piecewise continuously differentiable with jump discontinuities at $\tau_1, \ldots, \tau_N$, $t_0 < \tau_1 < \ldots < \tau_N \leqslant t_1$. Then theorem 1 is valid provided (32) and (33) are replaced by the following conditions: There exist numbers β_j^k, $j = 1, \ldots, s$, $k = 1, \ldots, N$ such that

$$p_i(\tau_k^-) - p_i(\tau_k^+) = \sum_{j=1}^{s} \beta_j^k \frac{\partial g_j(x^*(\tau_k), \tau_k)}{\partial x_i}; \qquad i = 1, \ldots, n, \ k = 1, \ldots, N.$$

(If $\tau_N = t_1$, put $p_i(\tau_N^+) = p_i(t_1)$.) $\qquad\qquad\qquad\qquad (37)$

$\beta_j^k \geqslant 0$. Moreover, (i) $\beta_j^k = 0$ if $g_j(x^*(\tau_k), \tau_k) > 0$ and (ii) $\beta_j^k = 0$ if $\tau_k \in (t_0, t_1)$, $g_j(x^*(\tau_k), \tau_k) = 0$ and $(\partial g_j(x^*(t), t)/\partial x) \cdot f(x^*(t), u^*(t), t)$ is discontinuous at τ_k, $j = 1, \ldots, s$, $k = 1, \ldots, N$. $\qquad (38)$

A proof of this result is given below.

Note that the discontinuity property in (38(ii)) is equivalent to the discontinuity of $(d/dt)g_j(x^*(t), t) = (\partial g_j(x^*(t), t)/\partial x) f(x^*(t), u^*(t), t) + \partial g_j(x^*(t), t)/\partial t$, since the last term is continuous.

Suppose x_i does not appear in $g_j(x_1, \ldots, x_n, t)$ for any $j = 1, \ldots, s$. Then we conclude from (37) that the corresponding adjoint variable $p_i(t)$ has no discontinuity.

Moreover, from (38(i)) we see that if the constraints (27) are not binding at any $t \in [0, T]$ along the optimal path $x^*(t)$, then $p_1(t), \ldots, p_n(t)$ are all continuous. This is in accordance with our previous results for problems in which (27) did not appear.

Finally, useful results on the continuity of $p(t)$ can also be found in note 3(e).

Note 2. Suppose $x(t)$ is also required to belong to a convex set $A(t)$ for all $t \in [t_0, t_1]$. If $x^*(t)$ belongs to the interior of $A(t)$ for all $t \in [t_0, t_1]$, then the conditions in theorem 1 are sufficient for optimality. $\hat{H}(x, p(t), t)$ needs to be concave only for $x \in A(t)$. The same result applies to the modification of theorem 1 given in note 1.

Proof of theorem 1　　The proof is similar to the proofs of theorem 4.5 and theorem 2.5, and the same compact notation is used:

$$\Delta = \int_{t_0}^{t_1} (f_0^* - f_0)dt = \int_{t_0}^{t_1} (H^* - H)dt + \int_{t_0}^{t_1} p \cdot (\dot{x} - \dot{x}^*)dt$$

$$\overset{(1)}{\geqslant} \int_{t_0}^{t_1} \left(-\frac{\partial H^*}{\partial x}\right) \cdot (x - x^*)dt + \int_{t_0}^{t_1} p \cdot (\dot{x} - \dot{x}^*)dt$$

$$\overset{(2)}{=} \int_{t_0}^{t_1} \left(\dot{p} + \lambda \frac{\partial g^*}{\partial x}\right) \cdot (x - x^*)dt + \int_{t_0}^{t_1} p \cdot (\dot{x} - \dot{x}^*)dt$$

$$\overset{(3)}{\geqslant} \int_{t_0}^{t_1} \left[\frac{d}{dt}(p \cdot (x - x^*))\right]dt = p(t_1^-) \cdot (x(t_1) - x^*(t_1)) - p(t_0) \cdot (x(t_0) - x^*(t_0))$$

$$= p(t_1) \cdot (x(t_1) - x^*(t_1)) + (p(t_1^-) - p(t_1)) \cdot (x(t_1) - x^*(t_1)) \geqslant 0.^{(4)}$$

Explanation

(1): The inequality follows if we can show that $H^* - H \geqslant -(\partial H^*/\partial x) \cdot (x - x^*)$. The latter inequality has already been established in the proof of theorem 2.5. (A more elementary argument can be given for the important special case in which H is concave in (x, u) and U is convex. See the proof of theorem 2.4.)

(2): Use $L = H + \lambda \cdot g$ and (31)

(3): Note that $\lambda_j g_j(x(t), t) \geqslant \lambda_j g_j(x^*(t), t) = 0$, by (27) and (30). Since $\lambda_j g_j(x, t)$ is quasi-concave in x, $\lambda_j(\partial g_j^*/\partial x) \cdot (x - x^*) \geqslant 0$. (See (B.4).)

(4): $p(t_1) \cdot (x(t_1) - x^*(t_1)) \geqslant 0$ by the arguments used subsequent to (2.121). From (32) we see that $(p(t_1^-) - p(t_1)) \cdot (x(t_1) - x^*(t_1)) = \sum_{j=1}^{s} \beta_j(\partial g_j^*/\partial x) \cdot (x(t_1) - x^*(t_1))$, and each term in this sum is $\geqslant 0$ by the same argument as we used in (3).　∎

Proof of the result in note 1 The argument above must be slightly adjusted due to the discontinuities of $p(t)$. Putting $r(t) = p(t) \cdot (x(t) - x^*(t))$, we obtain

$$\Delta \geq \int_{t_0}^{t_1} \left(\frac{d}{dt} r(t) \right) dt = \sum_{k=0}^{N-1} [r(\tau_{k+1}^-) - r(\tau_k^+)], \tag{39}$$

where $\tau_0 = t_0$ and $\tau_N = t_1$ (an assumption we can always make). Expanding the sum in (39) and rearranging, it is easy to verify that the sum is equal to

$$\sum_{k=1}^{N} (r(\tau_k^-) - r(\tau_k^+)) + r(\tau_N^+) - r(\tau_0^+).$$

Now, $r(\tau_N^+) = r(t_1) = p(t_1) \cdot (x(t_1) - x^*(t_1)) \geq 0$ as in (4) above, while $r(\tau_0^+) = r(t_0^+) = p(t_0)(x(t_0) - x^*(t_0)) = 0$. Moreover, applying (37)

$$r(\tau_k^-) - r(\tau_k^+) = (p(\tau_k^-) - p(\tau_k^+)) \cdot (x(\tau_k) - x^*(\tau_k))$$

$$= \sum_{j=1}^{s} \beta_j^k (\partial g_j(x^*(\tau_k), \tau_k)/\partial x)(x(\tau_k) - x^*(\tau_k)).$$

Each term in the latter sum is ≥ 0 by the quasiconcavity of $\beta_j^k g_j$ (by an argument similar to that in (3) above). We conclude that $\Delta \geq 0$, so (x^*, u^*) is optimal. ∎

Example 2 Recall the problem in example 1. The solution we suggested with $p(t) = ht + c - ht'$ in $[0, t']$, $p(t) = c$ in (t', T), $p(T) = 0$ $(t' = (2/a)^{1/2})$, satisfies all the conditions in theorem 1. In particular, (32) and (33) are satisfied with $\beta = c$.

The next example is somewhat more complicated than the previous one and it has an interesting economic interpretation.

Example 3 (Growth that pollutes) Consider the following control problem:

$$\max \int_0^T (1 - s(t))u(t)k(t)dt \tag{40}$$

$$\dot{k}(t) = s(t)u(t)k(t), \qquad k(0) = k_0, \qquad k(T) \text{ is free} \tag{41}$$

$$\dot{z}(t) = u(t)k(t) - az(t), \qquad z(0) = z_0, \qquad z(T) \text{ is free} \tag{42}$$

$$g(k(t), z(t), t) = Z - z(t) \geq 0 \tag{43}$$

$$0 \leqslant s(t) \leqslant 1, \qquad 0 \leqslant u(t) \leqslant 1 \tag{44}$$

k_0, z_0, Z, T, and a are positive constants, $z_0 < Z$, $T > 1$,

$$(z_0 - k_0/a)e^{-aT} + k_0/a < Z. \tag{45}$$

This problem is closely related to the problem in example 2.4 and the variables have the same interpretation. In the present problem the disutility of pollution does not enter directly in the criterion functional, but instead there is a fixed ceiling Z on the stock of pollution. In addition to having the savings rate $s(t)$ as a control variable, we introduce in the present problem a second control variable $u(t)$ which is the capacity utilization rate. (In example 2.4, $u(t) \equiv 1$.) Note that in (42) we have assumed that the stock of pollution, for $u(t) \equiv 0$, decays (evaporates) exponentially, with a as the rate of decay. By the final condition in (45) we assume, in effect, that if there is no capital accumulation $(s(t) \equiv 0)$, then the stock of pollution never reaches Z even if the capacity utilization rate is $u(t) \equiv 1$.

We have here a control problem with two state variables (k and z) and two control variables (u and s). Note that the control region is $U = [0, 1] \times [0, 1]$.

We shall make use of theorem 1. We put $p_0 = 1$ and define the Lagrangian function L by

$$L(k, z, u, s, p_1, p_2, \lambda, t) = H + \lambda(Z - z) \tag{46}$$

where the Hamiltonian H is

$$H(k, z, u, s, p_1, p_2, t) = [1 + (p_1 - 1)s + p_2]uk - ap_2z. \tag{47}$$

Here p_1 and p_2 are the adjoint variables associated with (41) and (42), and λ is the multiplier associated with (43).

We want to find a solution candidate $(k^*(t), z^*(t), u^*(t), s^*(t))$ which satisfies all the conditions in theorem 1. For this problem conditions (29)–(34) in theorem 1 are as follows:

$u^*(t)$ and $s^*(t)$ maximize $\{[1 + (p_1(t) - 1)s + p_2(t)]uk^*(t) - ap_2(t)z^*(t)\}$
in $U = [0, 1] \times [0, 1]$, for v.e. t. $\tag{48}$

$$\lambda(t) \geqslant 0 \, (= 0 \text{ if } z^*(t) < Z) \tag{49}$$

$$\dot{p}_1(t) = -\frac{\partial L^*}{\partial k} = -[1 + (p_1(t) - 1)s^*(t) + p_2(t)]u^*(t), \qquad p_1(T) = 0 \tag{50}$$

$$\dot{p}_2(t) = -\frac{\partial L^*}{\partial z} = ap_2(t) + \lambda(t), \qquad p_2(T) = 0 \tag{51}$$

$$p_1(T^-)-p_1(T)=\beta \cdot 0=0 \tag{52}$$

$$p_2(T^-)-p_2(T)=\beta \cdot (-1)= -\beta \tag{53}$$

$$\beta \geq 0 \,(=0 \text{ if } z^*(T)<Z). \tag{54}$$

From (52) we see that $p_1(t)$ is continuous at $t=T$. Moreover, (41), (44), and (45) imply $k^*(t) \geq k_0 >0$ everywhere.

For v.e. t, we see from (48) that $(u^*(t), s^*(t))$ maximizes $F(u, s, t)=[1+(p_1(t)-1)s+p_2(t)]u$ for $u \in [0, 1]$, $s \in [0, 1]$. Now, in general,

$$\max_{u,s} F(u,s,t)=\max_u (\max_s F(u,s,t))=\max_s (\max_u F(u,s,t))$$

Specifically for F as defined above,

$$\max_s F(u,s,t)=u[\max_s [1+(p_1(t)-1)s]+p_2(t)]$$

$$=u[\max(1,p_1(t))+p_2(t)] \tag{55}$$

where we obtained $\max_s[1+(p_1(t)-1)s]=\max(1, p_1(t))$ be considering each of the cases $p_1(t)=1$, $p_1(t)>1$ and $p_1(t)<1$ separately. Thus we see that

$$\max_{u,s} F(u,s,t)=\max_u u[\max(1,p_1(t))+p_2(t)]$$

It follows that

$$\left.\begin{array}{l}\max (1,p_1(t))+p_2(t)>0 \Rightarrow u^*(t)=1 \\ \max (1,p_1(t))+p_2(t)<0 \Rightarrow u^*(t)=0\end{array}\right\} \quad (t<T) \tag{56}$$

Since $s^*(t)$ maximizes $F(u^*(t), s, t)$, we see that if $u^*(t)=0$, $s^*(t)$ is not determined by (48). If $u^*(t)>0$, it follows that

$$p_1(t)>1 \Rightarrow s^*(t)=1$$

$$p_1(t)<1 \Rightarrow s^*(t)=0. \tag{57}$$

From (56), the maximum value of $u[\max(1, p_1(t))+p_2(t)]$ for $u \in [0, 1]$ is $r(t) \equiv \max[0, \max(1, p_1(t))+p_2(t)]$. The function $\hat{H}$, defined in (35), is therefore

$$\hat{H}(k,z,p_1(t),p_2(t),t)=r(t)k-ap_2(t)z, \tag{58}$$

which is a linear, and hence concave, function of k and z for every $t<T$.

Consider now the differential equation (50) for $p_1(t)$, which can be written as

$$\dot{p}_1(t)=-r(t), \qquad p_1(T)=0. \tag{59}$$

Hence $\dot{p}_1(t) \leqslant 0$, and since $p_1(t)$ is continuous with $p_1(T) = 0$, $p_1(t)$ is non-increasing and non-negative for all $t \in [0, T]$. Define $t^* = \sup\{t : p_1(t) \geqslant 1\}$. (If the set is empty, let $t^* = 0$.) If $t^* > 0$, then $p_1(t^*) = 1$.

Note next that since $p_2(T^-) = -\beta \leqslant 0$ and $\dot{p}_2(t) = ap_2(t) + \lambda(t)$ with $\lambda(t) \geqslant 0$, it follows by (A.20) that $p_2(t) \leqslant 0$ in $[0, T]$.

If Z is large enough, we could have $z^*(t) < Z$ for all $t \in [0, T]$. This case is referred to as case A. Case B is the one in which $z^*(t)$ touches Z at some point t', t' chosen as the smallest t with this property. One possibility is that $t' = T$, (case B_1). On the other hand, if $t' < T$, (the case B_2), we expect that $z^*(t) = Z$ in $[t', T]$, since it does not seem reasonable that we decrease the stock of pollution once it has reached the ceiling. Thus

(A) $z^*(t) < Z$ for all $t \in [0, T]$.

(B_1) $z^*(t) < Z$ for all $t \in [0, T)$, $z^*(T) = Z$

(B_2) $z^*(t) < Z$ for all $t \in [0, t')$, $z^*(t) = Z$ for $t \in [t', T]$, $t' \neq T$.

(A) If case A prevails, then by (49), $\lambda(t) \equiv 0$. From (54) and (53) $p_2(T^-) = p_2(T)$ and from (51) $\dot{p}_2(t) = ap_2(t)$ with $p_2(T) = 0$, so $p_2(t) \equiv 0$. It follows from (56) that $u^*(t) \equiv 1$. But then the results of the closely related model in exercise 2.3.3 suggest that $s^*(t) = 1$ in $[0, T-1]$, $s^*(t) = 0$ in $(T-1, T]$, and t^* is equal to $T-1$ in this case. Moreover, by (41) we find

$$k^*(t) = \begin{cases} k_0 e^t & \text{in } [0, t^*] \\ k_0 e^{T-1} & \text{in } (t^*, T] \end{cases} \quad (t^* = T-1) \tag{60}$$

and then, by (42), with $t^* = T-1$,

$$z^*(t) = \begin{cases} \left(z_0 - \dfrac{k_0}{a+1}\right)e^{-at} + \dfrac{k_0}{a+1}e^t & \text{in } [0, t^*] \\[3mm] \left[z_0 - \dfrac{k_0}{a+1} - \dfrac{k_0}{a(a+1)}e^{t^*(a+1)}\right]e^{-at} + \dfrac{k_0}{a}e^{t^*} & \text{in } (t^*, T] \end{cases} \tag{61}$$

If $z^*(t)$ as defined in (61) satisfies $z^*(T) < Z$ for all $t \in [0, T]$, then we have revealed a candidate for optimality. The associated $p_1(t)$ is easily found to be $p_1(t) = e^{T-t-1}$ for $t \in [0, T-1]$, $p_1(t) = T-t$ for $t \in (T-1, T]$.

Before we turn to cases B_1 and B_2, let us show that if A is not satisfied, then $t^* > 0$. Assume on the contrary that $t^* = 0$. Then $p_1(t) < 1$ for $t > 0$, so by (41) and (57), $\dot{k}^*(t) \equiv 0$, and then $k^*(t) \equiv k_0$. From (42), $\dot{z}^*(t) = u^*(t)k_0 - az^*(t)$. Clearly, $z^*(t) \leqslant \bar{z}(t)$, where $\bar{z}(t)$ is the solution to $\dot{\bar{z}} = k_0 - a\bar{z}$, $\bar{z}(0) = z_0$, i.e. $\bar{z}(t) = (z_0 - k_0/a)e^{-at} + k_0/a$. Now, if $z_0 \geqslant k_0/a$, then $\bar{z}(t)$ is non-increasing and thus $\bar{z}(t) \leqslant z_0 < Z$, by (45). On the other hand, if

$z_0 < k_0/a$, then $\bar{z}(t)$ is strictly increasing and thus $\bar{z}(t) < \bar{z}(T) = (z_0 - k_0/a)e^{-aT} + k_0/a < Z$ by (45). It follows that $z^*(t) \leq \bar{z}(t) < Z$ for all t, which is case A. We conclude that in cases B_1 and B_2, $t^* > 0$.

(B_1) In this case we see that in the interval $[0, T)$, $\lambda(t) = 0$ and $p_2(t) = p_2(T^-)e^{-a(T-t)} \leq 0$. It is reasonable to suggest that $u^*(t) = 1$ in $[0, T]$. From (56) it follows that for $t < T$, $\max(1, p_1(t)) + p_2(t) \geq 0$. In particular, $p_2(T^-) \geq -\max(1, p_1(T^-)) = -1$ since $p_1(T^-) = p_1(T) = 0$. Hence, in $[0, T)$, $0 \geq p_2(t) \geq -1$ and $p_2(t)$ is non-increasing. At T, $p_2(t)$ jumps up from $p_2(T^-)$ to 0.

Let us turn to $p_1(t)$. In $[t^*, T]$, $p_1(t) \leq 1$ so $\max(1, p_1(t)) + p_2(t) = 1 + p_2(t) \geq 0$. Hence from (59), $\dot{p}_1(t) = -1 - p_2(t) = -1 - p_2(T^-)e^{-a(T-t)}$, $p_1(T) = 0$. By integration we get

$$p_1(t) = T - t + \frac{p_2(T^-)}{a}(1 - e^{-a(T-t)}) \qquad t \in [t^*, T]. \tag{62}$$

Since $p_1(t^*) = 1$, the following relationship between $p_2(T^-)$ and t^* is obtained from (62):

$$p_2(T^-) = \frac{a(t^* - (T-1))}{1 - e^{-a(T-t^*)}}. \tag{63}$$

Moreover, in $[0, t^*)$, $\dot{p}_1(t) = -p_1(t) - p_2(t) = -p_1(t) - p_2(T^-)e^{-a(T-t)}$. Solving this linear differential equation with $p_1(t^*) = 1$, we obtain in $[0, t^*)$:

$$p_1(t) = \left[1 + \frac{p_2(T^-)}{a+1}e^{-a(T-t^*)}\right]e^{-(t-t^*)} - \frac{p_2(T^-)}{a+1}e^{-a(T-t)}. \tag{64}$$

Again the corresponding $k^*(t)$ and $z^*(t)$ are as given in (60) and (61), although t^* is not necessarily equal to $T-1$ in the present case. The requirement $z^*(T) = Z$, now gives the relationship

$$\left[z_0 - \frac{k_0}{a+1} - \frac{k_0}{a(a+1)}e^{t^*(a+1)}\right]e^{-aT} + \frac{k_0 e^{t^*}}{a} = Z,$$

that is

$$e^{t^*} - \frac{1}{a+1}e^{t^*(a+1)}e^{-aT} = \frac{a}{k_0}\left[Z - \left(z_0 - \frac{k_0}{a+1}\right)e^{-aT}\right]. \tag{65}$$

Finally, (57) suggests that we put $s^*(t) = 1$ in $[0, t^*]$, $s^*(t) = 0$ in $(t^*, T]$.

We have now come up with a unique proposal $u^*(t)$, $s^*(t)$, $k^*(t)$, $z^*(t)$, $p_1(t)$, $p_2(t)$, and $\lambda(t)$. Of course, we have found a real candidate for a B_1-type solution only provided the parameters in the problem are such that t^* as

defined by (65) belongs to $(0, T)$ and provided $z^*(t) < Z, t < T, p_2(T^-) \geq -1$.
(B_2) In this case it seems reasonable to assume that $u^*(t) = 1$ in $[0, t']$.
During that period $z^*(t)$ increased from z_0 to Z. In the interval $[t', T]$,
$z^*(t) = Z$, so $\dot{z}(t) \equiv 0$ in $(t', T]$. It is reasonable to suggest that the capacity
utilization rate $u^*(t)$ is < 1 in $(t', T]$ in order to keep $z^*(t)$ at the level Z.
From (42) we then get $u^*(t) = aZ/k^*(t) \in (0, 1)$. It follows from (56) that max
$(1, p_1(t)) + p_2(t) = 0$, and hence from (59), $\dot{p}_1(t) = 0$ in (t', T). Since $p_1(t)$ is
continuous and $p_1(T) = 0$, we conclude that $p_1(t) \equiv 0$ and $p_2(t) = -\max (1,$
$p_1(t)) \equiv -1$ in (t', T). By (51), $\lambda(t) = a$ on (t', T).

On the interval $[0, t')$, $z^*(t) < Z$, so $\lambda(t) = 0$. From (51) it follows that
$p_2(t) = p_2(t')e^{-a(t'-t)}$. Since p_2 is -1 on (t', T) and continuous at t', $p_2(t') = 1$,
and thus

$$p_2(t) = \begin{cases} -e^{-a(t'-t)} & \text{in } [0, t'] \\ -1 & \text{in } (t', T). \end{cases} \tag{66}$$

On $(t', T]$ we saw above that $p_1(t) = 0$. Since $p_1(t^*) = 1$, $t^* < t'$. On the
interval $[0, t^*]$, $p_1(t) \geq 1$ and $p_2(t) = -e^{a(t-t')}$, so max $(1, p_1(t)) + p_2(t) =$
$p_1(t) + p_2(t) \geq 0$; hence from (59), $\dot{p}_1(t) = -p_1(t) - p_2(t) = -p_1(t) + e^{-a(t'-t)}$.
On (t^*, t'), $\dot{p}_1(t) = -1 + e^{-a(t'-t)}$. Hence, we easily derive that

$$p_1(t) = \begin{cases} \left(1 - \dfrac{1}{a+1}e^{-a(t'-t^*)}\right)e^{-(t-t^*)} + \dfrac{1}{a+1}e^{-a(t'-t)} & \text{in } [0, t^*] \\[2mm] t' - t - \dfrac{1}{a}(1 - e^{-a(t'-t)}) & \text{in } (t^*, t'] \quad (67) \\[2mm] 0 & \text{in } (t', T]. \end{cases}$$

Again (57) suggests $s^*(t) = 1$ in $[0, t^*]$, $s^*(t) = 0$ in $(t^*, T]$. Since $p_1(t^*) = 1$,
(67) implies

$$t' - t^* + \frac{1}{a}e^{-a(t'-t^*)} = 1 + \frac{1}{a}. \tag{68}$$

Put $\psi(w) = w + e^{-aw}/a$. Then $\psi(0) = 1/a$, $\psi'(w) = 1 - e^{-aw}$ which is > 0 for
$w > 0$. Thus ψ is strictly increasing for $w \geq 0$. Moreover,
$\psi(1) = 1 + e^{-a}/a < 1 + 1/a$. The situation is as indicated in fig. 5.3. We see that
the equation $\psi(w) = 1 + 1/a$ has a unique positive solution, $\bar{w} > 1$. Hence (68)
has a solution $t' - t^* > 1$.

As in case B_1, $k^*(t)$ develops according to (60) (t^* usually $\neq T - 1$), and
$z(t)$ develops according to (61) in the interval $[0, t^*]$ and in $(t^*, t']$, while in
$[t', T]$, $z^*(T) = Z$. In the present case, we obtain a new relationship between

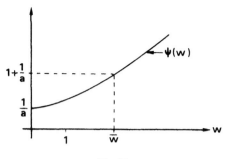

Fig. 5.3.

t^* and t' by the requirement $z^*(t')=Z$:

$$\left[z_0 - \frac{k_0}{a+1} - \frac{k_0}{a(a+1)} e^{t^*(a+1)}\right]e^{-at'} + \frac{k_0 e^{t^*}}{a} = Z. \tag{69}$$

We have again come up with a unique proposal $u^*(t)$, $s^*(t)$, $k^*(t)$, $z^*(t)$, $p_1(t)$, $p_2(t)$ and $\lambda(t)$. But only if t^* and t' defined by the two eqs (68) and (69) satisfy $0 < t^* < t' < T$ do we have a B_2-solution.

As noted above (see (58)), the concavity condition (35) is satisfied in the present problem. Moreover, the pure state constraint-function in (43) is linear, and hence quasi-concave. It therefore follows from theorem 1 that if, for a given set of problem data, we find a quadruple $P = (k^*(t), z^*(t), u^*(t), s^*(t))$ with associated multipliers $p_1(t)$, $p_2(t)$ and $\lambda(t)$ satisfying (48)–(54), then P is optimal.

We have proposed solution candidates in each of the three cases A, B_1 and B_2. However, a number of problems remain to be sorted out.

First we must prove that for all values of the problem data there *exist* a quadruple $(k^*(t), z^*(t), u^*(t), s^*(t))$ and associated multipliers $p_1(t)$, $p_2(t)$ and $\lambda(t)$ such that A, B_1 or B_2 holds.

Consider first the solution suggested in case A. If the problem data are such that $z^*(T)$ (see (61)) is $< Z$, then all the conditions in theorem 1 are satisfied, and the optimal solution is found. It only remains to check that $z^*(t) < Z$ for all $t \in [0, T]$.

Suppose next that A fails such that $z^*(T) \geqslant Z$. Recall that in any case $p_2(T^-) \leqslant 0$; in case B_1, $p_2(T^-) \in [-1, 0]$, while in case A, $p_2(T^-) = 0$. Observe next that if $p_2(T^-) = 0$, the solution suggested in B_1, coincides with the solution in A. (From (63), $t^* = T - 1$ etc.).

Consider condition (63). Denote the right-hand side by $g(t^*)$. Put $p_2(T^-) = -1$ and consider the equation $-1 = g(t^*)$. Since $g(t^*) \to -\infty$, as

$t^* \to -\infty$ and $g(t^*) = 0$ for $t^* = T - 1$, $-1 = g(t^*)$ has a solution in $(-\infty, T-1)$. Let t_1^* denote the largest of these solutions. We claim that $g(t^*)$ is strictly increasing in $(t_1^*, T-1)$. Note that

$$g'(t^*) = \frac{a(1 - e^{-a(T-t^*)}) + a^2(t^* - (T-1))e^{-a(T-t^*)}}{(1 - e^{-a(T-t^*)})^2}.$$

By the choice of t_1^*, $g(t^*) > -1$ in $(t_1^*, T-1)$, so $g(t^*) \in (-1, 0)$ in this interval. Hence, for some $\gamma \in (0, 1)$, $a(t^* - (T-1)) = -\gamma(1 - e^{-a(T-t^*)})$. Thus the denominator in the expression for $g'(t^*)$ is $a(1 - e^{-a(T-t^*)}) \cdot (1 - \gamma e^{-a(T-t^*)}) > 0$.

Since $g(t^*) \in (-1, 0)$ is strictly increasing in $(t_1^*, T-1)$, we see that if $p_2(T^-)$ decreases in $[-1, 0]$, t^* decreases. From (61) we find $dz^*(T)/dt^* = (k_0/a)e^{t^*}(1 - e^{-a(T-t^*)}) > 0$, so $z^*(T)$ decreases as t^* decreases. Let us decrease $p_2(T^-)$ from 0 towards -1. If $z^*(T)$, which by assumption starts out $\geqslant Z$, becomes equal to Z for some value of $p_2(T^-) \in [-1, 0]$, then we have a real B_1 candidate for optimality. (It only remains to check that $z^*(t) < Z$ for $t < T$.)

If $z^*(T) > Z$ for $p_2(T^-) = -1$, we turn to case B_2: For $p_2(T^-) = -1$, if we choose $t' = T$ in B_2, the solution suggested in B_2 coincides with the solution in B_1. (Note, in particular, that (63) with $p_2(T^-) = -1$ and $t^* = t_1^*$, reduces to (68) with $t' = T$.) Let t' decrease. Since $t' - t^* = \bar{w}$ when $t^* > 0$, t^* decreases as well. For $t' = T$, $z^*(T) > Z$ and for $t' = \bar{w}$, $t^* = 0$ with $z^*(t') < Z$. Thus for some $t' \in (0, T)$, $z^*(t') = Z$, and we have a real type-B_2 solution. (It only remains to check that $z^*(t) < Z$ for $t < t'$ and that $u^*(t) < 1$ for $t > t'$.)

In each of these cases we have specified the behaviour of $k^*(t)$, $z^*(t)$, $u^*(t)$, $s^*(t)$, $p_1(t)$, $p_2(t)$ and $\lambda(t)$ and the conditions of theorem 1 are all satisfied. It remains only to show that

when $z^*(T) < Z$, then $z^*(t) < Z$ for all $t \in [0, T]$.   (70(i))

when $z^*(T) = Z$ for some $p_2(T^-) \in [-1, 0]$, then $z^*(t) < Z$ for

all $t \in [0, T)$.   (70(ii))

when $z^*(t') = Z$ for some $t' \in [0, T]$, then $z^*(t) < Z$ for $t < t'$

and $u^*(t) = aZ/k^*(t) \in (0, 1)$ for $t > t'$.   (70(iii))

One way of establishing (70) is shown in example 5, and that verification completes the analysis of this model.

Exercise 5.2.1 Consider the following problem:

$$\max \int_0^2 (1-x(t))dt, \qquad \dot{x}(t)=u(t), \qquad x(0)=1, \qquad x(2) \text{ free} \tag{71}$$

$$g(x(t), t)=x(t) \geqslant 0 \tag{72}$$

$$u(t) \in [-1, 1] \tag{73}$$

(i) Try to guess at the optimal solution. (It pays to keep $x(t)$ low. Start by reducing $x(t)$ as fast as possible until $x(t)=0$ etc.)

(ii) Use theorem 1 to prove optimality.

Exercise 5.2.2 Solve the following problems:

(i) $\max \int_0^\pi -u \sin t \, dt, \qquad \dot{x}=u,$

$\qquad x(0)=0, \qquad x(\pi)=0, \qquad x \geqslant 0, \qquad u \in [-1, 1].$

(ii) $\max \int_0^3 (4-t)u \, dt, \qquad \dot{x}=u,$

$\qquad x(0)=0, \qquad x(3)=3, \qquad x \leqslant 1+t, \qquad u \in [0, 2].$

Exercise 5.2.3 (**Optimal maintenance**) Consider the following problem:

$$\max \int_0^T [ax(t)-u(t)]dt \tag{74}$$

$$\dot{x}(t)= -\alpha x(t)+u(t), \qquad x(0)=x_0 >0, \qquad x(T) \text{ free} \tag{75}$$

$$g(x(t), t)=\bar{x}-x(t) \geqslant 0 \tag{76}$$

$$u(t) \in [0, \bar{u}] \tag{77}$$

$\qquad T, a, \alpha$ and $\bar{u}$ are positive, $\qquad \bar{u}>\alpha\bar{x}, \qquad a>\alpha, \qquad$ and $\bar{x}>x_0.$ (78)

[Indication of an economic interpretation: A machine with quality $x(t)$ at time t earns a revenue $ax(t)$. The quality decays at a rate of α but it can be improved by spending an amount $u(t)$ on maintenance. The planning period $[0, T]$ is fixed.]

(i) Solve the problem with (76) deleted. (You will need to distinguish between the cases $\alpha T \leqslant$ and $> \ln a - \ln(a - \alpha)$.)
(ii) Consider the full problem (74)–(78). Explain why theorem 1 applies and write down all the conditions supplied by that theorem.
(iii) Solve the problem.

Exercise 5.2.4 (Growth with foreign lending) Consider exercise 4.4.5 and replace condition (4.145) by $u \leqslant \bar{u}$, $\bar{u}$ a fixed positive constant. Add the pure state constraint $B \geqslant -c$, c a positive constant, ($-c$ representing a lower limit on the country's credit position). Assume finally that $bK_0 - rc - \bar{u} > 0$.

(i)   Show (4.149)–(4.151) in exercise 4.4.5, replacing the equality in (4.151) by $u^*(t) = \bar{u}$.
(ii) Show that (35) in theorem 1 is satisfied.
(iii) Assuming that a solution of the sufficient conditions in that theorem exists, show that such a solution has the following form: There exist four numbers $t_1 \leqslant t_2 < t' < t_3$, $[t_1, t_3] \neq [0, T]$ such that $s^*(t) = 1$ on $[0, t']$, $s^*(t) = 0$ on $(t', T]$, $u^*(t) = \underline{u}$ on $[0, t_1]$, $u^*(t) = 0$ on $(t_1, t_2]$, $u^*(t) = \bar{u}$ on $(t_2, t_3]$, $u^*(t) = \underline{u}$ on $(t_3, T]$. The numbers t_1, t_2, t', t_3 satisfy the following relations (a): $t' = \max(t'', T - 1/b)$, where t'' satisfies $K^*(t'') = K_T$. Either: (α) $t_1 = t_2$, $t_3 = T$. Then $t_1 = t_2 = \bar{u}T/(\bar{u} - \underline{u})$ and $p_1(t_1) = p_2(t_1)$. This is an acceptable solution if $B^*(t)$ does satisfy $B^*(t_1) \geqslant -c$ and $p_2(T) \geqslant 1$; or (β) $t_1 = t_2$, $t_3 < T$. Then $t_1 = t_2$ and t_3 satisfy the relations (b): $p_2(t_3) = 1$, either $t_1 = 0$ or $t_1 > 0$ and $p_1(t_1) = p_2(t_1)$ and (c): $\underline{u}t_1 + \bar{u}(t_3 - t_1) + \underline{u}(T - t_3) = 0$. Test for acceptance: $p_2(T) \geqslant 0$, $B^*(t_1) \geqslant -c$; or (γ) $t_1 < t_2$, $t_3 = T$. Then $t_1 = -c/\underline{u}$, $t_2 = T - c/\bar{u}$. ($p_1 = p_2$ in (t_1, t_2)). Test for acceptance: $p_2(T) \geqslant 1$. (δ) $t_1 < t_2$, $t_3 < T$. Then $t_1 = -c/\underline{u}$, and t_2 and t_3 satisfy (b) and (d): $c + \bar{u}(t_3 - t_2) + \underline{u}(T - t_3) = 0$. ($p_1 = p_2$ in (t_1, t_2)).
(iv) Try to use an argument similar to that in (v) and (vi) in exercise 4.4.5 to show that the sufficient conditions do have a solution.
     (Hint: Use $p_1(T)$ and $p_2(T)$ as parameters, and see how t_1, t_2, and t_3 vary with $p_2(T)$ for each fixed $p_1(T)$.)

Exercise 5.2.5 An example in which $g(x^*(t), t)$ behaves as in fig. 5.2 at τ_k is the following:

$$\max \int_0^4 -(u - 1)^2 e^{-t} dt, \qquad \dot{x} = u, \qquad x(0) = 0, \qquad x(4) \text{ free},$$

$$u \in (-\infty, \infty), \qquad x \leqslant 2 + e^{-3}.$$

Solve the problem.
(Hint: $p(t)$ is continuous. See note 3(e).)

Exercise 5.2.6 (**Inventory policy**) Consider the problem in example 1.

(i) Change the criterion to min $\int_0^T [cu(t) + b(u(t))^2 + hx(t)] dt$, and drop (14).
 Assume that $b > 0$. Solve the problem when (i) $h > 2ab$ and
 $aT^2(1 - 2ab/h) > 2$, (ii) $h < 2ab$ and $aT^2 > 2$.

(ii) Change the criterion to min $\int_0^T [c(u(t))^\gamma + hx(t)] dt$, $\gamma \in (0, 1)$. Try to explain
 why no optimum exists in this problem.
 (Hint: The convexity condition in the standard existence theorem
 (theorem 5) fails.)

3. Necessary conditions

We begin in this section our discussion of necessary conditions for optimality in the pure state constraint problem (23)–(27). Briefly formulated the
problem is:

$$\max \int_{t_0}^{t_1} f_0(x, u, t) dt, \qquad \dot{x} = f(x, u, t), \qquad x(t_0) = x^0, \tag{79}$$

$$g(x, t) \geqslant 0, \qquad u \in U, \tag{80}$$

with the usual terminal conditions (2.53) ($= (25)$).

It would be convenient to have the necessary conditions as close as
possible to the sufficient conditions. In particular, we might ask if conditions (29)–(31), (34), (37) and (38), while allowing for the additional
possibility that $p_0 = 0$, are necessary for the optimality of $(x^*(t), u^*(t))$. The
answer is "not quite". In fact, these conditions are referred to below as
"almost necessary conditions" (ANC) since the properties connected with
the multipliers $p_i(t)$, $\lambda_i(t)$ and β_j^k are not quite necessary for optimality.

In theorem 2 we shall formulate precise necessary conditions in which the
g_j-constraints are represented in the generalized Lagrangian by an alternative expression. Example 5 will show how to use these conditions in an
economic control problem. It turns out that the conditions in theorem 2 are
somewhat more complicated to apply than the ANC conditions. Since the
ANC conditions are very close to being necessary, in the sense that an
optimal solution highly unlikely fails to satisfy ANC, we recommend using
ANC conditions in actual control problems, at least at the explorative
stage.

We prepare the approach used in theorem 2 in the following sub-section

Preliminary considerations

Let $(x(t), u(t))$ be any admissible pair. At any point $t \in (t_0, t_1)$ at which $g_j(x(t), t)=0$, the function $\tau \to g_j(x(\tau), \tau)$ has a minimum. Then $(d/dt) \, g_j(x(t), t)=0$, provided the derivative exists. This total derivative is calculated as follows:

$$\frac{d}{dt} \, g_j(x(t), t) = \sum_{i=1}^{n} \frac{\partial g_j(x(t), t)}{\partial x_i} \, \dot{x}_i(t) + \frac{\partial g_j(x(t), t)}{\partial t}$$

$$= \sum_{i=1}^{n} \frac{\partial g_j(x(t), t)}{\partial x_i} \, f_i(x(t), u(t), t) + \frac{\partial g_j(x(t), t)}{\partial t}. \tag{81}$$

Notice that this derivative surely exists at continuity points of $u(t)$. Define the function h_j by

$$h_j(x, u, t) = \sum_{i=1}^{n} \frac{\partial g_j(x, t)}{\partial x_i} \, f_i(x, u, t) + \frac{\partial g_j(x, t)}{\partial t}. \tag{82}$$

With $h = (h_1, \ldots, h_s)$ and using a matrix notation, we have

$$h(x, u, t) = g'_x(x, t) \cdot f(x, u, t) + g'_t(x, t). \tag{83}$$

Thus at continuity points t of u, $h_j(x(t), u(t), t)=0$ whenever $g_j(x(t), t)=0$. Hence, for v.e. t, $(x(t), u(t))$ satisfies the constraint

$$h_j(x, u, t) = 0 \qquad \text{whenever} \qquad g_j(x, t) = 0, \qquad j = 1, \ldots, s. \tag{84}$$

In the necessary conditions presented in theorem 2 multipliers $q_j(t)$ are associated with the constraint (84). From our considerations above, this should not come as a complete surprise, but the precise explanation can only be found in proofs of the theorem.

Define the Lagrangian $\bar{L}$, with $q = (q_1, \ldots, q_s)$, by

$$\bar{L}(x, u, p, q, t) = H(x, u, p, t) - q \cdot h(x, u, t)$$
$$= p_0 f_0(x, u, t) + p \cdot f(x, u, t) - q \cdot h(x, u, t). \tag{85}$$

Theorem 2 (Necessary conditions. Pure state constraints)[1] Let $(x^*(t), u^*(t))$ be an admissible pair solving problem (23)–(27) with $g(x, t)$ a C^2-function of (x, t). Then there exist a number p_0, a vector function

---

[1] Neustadt (1976), Chapter V.3. Be aware of the fact that property (89(ii)) is a consequence of the other properties of the theorem. However, it is a useful property, explicitly stated in Hestenes (1966), Chapter 8. (See exercise 5.3.4.) Some of the comments in the subsequent note 3 are inspired by the reading of the manuscript of Feichtinger and Hartl (1986).

$p(t) = (p_1(t), \ldots, p_n(t))$ with one-sided limits everywhere, and a non-decreasing vector function $q(t) = (q_1(t), \ldots, q_s(t))$ such that

$$p_0 = 0 \quad \text{or} \quad p_0 = 1, \tag{86}$$

$$(p_0, p(t), q(t_1) - q(t_0)) \neq (0, 0, 0) \quad \text{for all } t \tag{87}$$

$u^*(t)$ maximizes $H(x^*(t), u, p(t), t)$ for $u \in U$ and for v.e. $t \in (t_0, t_1)$

where $H = p_0 f_0(x, u, t) + p \cdot f(x, u, t)$. $\tag{88}$

$q_j(t)$ is constant on any interval where $g_j(x^*(t), t) > 0$ $\tag{89(i)}$

$q_j(t)$ is continuous at all $t \in (t_0, t_1)$ where $g_j(x^*(t), t) = 0$

and $(\partial g_j(x^*(t), t)/\partial x) \cdot f(x^*(t), u^*(t), t)$ is discontinuous. $\tag{89(ii)}$

If we define

$$p^*(t) = p(t) + q(t) \cdot g'_x(x^*(t), t), \tag{90}$$

then $p^*(t)$ is continuous, and has a continuous derivative, $\dot{p}^*(t)$, at all points of continuity of $u^*(t)$ and $q(t)$, and we have:

$$\dot{p}^*(t) = -\frac{\partial \bar{L}^*}{\partial x} \quad (\partial \bar{L}/\partial x \text{ evaluated at } (x^*(t), u^*(t), p^*(t), q(t), t))$$

where $\bar{L} = H - q \cdot (g'_x(x, t) \cdot f(x, u, t) + g'_t(x, t))$ $\tag{91}$

$p_i(t_1)$ no conditions $\qquad\qquad i = 1, \ldots, l$ $\tag{92a}$

$p_i(t_1) \geq 0 (= 0 \text{ if } x_i^*(t_1) > x_i^1) \quad i = l+1, \ldots, m$ $\tag{92b}$

$p_i(t_1) = 0 \qquad\qquad\qquad i = m+1, \ldots, n.$ $\tag{92c}$

Note 3

(a) The conditions in the theorem are also satisfied if $q(t)$ is replaced by $q(t) + c$, where c is an arbitrary constant vector. Hence

$q(t)$ can be normalized by choosing, say, $q(t_0) = 0$

or $q(t_1) = 0$. $\tag{93}$

When $q(t)$ is replaced by $q(t) + c$, $p^*(t)$ changes to $p^*(t) + cg'_x(x^*(t), t)$. An easy calculation shows that this function satisfies $\dot{p} = -\bar{L}'_x(x^*(t), u^*(t), p, q(t) + c, t)$.

(b) For all $t \in (t_0, t_1)$, $u^*(t^+)$ and $u^*(t^-)$ maximize both $H(x^*(t), u, p(t^+), t)$ and $H(x^*(t), u, p(t^-), t)$.

(c) For all $t \in (t_0, t_1)$, $\beta_t \cdot g'_x(x^*(t), t) \cdot f(x^*(t), u^*(t^+), t) = \beta_t \cdot g'_x(x^*(t), t) \cdot f(x^*(t),$

$$u^*(t^-), t) = -\beta_t g_t'(x^*(t), t) = H^*(t^-) - H^*(t^+), \ \beta_t = q(t^+) - q(t^-).$$

(d) At any point $\tau \in (t_0, t_1)$, if either $H(x^*(\tau), u, p(\tau^-), \tau)$ or $H(x^*(\tau), u, p(\tau^+), \tau)$ has a unique maximum in U, then $u^*(t)$ is continuous at $t = \tau$.

(e) When $u^*(t)$ is continuous at $t = \tau \in (t_0, t_1)$, then one can often immediately show that $q(t)$, and hence $p(t)$, are continuous at τ. A general set of conditions ensuring the continuity of $q(t)$ at τ is (i) $u^*(t)$ continuous at τ, (ii) $U = \{u: F_j(u) \geqslant 0, \ j=1,\ldots,j'\}$, $F_j(u)$ and $f C_-^1$ functions, (iii) (6.92) is satisfied for $t = \tau$ if we define $g_j(x, u, t) = F_j(u)$, $j = 1,\ldots, s', j' = s'$; $g_j(x, u, t) = \bar{g}_j(x, t) = g_{j-s'}(x, t), \ j = s' + 1,\ldots, s + s'$.

(f) If f_0 and f have continuous partial derivatives w.r.t. t, then $(d/dt)H^*(t) = \partial H^*(t)/\partial t$ v.e..

Note 4 (**Non-triviality result**) In the main result of Chapter 4 (theorem 4.1) we imposed a constraint qualification which prevented $(p_0, p(t))$ from being anywhere equal to $(0, 0)$. Note that if the multipliers are trivial in the sense that $(p_0, p(t)) = (0, 0)$ for all t in some interval I, the Hamiltonian vanishes in I, and the maximum condition gives no information about $u^*(t)$ for $t \in I$. In theorem 2 no constraint qualification was imposed and as a consequence non-triviality of $(p_0, p(t))$ is not ensured.

It is advantageous to have available some general results about the non-triviality of the multipliers rather than being forced to derive such results for each particular problem. In this section we shall mention only one such result.

Suppose the following two conditions are satisfied:

$$G^*(t_0) = \phi, \qquad G^*(t_1) = \phi, \qquad \text{where } G^*(t) = \{j: g_j(x^*(t), t) = 0\} \qquad (94)$$

For all $t \in (t_0, t_1)$, either $G^*(t) = \phi$ or

there exists an $x_t \in R^n$ such that $[\partial g_j(x^*(t), t)/\partial x] x_t > 0$ for

all $j \in G^*(t)$. $\hspace{10cm}$ (95)

Then one can prove that[2]

$$(p_0, p(t)) \neq (0, 0) \text{ for all } t \text{ in some interval contained in } (t_0, t_1). \qquad (96)$$

We refer to (94), (95) as constraint qualifications (c.q.). When theorem 2 is applied, the c.q. enters the solution procedure as follows: (i) Find all admissible pairs for which the c.q. fails. (ii) Find all admissible pairs

---

[2]Neustadt (1976), Chapter V.3, 69 (69 has a misprint: Replace $=$ by $\neq$.) Condition (96) is a consequence of the other properties in theorem 2.

satisfying both the c.q. and the necessary conditions including the non-triviality result (96).

In the next section we shall discuss other non-triviality results and, in particular, shall consider the case $G^*(t_0) \neq \phi$, $G^*(t_1) \neq \phi$.

"*Almost necessary*" conditions

In the majority of cases the conditions related to the multipliers in theorem 2 are more complicated to apply than are the corresponding conditions in theorem 1 and note 1. It turns out that *if $q(t)$ is well-behaved* in the sense of having a finite number of jump discontinuities and being C^1 elsewhere, then these multiplier conditions are equivalent. Since $q(t)$ normally turns out to be well-behaved, we formulate the following "result":[3]

("Almost necessary" conditions (ANC) Pure state constraints) If $(x^*(t)$, $u^*(t))$ solves problem (23)–(28), then the conditions in theorem 2 are satisfied with $q(t)$ having jumps $\beta^k = q(\tau_k^+) - q(\tau_k^-)$ at a finite number of points τ_k, $k = 1, \ldots, N$, $t_0 \leqslant \tau_1 < \ldots < \tau_N \leqslant t_1$, and having a piecewise continuous derivative $\dot{q}(t) = \lambda(t)$ elsewhere, and with conditions (89)–(91) replaced by (30), (31), (37) and (38). (In (37), let $p(\tau_1^-) = p(t_0)$ if $\tau_1 = t_0$.) ∎

---

[3] Let us briefly explain the relationship between ANC and the various necessary and sufficient conditions.

  (i) In theorem 2, definition (90) and property (91) can be replaced by

$$p(t^-) = p(t_1) + \int_t^{t_1} (\partial H^*/\partial x) dt$$

$$+ \int_{[t,t_1]} (\partial g^*/\partial x) dq(t), \qquad t \in [t_0, t_1],$$

where the latter integral is a Lebesgue–Stieltjes integral. In this case we need only assume that g and $\partial g/\partial x$ exist and are continuous. Note that integration by parts in the latter integral in the above equation immediately gives (91) when definition (90) is used.

  (ii) In the sufficient conditions of note 1, (37) and (31) can be replaced by

$$p(t^-) = p(t_1) + \int_t^{t_1} (\partial L^*/\partial x) dt + \sum_{\tau_k \geqslant t} \beta^k \partial g^*(\tau_k)/\partial x.$$

If we define $q(t_0) = 0$ and $q(t) = \int_{t_0}^t \lambda(t) dt + \sum_{\tau_k > t} \beta^k$ for $t > t_0$, (or, equivalently, as in ANC), then the two formulas for $p(t^-)$ above coincide.

We recommend using ANC, rather than theorem 2 when solving control problems. Only in rare cases are ill-behaved $q(t)$'s needed. (An ill-behaved $q(t)$ may have a countable number of jump points, $\dot{q}(t)$ may exist only a.e., $q(t)$ is not necessarily an integral of $\dot{q}(t)$.) If an ill-behaved $q(t)$ is needed in a problem, then one will become aware of this fact when trying to apply ANC.

Note that even when we use ANC, non-triviality results as discussed above are important.

Sufficient conditions

In the previous section we stated a sufficiency theorem for the pure state constraint problem. Let us modify that theorem and bring it on a form which makes it easier to use in conjunction with theorem 2.[4]

Theorem 3 (Sufficient conditions corresponding to theorem 2) Let $(x^*(t), u^*(t))$ be an admissible pair for problem (23)–(27) with associated multipliers $p(t)$ and $q(t)$, satisfying the conditions in theorem 2 with $p_0 = 1$. Assume further that

$$\hat{H}(x, p(t), t) = \max_{u \in U} H(x, u, p(t), t)$$

is concave in $x \in R^n$ for v.e. t,   (97)

(assuming that the maximum value is attained).

$g(x, t)$ is quasi-concave in x.   (98)

Then $(x^*(t), u^*(t))$ is an optimal pair. ∎

Free final time problems

In theorem 2 the final time t_1 was assumed to be fixed. The corresponding result for a free final time problem is as follows.[5]

---

[4] Seierstad (1985b).
[5] This theorem can be derived in a straightforward way by using the standard trick of treating t as a state variable. See Ioffe and Tihomirov (1979), Chapter 5. (Their eq. (20) (which has a misprint) implies (99).)

Theorem 4 (**Necessary conditions. Free final time**)   Let $(x^*(t), u^*(t), t_1^*)$ be an admissible triple solving problem (23)–(27) with $t_1 \in [T_1, \ T_2]$, $t_0 \leqslant T_1 < T_2, T_1, T_2$ fixed. In addition to the standard assumption (2.6) on f_i, assume that $\partial f_i / \partial t, i = 0, \ldots, n$ exist and are continuous. Assume also that $g(x, t)$ is a C^2-function in (x, t). Then $(x^*(t), u^*(t))$ satisfies conditions (86)–(92) for $t_1 = t_1^*$. In addition,

$$H^*(t_1^{*-}) + (q(t_1^*) - q(t_1^{*-}))(\partial g(x^*(t_1^*), t_1^*)/\partial t) \begin{cases} \leqslant 0 \text{ if } t_1^* = T_1 \\ = 0 \text{ if } T_1 < t_1^* < T_2 \\ \geqslant 0 \text{ if } t_1^* = T_2 \end{cases} \quad (99)$$

where $H^*(t) = H(x^*(t), u^*(t), p(t), t)$. ∎
   (For $t_1^* = T_1 = t_0$, a comment similar to that of note 6.8 can be made).

Sufficient conditions corresponding to theorem 4 can be found by specializing theorem 6.7 to the present case.

Note 5   All the results in notes 3 and 4, as well as the non-triviality results in the next section, are valid for problems covered by theorem 4, for $t_1 = t_1^*$.

Existence

Theorem 5 (**Existence**)[6]   Consider problem (23)–(27), where t_1 is subject to choice in $[T_1, T_2], t_0 \leqslant T_1 \leqslant T_2$. Assume (merely) that f_0, f and g are continuous. Assume, in addition, that there exist admissible pairs, that U is compact and that for some $b > 0$, $||x(t)|| \leqslant b$ for all t, for all admissible pairs $(x(t), u(t))$. Suppose finally that the set $N(x, U, t) = \{(f_0(x, u, t) + \gamma, f(x, u, t)):$ $u \in U, \ \gamma \leqslant 0\}$ is convex for all $(x, t) \in R^n \times [t_0, T_2]$. Then there exists an optimal pair $(x^*(t), u^*(t))$, (where $u^*(t)$ is measurable). ∎

Example 4   Consider the following problem:

$$\max \int_0^4 \left(x + \frac{1}{u^2 + 1}\right) dt, \quad (100)$$

$$\dot{x} = 1 - \frac{1}{u^2 + 1}, \qquad x(0) = 0, \qquad x(4) \text{ free} \quad (101)$$

[6]Cesari (1983), Chapter 9. On necessary conditions for measurable control functions, see footnote 11.

$$g(x, t) = t - 2x \geqslant 0 \tag{102}$$

$$u = u(t) \in (-\infty, \infty). \tag{103}$$

We solve the problem by using theorem 3. The Hamiltonian, with $p_0 = 1$, is

$$H = \left(x + \frac{1}{u^2 + 1}\right) + p\left(1 - \frac{1}{u^2 + 1}\right) = x + p + \frac{1}{u^2 + 1}(1 - p)$$

and the Lagrangian $\bar{L}$ is,

$$\bar{L} = H - q \cdot (g'_x \cdot f + g'_t) = H - q\left(-2\left(1 - \frac{1}{u^2 + 1}\right) + 1\right)$$

$$= H - q\left(\frac{2}{u^2 + 1} - 1\right).$$

Choose $q(4) = 0$, (see note 3(a)). The condition that $u^*(t)$ maximizes the Hamiltonian v.e. is equivalent to the condition that

$$u^*(t) \text{ maximizes } \frac{1}{u^2 + 1}(1 - p(t)) \text{ for } u \in (-\infty, \infty)$$

for v.e. $t \in (0, 4)$. \tag{104}

Observe that no maximum in (104) can exist if $(1 - p(t)) < 0$, so we have to construct a function $p(t)$ which is $\leqslant 1$ everywhere in $(0, 4)$. If we look at (89(ii)), in the present example $g'_x(x^*, t)f(x^*, u^*, t) = -2[1 - 1/((u^*(t))^2 + 1)]$, which is discontinuous when $(u^*(t))^2$ is discontinuous. Hence (89) implies

$q(t)$ is constant on intervals where $x^*(t) < t/2$ \tag{105(i)}

$q(t)$ is continuous at any $t' \in (0, 4)$ if $x^*(t') = t'/2$ and
$(u^*(t))^2$ is discontinuous at t'. \tag{105(ii)}

The function $p^*(t)$ of (90) is here

$$p^*(t) = p(t) + q(t)(-2) = p(t) - 2q(t), \tag{106}$$

and we know that this function is continuous. Moreover,

$$\dot{p}^*(t) = -\frac{\partial \bar{L}}{\partial x} = -1 \text{ v.e.} \tag{107}$$

Since $p(4) = 0$ and $q(4) = 0$, we get from (106), $p^*(4) = 0$. This fact, together with (107) and the continuity of $p^*(t)$, imply

$$p^*(t)=4-t. \tag{108}$$

The first-order condition for (104) tells us that

$$-\frac{2u^*(t)}{(u^*(t))^2+1}(1-p(t))=0 \qquad \text{for v.e. } t,$$

and therefore,

$$u^*(t)=0 \qquad \text{or} \qquad p(t)=1 \qquad \text{for v.e. } t \tag{109}$$

In $(0, 4)$, $p(t)\equiv 1$ is impossible: In that case from (106) and (108), $q(t)=\frac{1}{2}(t-3)$ in $(0, 4)$ so $q(4^-)=1/2$. Since $q(4)=0$, this contradicts the fact that $q(t)$ is non-decreasing. Thus $p(s)<1$ for some $s\in(0, 4)$. Let t^* be the infimum of such s's. Let us show that $p(\bar{t})<1$ for all $\bar{t}>t^*$: There exists a point $s<t$, where $p(s)<1$. At a continuity point s' of $u^*(t)$ close to s, $s'<\bar{t}$, both $p(s')<1$ and (109) hold. By (109), $u^*(s')=0$. Since $\dot{x}^*(s')=0$, $x(t)<t/2$ in some (perhaps small) interval (a, b) around s'. (The function $y(t)=x(t)-t/2$ is everywhere ≤ 0, and $\dot{y}(s')=-1/2$.) By (106), (107) and the fact that $q(t)$ is constant on (a, b), we get that $p(t)$ decreases in (a, b). Hence $p(b)\leq p(s')<1$, $\dot{x}^*(t)=0$ v.e. in (s', b), and $x(b)<b/2$. Since $x(t)<t/2$ also just to the right of b, $p(t)$ continues to decrease beyond b. In fact, the interval (s', b) can be extended to $(s', 4)$, with $p(t)<1$ holding for all $t>s'$, in particular for $t=\bar{t}$.

Thus we have to end up with the following suggestion for $p(t)$: For some $t^*<4$, $p(t)=1$ in $(0, t^*)$, while $p(t)<1$ in $(t^*, 4]$.

On $(t^*, 4]$, $u^*(t)=0$, so $x^*(t)=x^*(t^*)=t^*/2$ and $x^*(t)<t/2$. Hence $q(t)=0$, and $p(t)=p^*(t)=4-t$ on $(t^*, 4]$.

We claim that $t^*=3$. If $t^*<3$, $p(t)=4-t>1$ near t^*, a contradiction. Note next that by (106) and the continuity of $p^*(t)$, $p(t^{*+})-p(t^{*-})=2(q(t^{*+})-q(t^{*-}))\geq 0$. If $t^*>3$, then $p(t^{*-})\leq p(t^{*+})=4-t^*<1$, contradicting the definition of t^*. (Another way of proving that $p(t^{*-})\leq p(t^{*+})$ is to note that by (105(ii)), $q(t)$, and hence $p(t)$, actually are continuous at t^*.)

In $(0,3)$, $p(t)\equiv 1$, so by (106) and (108), $q(t)=\frac{1}{2}(t-3)<0$, while $q(t)=0$ in $(3, 4)$. Hence $q(t)$ is non-decreasing. Thus we have the following suggestion for an optimal solution:

$$u^*(t)=\begin{cases} \pm 1 & \text{on } [0,3] \\ 0 & \text{on } (3,4] \end{cases}, \qquad x^*(t)=\begin{cases} t/2 & \text{on } [0,3] \\ 3/2 & \text{on } (3,4] \end{cases} \tag{110}$$

with

$$p(t) = \begin{cases} 1 & \text{on } [0,3] \\ 4-t & \text{on } (3,4] \end{cases}, \quad q(t) = \begin{cases} \frac{1}{2}(t-3) & \text{on } [0,3] \\ 0 & \text{on } (3,4] \end{cases}. \tag{111}$$

It is easy to verify that all the conditions in theorem 3 are now satisfied. In particular, the maximized Hamiltonian as well as the constraint function $g(x, t) = t - 2x$ are linear in x, so conditions (97) and (98) are trivially satisfied. Hence (110) solves our problem.

Example 5 We want to apply theorem 2 to the problem in example 3. Thus we assume that $(x^*(t), u^*(t))$ is an optimal solution and we try to deduce its properties. We normalize $q(t)$ by requiring $q(T) = 0$. First we want to prove that the necessary conditions cannot be satisfied with $p_0 = 0$. Suppose $p_0 = 0$. From (90) we have in the present case,

$$p_1^*(t) = p_1(t) + q(t) \cdot \frac{\partial g}{\partial k} = p_1(t),$$

$$p_2^*(t) = p_2(t) + q(t) \frac{\partial g}{\partial z} = p_2(t) - q(t). \tag{112}$$

In particular, $p_1^*(T) = p_1(T) = 0$ and $p_2^*(T) = p_2(T) - q(T) = 0$ as a result of (92). The Lagrangian $\bar{L}$ is

$$\bar{L} = p_0(1-s)uk + p_1\, suk + p_2(uk - az) - q(-1)(uk - az), \tag{113}$$

and with $p_0 = 0$, we obtain from (112) and (91):

$$\dot{p}_1(t) = -\frac{\partial \bar{L}^*}{\partial k} = -p_1(t)s^*(t)u^*(t) - p_2^*(t)u^*(t) - q(t)u^*(t)$$

$$= -p_1(t)s^*(t)u^*(t) - p_2(t)u^*(t) \tag{114(i)}$$

$$\dot{p}_2^*(t) = -\frac{\partial \bar{L}^*}{\partial z} = a(p_2^*(t) + q(t)). \tag{114(ii)}$$

From (114(ii)), using (A.20) with $t_0 = T$, $p_2^*(T) = 0$, we obtain for $t \leqslant T$,

$$p_2^*(t) = -a \int_t^T q(v)e^{a(t-v)}\, dv.$$

Since, moreover,

$$\int_t^T ae^{a(t-v)}\,dv = 1 - e^{a(t-T)},$$

$$\int_t^T aq(t)e^{a(t-v)}\,dv = q(t) - q(t)e^{a(t-T)},$$

so

$$p_2(t) = p_2^*(t) + q(t) = -\int_t^T aq(v)e^{a(t-v)}\,dv + \int_t^T aq(t)e^{a(t-v)}\,dv + q(t)e^{a(t-T)}$$

$$= -\int_t^T a(q(v)-q(t))e^{a(t-v)}\,dv + q(t)e^{a(t-T)} \leq 0, \tag{115}$$

in as much as $q(t) \leq q(v) \leq q(T) = 0$, since $q(t)$ is non-decreasing. Note that if $q(t'') < 0$ for some t'', then $p_2(t'') < 0$. Furthermore, since $p_0 = p_1(T) = p_2(T) = 0$ by (87), $q(0) = 0$ is impossible. Hence $q(0) < 0$. Note also that since $p_2(t) \leq 0$ for all t, by using the bouncing off result (A.15) on $\dot{p}_1 = -p_1 s^*(t)u^*(t) - p_2(t)u^*(t)$ in $(0, T]$, with $\bar{p}_1 = p_1(T) = 0$, we can conclude that $p_1(t) \leq 0$ for all $t \in (0, T]$.

Let t' be the largest point t such that $z(v) < Z$ for $v \in [0, t')$, (remember $z_0 < Z$). Now $q(v)$ is constant on $[0, t')$, $0 > q(0) = q(t)$ for all $t \in [0, t')$, i.e. $p_2(t) < 0$ for all $t \in [0, t')$. The relations $p_0 = 0$, $p_1(t) \leq 0$, $p_2(t) < 0$ on $[0, t')$ gives that only $u = 0$ maximizes the Hamiltonian. But $u = 0$ means $z(t') \leq z_0 < Z$. This contradicts the definition of t', unless $t' = T$. Thus $z(t) < Z$ everywhere, and therefore $q(t) \equiv q(0) < 0$, which contradicts $q(T) = 0$. Hence $p_0 = 1$.

Next, knowing that $p_0 = 1$, we want to prove that any optimal solution must be of the types A, B_1 and B_2 described in example 3. Now, (114(ii)) also holds if $p_0 = 1$, and thus $p_2(t) \leq 0$ in this case as well.

Let us state two useful properties. Let $(k(t), z(t), u(t), s(t))$ be an admissible quadruple. Then:

Let $\beta \in (0, T]$. If $z(t) < z(\beta)$ for some $t < \beta$, then $k(\beta) > az(\beta)$. $\quad$ (116a)

If $u(t)=1$ in some interval (α, β), and if $z(\beta)>z_0$, then $z(t)<z(\beta)$ for all $t\in[\alpha, \beta)$.   (116b)

Furthermore:

If $(k^*(t), z^*(t))$ satisfies the necessary conditions and $z^*(t')=Z$ for some $t'\leqslant T$, while $z^*(\tau)<Z$ for some τ arbitrarily near $t', \tau<t'$, then

(∗) $\max(1, p_1(t))+p_2(t)>0$ for $t<t'$

and $u^*(t)\equiv1$ in $[0,t')$. Furthermore, $z^*(t)<Z$ for all $t<t'$.   (117)

Before proving these properties, let us use them to show that an optimal quadruple $(k^*(t), z^*(t), u^*(t), s^*(t))$ must be one of the types A, B_1, or B_2. First note that if $z^*(T)<Z$, then $z^*(t')=Z$ for some $t'<T$ is impossible. To prove this, assume the contrary. Let t' be the largest t' such that $z^*(t')=Z$. Then $q(t)$ is constant in $(t', T]$, and since $q(T)=0$, $q(t)\equiv0$ in $(t', T]$. From (112) and (114(ii)), $\dot{p}_2(t)=ap_2(t)$ with $p_2(T)=0$, so $p_2(t)\equiv0$ in $(t', T]$. Using (56) we see that $u^*(t)\equiv1$ in $(t', T]$. From (116a), since $z(0)<z(t')$, $k^*(t')>az^*(t')=aZ$. But $k^*(t)$ is non-decreasing, so $k^*(t)\geqslant k^*(t')>aZ$ for all $t\in(t', T]$, and thus $\dot{z}^*(t)>k^*(t)-aZ>0$. It follows that $z^*(T)>z^*(t')=Z$, a contradiction. We conclude that if $z^*(T)<Z$, $z^*(t)<Z$ for all t, which proves (70(i)), and takes care of case A.

Next, consider the case $z^*(T)=Z$. Define $t'=\inf\{t': z^*(t)=Z \text{ in } [t', T]\}$. (Perhaps $t'=T$.) There exist τ arbitrarily near t', $\tau<t'$, where $z^*(\tau)<z^*(t')=Z$. Then (117) states that $z^*(t)<Z$ for all $t<t'$. Hence we get case B_1 if $t'=T$ and case B_2 if $t'<T$. We can now make sure that the guesses made in cases B_1 and B_2 in example 3 were correct: From (117), $u^*(t)=1$ in $[0, t']$. Furthermore, from the fact that $z^*(t)<Z$ for $t<t'$, we see that from the time $z^*(t)$ first attains the value Z we continue to have $z^*(t)=Z$. (Thus the present definition of t' coincides with that in example 3.) Finally, in $(t', T]$, $\dot{z}^*\equiv0$, so for $t\in(t', T]$, $u^*(t)=aZ/k^*(t)\leqslant aZ/k^*(t')<1$, where we used (116a) with $\beta=t'$.

Thus we see that if there exists an optimal solution, then it must be one of the types A, B_1 or B_2 with $u^*(t)$ as described. In each of these cases we derived in example 3 the behaviour of $s^*(t)$ and the appropriate conditions on t^* and t'. Finally, in the present problem, theorem 5 is easily seen to guarantee the existence of an optimal solution. Thus it is unnecessary to prove that one of the solutions of type A, B_1 or B_2 actually satisfies the necessary conditions. (The latter verification is essential if we use sufficient conditions.)

It remains to prove properties (116) and (117). In order to prove (116a) assume by contradiction that $k(\beta) \leqslant az(\beta)$. Consider the differential equation $\dot{z} = h(z, s) = u(s)k(s) - az$ on $(t, \beta]$. If $s \in (t, \beta]$, since k is non-decreasing, $k(s) \leqslant k(\beta) \leqslant az(\beta)$. Thus, since $u(s) \leqslant 1$, $h(z(\beta), s) = u(s)k(s) - az(\beta) \leqslant k(s) - az(\beta) \leqslant 0$ for all $s \in (t, \beta]$. By the bouncing-off result (A.14), $z(s) \geqslant z(\beta)$ in $(t, \beta]$. By continuity, also $z(t) \geqslant z(\beta)$, a contradiction.

Next, let us prove (116b). Assume by contradiction that $z(t') \geqslant z(\beta)$ for some $t' \in [\alpha, \beta)$. As a continuous function on $[\alpha, \beta]$, $z(t)$ has a maximum at some point $t_1 \in [\alpha, \beta]$, and we can choose $t_1 < \beta$. We have that $z(t_1) \geqslant z(t') \geqslant z(\beta) > z(0) = z_0$. Using (116a), we see that $k(t_1) > az(t_1)$. But since t_1 is a maximum point, $\dot{z}(t_1^+) = u(t_1^+)k(t_1) - az(t_1) = k(t_1) - az(t_1) \leqslant 0$, a contradiction.

Finally, to prove (117), let τ be any point $< t'$ where $z^*(\tau) < Z$. Let (α, β) be the largest interval containing τ such that $z^*(t) < Z$ for $t \in (\alpha, \beta)$. Then $z^*(\beta) = Z$, and there exist points t arbitrarily close to β, $t < \beta$, such that $\dot{z}^*(t) \geqslant 0$, and by the continuity of $z^*(t)$, $z^*(t) > 0$ for t close to β. For such values of t, it follows from $\dot{z}^*(t) = u^*(t)k^*(t) - az^*(t) \geqslant 0$ that $u^*(t) > 0$. From (56) it follows that

$$\max (1, p_1(t)) + p_2(t) \geqslant 0. \tag{118}$$

Choosing a sequence of such t's converging to β^-, we see that (118) is valid for $t = \beta^-$ as well.

We next claim that $p_2(t)$ as well as $p_1(t)$, and hence $\max (1, p_1(t))$, are non-increasing in (α, β). Note first that from (89(i)) we see that $q(t)$ is a constant $\bar{q}$ in (α, β). From the facts that $p_2(t) = p_2^*(t) + \bar{q}$ (see (112)) and $p_2^*(t)$ is continuous, it follows that $p_2(t)$ is continuous and $\dot{p}_2(t) = \dot{p}_2^*(t)$ in (α, β). Hence from (114(ii)), $\dot{p}_2(t) = ap_2(t)$ which is $\leqslant 0$ from (115), ((115) is valid for $p_0 = 1$ as well). We conclude that $p_2(t)$ is non-increasing in (α, β). Note also that if $p_2(\beta^-) < 0$, then $p_2(t)$ is strictly decreasing.

As for $p_1(t)$, with $p_1(t) = p_1^*(t)$, the differential equation for $p_1(t)$ (obtained from $\dot{p}_1^* = -\partial \bar{L}^*/\partial k$) is the same as the corresponding equation for $p_1(t)$ in example 3, and again (see (59)), $\dot{p}_1(t) \leqslant 0$, so $p_1(t)$ is non-decreasing.

Since we have shown that $\max (1, p_1(t))$ and $p_2(t)$ are both non-increasing in (α, β), it follows that if (118) holds with strict inequality for $t = \beta^-$, then inequality (*) in (117) holds in (α, β). If (118) holds with equality for $t = \beta^-$, then $p_2(\beta^-) < 0$, and since we observed above that then $p_2(t)$ is strictly decreasing in (α, β), inequality (*) in (117) will again hold in (α, β). Thus $u^*(t) = 1$ in (α, β) by (56), and by (116b), $z(\alpha) < z(\beta) = Z$. Hence, $\alpha = 0$ by the maximality of the interval (α, β).

Note that $u^*(t) = 1$ in $(0, \tau) \subset (0, \beta)$, and that inequality (*) in (117) holds in $(0, \tau)$. There exists a sequence of points $\tau = \tau_n$ of the type above which converges to t', hence we obtain that $u^*(t)=1$ on $(0,t')$, and that $(*)$ in (117) also holds in this interval. From (116b), $Z=z(t')>z_0$ implies $z(t)<Z$ for all $t<t'$. This concludes the proof of (117).

Note 6   It is instructive to find the candidates for optimality in example 5 by using ANC. Note, in particular, that the argument for $p_2(t)\leqslant 0$ becomes much easier.

Exercise 5.3.1   Consider the model in example 4. Denote the criterion functional by V.

(i) Compute V for the admissible controls $u^*(t)\equiv 0$ and $u^*(t)\equiv 1$.
(ii) Let $u^*(t)=1$ for $t\in[0,t^*]$ and $u^*(t)=0$ for $t\in(t^*,4]$. V will then depend on t^*, $V=V(t^*)$. Find the value of t^* which maximizes $V(t^*)$. Do you get a partial confirmation of the result in example 4?

Exercise 5.3.2   Consider the problem

$$\max \int_0^4 [(x(t)-2)^2 + u(t)]dt, \qquad \dot{x}=-u,$$

$$x(0)=1, \qquad x(4)=1, \qquad u(t)\in[-1,1], \qquad x(t)\geqslant 0.$$

Prove that $p_0=0$ is impossible, and solve the problem.

Exercise 5.3.3*   Consider theorem 2. By using the continuity of $p^*(t)$ and taking left and right limits at a discontinuity point τ_k of $p(t)$, prove (37) with $\beta_j^k=q_j(\tau_k^+)-q_j(\tau_k^-)$, with the aid of (90).

Exercise 5.3.4*   Derive condition (89(ii)) from (88) and (89(i)).
(Hint: $H^*(t^+)\geqslant H(x^*(t), u, p(t^+), t)$ when $u=u^*(t^-)$, $H^*(t^-)\geqslant H(x^*(t), u, p(t^-), t)$ when $u=u^*(t^+)$. Use also the result in exercise 5.3.3 and the fact that if $g_j^*(t)=0$, then $(d^-/dt)g_j^*(t)\leqslant 0$, $(d^+/dt)g_j^*(t)\geqslant 0$, and $g_{jx}'^*(t)(f^*(t^+)-f^*(t^-))\geqslant 0$.)

Exercise 5.3.5*   Consider note 3. Prove the first two equalities in (c) using (89) and the fact that $(d/dt)g_j^*(t)=0$ if $g_{jx}'^* f^*$ is continuous at t and $g_j^*(t)=0$. Using these two equalities, prove (b) and the third equality in (c). Use (b) to prove (d) and (d) to prove (e). Prove (f) using t as a new state variable.

4. Non-triviality results*

In this section we extend the non-triviality result in note 4 in several directions. Non-triviality results play, essentially, the same role in optimal control theory as do constraint qualifications in non-linear programming. In the absence of conditions securing non-triviality, the maximum condition might give no information about the optimal control. (See e.g. exercise 5.4.1.)

Consider problem (23)–(27). If the g-constraints are inactive at t_0 as well as at t_1, condition (95) in note 4 implies that the multipliers are non-trivial in an interval contained in (t_0, t_1). We shall now consider some non-triviality results covering cases in which some of the state constraints are active at t_0 and/or at t_1.

In addition to (94), define

$$G^*(t_0^+) = \{j: \text{there exist numbers } s > t_0, \text{ arbitrarily near } t_0$$
$$\text{such that } g_j(x^*(s), s) = 0\}$$

$$G^*(t_1^-) = \{j: \text{there exist numbers } s < t_1, \text{ arbitrarily near } t_1$$
$$\text{such that } g_j(x^*(s), s) = 0\}. \tag{119}$$

Note that $G^*(t_0)$ contains $G^*(t_0^+)$ and $G^*(t_1)$ contains $G^*(t_1^-)$.

Call the restrictions, here $x(t_0) = x^0$, $g \geq 0$, *first-order compatible at* $(x^*(t_0), t_0)$ if either (i) $G^*(t_0) = \phi$ or (ii) $G^*(t_0) \neq \phi$ and:

For some $v \in \text{co}\{f(x^*(t_0), u, t_0): u \in U\}$

$$v \cdot \frac{\partial g_j(x^*(t_0), t_0)}{\partial x} + \frac{\partial g_j(x^*(t_0), t_0)}{\partial t} > 0 \text{ for all } j \in G^*(t_0^+). \tag{120}$$

(As usual, $\text{co}\{A\}$ denotes the convex hull of A.)

Call the restrictions, here (25) and $g \geq 0$, *first-order compatible at* $(x^*(t_1), t_1)$ if either (i) $G^*(t_1) = \phi$ or (ii) $G^*(t_1) \neq \phi$, $l = n$ and (121) below holds, or (iii) $G^*(t_1) \neq \phi$, $l < n$ and (122) below holds.

For some $v \in \text{co}\{f(x^*(t_1), u, t_1): u \in U\}$

$$v \cdot \frac{\partial g_j(x^*(t_1), t_1)}{\partial x} + \frac{\partial g_j(x^*(t_1), t_1)}{\partial t} < 0 \text{ for all } j \in G^*(t_1^-). \tag{121}$$

There exists a vector $a \in \Gamma_1 = \{a: a \in R^n, a_i = 0 \text{ for } i = 1, \ldots, l,$
$a_i \geq 0 \text{ for } i = l+1, \ldots, m\}$ such that $a \cdot [\partial g_j(x^*(t_1), t_1)/\partial x] > 0$
for all $j \in G^*(t_1)$. \tag{122}

We have now the following result:

Theorem 6[7]   If

the restrictions are first-order compatible at t_0 and at t_1,
and (95) holds, $\qquad\qquad\qquad\qquad\qquad\qquad\qquad\qquad\qquad\qquad$ (123)

then

$(p_0, p(t)) \neq (0, 0)$ for all t in some interval contained in (t_0, t_1).   (124)

Suppose (123) holds and

for all $t \in (t_0, t_1]$ there exists a $v \in \text{co}\{f(x^*(t), u, t): u \in U\}$ such
that

$$v \cdot \frac{\partial g_j(x^*(t), t)}{\partial x} + \frac{\partial g_j(x^*(t), t)}{\partial t} > 0 \qquad \text{for all } j \in G^*(t), \qquad \text{(125a)}$$

for all $t \in [t_0, t_1)$ there exists a $v \in \text{co}\{f(x^*(t), u, t): u \in U\}$
such that

$$v \cdot \frac{\partial g_j(x^*(t), t)}{\partial x} + \frac{\partial g_j(x^*(t), t)}{\partial t} < 0 \text{ for all } j \in G^*(t). \qquad \text{(125b)}$$

Then

$(p_0, p(t)) \neq (0, 0)$ for $t = t_0$, $t = t_0^+$, $t = t_1$, $t = t_1^-$ and for all
$t = t^+$, $t = t^-$ with $t \in (t_0, t_1)$. $\qquad\qquad\qquad\qquad\qquad\qquad$ (126)

Note 7   If the terminal point $x(t_1)$ is fixed ($l = n$), and the restrictions are
first-order compatible at $(x^*(t_1), t_1)$, then in theorem 2 we can put $q(t_1^-) = q(t_1)$, $p(t_1^-) = p(t_1)$. In this connection the following result is useful: If the
restrictions are first-order compatible at t_0 and t_1 and (125a) holds, then in
theorem 2, $(p_0, p(t_1^-)) \neq (0, 0)$.

Similarly, if the initial point $x(t_0)$ is fixed (as we have assumed so far), and
the restrictions are first-order compatible at $(x^*(t_0), t_0)$, then we can put
$q(t_0^+) = q(t_0)$, $p(t_0^+) = p(t_0)$. Moreover, if the restrictions are first-order compatible at t_0 and t_1, and (125b) holds, then $(p_0, p(t_0^+)) \neq (0, 0)$.[8]

Note 8   Assume that $U = R^r$ and that $\partial f_0/\partial u$ and $\partial f/\partial u$ exist and are
continuous. Then property (126) holds if for each $t \in [t_0, t_1]$ either $G^*(t) = \phi$

---

[7] For references, see footnote 10. A technical point is discussed in footnote 11.
[8] For references, see footnote 11.

or $G^*(t) \neq \phi$ and

the matrix with elements $\alpha_{jk} = \sum_{i=1}^{n} \dfrac{\partial g_j(x^*(t), t)}{\partial x_i} \dfrac{\partial f_i(x^*(t), u, t)}{\partial u_k}$

$k = 1, \ldots, r, \, j \in G^*(t)$ has a rank equal to the number of elements in $G^*(t)$ both for $u = u^*(t^+)$ and $u = u^*(t^-)$. (Only one of these conditions applies when $t = t_0$ and $t = t_1$.) $\qquad$ (127)

Suppose that f_0 and f are C^2. Then the maximum condition implies:

$$\partial H(x^*(t), u^*(t), p(t), t)/\partial u = 0 \qquad \text{v.e. in } (t_0, t_1).\qquad (128)$$

Assume that $u^*(t)$ is piecewise continuous. Then condition (127) and the property (128) imply that the discontinuity points of $q(t)$ are included among those of $u^*(t)$. If $u^*(t)$ has a uniformly continuous derivative between its jump points, then $q(t)$ also has this property, making the "almost necessary" conditions of section 3 truly necessary.[9]

Note 9 (Non-triviality when $x(t_0)$ is not necessarily fixed) Suppose we replace the condition $x(t_0) = x^0$ by the initial conditions (3.29). Then the necessary conditions in theorem 2 and ANC remain valid, and in addition the transversality condition (3.30) with $S_0 \equiv 0$ holds. It turns out that all the non-triviality results of this section hold if we redefine the concept "first-order compatible at $(x^*(t_0), \, t_0)$" to mean: Either (i) $G^*(t_0) = \phi$ or (ii) $G^*(t_0) \neq \phi$, $N = \{1, \ldots, n\}$ and (120) hold or (iii) $G^*(t_0) \neq \phi$, $N \neq \{1, \ldots, n\}$ and

There exists a vector $a \in \Gamma_0 = \{a : a \in R^n, \, a_i = 0$
for $i \in N$, $a_i \geq 0$ for $i \in N'\}$
such that $a \cdot (\partial g_j(x^*(t_0), t_0)/\partial x) > 0$ for all $j \in G^*(t_0)$. $\qquad$ (129)

(Sufficient conditions for this case are obtained by adding (3.30) (with $S_0 \equiv 0$) to the conditions of theorems 1 and 3 and of note 1, allowing $t_0 \leq \tau_1$ in this note.)

In the case where $N \neq \{1, \ldots, n\}$ the following result is useful: If the restrictions are first-order compatible at $(x^*(t_1), t_1)$ and (125b) holds, then $(p_0, \, p(t_0)) \neq (0, \, 0)$. If we add compatibility also at $(x^*(t_0), \, t_0)$, then $(p_0, \, p(t_0^+)) \neq (0, \, 0)$, as well.

Finally, in the case $l < n$, the following symmetric result is useful: If the

---

[9] The case treated in this note is closely related to that of Hestenes (1966), Chapter 8.

restrictions are first-order compatible at $(x^*(t_0), t_0)$ and (125a) holds, then $(p_0, p(t_1)) \neq (0, 0)$. If we add compatibility also at $(x^*(t_1), t_1)$, even $(p_0, p(t_1^-)) \neq (0, 0)$.[10]

Example 6 The last non-triviality result can be used to see that $p_0 \neq 0$ in example 5. In fact, in that example $z^*(0) = z_0 < Z$, so $G^*(0) = \phi$, and (125a) is satisfied if we put $s = u = 0$, $v = (0, -aZ)$, since

$$v \cdot \left(\frac{\partial g^*}{\partial k}, \frac{\partial g^*}{\partial z} \right) + \frac{\partial g^*}{\partial t} = (0, -aZ) \cdot (0, -1) = aZ > 0. \tag{130}$$

Note 10 All the results in this chapter, except theorem 4 and Notes 3(b)–(f), are also valid if the pure state restrictions are

$$g_j(x, t) \geqslant 0 \qquad \text{for all } t \in [a_j, b_j], \tag{131}$$

where $[a_j, b_j], j = 1, \ldots, s$ are fixed subintervals of $[t_0, t_1]$. The only change needed is that the set $G^*(t)$ has to be defined as $\{j : t \in [a_j, b_j], g_j(x^*(t), t) = 0\}$, and that the sets $G^*(t_0^+)$ and $G^*(t_1^-)$ are similarly redefined. Furthermore, the function $q_j(t)$ is constant on (t_0, a_j) and on (b_j, t_1), and (89(ii)) does not hold for $t = a_j$ or $t = b_j$.

Note 11 (**Weaker regularity conditions**) Assume that f_0 and f satisfy the conditions in (a) and (b) in note 2.6. Then all the results in this chapter are valid, except theorem 4, property (89(ii)) and note 3(b)–(f). Moreover, condition (38(ii)) should be deleted and, finally, condition (125) has to be modified in the following way: In (a), v must belong to $\text{co}\{f(x^*(t), u, t^-): u \in U\}$, and in (b), v must belong to $\text{co}\{f(x^*(t), u, t^+): u \in U\}$.[11]

---

[10]The proof of (124) in the case where $G^*(t) \neq \phi$ and $x(t)$ is fixed for $t = t_0$ and/or for $t = t_1$, is given in Seierstad (1986b). (A modification of the standard proofs is needed.)

  In that case (124) *is not* a consequence of the conditions in theorem 2. In all other cases (124) follows from the conditions in theorem 2 as shown in Neustadt (1976), Chapter V.3.

  If (124) is known to hold, then, assuming (125), property (126) follows from the conditions in theorem 2, as does the non-triviality results in note 7 and the last two sections of the present note.

[11]With (131) the results in section 3 are essentially valid if we also allow for bounded, measurable controls. (See footnote 9, Chapter 2.) The slight modifications needed are the following: (88) holds a.e., $p^*(t)$ is absolutely continuous with a derivative a.e. which a.e. satisfies (91). In (99), $H^*(t_1^{*-})$ has to be replaced by $\sup_{u \in U} H(x^*(t_1^*), u, p(t_1^{*-}), t_1^*)$ which *is* finite.)

  Note also the following delicate technical point: If we consider the special case in which $G^*(t) \neq \phi$ and $x(t)$ is fixed for $t = t_0$ and/or $t = t_1$, then the non-triviality conditions can only be derived if we allow the admissible controls to have an infinite number of jump points, and not necessarily with one-sided limits at t_0 and t_1. If we insist on allowing only piecewise continuous admissible controls, then the symbol "co" in (120) and in (121) must be deleted in order for the results to be correct. (See Seierstad, (1986b).)

In note 2, f_0 and f need only be defined for $x \in A(t)$ and have the regularity properties described in note 2.12 for the Arrow case (resp. Mangasarian case).

Exercise 5.4.1 In this exercise the use of non-triviality conditions is illustrated.

(i) Maximize or minimize $\int_0^1 x(t)dt$, $\dot{x}(t)=u(t) \in [0,\ 1]$, $x(0)=0$, $x(1)$ free, $x(t) \geqslant 0$. Use note 9 to prove that $(p_0,\ p(1)) \neq (0,\ 0)$, and conclude that $p_0=1$.
(ii) Require in (i) that $x(1) \geqslant 0$. Use note 9 to prove that $(p_0,\ p(1^-)) \neq (0,\ 0)$, and then argue why $p_0=1$. (Use (88).)
(iii) Replace $\dot{x}=u$ by $\dot{x}=(t-1/2)^2 u$ in (i). Use (124) to prove that $p_0=1$.
(iv) In (i) let $u \in [-1,\ 1]$ and require $x(1)=0$. Use (126) to show that $p_0=1$. (Hint: $p_0=0$ implies $u^*=\pm 1$. At some point $\check{t} \in (0,1)$ a switch must occur i.e. $p(\check{t}^+) \neq p(\check{t}^-)$, contradicting (89).)

5. Miscellaneous*

We shall conclude this chapter with a discussion of two different topics. First, we compare our approach to pure state constraint problems with an alternative formulation in the literature. Secondly, we give an heuristic argument indicating that the necessary conditions in theorem 2 give enough information.

An alternative approach

When solving pure state constraint problems it is often assumed that the optimal solution has the following property: There exists a finite partition of (t_0, t_1) into intervals (a_i, a_{i+1}), $i=1, \ldots, i^*$ such that on any one of them the same set of constraints is binding on the entire interval, i.e. $G^*(t)$ defined in (94) is constant on each interval (a_i, a_{i+1}), $i=1, \ldots, i^*$, where $a_1=t_0$, $a_{i^*+1}=t_1$.

This assumption is, for example, in Hadley and Kemp (1971), and it makes it possible to reformulate our necessary conditions in the following way:

Consider theorem 2. Define, $q_j^1(t)=q_j(t)$ if $j \notin G^*(t)$, $q_j^1(t)=0$ if $j \in G^*(t)$ and let $q^0(t)=q(t)-q^1(t)$. Define moreover, $\hat{p}(t)=p^*(t)-q^1(t)g_x'(x^*(t),\ t)$. By our assumption on $G^*(t)$, $q^1(t)$ is piecewise constant. In fig. 5.4 we have indicated

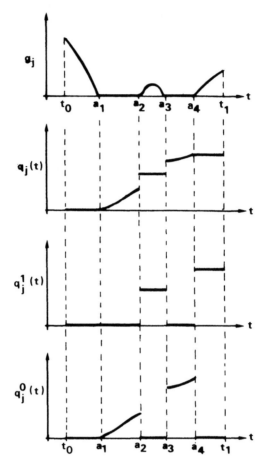

Fig. 5.4 $dg_j(x^*(t), t)/dt = (\partial g_j^*/\partial x) f(x^*, u^*, t) + \partial g_j^*/\partial t$ is discontinuous at a_1 and at a_4, so according to (89(ii)), $q_j(t)$ is continuous at these points. (We have normalized $q(t)$ by putting $q(t_0) = 0$.)

a possible development for $\varphi(t) = g_j(x^*(t), t)$ and for $q_j(t)$ (which, according to theorem 2, is non-decreasing and constant if $\varphi(t) > 0$ and we have also shown the corresponding graphs for q_j^1 and q_j^0.

As in note 3 we see that $\hat{p}(t)$ satisfies $\dot{\hat{p}} = -\bar{L}_x'(x^*(t), u^*(t), \hat{p}(t), q^0(t), t)$ v.e. The function $u^*(t)$ maximizes $H(x^*(t), u, p(t), t) = p_0 f_0(x^*(t), u, t) + p(t)f(x^*(t), u, t)$. Since $p(t) = p^*(t) - q(t)g_x'^* = \hat{p}(t) + q^1(t)g_x'^* - q(t)g_x'^* = \hat{p}(t) - q^0(t)g_x'^*$, $u^*(t)$ maximizes $p_0 f_0(x^*(t), u, t) + \hat{p}(t)f(x^*(t), u, t) - q^0(t)g_x'^*f(x^*(t), u, t) = H(x^*(t), u, \hat{p}(t), t) - q^0(t)g_x'^*f(x^*(t), u, t)$.

By using the continuity of $p^*(t) = \hat{p}(t) + q^1(t)g'^*_x$, we obtain for $i = 2, \ldots, i^*$ the jump conditions:

$$\hat{p}(a_i^+) = \hat{p}(a_i^-) + \sum_{j=1}^{s} q_j^1(a_i^-)\partial g_j^*/\partial x - \sum_{j=1}^{s} q_j^1(a_i^+)\partial g_j^*/\partial x, \qquad (132)$$

where the *'s indicate evaluation at $(x^*(a_i), a_i)$.

Consider any of the points a_i, $i = 2, \ldots, i^*$. Let us agree to define j as an *entering (leaving) index* at a_i if j enters (leaves) the set of active indices $G^*(t)$ when t passes a_i. Then (132) can be reformulated as follows:

$$\hat{p}(a_i^+) = \hat{p}(a_i^-) + \sum_{\substack{\text{entering}\\ j\text{'s}}} q_j(a_i^-)\partial g_j^*/\partial x - \sum_{\substack{\text{leaving}\\ j\text{'s}}} q_j(a_i^+)\partial g_j^*/\partial x. \qquad (133)$$

To derive (133), we make use of the following facts: For an entering index at a_i (see a_1 and a_3 in fig. 5.4), $q_j^1(a_i^-) = q_j(a_i^-)$ and $q_j^1(a_i^+) = 0$. For a leaving index j at a_i (see a_2 and a_4 in fig. 5.4), $q_j^1(a_i^-) = 0$ and $q_j^1(a_i^+) = q_j(a_i^+)$. Moreover, if at a_i some g_j is inactive on both sides, $q_j^1(a_i^-) = q_j(a_i^-) = q_j(a_i^+) = q_j^1(a_i^+)$, so the corresponding terms in (132) cancel. Finally, if at a_i some g_j is active on both sides of a_i, $q_j^1(a_i^-) = q_j^1(a_i^+) = 0$.

The following necessary conditions are now valid (applying, implicitly, the normalization $q(t_0) = 0$):

If $(x^*(t), u^*(t))$ solves the problem, then there exist a number p_0, $(p_0 = 0$ or $p_0 = 1)$, and a piecewise continuous function $\hat{p}(t)$, differentiable v.e. and a non-negative function $q^0(t)$, non-decreasing on each interval (a_i, a_{i+1}), such that

$$\dot{\hat{p}}(t) = -\bar{L}'_x(x^*(t), u^*(t), \hat{p}(t), q^0(t), t) \qquad \text{v.e.} \qquad (134)$$

For v.e. $t \in (t_0, t_1)$,

$$u^*(t) \text{ maximizes } H(x^*(t), u, \hat{p}(t), t) - q^0(t)g'_x(x^*(t), t)f(x^*(t), u, t),$$
$$\text{for } u \in U. \qquad (135)$$

Moreover, for some non-negative numbers δ_j^{i-}, δ_j^{i+}, $i = 2, \ldots, i^*$,

$$\hat{p}(a_i^+) = \hat{p}(a_i^-) + \sum_{\substack{\text{entering}\\ j\text{'s}}} \delta_j^{i-} \partial g_j^*/\partial x - \sum_{\substack{\text{leaving}\\ j\text{'s}}} \delta_j^{i+} \partial g_j^*/\partial x. \qquad (136)$$

The δ's satisfy the relations: $\delta_j^{i-} = \delta_j^{(i-1)+}$, $\delta_j^{i+} \geqslant \delta_j^{i-}$, $i = 2, \ldots, i^*, j = 1, \ldots, s$, $\delta_j^{i-} = \delta_j^{i+}$ if $j \notin G^*(a_i)$, $i = 2, \ldots, i^*$ and $\delta_j^{1+} = 0$ if $j \notin G^*(a_1)$. Moreover, for $i = 2, \ldots, i^*$, $\delta_j^{i+} \geqslant q_j^0(a_i^-)$, $\delta_j^{i-} \leqslant q_j^0(a_i^+)$ if $a_i \in G^*(a_i^+)$ and for an entering index j, if $\varphi^* = (\partial g_j^*/\partial x) f^*$ is discontinuous at a_i, then $\delta_j^{i-} = q_j^0(a_i^+)$. For a leaving index j, if φ^* is discontinuous at a_i, then $\delta_j^{i+} = q_j^0(a_i^-)$. $\qquad$ (137)

The function $q^0(t)$ also satisfies

if $t \neq a_i$ and $j \notin G^*(t)$, then $q_j^0(t) = 0$. $\qquad$ (138)

Finally, (92) is valid with $p(t_1)$ replaced by $\hat{p}(t_1) - q^0(t_1) g_x'(x^*(t_1), t_1)$. ∎

(We omit a translation of (87) into the present setting. See instead note 13 below.)

We obtained the conditions above from our previous discussion by putting $\delta_j^{i-} = q_j(a_i^-)$ and $\delta_j^{i+} = q_j(a_i^+)$, for entering and leaving indices, respectively.

The equations $\delta_j^{i-} = q_j^0(a_i^+)$ and $\delta_j^{i+} = q_j^0(a_i^-)$ hold for all entering and leaving indices j (resp.) if q_j is continuous at a_i, $i = 2, \ldots, i^*$, and conditions implying the last fact was stated in note 3, (d) and (e).

Condition (136) is the same as the crucial jump condition (5–167) in Hadley and Kemp (1971).

Essentially, the above necessary conditions were obtained by subtracting different constants from $q(t)$ on the various subintervals (a_{i-1}, a_i). The reader can find still other necessary conditions in the literature, which are obtained by subtracting other constants from $q(t)$. For example, this can always be done in such a way that $q_j^0(t)$ is continuous at all points where j leaves the set of active indices.

Note 12 For any j, suppose either that $(\partial g_j^*/\partial x) f^*$ is discontinuous – or that the conditions of note 3(d) and (e) are satisfied – at all points where j enters or leaves the set of active indices. (This implies that $q_j(t)$ is continuous at such points.) Then the following facts should be noted:

Suppose j is an inactive index which enters $G^*(t)$ at $t = a_i$, let a_{i_j}, $i_j < i$, be the last point at which j was active. Then $\delta_j^{i-} = \delta_j^{(i-1)+} = \delta_j^{(i-1)-} = \delta_j^{(i-2)+} = \ldots = \delta_j^{i_j+} = q_j^0(a_i^-)$. If j was never active before, put $i_j = 1$. In this case we get again $\delta_j^{i-} = \ldots = \delta_j^{1+} = 0 = q_j^0(a_1^+)(= q_j^0(a_1^-))$, by convention, as j is inactive at t_0). Thus we see that (136) in this case can be written

$$\hat{p}(a_i^+) = \hat{p}(a_i^-) + \sum_{\substack{\text{entering} \\ j\text{'s}}} q_j^0(a_i^-) \partial g_j^*/\partial x - \sum_{\substack{\text{leaving} \\ j\text{'s}}} q_j^0(a_i^-) \partial g_j^*/\partial x. \qquad (139)$$

Note that $\hat{p}(a_i^+)$ is determined by the values of $x^*(t)$, $\hat{p}(t)$ and $q^0(t)$ on $[t_0, a_i]$.

Note 13 (**Non-triviality**) In the present case, if the restrictions are first-order compatible at t_0 and t_1 and if (125) holds, then

$$(p_0, \hat{p}(t) - q^0(t)g'_x(x^*(t), t)) \neq (0, 0) \text{ for } t = t_0^+, t = t_1, t = t_1^-$$

and for $t = t^+$ and $t = t^-$, for all $t \in (t_0, t_1)$. $\qquad$ (140)

In this case we can assume that $q^0(t_0) = q^0(t_0^+)$. Also, $\delta_j^{1+} = 0$ for all j. In particular, this non-triviality result is valid if $U = R^r$ and (127) holds. In this case $q^0(t)$ is continuous at all points $t \notin \{a_1, \ldots, a_{i^*+1}\}$ at which $u^*(t)$ is continuous.

Why the necessary conditions contain enough information

As in the previous chapters we shall give heuristic arguments indicating that the necessary conditions contain enough information. In fact, we check in a rough way that the necessary conditions give the sufficient number of conditions ("equations") for determining the optimal solution. The arguments we use partly correspond to those used by some authors in presenting the necessary conditions.

We restrict our attention to fixed initial, fixed terminal point problems, where $G^*(t_0) = G^*(t_1) = \phi$. We assume that the conditions mentioned in the beginning of note 12 hold, such that $q_j(t)$ is continuous at all points where j leaves or enters the set of active indices. We then saw that $\delta_j^{i-} = q_j^0(a_i^+)$ and $\delta_j^{i+} = q_j^0(a_i^-)$ for entering and leaving indices, respectively. As in the preceding section we assume that $G^*(t)$ is constant on each interval (a_i, a_{i+1}), $i = 1, \ldots, i^*$, $a_1 = t_0$, $a_{i^*+1} = t_1$.

We begin by studying how $u^*(t)$ is determined on an interval (a_i, a_{i+1}). For all active indices j, $(d/dt)g_j(x^*(t), t) = (\partial g_j^*/\partial x)f^* + \partial g_j^*/\partial t = 0$. Thus for all $t \in (a_i, a_{i+1})$, $u^*(t) \in U'(t)$ where

$$U'(t) = \{u \in U : (\partial g_j^*/\partial x)f(x^*(t), u, t) + \partial g_j^*/\partial t = 0, j \in G^*(t)\}.$$

In order to determine the value that $u^*(t)$ must take in $U'(t)$, we use the maximum condition (135), which surely holds for $u \in U'(t) \subset U$. Note first that if $u \in U'(t)$, then for all $j \in G^*(t)$, $q_j^0(t)(\partial g_j^*/\partial x)f(x^*(t), u, t) = -q_j^0(t)\partial g_j^*/\partial t$. Since $q_j^0(t) = 0$ for $j \notin G^*(t)$, it follows that for $u \in U'(t)$, $q^0(t)g_x'^* f(x^*(t), u, t) = -q^0(t)g_t'^*$. Since the latter expression is independent of u, it follows from (135) that for v.e. t,

$$u^*(t) \text{ maximizes } H(x^*(t), u, \hat{p}(t), t) \text{ in } U'(t). \qquad (141)$$

This fact gives us a clue as to how to proceed. Consider the problem

$$\max_{u} H(x, u, p, t) \text{ for}$$

$$u \in \{u \in U: (\partial g_j(x, t)/\partial x)f(x, u, t) + \partial g_j(x, t)/\partial t = 0, j \in J\} \tag{142}$$

where J is a subset of $\{1, \ldots, s\}$.

In the rest of our argument we shall assume that $U = R^r$ or $U = \{u \in R^r: F_i(u) \geqslant 0, i = 1, \ldots, i'\}$. In both these cases it is reasonable to expect that problem (142) usually determines the maximum point u as a unique, well-behaved function of (x, p, t).

Suppose $U = R^r$, and imagine that we write down the first-order conditions for problem (142) with $-q_j^0, j \in J$ as Lagrange multipliers, i.e. using the Lagrange function $H - \Sigma^* q_j^0 [(\partial g_j/\partial x)f + \partial g_j/\partial t]$, where * indicates summation over all $j \in J$. We then get:

$$H'_u(x, u, p, t) - \sum^{*} q_j^0 (\partial g_j/\partial x) \cdot f'_u(x, u, t) = 0. \tag{143}$$

Let us assume that (143) and the constraints in (142) define u and q_j^0 as unique well-behaved functions of x, p and t, $q_j^0 = q_j^0(x, p, t)$. Put $q_j^0(x, p, t) = 0$ if $j \notin J$.

We claim that for $J = G^*(t)$, $x = x^*(t)$, $p = \hat{p}(t)$, the function $q_j^0(x, p, t)$ coincides with $q_j^0(t)$ as previously defined: Since $u^*(t)$ maximizes $H(x^*(t), u, p(t), t)$ for $u \in U = R^r$, $H'_u(x^*(t), u^*(t), p(t), t) = p_0 \partial f_0^*/\partial u + p(t)\partial f^*/\partial u = 0$. Replacing $p(t)$ with $\hat{p}(t) - \Sigma^* q_j^0(t)\partial g_j^*/\partial x$, where * indicates summation over all active indices, we obtain condition (143) with $x = x^*(t)$, $p = \hat{p}(t)$, $u = u^*(t)$. By the uniqueness assumption, $q_j^0(t) = q_j^0(x^*(t), \hat{p}(t), t)$.

Suppose next that $U = \{u \in R^r: F_i(u) \geqslant 0, i = 1, \ldots, i'\}$. In this case the first-order conditions for problem (142) involve the introduction of multipliers associated with each F_i-constraint as well as with each of the constraints $(\partial g_j/\partial x) \cdot f + \partial g_j/\partial t = 0, j \in J$. As in the previous case we assume that we obtain a unique maximum point and unique multipliers $-q_j^0$, which are well-behaved functions of (x, p, t).

The maximization of $H(x^*(t), u, p(t), t)$ in U leads to first-order conditions involving multipliers corresponding to each F_i. Replacing $p(t)$ by $\hat{p}(t) - \Sigma^* q_j^0(t)(\partial g_j^*/\partial x)$, it is easy to see that the first-order conditions look exactly the same as the first-order conditions for problem (142) when $J = G^*(t)$, $x = x^*(t)$ and $p = \hat{p}(t)$. By the uniqueness assumption we again obtain $q_j^0(t) = q_j^0(x^*(t), \hat{p}(t), t)$.

We now construct a quadruple of functions $P = (x(t; \hat{p}_0), u(t; \hat{p}_0), \hat{p}(t; \hat{p}_0), q^0(t; \hat{p}_0))$ by a stepwise procedure, moving from left to right on the sequence of intervals (a_i, a_{i+1}), $i = 1, \ldots, i^*$. The quadruple is constructed in such a way that it satisfies the necessary conditions, except the terminal conditions. The symbol $\hat{p}^0 (= \hat{p}(t_0))$ denotes an unknown vector of parameters which only in the end will be determined.

Since $G^*(t_0) = \phi$, the set J_1 of active indices is empty in an interval to the right of a_1. Solving problem (142) with $J = J_1$, in this interval, we get a function $u(x, \hat{p}, t)$. Since none of the g_j-constraints are active, we put $q^0(x, \hat{p}, t) \equiv 0$ in the interval. Next, we find solutions $x(t; \hat{p}^0)$, $\hat{p}(t; \hat{p}^0)$ of the system

$$\dot{x} = f(x, u(x, \hat{p}, t), t), \qquad \dot{\hat{p}} = -L'_x(x, u(x, \hat{p}, t), \hat{p}, q^0(x, \hat{p}, t), t) \qquad (144)$$

with initial conditions $x = x(t_0)$, $\hat{p} = \hat{p}^0$ at $t = t_0$. Define $u(t; \hat{p}^0) = u(x(t; \hat{p}^0), \hat{p}(t; \hat{p}^0), t)$, $q^0(t; \hat{p}^0) = q^0(x(t; \hat{p}^0), \hat{p}(t; \hat{p}^0), t) \equiv 0$. We restrict the definition of the quadruple $P = (x(t; \hat{p}^0), u(t; \hat{p}^0), \hat{p}(t; \hat{p}_0), q^0(t; \hat{p}^0))$ to the interval $[a_1, a_2]$ where a_2 is the first point at which the set of active indices changes, i.e. the first point at which some index becomes active.

We assume for simplicity that at each of the points $a_2, a_3, \ldots$, to be defined, either a single index enters the set of active indices (simultaneously a single index might leave), or that the sole change in the set of active indices is that one index leaves the set. Moreover, we make the same assumption as in the beginning of note 12.

Now, while passing the point a_2 a new set of active indices J_2 (containing a single index) is determined, and J_2 is in action on some interval to the right of a_2. We next solve the problem (142) with $J = J_2$, obtaining functions $u(x, \hat{p}, t)$, $q_j^0(x, \hat{p}, t)$, $j \in J_2$ and put $q_j^0(x, \hat{p}, t) = 0$ for $j \notin J_2$. Again we insert these functions into the differential equations in (144) and solve these equations to the right of a_2, with initial conditions $x = x(a_2; \hat{p}^0)$, $\hat{p} = \hat{p}(a_2^+, \hat{p}^0)$, the latter vector being determined by (139). We then define $u(t; \hat{p}^0) = u(x(t; \hat{p}^0), \hat{p}(t; \hat{p}^0), t)$ and $q^0(t; \hat{p}^0) = q^0(x(t; \hat{p}^0), \hat{p}(t; \hat{p}^0), t)$, and use this definition of the quadruple P on an interval (a_2, a_3), where a_3 is defined in the following way: When $x(t; \hat{p}^0)$ is extended to the right of a_2, either a new index becomes active at some point a_3 (at the same time the old active index might become inactive), or, before this happens we may want the index in J_2 to become inactive. In the latter case, a_3 is an unknown parameter. In either case we have defined a new set of active indices J_3 which is going to be in action on an interval (a_3, a_4).

The process is continued in the same way: We construct the quadruple P successively on intervals $(a_3, a_4), (a_4, a_5), \ldots$, and we obtain corresponding sets of active indices $J_4, J_5, \ldots$. (Except in the beginning and near the end,

J_k may contain several indices.) The points a_i, $i \geq 4$, are determined in the same way as a_3: When $x(t; \hat{p})$ is extended to the right of a_{i-1}, either a new index becomes active at a_i (and perhaps some other index becomes inactive at a_i), or, we might decide to let a_i be some earlier point at which we want an index to become inactive. In the latter case the index which is going to be inactive is subject to choice, and a_i is an unknown parameter. (In the former case a_i is the first point to the right of a_{i-1} at which some new index becomes active, and a_i is determined by $x(t, \hat{p}^0)$.)

In general, there will be a number of different sequences of sets of active indices, J_i. These sequences will give rise to different quadruples P. However, many of these quadruples will lead to infeasible solution candidates. Firstly, all $q_j^0(t)$, $j \in J_i$, should be non-negative and non-decreasing. A violation of this condition for some j must therefore lead to the exclusion of j from the set of active indices. Secondly; if $j \notin J_i$ but $g_j(x(t; \hat{p}^0), t)$ becomes <0 to the right of a_i, then j ought to be included in the set of active indices. Thirdly, many sequences $\{J_i\}$ give rise to infeasible candidates because condition (135) or other necessary conditions are violated.

The crucial point we now want to make is that when a sequence which gives rise to a feasible solution is fixed, we then have enough conditions (equations) to determine the unknowns. The conditions can be counted in the following somewhat indirect way: Assume first that the number of elements in J_i, denoted n_i, increases from zero to some maximum $n_{i'}$ and then decreases to zero. Imagine that we reverse the direction of time and construct solutions $\check{x}(t, \hat{p}^1)$, $\check{p}(t, \hat{p}^1)$ starting at t_1, with $p(t_1) = \hat{p}^1$ as unknown. Then all leaving indices in the former construction become entering indices in the latter construction, determined by $\hat{p}^1$. Take any t in $(a_{i'}, a_{i'+1})$. Then we can determine $\hat{p}^0$ and $\hat{p}^1$ by the equations $x(t, \hat{p}^0) = \check{x}(t, \hat{p}^1)$, $p(t, \hat{p}^0) = \check{p}(t, \hat{p}^1)$. This also determines all entering and leaving time points a_i.

Next, if n_i exhibits two peaks, with a minimum $n_{i''}$ inbetween, the above process can be carried out on $[t_0, t'']$ and $[t'', t_1]$ separately, t'' some fixed point in $(a_{i''}, a_{i''+1})$, using $x(t'')$ as an unknown parameter. The unknown $x(t'')$ is then determined by the condition that the right end $x(t'')$ of the solution $x(.)$ obtained on $[t_0, t'']$ must equal the left end $x(t'')$ of the solution $x(.)$ obtained on $[t'', t_1]$. The case with several peaks is treated in a similar manner.

Note that the procedure described above is not to be regarded as a standard solution procedure. However, sometimes it may give clues as to how a problem could be attacked.

CHAPTER 6

MIXED AND PURE STATE CONSTRAINTS

In this final chapter we shall consider very general optimal control problems with mixed and pure state constraints and with general initial and terminal conditions. It is important to examine such problems since a number of interesting economic control applications ultimately do (or should) involve both types of constraints.

Besides generalizing most of the results in the preceding chapters we shall also tie up various loose ends from the previous discussions. In particular, we shall study in more detail infinite horizon models, scrap value functions, and more general initial and terminal conditions.

1. The main problem

The main problem we examine is the following:

$$\max \int_{t_0}^{t_1} f_0(x(t), u(t), t) \mathrm{d}t \qquad (t_0, t_1 \text{ fixed}) \tag{1}$$

subject to

$$\dot{x}(t) = f(x(t), u(t), t), \qquad x(t_0) = x^0 \ (x^0 \text{ fixed in } R^n), \tag{2}$$

the terminal conditions

$$x_i(t_1) = x_i^1, \qquad i = 1, \dots, l \qquad (x_i^1 \text{ all fixed}) \tag{3a}$$

$$x_i(t_1) \geqslant x_i^1, \qquad i = l+1, \dots, m \qquad (x_i^1 \text{ all fixed}) \tag{3b}$$

$$x_i(t_1) \text{ free}, \qquad i = m+1, \dots, n \tag{3c}$$

the control variable restriction

$$u(t) \in U, \ U \text{ a fixed set in } R^r \tag{4}$$

357

and the constraints

$$g_j(x(t), u(t), t) \geqslant 0, \qquad\qquad j = 1, \ldots, s'$$
$$g_j(x(t), u(t), t) = \bar{g}_j(x(t), t) \geqslant 0 \qquad j = s' + 1, \ldots, s. \qquad (5)$$

From (5) it is seen that we represent mixed as well as pure state constraints by $g_j(x, u, t) \geqslant 0$, but for $j = s' + 1, \ldots, s$, g_j is independent of u. As before, x is an n-dimensional state vector, u is an r-dimensional control vector and f is an n-dimensional vector function. Throughout this chapter we shall assume that

f_0, f, $g = (g_1, \ldots, g_s)$ and their derivatives w.r.t. x and u exist and are continuous.   (6)

Of course, in the general case, (1)–(5) is a very complicated optimal control problem. Therefore we begin by considering some reasonably simple results which nevertheless are powerful enough to decide optimality in a number of interesting cases.

2. Some Mangasarian-type sufficiency results

In order to solve problems of the type (1)–(5) we invariably associate an adjoint function $p(t) = (p_1(t), \ldots, p_n(t))$ with (2) and we associate a multiplier function $q(t) = (q_1(t), \ldots, q_s(t))$ with the constraints (5). If there are pure state constraints imposed, the adjoint function $p(t)$ has, in general, discontinuities. In our first sufficiency result we are concerned with the case in which $p(t)$ is continuous. It generalizes theorem 4.5 and the Mangasarian version of theorem 5.1.

Again *the Hamiltonian* is defined by

$$H(x, u, p, t) = p_0 f_0(x, u, t) + p \cdot f(x, u, t) \qquad (7)$$

and *the Lagrangian* by

$$L(x, u, p, q, t) = H(x, u, p, t) + q \cdot g(x, u, t). \qquad (8)$$

Then the following theorem can be proved:

Theorem 1 (**Sufficient conditions, p(t) continuous**) Let $(x^*(t), u^*(t))$ be an admissible pair for problem (1)–(6). Assume that there exist a continuous and piecewise continuously differentiable function $p(t) = (p_1(t), \ldots, p_n(t))$, a piecewise continuous function $q(t) = (q_1(t), \ldots, q_s(t))$, and a vector $\beta = (\beta_1, \ldots,$

β_s) such that the following properties hold with $p_0 = 1$:

$$(\partial L^* / \partial u) \cdot (u - u^*(t)) \leqslant 0 \qquad \text{for all } u \in U, \text{ for v.e. } t. \tag{9}$$

(The *, here and below, indicates that the derivative is evaluated along $(x^*(t), u^*(t), p(t), q(t), t.)$

$$\dot{p}_i(t) = -\frac{\partial L^*}{\partial x_i}, \qquad i = 1, \ldots, n \qquad \text{v.e.} \tag{10}$$

$$p_i(t_1) - \sum_{j=1}^{s} \beta_j \frac{\partial g_j^*(t_1)}{\partial x_i} \begin{cases} \text{no condition} & i = 1, \ldots, l \\ \geqslant 0 \ (=0 \text{ if } x_i^*(t_1) > x_i^1) & i = l+1, \ldots, m \\ = 0 & i = m+1, \ldots, n \end{cases} \tag{11}$$

$$\beta_j = 0 \qquad\qquad\qquad\qquad\qquad j = 1, \ldots, s'$$
$$\beta_j \geqslant 0 \ (=0 \text{ if } g_j(x^*(t_1), u^*(t_1), t_1) > 0), \qquad j = s'+1, \ldots, s \tag{12}$$

$$q_j(t) \geqslant 0 \ (=0 \text{ if } g_j(x^*(t), u^*(t), t) > 0), \qquad j = 1, \ldots, s. \tag{13}$$

$$g_j(x, u, t) \text{ is quasi-concave in } (x, u) \text{ for each } t, \qquad j = 1, \ldots, s. \tag{14}$$

$$H(x, u, p(t), t) \text{ is concave in } (x, u) \text{ for each } t. \tag{15}$$

Then $(x^*(t), u^*(t))$ is optimal. ∎

For a proof of this result see the proof of the more general theorem 13.

If $U = R^r$, $g_j = h_j$, and $s' = s$, we see that theorem 1 reduces to theorem 4.5. In particular, since $U = R^r$, (9) does reduce to (4.31), and since $\beta_j = 0$ for all j, (11) reduces to (4.35).

On the other hand, if we put $s' = 0$ and $\bar{g}_j = g_j, j = 1, \ldots, s$, and U is convex, theorem 1 reduces to theorem 5.1 with (5.35) replaced by the stronger condition, (15). Note, in particular, that in this case $\partial L / \partial u = \partial H / \partial u$, and (9) is equivalent to (5.29).

In theorem 1 $p(t)$ is continuous also at t_1, so that $p_i(t_1^-) = p_i(t_1)$. However, if we redefine $p_i(t_1)$ to be equal to the expression on the left-hand side of (11), we see that (11) reduces to (5.34).

Theorem 1 is valid without any constraint qualification. However, the multipliers will normally satisfy the conditions in the theorem only if certain constraint qualifications hold. When attempting to solve an actual control problem we suggest using theorem 1 without bothering about constraint qualifications. If such an attempt fails, the reason might be that constraint qualifications are not satisfied, and hence that $p(t)$ is discontinuous. Then one can turn to the next theorem which is specially designed for such cases.

Example 1

$$\max \int_0^3 (-x(t)+u(t))dt, \qquad \dot{x}(t)=1-u(t),$$

$$x(0)=2, \qquad x(3) \text{ free}, \qquad u(t)\in U=[0,3], \tag{16}$$

$$g_1(x(t),u(t),t)=4-x(t)-u(t)\geqslant 0, \qquad \bar{g}_2(x(t),t)=x(t)\geqslant 0. \tag{17}$$

This is obviously a problem of the type (1)–(6). We shall start by trying to anticipate a solution. In order to get a high value of the integral, we should try to keep $x=x(t)$ small and $u=u(t)$ large, while still fulfilling $x+u\leqslant 4$. Since $x(0)=2$, we can at most put $u(0)=2$. We suggest putting $u=4-x$ in the beginning. Then $\dot{x}=x-3$. Solving this differential equation with $x(0)=2$, we get $x=3-e^t$. When $3-e^t=1$, i.e. when $t=t_1=\ln 2$, u has increased to 3 and since u must belong to $[0,3]$, we switch to $u(t)=3$ from t_1 and on. Then x continues to decrease. Since $x(t_1)=1$ and $\dot{x}=-2$, $x=1+2\ln 2-2t$. Because of the restriction $x\geqslant 0$, we can keep $u=3$ only until $1+2\ln 2-2t=0$, i.e. until $t=t_2=1/2+\ln 2$. For $t>t_2$ we use a control such that $x(t)\equiv 0$, (i.e. $u(t)\equiv 1$). Hence our suggestion for an optimal solution is, with $t_1=\ln 2$, $t_2=1/2+\ln 2$:

$$u^*(t)= \begin{cases} 1+e^t & t\in[0,t_1] \\ 3 & t\in(t_1,t_2] \\ 1 & t\in(t_2,3] \end{cases} \tag{18a}$$

$$x^*(t)= \begin{cases} 3-e^t & t\in[0,t_1] \\ 1+2\ln 2-2t & t\in(t_1,t_2] \\ 0 & t\in(t_2,3] \end{cases} \tag{18b}$$

To prove the optimality of (18) we find an adjoint function $p(t)$ and multipliers $q_1(t)$ and $q_2(t)$, associated with the two constraints in (17), such that all the conditions in theorem 1 are satisfied.

Note first that the Hamiltonian and the Lagrangian are ($p_0=1$)

$$H=-x+u+p(1-u), \qquad L=H+q_1(4-x-u)+q_2 x \tag{19}$$

so that

$$\frac{\partial L}{\partial u}=1-p-q_1, \qquad \frac{\partial L}{\partial x}=-1-q_1+q_2.$$

Let us consider the interval $(t_2, 3]$. In this interval the first constraint in (17) is inactive, so by (13), $q_1(t) = 0$. Moreover, in $(t_2, 3]$, $u^*(t) = 1$, which is an interior point in $U = [0, 3]$, and thus from (9), $\partial L^*/\partial u = 1 - p(t) - q_1(t) = 1 - p(t) = 0$. Hence we must have $p(t) = 1$ in $(t_2, 3]$. According to (10), $\dot{p}(t) = 1 + q_1(t) - q_2(t) = 1 - q_2(t)$, which is valid provided we put $q_2(t) = 1$, and this choice agrees with (13). According to (11), $p(3) - \beta_1(-1) - \beta_2 = 0$, since $x(3)$ is free. (12) gives that $\beta_1 = 0$ and $\beta_2 \geq 0$. Since $p(3) = 1$, by putting $\beta_2 = 1$, (11) and (12) are satisfied.

Consider next the interval $(t_1, t_2]$. Here both constraints in (17) are inactive and thus $q_1(t) = q_2(t) = 0$. From (10), $\dot{p}(t) = 1$. Since $p(t_2^+) = 1$, by the continuity of $p(t), p(t) = t + 1/2 - \ln 2$. Hence $\partial L^*/\partial u = 1 - p(t) = 1/2 + \ln 2 - t \geq 0$ in $(t_1, t_2]$, so $(\partial L^*/\partial u)(u - 3) \leq 0$ for all $u \in [0, 3]$. Thus, (9) is satisfied.

Finally, consider $[0, t_1]$. For $t < t_1$, $u^*(t) \in (0, 3)$ and thus (9) requires $0 = \partial L^*/\partial u = 1 - p(t) - q_1(t)$, i.e. $q_1(t) = 1 - p(t)$. On the other hand, since $q_2(t) = 0$ (as $x^*(t) > 0$), (10) gives $\dot{p}(t) = 1 + q_1(t) = 2 - p(t)$. Solving this equation with $p(t_1) = 1/2$, we get $p(t) = 2 - 3e^{-t}$. Hence, $q_1(t) = 1 - p(t) = 3e^{-t} - 1 > 0$, agreeing with (13), since $x^*(t) + u^*(t) = 4$ in this interval.

We have now proved that (9)–(13) are all satisfied. Since the concavity conditions (14) and (15) are trivially fulfilled, we conclude that (18) solves the problem.

In pure state constraint problems jumps in $p(t)$ at interior points in $[t_0, t_1]$ may occur. Such jumps may occur even if there are mixed constraints only. (See exercise 6.2.5.) Therefore the following modification of theorem 1 is sometimes useful.

Theorem 2 (**Sufficient conditions. Jumps in** $p(t)$ **in** (t_0, t_1)) Suppose we replace the continuity assumptions on $p(t)$ in theorem 1 by the assumption that $p(t)$ is piecewise continuous and piecewise continuously differentiable. Suppose, further, that all the conditions in theorem 1, except (12), are satisfied. The vector $\beta = (\beta_1, \ldots, \beta_s)$ is now assumed to satisfy

$$\beta_j \geq 0 (= 0 \text{ if } g_j(x^*(t_1), u^*(t_1), t_1) > 0) \qquad j = 1, \ldots, s \qquad (20(\text{i}))$$

$$\beta \cdot \partial g(x^*(t_1), u^*(t_1), t_1)/\partial u \cdot [u - u^*(t_1)] \leq 0 \text{ for each } u \in U. \qquad (20(\text{ii}))$$

Let $\tau_k, t_0 < \tau_1 < \ldots < \tau_N < t_1$ be the jump points of $p(t)$. Assume that associated with each τ_k there exist vectors $\beta_k^+ = (\beta_{k1}^+, \ldots, \beta_{ks}^+)$ and $\beta_k^- = (\beta_{k1}^-, \ldots, \beta_{ks}^-)$ such that for all $i = 1, \ldots, n$ and all $k = 1, \ldots, N$,

$$p_i(\tau_k^-) - p_i(\tau_k^+) = \sum_{j=1}^{s} \beta_{kj}^+ \frac{\partial g_j(x^*(\tau_k), u^*(\tau_k^+), \tau_k)}{\partial x_i} +$$

$$\sum_{j=1}^{s} \beta_{kj}^{-} \frac{\partial g_j(x^*(\tau_k),\, u^*(\tau_k^{-}),\, \tau_k)}{\partial x_i} \tag{21}$$

where for all $k=1,\ldots,N$,

$$\beta_k^{+} \cdot \partial g(x^*(\tau_k), u^*(\tau_k^{+}), \tau_k)/\partial u \cdot [u - u^*(\tau_k^{+})] \leqslant 0$$

$$\text{for each } u \in U \tag{22}$$

and

$$\beta_k^{-} \cdot \partial g(x^*(\tau_k),\, u^*(\tau_k^{-}), \tau_k)/\partial u \cdot [u - u^*(\tau_k^{-})] \leqslant 0$$

$$\text{for each } u \in U. \tag{23}$$

Moreover, assume for each $k=1,\ldots,N$ and each $j=1,\ldots,s$,

$$\beta_{kj}^{+} \geqslant 0 \ (=0 \text{ if } g_j(x^*(\tau_k), u^*(\tau_k^{+}), \tau_k) > 0) \tag{24}$$

$$\beta_{kj}^{-} \geqslant 0 \ (=0 \text{ if } g_j(x^*(\tau_k),\, u^*(\tau_k^{-}), \tau_k) > 0) \tag{25}$$

Then $(x^*(t), u^*(t))$ is optimal. ∎

For a proof of this result see the proof of the more general theorem 14.

Note 1   Define the sets $U^*(t^{+})$ and $U^*(t^{-})$ by

$$U^*(t^{+}) = \left\{ u \in U : \frac{\partial g_j^*(t^{+})}{\partial u} [u - u^*(t^{+})] \geqslant 0, \quad j=1,\ldots,s' \right\}$$
$$U^*(t^{-}) = \left\{ u \in U : \frac{\partial g_j^*(t^{-})}{\partial u} [u - u^*(t^{-})] \geqslant 0, \quad j=1,\ldots,s' \right\} \tag{26}$$

(We write $\partial g_j^*/\partial u$ for $\partial g_j/\partial u$ evaluated at $(x^*(t), u^*(t), t)$.)

The following consequences of the conditions of theorem 2 are very useful:

$$\beta_{kj}^{+} = 0 \quad \text{if } \frac{\partial g_j^*(\tau_k^{+})}{\partial u} \cdot [u - u^*(\tau_k^{+})] > 0 \text{ for some } u \in U^*(\tau_k^{+}),$$

$$j=1,\ldots,s' \tag{27}$$

$$\beta_{kj}^{-} = 0 \quad \text{if } \frac{\partial g_j^*(\tau_k^{-})}{\partial u} \cdot [u - u^*(\tau_k^{-})] > 0 \text{ for some } u \in U^*(\tau_k^{-}),$$

$$j=1,\ldots,s'. \tag{28}$$

(To prove (27), take any j, $j=1,\ldots,s'$ and suppose $\partial g_j^*(\tau_k^+)/\partial u \cdot [\bar{u}-u^*(\tau_k^+)]>0$ for some $\bar{u} \in U$ such that $\partial g_i^*(\tau_k^+)/\partial u \cdot [\bar{u}-u^*(\tau_k^+)] \geqslant 0$ for all $i=1,\ldots,s'$. We want to prove that $\beta_{kj}^+=0$. Suppose $\beta_{kj}^+>0$. Then $\beta_{kj}^+ \partial g_j^*(\tau_k^+)/\partial u \cdot (\bar{u}-u^*(\tau_k^+))>0$, and thus one of the terms in the sum $\Sigma_{i=1}^s \beta_{ki}^+ \partial g_i^*(\tau_k^+)/\partial u \cdot [\bar{u}-u^*(\tau_k^*)]$ is >0, while all the other terms are $\geqslant 0$. Hence the sum is >0, contradicting (22). The statement in (28) is proved similarly.)

In the same way we prove that $\beta_j^-=0$ if the inequality in (28) holds for τ_k^- replaced by t_1 for some $u \in U^*(t_1^-)$. Since (27) and (28) tell us nothing about β_{kj}^+ and β_{kj}^- when $j>s'$, the following result is useful:

$$\beta_{kj}^+=\beta_{kj}^-=0 \quad \text{if } j>s', \ \bar{g}_j(x^*(\tau_k),\tau_k)=0 \quad \text{and}$$
$$(\partial \bar{g}_j^*(x^*(t),t)/\partial x)f(x^*(t),u^*(t),t) \text{ is discontinuous at } t=\tau_k \tag{29}$$

Other conditions implying the vanishing of β_{kj}^+ and β_{kj}^- are those of note 4(c).

In order to see the relationship between (21) and (5.37), note that for $j>s'$, g_j is independent of u, so the sum of the terms corresponding to $j>s'$ in (21) can be joined into the sum $\Sigma_{j=s'+1}^s (\beta_{kj}^+ + \beta_{kj}^-)\partial g_j^*(\tau_k)/\partial x_i$.

Note 2* In theorems 1 and 2 the quasi-concavity of $g_j(x,u,t)$ w.r.t. (x,u) can be replaced by some weaker conditions. If we define

$$\Gamma(t)=\{(x,u): g(x,u,t) \geqslant 0, \ u \in U\}, \tag{30}$$

these weaker conditions are the following three:

$(q(t)g_x'(x^*(t),u^*(t),t), q(t)g_u'(x^*(t),u^*(t),t))$ is a support
functional at $(x^*(t),u^*(t))$ of $\Gamma(t)$ for v.e. t. $\tag{31}$

For $k=1,\ldots,N$,

$(\beta_k^+ g_x'(x^*(\tau_k),u^*(\tau_k^+),\tau_k), \beta_k^+ g_u'(x^*(\tau_k),u^*(\tau_k^+),\tau_k))$ is a support
functional at $(x^*(\tau_k),u^*(\tau_k^+),\tau_k)$ of $\Gamma(\tau_k)$ $\tag{32}$

and

$(\beta_k^- g_x'(x^*(\tau_k),u^*(\tau_k^-),\tau_k), \beta_k^- g_u(x^*(\tau_k),u^*(\tau_k^-),\tau_k))$ is a support
functional at $(x^*(\tau_k),u^*(\tau_k^-),\tau_k)$ of $\Gamma(\tau_k)$ $\tag{33}$

(See Appendix B.2 for the definition of support functional.)

In the next example we use theorem 2 and check to see if $p(t)$ has jumps.

Example 2

$$\max \int_0^2 (ax(t)+u(t))dt, \qquad \dot{x}(t)=-2u(t),$$

$$x(0)=1, \qquad x(2) \text{ free}, \qquad U=[-2,2], \tag{34}$$

$$g_1(x(t),u(t),t)=x(t)-(u(t))^2 \geqslant 0,$$
$$g_2(x(t),u(t),t)=\bar{g}_2(x(t),t)=1-x(t) \geqslant 0 \tag{35}$$

where a is a positive constant.

In this example, with $p_0=1$, the Lagrangian and its partial derivatives w.r.t. x and u are

$$L=ax+u-2up+q_1(x-u^2)+q_2(1-x),$$

$$\frac{\partial L}{\partial x}=a+q_1-q_2, \qquad \frac{\partial L}{\partial u}=1-2p-2uq_1. \tag{36}$$

If $(x^*(t),u^*(t))$ is a solution candidate, and theorem 2, is applied, then according to (9),

$$\frac{\partial L^*}{\partial u}(u-u^*(t))=[1-2p(t)-2q_1(t)u^*(t)]\cdot(u-u^*(t)) \leqslant 0$$

$$\text{for each } u \in [-2,2]. \tag{37}$$

According to (10),

$$\dot{p}(t)=-\frac{\partial L^*}{\partial x}=-a-q_1(t)+q_2(t), \tag{38}$$

while (11) and (20) give

$$p(2)-\beta_1+\beta_2=0 \tag{39(i)}$$
$$\beta_1 \geqslant 0 \; (=0 \text{ if } x^*(2)>(u^*(2))^2); \qquad \beta_2 \geqslant 0 \; (=0 \text{ if } x^*(2)<1) \tag{39(ii)}$$
$$\beta_1 u^*(2)[u-u^*(2)] \geqslant 0 \text{ for all } u \in [-2,2] \tag{39(iii)}$$

From (13),

$$q_1(t) \geqslant 0 \; (=0 \text{ if } x^*(t)>(u^*(t))^2), \qquad q_2(t) \geqslant 0 (=0 \text{ if } x^*(t)<1) \tag{40}$$

If there is a jump at τ_k, (21) gives that

$$p(\tau_k^-)-p(\tau_k^+)=\beta_{k1}^+ +\beta_{k1}^- -\beta_{k2}^+ -\beta_{k2}^-. \tag{41}$$

Conditions (22) and (23) reduce to

$$\beta_{k1}^+ u^*(\tau_k^+)\cdot(u-u^*(\tau_k^+))\geqslant 0 \qquad \text{for all } u\in[-2,2] \tag{42}$$

$$\beta_{k1}^- u^*(\tau_k^-)\cdot(u-u^*(\tau_k^-))\geqslant 0 \qquad \text{for all } u\in[-2,2], \tag{43}$$

and according to (24), and (25),

$$\beta_{k1}^+ \geqslant 0 \ (=0 \text{ if } x^*(\tau_k)>(u^*(\tau_k^+))^2, \qquad \beta_{k2}^+\geqslant 0 \ (=0 \text{ if } x^*(\tau_k)<1) \tag{44}$$

$$\beta_{k1}^- \geqslant 0 \ (=0 \text{ if } x^*(\tau_k)>(u^*(\tau_k^-))^2, \qquad \beta_{k2}^-\geqslant 0 \ (=0 \text{ if } x^*(\tau_k)<1) \tag{45}$$

Since the concavity conditions (14) and (15) are trivially satisfied, we see from theorem 2 that if the pair $(x^*(t),u^*(t))$ satisfies (34)–(45), then that pair solves problem (34), (35).

We know that $x^*(0)=1$. Let t' be the maximal value of t for which $x^*(s)=1$ on $[0,t]$. We guess that $t'>0$. Then from (34) $u^*(t)\equiv 0$ and since $g_1\geqslant 0$ becomes inactive, from (40), $q_1(t)\equiv 0$ on $[0,t']$. Since $u^*(t)\equiv 0$, (37) implies that $\partial L^*/\partial u=0$, i.e. $p(t)=1/2$. From (38), $0=-a+q_2(t)$, so $q_2(t)\equiv a$.

Note that $t'=2$ is not possible, since by (39), $p(2)-\beta_1+\beta_2=0$ with $p(2)=1/2$, $\beta_1=0$, and $\beta_2\geqslant 0$, a contradiction. $x^*(t)$ leaves the constraint $x\leqslant 1$ at t'. Conceivably, $u^*(t)>0$ in some interval $I=[t',t'+\alpha)$, $\alpha>0$. Since $(u^*(t))^2\leqslant x^*(t)<1$ in I, $q_2(t)=0$ and $0<u^*(t)<1$ in I. By (38), (41), $p(t)$ is decreasing on I. From (37), $\partial L^*/\partial u=1-2p(t)-2q_1(t)u^*(t)=0$, since $u^*(t)\in(0,1)$, so $2q_1(t)u^*(t)=1-2p(t)\geqslant 0$ in I, with strict inequality holding for $t\in(t',t'+\alpha)$, since $p(t)$ is decreasing in I. Thus $q_1(t)>0$ in I and so g_1 is active in I.

A priori, $p(t)$ could have a jump at $\tau_1=t'$. We claim, however, that $p(t)$ is continuous at t'. Note first that $\partial \bar{g}_2^*/\partial x\cdot f^*=(-1)(-2u^*(t))=2u^*(t)$ is discontinuous at $\tau_1=t'$, so by (29), $\beta_{12}^+=\beta_{12}^-=0$. Since $u^*(t'^+)=(x^*(t'))^{1/2}=1\in(0,2)$, from (42), $\beta_{11}^+=0$. Moreover, from (45), $\beta_{11}^-=0$. It follows from (41) that $p(t'^+)=p(t'^-)$, so $p(t)$ is indeed continuous at $\tau_1=t'$.

We proceed by anticipating that $x^*(t)=(u^*(t))^2$ with $u^*(t)>0$ on all of $[t',2]$. From (34) we then obtain $\dot{x}^*(t)=-2(x^*(t))^{1/2}$, and solving this equation with $x^*(t')=1$, we get $x^*(t)=(1+t'-t)^2$, so $u^*(t)=1+t'-t$. Since $u^*(t)\in(-2,2)$, from (37) we get $\partial L^*/\partial u=0$, i.e. $1-2p(t)-2u^*(t)q_1(t)=0$. From (38), $\dot{p}(t)=-a-q_1(t)$, since $q_2(t)=0$, because $x^*(t)<1$. Hence $1-2p(t)+2a(1+t'-t)+2(1+t'-t)\dot{p}(t)=0$. Put $\varphi(t)=2(1+t'-t)p(t)$. Then $\varphi(t')=2p(t')=1$ and $\dot{\varphi}(t)=-2p(t)+2(1+t'-t)\dot{p}(t)$, so $1+2a(1+t'-t)+\dot{\varphi}(t)=0$. Hence, $\dot{\varphi}(t)=-1-2a(1+t'-t)$, $\varphi(t')=1$, which gives $\varphi(t)=1+t'-t+a(1+t'-t)^2-a$, and thus $p(t)=\frac{1}{2}(1+a(1+t'-t))-\frac{1}{2}a(1+t'-t)^{-1}$. For $t\in[t',t'+1)$ we find $\dot{p}(t)=-\frac{1}{2}a-\frac{1}{2}a(1+t'-t)^{-2}\leqslant -a$. The function $t\to\frac{1}{2}(1+a(1+t'-t))-\frac{1}{2}a(1+t'-t)^{-1}$ is graphed in fig. 6.1 for

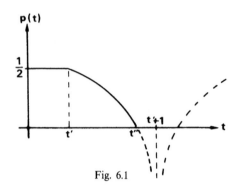

Fig. 6.1

$t \geqslant t'$. It is easy to see that this function has a zero at $t'' = t' + 1 + 1/2a - (1 + 1/4a^2)^{1/2} \in (t', t'+1)$.

Since $x^*(2) = (t'-1)^2 < 1$ for $t' \in (0, 2)$, we see from (39) that $\beta_2 = 0$ and $p(2) = \beta_1 \geqslant 0$. Hence $t'' \geqslant 2$, and then $t' = t'' - 1 - 1/2a + (1 + 1/4a^2)^{1/2} > 1$. Since $u^*(2) = 1 + t' - 2 \in (-2, 2)$, from (39(iii)) it follows that $\beta_1 = 0$, and therefore $p(2) = 0$. Hence we put $t'' = 2$, i.e. $t' = 1 + v$, where $v = -1/2a + (1 + 1/4a^2)^{1/2}$. We have thus arrived at the following suggestion for an optimal solution:

$$u^*(t) = \begin{cases} 0 & \text{on } [0, 1+v] \\ 2+v-t & \text{on } (1+v, 2] \end{cases}$$

$$x^*(t) = \begin{cases} 1 & \text{on } [0, 1+v] \\ (2+v-t)^2 & \text{on } (1+v, 2] \end{cases} \tag{46}$$

The associated adjoint function and multipliers are: In $[0, 1+v]$, $p(t) = 1/2$, $q_1(t) = 0$, $q_2(t) = a$, while in $(1+v, 2]$, $p(t) = \frac{1}{2}(1 + a(2+v-t)) - \frac{1}{2}a(2+v-t)^{-1}$ (with $\dot{p}(t) \leqslant -a$), $q_1(t) = -a - \dot{p}(t) \geqslant 0$, $q_2(t) = 0$. Finally, $\beta_1 = \beta_2 = 0$ and $p(t)$ has no jumps, so we can choose all β_{kj}^+ and β_{kj}^- equal to 0. It is now an easy matter to check that conditions (37)–(45) are all satisfied. Hence (46) defines an optimal solution to the problem.

In exercise 6.2.5 we consider some optimal control problems in which $p(t)$ has jumps at interior points of $[t_0, t_1]$.

Exercise 6.2.1 Solve the problem

$$\max \int_0^1 (v-x)dt, \qquad \dot{x} = u, \qquad x(0) = 1/16, \qquad x(1) \text{ free,}$$

$$u \in [-1, 1], \qquad v^2 \leqslant x, \qquad x \leqslant 1/8.$$

(*u* and *v* are controls.)

Exercise 6.2.2 Solve the problem

$$\max \int_0^2 (2u - x)dt, \qquad \dot{x} = -u, \qquad x(0) = e, \qquad x(2) \text{ free},$$

$$u \in [-3, 3], \qquad u \leqslant x, \qquad t \leqslant x.$$

Exercise 6.2.3 Solve the problem

$$\max \int_a^1 (-u^2 - x)dt, \qquad \dot{x} = -u, \qquad x(a) = 19/16, \qquad x(1) \text{ free},$$

$$u \leqslant x, \qquad t/8 + 11/64 \leqslant x, \qquad \text{with } a = -1 - \ln 3.$$

Exercise 6.2.4 Solve the problem

$$\max \int_0^2 u dt, \qquad \dot{k} = v, \qquad k(0) = 1, \qquad k(2) \text{ free},$$

$$\dot{z} = u + v, \qquad z(0) = 1, \qquad z(2) \text{ free}$$
$$0 \leqslant u, \qquad 0 \leqslant v, \qquad u + v \leqslant k, \qquad z \leqslant 2 + t$$

(*u* and *v* are the controls.)

Exercise 6.2.5 Consider the pure state constraint problem

$$\max \int_0^2 |1 - t|(2u - u^2)dt, \qquad \dot{x} = (1 - t)u, \qquad x(0) = 0, \qquad x(2) \text{ free}$$

$$x \leqslant 0, \qquad u \in (-\infty, \infty).$$

(i) Show that $u^*(t) = 0$, $x^*(t) = 0$, $p(t) = -1$, $q(t) = 0$ for $t \in [0, 1]$, while $u^*(t) = 1$, $x^*(t) = (-1/2)t^2 + t - 1/2$, $p(t) = 0$, $q(t) = 0$ for $t \in [1, 2]$ solve the problem. (Note that $p(t)$ has a jump at $t = 1$.) Show that the condition in (29) fails.

(ii) Introduce a new control variable $v \in (-\infty, +\infty)$, and replace the criterion by $\int_0^2 |1-t|(2u-u^2-v^2)dt$ and the constraint $x \leq 0$ by $x+v^2 \leq 0$. Verify that the same choice of u^*, x^*, p, and q as in (i), with $v^*(t) \equiv 0$, solve our problem. Show that conditions (27) and (28) fail.

Exercise 6.2.6 Solve the problem

$$\max \int_0^T (-u-x)dt, \qquad \dot{x} = -u,$$

$$x(0) = 1, \qquad T \text{ fixed} > 1, \qquad 0 \leq u \leq x, x \geq e^{-2}.$$

3. An Arrow-type sufficiency result

In this section we briefly discuss an Arrow-type version of theorems 1 and 2. It follows from our comments on theorem 4.6 that we cannot obtain a generalization of theorems 1 and 2 merely by replacing concavity of the Hamiltonian by concavity of a maximized Hamiltonian w.r.t. all "relevant values" of x.

We consider problem (1)–(6), but we assume that $U = R^r$. In theorems 1 and 2 we required $g(x, u, t)$ to be quasi-concave in (x, u). When using an Arrow-type concavity condition, the quasi-concavity condition on $g_j(x, u, t)$, $j = 1, \ldots, s'$ can be replaced by a constraint qualification.

In order to formulate the constraint qualification, if $(x^*(t), u^*(t))$ is an admissible pair which we want to test for optimality, define for each $t \in [t_0, t_1]$ the sets I_t^- and I_t^+ by

$$I_t^- = \{j : j \leq s', \quad g_j(x^*(t), u^*(t^-), t) = 0\}$$
$$I_t^+ = \{j : j \leq s', \quad g_j(x^*(t), u^*(t^+), t) = 0\} \tag{47}$$

The constraint qualification, often referred to as a "full rank"-condition, reads:

If $I_t^- \neq \phi$, the matrix $[\partial g_j(x^*(t), u^*(t^-), t)/\partial u_i]$, $i = 1, \ldots, r$, $j \in I_t^-$, has a rank equal to the number of elements in I_t^-.   (48(i))

If $I_t^+ \neq \phi$, the matrix $[\partial g_j(x^*(t), u^*(t^+), t)/\partial u_i]$, $i = 1, \ldots, r$, $j \in I_t^+$ has a rank equal to the number of elements in I_t^+.   (48(ii))

(If $t = t_0$ in (48), drop (i). If $t = t_1$, drop (ii).)

Theorem 3 (Sufficient conditions, Arrow-type.)[1] Theorems 1 and 2 remain valid if it is assumed that

$$U = R^r \tag{49}$$

$$\hat{H}(x^*(t), p(t), t) = H(x^*(t), u^*(t), p(t), t) \qquad \text{for v.e.} t, \tag{50}$$

where

$$\hat{H}(x, p, t) = \max \{ H(x, u, p, t): u \in U(x, t) \}$$

where $U(x, t) = \{ u \in R^r: g_j(x, u, t) \geqslant 0, j \leqslant s' \}$, (assuming that the max value is attained for all points $(x, p(t), t)$) (51)

the constraint qualification (48) holds for v.e.t, (52)

and if, in addition, the concavity assumptions are modified in the following way: (14) is replaced by:

$$g_{s'+1}, \ldots, g_s \text{ are quasi-concave in } x \in R^n \text{ for each } t, \tag{53}$$

and (15) is replaced by:

Let $\tilde{A}_1(t) = \{ x: \text{ for some } u \in R^r, g_j(x, u, t) \geqslant 0 \text{ for } j = 1, \ldots, s' \}$. The function $x \to \hat{H}(x, p(t), t)$ is concave on $\tilde{A}_1(t)$, provided $\tilde{A}_1(t)$ is convex. If $\tilde{A}_1(t)$ fails to be convex, the function $x \to \hat{H}(x, p(t), t)$ has a concave extension to co $\tilde{A}_1(t)$ (the convex hull of $\tilde{A}_1(t)$). (54)

The numbers $\beta_{kj}^-, \beta_{kj}^+, j = 1, \ldots, s'$ can only be non-zero if (48) fails to hold for $t = \tau_k$, and $\beta_j, j = 1, \ldots, s'$, can only be non-zero if (48(i)) fails to hold for $t = t_1$. ∎

Note 3 Suppose $A(t)$ is a given set in R^n and impose on the admissible pair $(x(t), u(t))$ the additional restriction $x(t) \in A(t)$. Then theorem 3 still holds provided $x^*(t)$ belongs to the interior of $A(t)$ for each t. In this case $x \to \hat{H}(x, p(t), t)$ needs to exist and be concave only for $x \in \text{co}(A(t) \cap \tilde{A}_1(t))$ for v.e. t.

*Optimal control without differentiability**

Starting with Clarke (1976), (see also Russel (1976)) there have appeared a great number of results on problems where $\partial f_0 / \partial x$ and $\partial f / \partial x$ do not exist,

---

[1] Seierstad and Sydsæter (1977).

and a number of necessary conditions have been proved, involving some notion of a generalized derivative.

We shall only formulate some useful sufficient conditions for this case, which involve the simple notions of supergradients and support functionals (see Appendix B).

Consider again problem (1)–(5), assuming that f_0 and f are continuous, or has discontinuities of the type described in note 2.6(b). Then we have the following theorem.

Theorem 4 **(Sufficient conditions without differentiability)** Let $(x^*(t), u^*(t))$ be an admissible pair in problem (1)–(5). Assume that there exists a piecewise continuous and piecewise continuously differentiable function $p(t)$ with jump points $t_0 < \tau_0, \ldots, \tau_N \leqslant t_1$ such that the following properties hold for v.e.t:

$$H(x^*(t), u^*(t), p(t), t) - H(x, u, p(t), t) \geqslant \dot{p}(t)(x - x^*(t))$$

for all $(x, u) \in \Gamma(t) = \{(x, u): g(x, u, t) \geqslant 0, u \in U\}$ \hfill (55)

$(p(\tau_i^-) - p(\tau_i^+))(x - x^*(\tau_i)) \geqslant 0$ for all $(x, u) \in \Gamma(t)$.

(If $\tau_N = t_1$, $p(\tau_N^+)$ means $p(t_1)$.) \hfill (56)

$p(t)$ satisfies (2.60). \hfill (57)

Then $(x^*(t), u^*(t))$ solves the problem. ∎

The conditions (55) and (56) are satisfied if there exist piecewise continuous vector functions $c^x(t)$, $c^u(t)$, $d^x(t)$, $d^u(t)$, and vectors b_k^+ and b_k^- in R^n, corresponding to the jumps point τ_k, $(b_N^+ = 0$ if $\tau_N = t_1)$, such that

(i) $(b_k^+, 0, \ldots, 0)$ is a support functional at $(x^*(\tau_k), u^*(\tau_k^+))$ for the set $\Gamma(\tau_k)$.

(ii) An analogous property is satisfied by $(b_k^-, 0, \ldots, 0)$ at $(x^*(\tau_k), u^*(\tau_k^-))$

(iii) For v.e. t, $(d^x(t), d^u(t))$ is a support functional at $(x^*(t), u^*(t))$ for the set $\Gamma(t)$.

(iv) For v.e. t, $(c^x(t), c^u(t))$ is a supergradient to $(x, u) \rightarrow H(x, u, p(t), t)$ at $(x^*(t), u^*(t))$ w.r.t. R^{n+r}, where $p(t)$ is a piecewise continuous and piecewise continuously differentiable function satisfying $\dot{p} = -c^x(t) - d^x(t)$ v.e., $p(\tau_k^-) - p(\tau_k^+) = b_k^+ + b_k^-$.

(v)   For v.e.t, for all $(x, u) \in \Gamma(t)$, $(c^u(t) + d^u(t))(u - u^*(t)) \leqslant 0$.

Exercise 6.3.1 **(Growth with foreign lending)** Consider problem (4.142)–(4.147) in exercise 4.4.5 and add the pure state constraint

$$B \geqslant -c \ (c \text{ positive constant}).$$

(See also exercise 5.2.4.)

(i) Prove that (4.149)–(4.151) in exercise 4.4.5 still hold.

(ii) *Assume that we have a solution* $(x^*(t), u^*(t))$ of conditions (9)–(13) in theorem 1. Show that this solution has the following form: There exist four numbers $t_1 \leqslant t_2 < t' < t_3$, $[t_1, t_3] \neq [0, T]$ such that $s^*(t) = 1$ on $[0, t']$, $s^*(t) = 0$ on $(t', T]$, $u^*(t) = \underline{u}$ on $[0, t_1]$, $u^*(t) = 0$ on $(t_1, t_2]$, $u^*(t)$ satisfies $bK^*(t) + rB^*(t) - u^*(t) = d$ on $(t_2, t_3]$, $u^*(t) = u$ on $(t_3, T]$. $B^*(T) = 0$. t' is $\geqslant \max(t'', T - 1/b)$, $K^*(t'') = K_T$. On $[t_2, t']$, $K^*(t) = K^*(t_2) + d(t - t_2)$, on $[t', T]$, $K^*(t) = K^*(t')$. On $[t_2, t_3]$, $B^*(t)$ is the solution of $\dot{B} = bK^*(t) + rB - d$, $B^*(t_2) = B^*(t_1)$. Write $B^*(t_3) = B^*(t_3, t_2, B^*(t_1))$, to show the dependence on t_2 and $B^*(t_1)$. Then one of the four cases holds:

(α) $t_1 = t_2$, $t_3 = T$. Then t_1 satisfies $B^*(T, t_1, \underline{u}t_1) = 0$, $p_2(T) \geqslant 1$.

(β) $t_1 = t_2$, $t_3 < T$. Then t_1 and t_3 satisfy the relations (b): $p_2(t_3) = 1$, and (c): $B^*(t_3, t_1, \underline{u}t_1) + u(T - t_3) = 0$. Furthermore, $p_1(t_1) = p_2(t_1)$ if $t_1 > 0$.

(γ) $t_1 < t_2, t_3 = T$. Then $t_1 = -c/\underline{u}$ and t_2 satisfies (d): $B^*(T, t_2, c) = 0$. $p_2(T) \geqslant 1$, $p_1 = p_2$ in (t_1, t_2)

(δ) $t_1 < t_2$, $t_3 < T$. Then $t_1 = -c/\underline{u}$ and t_2 and t_3 satisfy (b) and $B^*(t_3, t_2, c) + u(T - t_3) = 0$, $p_1 = p_2$ in (t_1, t_2).

(iii) Using arguments similar to those in exercise 4.4.5, show that the constraint $bK + rB - u \geqslant d$ is never active with $u = u$ along a solution of the type described in (ii). Use this to show that the constraint qualification (48) holds along the trajectory. Show that (53) and (54) hold.

(iv) Try to use an argument similar to that in (v) and (vi) in exercise 4.4.5 to show that the sufficient conditions in theorem 2 do have a solution. (Hint: Again use $p_1(T)$ and $p_2(T)$ as parameters, and see how t_1, t_2, and t_3 vary with $p_2(T)$ for each fixed $p_1(T)$.)

Exercise 6.3.2 Consider the problem

$$\max \int_0^1 u^2 \, dt, \qquad \dot{x} = -2u, \qquad x(0) = 4, \qquad x(1) \text{ free}$$

$$u \in (-\infty, \infty), \qquad u^2 \leqslant x, \qquad x \leqslant 4.$$

(i) Solve the problem without the restriction $x \leqslant 4$. Use theorem 3.

(ii) Solve the problem with the restriction $x \leqslant 4$. Use theorem 3.

4. Necessary conditions

In this section our main concern is the problem of establishing necessary conditions for optimality in problem (1)–(6). Ideally, the necessary conditions should be as close as possible to the sufficient conditions discussed in theorems 1 and 3. However, due to the presence of pure state constraints, the formulation of necessary conditions based on the multipliers introduced in theorems 1 and 3 is somewhat complicated. (See the introduction to section 3 in Chapter 5.)

In order to obtain correct necessary conditions which are reasonably simple, we proceed as suggested in Chapter 5, section 3, introducing the modified Lagrangian. With $H(x, u, p, t) = p_0 \cdot f_0(x, u, t) + p \cdot f(x, u, t)$, define

$$\bar{L}(x, u, p, q, \mu, t) = H(x, u, p, t) + \sum_{j=1}^{s'} q_j g_j(x, u, t)$$

$$- \sum_{j=s'+1}^{s} \mu_j h_j(x, u, t) \tag{58}$$

where

$$h_j(x, u, t) = [\partial \bar{g}_j(x, t)/\partial x] \cdot f(x, u, t)$$
$$+ \partial \bar{g}_j(x, t)/\partial t \qquad j = s'+1, \ldots, s \tag{59}$$

and where $q = (q_1, \ldots, q_{s'})$ and $\mu = (\mu_{s'+1}, \ldots, \mu_s)$. We also define

$$\check{L}(x, u, p, q, t) = H(x, u, p, t) + \sum_{j=1}^{s'} q_j g_j(x, u, t). \tag{60}$$

We shall assume that

$$g_j(x, u, t) \text{ is a } C^2\text{-function}, j = s'+1, \ldots, s. \tag{61}$$

One can then prove the following theorem:

Theorem 5 (Necessary conditions. Full rank constraint qualification)[2] Let $(x^*(t), u^*(t))$ be an admissible pair solving problem (1)–(6), (61) with $U = R^r$. Assume that the constraint qualification (48) holds for all $t \in [t_0, t_1]$. Then

---

[2] Neustadt (1976) Chapter VI, 3. The full rank condition together with the facts that $\mu(t)$ and $u^*(t)$ have one-sided limits everywhere imply that $q(t)$ has one-sided limits as well. For measurable controls the theorem holds subject to the following changes: $q(t)$ is bounded and measurable, (64), (65), and (68) hold a.e., $p^*(t)$ is absolutely continuous and the rank condition must be modified as in footnote 1 to Chapter 4.

there exist a number p_0, vector functions $p(t) = (p_1(t), \ldots, p_n(t))$ and $q(t) = (q_1(t), \ldots, q_{s'}(t))$ and a non-decreasing vector function $\mu(t) = (\mu_{s'+1}(t), \ldots, \mu_s(t))$, all having one-sided limits everywhere, such that the following conditions hold:

$$p_0 = 0 \text{ or } p_0 = 1 \tag{62}$$

$$(p_0, p(t), \mu(t_1) - \mu(t_0)) \neq (0, 0, 0) \text{ for all } t. \tag{63}$$

For v.e. $t \in (t_0, t_1)$,

$$H(x^*(t), u^*(t), p(t), t) \geq H(x^*(t), u, p(t), t) \text{ for all } u \text{ such that}$$
$$g_j(x^*(t), u, t) > 0 \text{ for } j = 1, \ldots, s'. \tag{64}$$

$$\partial \check{L}^* / \partial u = 0, \ * \text{ denotes evaluation at } (x^*(t), u^*(t), p(t), q(t), t). \tag{65}$$

$\mu_j(t)$ is constant on any interval where $\bar{g}_j(x^*(t), t) > 0$,
$j = s'+1, \ldots, s$ (66(i))

$\mu_j(t)$ is continuous at all $t \in (t_0, t_1)$ at which $\bar{g}_j(x^*(t), t) = 0$
and $(\partial \bar{g}_j(x^*(t), t) / \partial x) \cdot f(x^*(t), u^*(t), t)$ is discontinuous,
$j = s'+1, \ldots, s$. (66(ii))

Defining

$$p^*(t) = p(t) + \sum_{j=s'+1}^{s} \mu_j(t) \partial \bar{g}_j(x^*(t), t) / \partial x \tag{67}$$

$p^*(t)$ is continuous, and has a derivative v.e. given by

$$\dot{p}^*(t) = -\partial \bar{L}^* / \partial x \tag{68}$$

For all t,

$$q_j(t) \geq 0 \ (=0 \text{ if } g_j(x^*(t), u^*(t), t) > 0), \qquad j = 1, \ldots, s'. \tag{69}$$

Finally, $p(t)$ satisfies

$$\begin{array}{ll} p_i(t_1) \text{ no condition} & i = 1, \ldots, l \\ p_i(t_1) \geq 0 \ (=0 \text{ if } x_i^*(t_1) > x_i^1) & i = l+1, \ldots, m \\ p_i(t_1) = 0 & i = m+1, \ldots, n \end{array} \tag{70}$$

Note 4 (a) A fact often used in problem solving is that $\partial \check{L}^*(t^+) / \partial u = 0$ for all $t \in [t_0, t_1)$, and $\partial \check{L}^*(t^-) / \partial u = 0$ for all $t \in (t_0, t_1]$.

Notice also that for all $t\in(t_0,t_1)$, $u^*(t^+)$ and $u^*(t^-)$ maximize both $H(x^*(t),u,p(t^+),t)$ and $H(x^*(t),u,p(t^-),t)$ in the set $\bar{U}^+(t)$, (the closure of $U^+(t)$), where $U^+(t)=\{u:g_j(x^*(t),u,t)>0,\ j=1,\ldots,s'\}$.

(b) The conditions in the theorem are satisfied also if we replace $\mu(t)$ by $\mu(t)+c$, where c is an arbitrary constant vector. Hence,

$$\mu(t)\text{ can be normalized by choosing, }\mu(t_0)=0\text{ or }\mu(t_1)=0. \tag{71}$$

(c) If $H(x^*(t),u,p(t),t)$ for each $t\in(t_0,t_1)$ has a unique maximum in $\bar{U}^+(t)$, and the constraint qualification (92) below holds for all $t\in(t_0,t_1)$, then $u^*(t)$, $p(t)$, and $\mu(t)$ are continuous in (t_0,t_1).

Note 5 Define

$$G^*(t)=\{j:j>s',\ \bar{g}_j(x^*(t),\ t)=0\} \tag{72}$$

(a) The conditions in (64) and (65) give no information about $u^*(t)$ unless

$$(p_0,p(t))\neq(0,0)\text{ for all }t\text{ in some interval contained in }(t_0,t_1). \tag{73}$$

This property *is* satisfied if $G^*(t_0)=G^*(t_1)=\phi$ and the following condition holds:

For all $t\in(t_0,t_1)$ either $G^*(t)=\phi$, or there exists a vector $x_t\in R^n$ such that:

$$[\partial\bar{g}_j(x^*(t),t)/\partial x]\cdot x_t>0 \qquad\text{for all }j\in G^*(t). \tag{74}$$

(b)* If $G^*(t_0)$ and/or $G^*(t_1)$ are non-empty, the non-triviality results of Chapter 5, section 4 can be modified to cover the present problem: Replace g_j, $j=1,\ldots,s$ in the definitions of first-order compatibility in Chapter 5, section 4 by $\bar{g}_j$, $j=s'+1,\ldots,s$. Replace U by $U^+(t_0)$, $U^+(t_1)$, and $U^+(t)$ in conditions (5.120), (5.121) and (5.125), respectively. Then (73) holds, if the restrictions are first-order compatible at t_0 and t_1 and if (74) holds. If, in addition, (5.125) holds, then (5.126) is also valid.

Another property that entails (5.126) is condition (92). Note, finally, that the results in note 5.7 remain valid for the above redefinitions and the modification of (5.125)[3].

Note 6 **"Almost necessary conditions"** The results concerning "almost necessary conditions" in Chapter 5, section 3 can also be generalized to the present problem. In fact, if $\mu(t)$ in theorem 5 is well-behaved (in the sense

---

[3]Seierstad (1986a).

that it has a finite number of jump discontinuities and is C^1 elsewhere), then condition (68) and the continuity condition on $p^*(t)$ can be replaced by (10) and (21). The latter conditions are usually simpler to apply, but remember that there is no a priori guarantee that $\mu(t)$ is well-behaved. However, since "ill-behaved" $\mu(t)$ – functions are rare, we formulate the following "result":

("*Almost necessary*" conditions, (ANC). *Mixed and pure constraints*)

Assume that $(x^*(t), u^*(t))$ solves problem (1)–(6), (61), $U = R^r$, with (48) being satisfied for all $t \in [t_0, t_1]$. Then the conditions in theorem 5 are satisfied with $\mu_j(t)$ having jumps $\beta_{kj} = \mu_j(\tau_k^+) - \mu_j(\tau_k^-)$ at a finite number of points τ_k, $k = 1, \ldots, N$, $t_0 \leqslant \tau_1 < \ldots < \tau_N \leqslant t_1$, and having piecewise continuous derivatives $\dot{\mu}_j(t) = q_j(t)$, $j = s' + 1, \ldots, s$ for all $t \neq \tau_k$, and with the continuity condition on $p^*(t)$, (67) and (68) replaced by (10) and:

$$p(\tau_k^-) - p(\tau_k^+) = \sum_{j=s'+1}^{s} \beta_{kj} \frac{\partial \bar{g}_j(x^*(\tau_k), \tau_k)}{\partial x}, \qquad k = 1, \ldots, N. \tag{75}$$

(If $\tau_1 = t_0$, put $p(\tau_1^-) = p(t_0)$, $\mu(\tau_1^-) = \mu(t_0)$. If $\tau_N = t_1$, put $p(\tau_N^+) = p(t_1)$, $\mu(\tau_N^+) = \mu(t_1)$.) All $\beta_{kj} \geqslant 0$. Moreover,

$$\beta_{kj} = 0 \text{ if } \bar{g}_j(x^*(\tau_k), \tau_k) > 0 \tag{76a}$$

$\beta_{kj} = 0$ if $\tau_k \in (t_0, t_1)$, $\bar{g}_j(x^*(\tau_k), \tau_k) = 0$ and

$\partial \bar{g}_j(x^*(t), t)/\partial x \cdot f(x^*(t), u^*(t), t)$ is discontinuous at τ_k. ∎ (76b)

As in the corresponding case in Chapter 5, we recommend utilizing ANC rather than theorem 5. If we find one or more solutions to ANC, then we can be reasonably sure that there exists no other solution to the true necessary conditions. In addition, if for some candidate an ill-behaved $p(t)$ is needed for the true necessary conditions to be satisfied, this fact can be discovered when trying to apply ANC.

Note that the non-triviality results in note 5 are important also when using ANC. Furthermore, note that when the constraint qualification (48) fails, to obtain the appropriate ANC, we have to add the terms:

$$\sum_{j=1}^{s'} \beta_{kj}^- \frac{\partial g_j(x^*(\tau_k), u^*(\tau_k^-), \tau_k)}{\partial x} \quad \text{and} \quad \sum_{j=1}^{s'} \beta_{kj}^+ \frac{\partial g_j(x^*(\tau_k), u^*(\tau_k^+), \tau_k)}{\partial x} \tag{77}$$

to the right hand side of (75), where $\beta_k^- = (\beta_{k1}^-, \ldots, \beta_{ks'}^-)$ and $\beta_k^+ = (\beta_{k1}^+, \ldots, \beta_{ks'}^+)$ satisfy (24), (25), $\sum_{j=1}^{s'} \beta_{kj}^- (\partial g_j^*(\tau_k^-)/\partial u) = 0$, $\sum_{j=1}^{s'} \beta_{kj}^+ (\partial g_j^*(\tau_k^+)/\partial u) = 0$. If

$\tau_1 = t_0$, $(\tau_N = t_1)$, omit the former (latter) sum in (77). In this case (63) has to be replaced by the condition that $(p_0, p(t), q(t), \mu(t_1) - \mu(t_0), \beta_1^-, \ldots, \beta_N^-, \beta_1^+, \ldots, \beta_N^+) \neq 0$ for some t.

To show that $p_0 \neq 0$, sometimes also the following condition is useful: Define

$$p(\tau_k) = p(\tau_k^+) + \sum_{j \leqslant s'} \beta_{kj}^+ \frac{\partial g_j^*(\tau_k^+)}{\partial x}$$

for $\tau_k \in (t_0, t_1)$. Then condition (64) holds for both $t = t_0$ and $t = t_1$ and for all $\tau_k \in (t_0, t_1)$ for which $\beta_{kj} = 0$, $j = s' + 1, \ldots, s$, both for $u^*(\tau_k^-)$ and $u^*(\tau_k^+)$.

It is convenient to have a sufficiency theorem corresponding to the formulation of the necessary conditions in theorem 5.

Theorem 6 (Sufficient conditions. Arrow-type. Mixed and pure state constraints)[4] Let $(x^*(t), u^*(t))$ be an admissible pair for problem (1)–(6), (61) with $U = R^r$. Suppose that the full rank c.q. (48) holds for all $t \in [t_0, t_1]$, and assume that there exist functions $p(t)$, $q(t)$ and $\mu(t)$ such that for $p_0 = 1$, the conditions (65)–(70) are satisfied. Assume also that

$$\hat{H}(x^*(t), p(t), t) = H(x^*(t), u^*(t), p(t), t) \qquad \text{v.e.} \tag{78}$$

and that the concavity conditions (53) on $g_{s'+1}, \ldots, g_s$ and (54) on $\hat{H}$ are satisfied. Then $(x^*(t), u^*(t))$ is optimal. ∎

Note 7   The free final time problem   In the problem of theorem 5 suppose that t_1 is not fixed but free to vary in an interval $[T_1, T_2]$, $t_0 \leqslant T_1 < T_2$. In addition to the assumptions (6) and (61), we require f_0, f, and $g_j, j \leqslant s'$ to have continuous derivatives also w.r.t. t. If t_1^* denotes the optimal final time, then theorem 5 is still valid for $t_1 = t_1^*$. The following additional condition is satisfied:[5]

$$H^*(t_1^{*-}) + \sum_{j=s'+1}^{s} [\mu_j(t_1^*) - \mu_j(t_1^{*-})] \frac{\partial \bar{g}_j^*(t_1^*)}{\partial t}$$

$$\begin{array}{ll} \leqslant 0 & \text{if } t_1^* = T_1 \\ = 0 & \text{if } T_1 < t_1^* < T_2 \\ \geqslant 0 & \text{if } t_1^* = T_2 \end{array} \tag{79}$$

---

[4] Seierstad (1985b).
[5] See footnote 14.

with $H^*(t) = H(x^*(t), u^*(t), p(t), t)$, and $\partial \bar{g}_j^*(t)/\partial t = \partial \bar{g}_j(x^*(t), t)/\partial t$. Notice that notes 4 and 5 are valid for $t_1 = t_1^*$.

Note 8 If $t_1^* = T_1 = t_0$, read (79) as saying that for some vectors $\bar{\beta} = (\beta_{s'+1}, \ldots, \beta_s)$, $p(t_0)$ we have $\sup_{u \in U^+(t_0)} H(x_0, u, p(t_0^-), t_0) + \bar{\beta}(\partial \bar{g}^*(t_0)/\partial t) \leqslant 0$, where $p(t_0^-) = p(t_0) + \bar{\beta}(\partial \bar{g}^*(t_0)/\partial x)$, $\bar{\beta}$ and $p(t_0)$ satisfying (12) and (70), respectively, with $t_1 = t_0$. Finally, $(p_0, p(t_0), \bar{\beta}) \neq 0$, $(p_0 = 0$ or $1)$. This note is valid only if the set $U^+(t_0) \neq \phi$.

Next we give a sufficient condition for the free final time problem:

Theorem 7 (**Sufficient condition, free final time**)[6] Consider the problem (1) – (6) with $t_1 \in [T_1, T_2]$, t_1 free and $t_0 \leqslant T_1 < T_2$. Assume that for each $T \in [T_1, T_2]$ there exists an admissible pair $(x_T(t), u_T(t))$ defined on $[t_0, T]$ with associated multipliers $p_T(t)$, $q_T(t)$ and β^T satisfying all the sufficient conditions of theorem 1. Assume that $T \to x_T(T)$ is Lipschitz continuous, $T \to \beta^T$ and $T \to p_T(T)$ are piecewise continuous, $u_T(t)$ and $q_T(t)$ take values in fixed, bounded subsets of R^r and R^s that are independent of $T \in [T_1, T_2]$, and that $u_T(T)$ belongs to the closure of the set $\{u \in U : g_j(x_T(T), u, T) > 0$ for all $j \leqslant s'\}$ for all T. Assume, finally, that (for $p_0 = 1$), the function

$$F(T) = H(x_T(T), u_T(T), p_T(T), T) + \beta^T \, \partial g(x_T(T), u_T(T), T)/\partial t \qquad (80)$$

has the property that there exists a $T^* \in [T_1, T_2]$ such that

$$\begin{aligned} F(T) &\geqslant 0 \qquad \text{for } T < T^* \text{ if } T_1 < T^* \\ F(T) &\leqslant 0 \qquad \text{for } T > T^* \text{ if } T_2 > T^* \end{aligned} \qquad (81)$$

Then the pair $(x_{T^*}(t), u_{T^*}(t))$ defined on $[t_0, T^*]$ is optimal.

(If $T_1 = t_0$, then it is sufficient to test the properties of the theorem for $T > T_1$.) ■

Note 9 If t_1 is free in $[T_1, \infty)$ and there exists a $T^* \in [T_1, \infty)$ such that theorem 7 is satisfied for each interval $[T_1, T_2]$ containing T^*, then the triple $(x_{T^*}(t), u_{T^*}(t), T^*)$ is optimal.

Note 10 (**Arrow-type, free final time sufficient condition**) If $U = R^r$, then in theorem 7 the quintuples $(x_T(t), u_T(t), p_T(t), q_T(t), \beta_T)$ can be allowed to satisfy the modification of the sufficient conditions of theorem 1 described in theorem 3.

[6]Seierstad (1984a).

Some generalizations*

If an alternative constraint qualification is imposed, theorem 5 can be generalized to the case in which U is any convex subset of R^r. With I_t^- and I_t^+ as defined in (47), the c.q. is as follows: For all $t \in [t_0, t_1]$,

If $I_t^- \neq \phi$, then there exists a vector $u^- \in U$ such that

$$\frac{\partial g_j(x^*(t), u^*(t^-), t)}{\partial u} \cdot (u^- - u^*(t^-)) > 0 \text{ for all } j \in I_t^-. \tag{82(i)}$$

If $I_t^+ \neq \phi$, then there exists a vector $u^+ \in U$ such that

$$\frac{\partial g_j(x^*(t), u^*(t^+), t)}{\partial u} \cdot (u^+ - u^*(t^+)) > 0 \text{ for all } j \in I_t^+. \tag{82(ii)}$$

If $t = t_0$, drop (i); if $t = t_1$, drop (ii).

Theorem 8 (Necessary conditions. U a convex subset)[7]   Replace the condition $U = R^r$ in theorem 5 with the condition that U is a convex subset of R^r, and replace the full rank c.q. (48) by the constraint qualifications (82). Then theorem 5 remains valid with the following modifications: $q(t)$ is bounded and measurable, $p^*(t)$ is absolutely continuous, (65) is replaced by

$$\text{For a.e.} t, \ \partial \check{L}^*/\partial u \cdot (u - u^*(t)) \leq 0 \text{ for all } u \in U, \tag{83}$$

and (68) holds a.e. Finally, in (64), u must also belong to U.   ∎

In most cases $q(t)$ will have one-sided limits everywhere, but the conditions in the theorem no longer ensure this property. Observe that all the assertions in note 5 are correct even in the present situation provided $U^+(t)$ is defined as $\{u : u \in U, \ g_j(x^*(t), u. t) > 0 \text{ for all } j = 1, \ldots, s'\}$.

Exercise 6.4.1 Solve the problems

(i) $\max \int_0^3 (x - au)\,dt, \qquad \dot{x} = u, \qquad x(0) = 1, \qquad x(3) \text{ free},$

$$-1 \leq u, \qquad u \leq x^2, \qquad x \leq 2.$$

---

[7] Neustadt (1976) Chapter VI, 3. In the measurable case the following changes must be made: $p^*(t)$ is absolutely continuous and (64) holds a.e. Condition (82) must be replaced by the condition that for each $(\bar{u}, t) \in \Omega$ (see footnote 1 in Chapter 4), if $g_j(x^*(t), \bar{u}, t) = 0$, then there exists a $u \in U$ such that $(\partial g_j(x^*(t), \bar{u}, t)/\partial u) \cdot (u - \bar{u}) > 0$, u the same for all such $j < s'$.

Assume that $0 < a$ and $\frac{1}{2} < 3 - a$.

(ii) $\max \displaystyle\int_0^T (x^2 - u)dt,$ $\dot{x} = u,$ $x(0) = 1,$ $x(T)$ free,

$u \geq -1,$ $u \leq x,$ $x \geq 0, \; T > 1.$

(iii) $\max \displaystyle\int_0^T (x^2 - 4x)dt,$ $\dot{x} = u,$ $x(0) = 1,$ $x(T)$ free

$u \geq -1,$ $u \leq x,$ $x \geq 0.$

5. Infinite horizon problems

Infinite horizon problems without state variable restrictions were studied in considerable detail in Chapter 3, section 7. Some of the results from that section will now be generalized to problems *with* state variable restrictions.

We study a system whose evolution for $t \geq t_0$ is determined by the usual vector differential equation:

$$\dot{x}(t) = f(x(t), u(t), t), \qquad x(t_0) = x^0 \qquad (x^0 \text{ fixed in } R^n) \tag{84}$$

with the control variable restriction

$$u(t) \in U, \qquad U \text{ a fixed convex set in } R^r, \tag{85}$$

the constraints

$$\begin{aligned}
g_j(x(t), u(t), t) &\geq 0 & j &= 1, \ldots, s' \\
g_j(x(t), u(t), t) &= \bar{g}_j(x(t), t) \geq 0 & j &= s' + 1, \ldots, s
\end{aligned} \tag{86}$$

and the terminal conditions

$$\lim_{t \to \infty} x_i(t) \text{ exists and equals } x_i^1 \qquad i = 1, \ldots, l \tag{87a}$$

$$\lim_{t \to \infty} x_i(t) \geq x_i^1 \qquad i = l + 1, \ldots, m \tag{87b}$$

$$\text{no condition at } \infty \qquad i = m + 1, \ldots, n. \tag{87c}$$

Throughout this section we assume that

f_0, f, g_j for $j = 1, \ldots, s'$, and the derivatives of these functions w.r.t. x and u exist and are continuous. $\tag{88}$

$\bar{g}_j$ is a C^2-function for $j = s' + 1, \ldots, s.$ $\qquad$ (89)

Our optimality criteria are all concerned with the behaviour of an integral, $\int_{t_0}^{t} f_0(x(\tau), u(\tau), \tau) d\tau$, as t becomes large. A number of criteria were already discussed and compared in Chapter 3, section 7. According to the weakest criterion, $(x^*(t), u^*(t))$ is *piecewise optimal* (PW-optimal) if for every $T \geqslant t_0$, the admissible pair $(x^*(t), u^*(t))$ restricted to $[t_0, T]$ is optimal in the corresponding finite horizon problem with terminal condition $x(T) = x^*(T)$, and with max $\int_{t_0}^{T} f_0(x, u, t) dt$ as the optimality criterion. A somewhat stronger optimality criterion is the catching-up criterion. An admissible pair $(x^*(t), u^*(t))$ is *catching-up optimal* (CU-optimal) provided

$$\lim_{t \to \infty} \left[\int_{t_0}^{t} f_0(x^*(\tau), u^*(\tau), \tau) d\tau - \int_{t_0}^{t} f_0(x(\tau), u(\tau), \tau) d\tau \right] \geqslant 0 \qquad (90)$$

for all admissible pairs $(x(t), u(t))$.

Necessary conditions

In order to obtain necessary conditions for optimality, a constraint qualification on $g_1, \ldots, g_s$ is needed. Define

$$h_j(x, u, t) = \begin{cases} g_j(x, u, t) & j = 1, \ldots, s' \\ (\partial \bar{g}_j(x, t)/\partial x) \cdot f(x, u, t) + \partial \bar{g}_j(x, t)/\partial t & j = s' + 1, \ldots, s. \end{cases} \qquad (91)$$

The constraint qualification we use is of the full rank type $((x^*(t), u^*(t))$ being the candidate for optimality):

The matrix $M^+(t) = [\partial h_j(x^*(t), u^*(t^+), t)/\partial u_i]$,

$i = 1, \ldots, r, \ j \in E^+(t) = \{j : g_j(x^*(t), u^*(t^+), t) = 0, j = 1, \ldots, s\}$

has rank equal to the number of elements in $E^+(t)$. $\qquad$ (92(i))

The matrix $M^-(t) = [\partial h_j(x^*(t), u^*(t^-), t)/\partial u_i]$

$i = 1, \ldots, r, \ j \in E^-(t) = \{j : g_j(x^*(t), u^*(t^-), t) = 0, j = 1, \ldots, s\}$

has rank equal to the number of elements in $E^-(t)$. $\qquad$ (92(ii))

(Drop (ii) if $t = t_0$; (i) (resp. (ii)) holds vacuously if $E^+(t)$, (resp. $E^-(t)$), is empty.)

The necessary conditions in the next theorem are valid for any type of conditions on the behaviour of $x(t)$ as t goes to infinity, not just for conditions (87). Since the theorem gives necessary conditions for piecewise optimality (PW-optimality) of $(x^*(t), u^*(t))$, it is also valid if we assume $(x^*(t), u^*(t))$ to be SCU-, CU- or OT-optimal, since these are stronger criteria than PW-optimality. (See (3.177).)

Theorem 9 (**Necessary conditions. Mixed and pure state constraints. Infinite horizon**)[8] Let $(x^*(t), u^*(t))$ be piecewise optimal (or SCU-, CU-, or OT-optimal) in problem (84)→(86), (88), (89) with $U = R^r$. Suppose that the constraint qualification (92) is satisfied for all $t \geqslant t_0$. Then there exist a number p_0, vector functions $p(t) = (p_1(t), \ldots, p_n(t))$, and $q(t) = (q_1(t), \ldots, q_{s'}(t))$, and a non-decreasing vector function $\mu(t) = (\mu_{s'+1}(t), \ldots, \mu_s(t))$, all having one-sided limits everywhere, such that conditions (64)–(69) are satisfied with $t_1 = \infty$, together with the conditions $p_0 = 0$ or $p_0 = 1$ and

$$(p_0, p(t^+)) \neq (0, 0) \text{ for all } t \geqslant t_0, \ (p_0, p(t^-)) \neq (0, 0) \text{ for all } t > t_0.$$
$$\mu(t_0) = \mu(t_0^+) = 0. \quad \blacksquare \tag{93}$$

There are no transversality conditions in the theorem. Consequently, the theorem usually does not contain enough information to narrow down the set of candidates for optimality to only a few ones. As we already mentioned in connection with theorem 3.12, the "natural" transversality conditions associated with (87) are *not* necessary for optimality. For conditions which almost are necessary transversality conditions in certain situations, see theorem 11 and note 13.

Note 11*
(a) Consider the case in which there are only pure state constraints (i.e. $s' = 0$) and assume that U is an arbitrary subset of R^r. Then theorem 9 is valid with the following modifications: (i) Replace the c.q. (92) by (5.125b), with $t_1 = \infty$, (ii) replace (93) by the condition $(p_0, p(t_0)) \neq (0, 0)$, (iii) replace (64) by the condition that the Hamiltonian is maximized by $u^*(t)$ in U, and (iv) delete (65).

If, in addition, (5.125a) holds for all $t > t_0$ and if the restrictions are first-order compatible at t_0, then (93) holds. (If (5.125a) fails, but all the other conditions hold, $(p_0, p(t_0^+)) \neq (0, 0)$.)
(b) Assume that U is a convex subset of R^r and allow mixed as well as pure state constraints ($s' > 0$). Then theorem 8 remains valid with the following modifications: (i) For all $t \in [t_0, \infty)$ the c.q. (82) is required to hold, as well

[8]Seierstad (1986a).

as (5.125b) (with $t_1 = \infty$, and modified as in note 5(b) with $U^+(t) = \{u \in U:$ $g_j(x^*(t), u, t) > 0, j = 1, \ldots, s'\}$, (ii) Condition (93) is replaced by $(p_0, p(t_0)) \neq (0, 0)$.

If, in addition, the restrictions are first order compatible at t_0 and (5.125a) is valid for $t > t_0$ (modified as in note 5(b)), then (93) holds true. (If (5.125a) fails, but all the other conditions hold, then $(p_0, p(t_0^+)) \neq (0, 0)$.)

It is possible to construct necessary transversality conditions corresponding to (87), but the conditions become so messy that we refrain from working them out. As noted above, theorem 9 gives an incomplete set of conditions in the sense that it will usually allow an infinite number of candidates. A case in point is the problem in the next example.

Example 3 Consider the problem

$$\max \int_0^\infty (x(t) + u(t))e^{-rt}\,dt,$$

$$\dot{x}(t) = -bu(t), \qquad x(0) = 1, \qquad x(\infty) \text{ no condition,} \tag{94}$$

$$\bar{g}(x(t), t) = 1 - x(t) \geqslant 0, \tag{95}$$

$$u(t) \in U = [-1, 1]. \tag{96}$$

Here r and b are constant, $b > r > 0$. (The integral in (94) exists for all admissible pairs, see exercise 6.5.1.)

We apply the modification of theorem 9 given in note 11(a). According to (59), (58), and (60),

$$h(x, u, t) = \frac{\partial \bar{g}}{\partial x} \cdot f + \frac{\partial \bar{g}}{\partial t} = (-1)(-bu) = bu \tag{97}$$

$$\bar{L}(x, u, p, q, \mu, t) = p_0(x + u)e^{-rt} - pbu - \mu bu \tag{98}$$

$$\check{L}(x, u, p, q, t) = p_0(x + u)e^{-rt} - pbu \tag{99}$$

Suppose $(x^*(t), u^*(t))$ solves the problem. Then (5.125a, b) are trivially satisfied since $v \cdot \partial g/\partial x + \partial g/\partial t = -v$ and $\mathrm{co}\{f(x^*(t), u, t): u \in U\} = [-b, b]$. Moreover, the restrictions $x(0) = 1$ and $x(t) \leqslant 1$ are first-order compatible at 0 since $G^*(0) \neq \phi$ and (5.120) is trivially satisfied. Therefore there exist a number p_0, $p_0 = 0$ or $p_0 = 1$, a function $p(t)$, and a non-decreasing function $\mu(t)$ such that (using (93)), $(p_0, p(t^\pm)) \neq (0, 0)$ for all $t \geqslant 0$, and $\mu(0^+) = 0$. Moreover, for v.e. $t > 0$, $u^*(t)$ maximizes $p_0(x^*(t) + u)e^{-rt} - bp(t)u$ for $u \in [-1,$

1], i.e. for v.e. $t>0$,

$\qquad u^*(t)$ maximizes $(p_0 e^{-rt} - bp(t))u$ $\qquad$ for $u \in [-1, 1]$. $\qquad$ (100)

From (100), we conclude that for v.e. $t>0$,

$\qquad u^*(t)=1, \qquad =?, \qquad =-1$ according as

$\qquad\qquad p_0 e^{-rt} - bp(t)>0, \qquad =0, \qquad <0.$ $\qquad$ (101)

By (66) we have that

$\qquad \mu(t)$ is constant on intervals where $x^*(t)<1$ $\qquad$ (102(i))

$\qquad \mu(t)$ is continuous if $x^*(t)=1$ and $u^*(t)$ is discontinuous. $\qquad$ (102(ii))

By (67) we define $p^*(t)=p(t)-\mu(t)$, and $p^*(t)$ is continuous. Using (68) we find for v.e. t,

$\qquad \dot{p}^*(t)= -\partial \bar{L}^*/\partial x = -p_0 e^{-rt}, \qquad$ so $p^*(t)=(p_0/r)e^{-rt}+C.$ $\qquad$ (103)

$\qquad p(t)=(p_0/r)e^{-rt}+C+\mu(t)$ for some constant C. $\qquad$ (104)

These are all the conditions derived from theorem 9 as modified in note 11.

Consider the case $p_0=0$. Then $p(0^+)\neq 0$, $p^*(t)=C$ and $p(t)=C+\mu(t)$. Since $\dot{x}^*(0^+)= -bu^*(0^+)$, the constraint in (95) is violated if $u^*(0^+)<0$. Hence from (101), $-bp(0^+)\geqslant 0$, so $p(0^+)\leqslant 0$ and thus $p(0^+)<0$. It follows that $-bp(t)>0$ in some small interval $[0, \varepsilon)$, $\varepsilon>0$. From (101), $u^*(t)=1$ for v.e. $t\in[0, \varepsilon)$. Since (95) is inactive, $\mu(t)$ is constant and so is $p(t)$ in $(0, \varepsilon)$. Evidently, the interval $(0, \varepsilon)$ can be extended to $(0, \infty)$ with the same properties holding. (Define $\varepsilon^* = \sup\{\varepsilon>0: x^*(s)<1$ on $(0, \varepsilon)\}$. Then $p(t)$ is constant, $u^*(t)=1$, and $x^*(t)<1$ on $(0, \varepsilon^*)$. If $\varepsilon^* < \infty$, then $x^*(\varepsilon^*)<1$ since $\dot{x}^*(t)<0$, and we have a contradiction of the definition of ε^*.) The only candidate for optimality is $u^*(t)\equiv 1$.

Consider next the case $p_0=1$. We have $x^*(0)=1$. Define $\tau=\sup\{t: x^*(s)=1$ on $[0, t]\}$. Possibly $\tau=0$ or $\tau=\infty$. Suppose $\tau<\infty$. Then there exist points $t_1>\tau$, arbitrarily close to τ where $x^*(t_1)<1$. Let t' be a point in (τ, t_1) which satisfies $x^*(t')<1$, $\dot{x}^*(t')<0$ and for which (101) holds. (Such a point t' exists, since $\dot{x}(t)$ cannot be $\geqslant 0$ v.e. in (τ, t_1).) Choose (α, β) as an interval around t' in which $x(s)<1$. Then from (102(i)), $\mu(t)$ is constant in (α, β). Since $\dot{x}^*(t')= -bu^*(t')<0$, $u^*(t')>0$ and from (101) we conclude that $e^{-rt'} - bp(t')=e^{-rt'} - (b/r)e^{-rt'} -bC -b\mu(t')\geqslant 0$. Define $\psi(t)=e^{-rt} - (b/$

$r)e^{rt} - bC$. Then $\dot{\psi}(t) = (b-r)e^{-rt} > 0$, so ψ is strictly increasing. Since $\psi(t') - b\mu(t') \geqslant 0$ and $\mu(t)$ is constant in (α, β), it follows that $\psi(t) - b\mu(t) > 0$ in (t', β). We conclude from (101) that $u^*(t) = 1$ in (t', β). Hence $x^*(\beta) < x^*(t') < 1$. Let $\beta^* = \sup \{\beta: x^*(s) < 1$ in $(t', \beta)\}$. Letting β^* play the role of β in the argument above, we obtain $x^*(\beta^*) < x^*(t') < 1$, which contradicts the definition of β^*, unless $\beta^* = \infty$. Thus we have $u^*(t) = 1$ for all $t \in (t_1, \infty)$, and since t_1 is arbitrarily near τ, we have arrived at the following conclusion: If $u^*(t)$ is an optimal control, then $u^*(t) = u_\tau^*(t)$ for some τ, where

$$u_\tau^*(t) = \begin{cases} 0 & \text{on } [0, \tau] \\ 1 & \text{on } (\tau, \infty) \end{cases} \tag{105}$$

The theorem, however, gives no information about τ.

From (94) it is easy to find the $x_\tau^*(t)$ corresponding to $u_\tau^*(t)$: In $[0, \tau]$, $x_\tau^*(t) = 1$, while on (τ, ∞), $x_\tau^*(t) = -b(t-\tau) + 1$. Thus the value of the criterion functional as a function of τ becomes, after some easy calculations,

$$I(\tau) = \int_0^\tau e^{-rt}dt + \int_\tau^\infty [2 - b(t-\tau)]e^{-rt}dt = \frac{i}{r} - \frac{1}{r^2}e^{-r\tau}(b-r) \tag{106}$$

Note that $I'(\tau) = (1/r)e^{-r\tau}(b-r) > 0$ for all τ, so $I(\tau)$ is strictly increasing. Granted that an optimal solution exists (see exercise 6.5.1), we therefore conclude that the optimal solution is $u^*(t) \equiv 0$, $x^*(t) \equiv 1$.

An existence theorem*

We end this section by formulating an existence theorem for a special class of infinite horizon problems, where the criterion integral exists for each admissible pair with values in $[-\infty, \infty)$. Interesting existence theorems for problems with other types of optimality criteria are more difficult to construct.

Theorem 10 (Existence. Infinite horizon)[9] Consider problem (84)–(87) with $\int_{t_0}^\infty f_0(x(t), u(t), t)dt$ as the criterion functional, assuming that f_0, f and g are continuous and U closed. Assume that there exists an admissible pair, and that conditions (3.189), (3.190) are satisfied. Assume that $N(x, U, t) = \{(f_0(x, u, t) + \gamma, f(x, u, t)): \gamma \leqslant 0, u \in U, g_j(x, u, t) \geqslant 0, j \leqslant s'\}$ is convex for

---

[9]Baum (1976).

all x and all $t \geq t_0$. Assume, finally, that for each T there exists a constant b_T such that the set

$$U(x, t) = \{u \in U : g_j(x, u, t) \geq 0, j = 1, \ldots, s'\}$$

is contained in the ball $B(0, b_T)$ for all $t \leq T$, and all $x \in B(0, y(t))$, where $y(t)$ is the solution of $\dot{y} = a(t)y + b(t)$, $y(t_0) = ||x^0||$, where $a(t)$ and $b(t)$ are the non-negative functions appearing in (3.190).

Then there exists an optimal pair $(x^*(t), u^*(t))$. (The function $u^*(t)$ is possibly merely measurable). ∎

Further existence results for infinite horizon problems are given at the end of this chapter.

Exercise 6.5.1 Consider example 3. Note that the integral in (94) exists for all admissible $(x(t), u(t))$ (since $|x(t)| \leq 1 + bt$). Apply theorem 10 to prove the existence of an optimal solution.

6. Infinite horizon problems: Sufficient conditions

We turn now to a study of sufficient conditions for problems of the type discussed in section 5.

Catching-up optimality is probably the most frequently encountered optimality criterion for infinite horizon problems in which $\int_{t_0}^{\infty} f_0(x, u, t)dt$ does not converge for all admissible pairs (x, u). In the formulation of our sufficient conditions we shall restrict our attention to CU-optimality. A more extensive discussion of sufficient conditions including other types of optimality criteria, is given in Seierstad and Sydsæter (1977), section 9. In that paper the main argument for the following result is also presented.

Theorem 11 (Sufficient conditions. Mixed and pure state constraints. Infinite horizon)[10] A pair $(x^*(t), u^*(t))$ which satisfies (84)–(89) is CU-optimal according to the criterion in (90) provided there exist a piecewise continuous and piecewise continuously differentiable function $p(t)$ with jump points τ_k, $t_0 < \tau_1 < \ldots < \tau_N$, a piecewise continuous function $q(t)$ and vectors β_k^+, β_k^- such that conditions (9), (10), (13)–(15) and (21)–(25) are satisfied with $p_0 = 1$, and, in addition:

$$\lim_{t \to \infty} p(t) \cdot (x(t) - x^*(t)) \geq 0 \qquad \text{for all admissible } x(t). \quad ∎ \qquad (107)$$

[10] Seierstad and Sydsæter (1977).

In this theorem we allow for jumps in $p(t)$. However, since jumps in $p(t)$ are relatively rare, when applying theorem 11 it may be a good idea to start by guessing that there are no jumps. This is the approach we use in example 4 below.

Note 12 (Arrow-type sufficient conditions.) If $U = R^r$ and (14), (15) are replaced by (50)–(54), the conclusion in theorem 11 remains valid.

Note 13 The conditions in (3.185) again imply (107). Note, however, that (3.185) does not take into account that our admissible pair is "constrained at infinity" by $g(x, u, t) \geq 0$.

The following result is useful *in the case where g is concave in (x, u)*: The condition (107) can be replaced by the following conditions:[11]

There exists a vector $\beta^\infty = (\beta_1^\infty, \ldots, \beta_s^\infty)$ such that

$$\beta_k^\infty \geq 0\left(= 0 \text{ if } \overline{\lim_{t \to \infty}} \, g_k(x^*(t), u^*(t), t) > 0 \right) \tag{108(i)}$$

$$\overline{\lim_{t \to \infty}} \, \beta^\infty \cdot g_u'(x^*(t), u^*(t), t) \cdot [u(t) - u^*(t)] \leq 0 \tag{108(ii)}$$

for all piecewise continuous control functions $u(t)$ where $u(t) \in U$, and such that either (107) or (3.185) is satisfied when $p(t)$ is replaced by

$$\varphi(t) = p(t) - \beta^\infty g_x'(x^*(t), u^*(t), t). \tag{109}$$

We see that $\varphi(t)$ behaves at infinity in the same way as $p(t)$ does in problems without g-constraints. Thus, in problems with g-constraints but without additional restrictions on $x(t)$ at infinity $(m = 0)$, we can frequently expect that $\varphi(t) \to 0$ as $t \to \infty$. (It is instructive to compare the definition of $\varphi(t)$ and its properties with (11) and (12). Evidently, we have to modify the standard transversality conditions to account for the effects of the g-constraints at t_1.)

Example 4 We consider a simple example that illustrates the use of theorem 11 and note 13.

$$\text{CU-max} \int_0^\infty (u(t) + v(t)) \mathrm{d}t \tag{110}$$

---

[11] It is easy to see (107) is a consequence of (108) and (109).

$$\dot{x}_1(t) = -u(t), \qquad x_1(0) = 1, \qquad x_1(\infty) \text{ free}, \qquad u(t) \in [-1, 1] \tag{111}$$

$$\dot{x}_2(t) = -v(t), \qquad x_2(0) = 1, \qquad x_2(\infty) \text{ free}, \qquad v(t) \in [-1, 1] \tag{112}$$

$$g(x_1(t), x_2(t), u(t), v(t), t) = 2x_1(t) + x_2(t) \geqslant 0. \tag{113}$$

The Hamiltonian and the Lagrangian are ($p_0 = 1$)

$$H = u + v - p_1 u - p_2 v = (1 - p_1)u + (1 - p_2)v, \quad L = H + q(2x_1 + x_2). \tag{114}$$

If we guess that $p_1(t)$ and $p_2(t)$ have no jumps, the sufficient conditions in theorem 11 supplemented by note 13 are as follows: For v.e. $t \geqslant 0$, $(1 - p_1(t)$, $1 - p_2(t)) \cdot [(u, v) - (u^*(t), v^*(t))] \leqslant 0$ for all $u \in [-1, 1]$ and all $v \in [-1, 1]$, so it follows that for v.e. $t \geqslant 0$,

$$(1 - p_1(t))(u - u^*(t)) \leqslant 0 \qquad \text{for all } u \in [-1, 1] \tag{115(i)}$$

$$(1 - p_2(t))(v - v^*(t)) \leqslant 0 \qquad \text{for all } v \in [-1, 1]. \tag{115(ii)}$$

For v.e. $t > 0$,

$$\dot{p}_1(t) = -\frac{\partial L^*}{\partial x_1} = -2q(t), \qquad \dot{p}_2(t) = -\frac{\partial L^*}{\partial x_2} = -q(t), \tag{116}$$

$$q(t) \geqslant 0 (=0 \text{ if } 2x_1^*(t) + x_2^*(t) > 0). \tag{117}$$

Finally, there exists a number β^∞ where

$$\beta^\infty \geqslant 0 \left(=0 \text{ if } \overline{\lim_{t \to \infty}} (2x_1^*(t) + x_2^*(t)) > 0 \right) \tag{118}$$

$$\overline{\lim_{t \to \infty}} \beta^\infty \cdot (0, 0) \cdot [(u(t), v(t)) - (u^*(t), v^*(t))] \leqslant 0 \qquad \text{for all } u(t), v(t), \tag{119}$$

and such that (107) is valid with $p(t)$ replaced by $(\varphi_1(t), \varphi_2(t)) = (p_1(t) - 2\beta^\infty, p_2(t) - \beta^\infty)$, i.e.

$$\overline{\lim_{t \to \infty}} (p_1(t) - 2\beta^\infty) \cdot (x_1(t) - x_1^*(t)) + (p_2(t) - \beta^\infty) \cdot (x_2(t) - x_2^*(t)) \geqslant 0. \tag{120}$$

The Hamiltonian as well as the constraint function g are linear in (x_1, x_2, u, v), so (14) and (15) are trivially satisfied. Thus, if we can find functions $x_1^*(t)$, $x_2^*(t)$, $u^*(t)$, $v^*(t)$, $p_1(t)$, $p_2(t)$, and $q(t)$, together with a number β^∞, satisfying (115)–(120), then theorem 11 guarantees that $(x_1^*(t), x_2^*(t), u^*(t), v^*(t))$ is optimal.

Since $2x_1^*(0) + x_2^*(0) = 3 > 0$, there is a maximal interval $[0, t')$ on which (113) is inactive. Then by (117) $q(t) = 0$ on $[0, t']$, so from (116) $p_1(t)$ and

$p_2(t)$ are constants on $[0, t']$. Proceed by assuming that (113) is active on (t', ∞), so that $2x_1^*(t) + x_2^*(t) = 0$. Then $2u^*(t) + v^*(t) = -2\dot{x}_1^*(t) - \dot{x}_2^*(t) = 0$ on (t', ∞).

Suppose $p_1(t) < 1$ for some $t \in (t', \infty)$ for which (115) holds. Then by (115(i)), $u - u^*(t) \leq 0$ for all $u \in [-1, 1]$, so $u^*(t) = 1$, and then $v^*(t) = -2u^*(t) = -2$, a contradiction. In the same way we prove that $p_1(t) > 1$ for some $t \in (t', \infty)$ for which (115) holds, leads to a contradiction. Hence $p_1(t) = 1$ v.e. on (t', ∞). Since p_1 is assumed to be continuous everywhere, and since $p_1(t)$ is constant on $[0, t']$ and $p_1(t)$ is 1 v.e. in (t', ∞), we conclude that $p_1(t) \equiv 1$ on $[0, \infty)$. It follows from (116) that $q(t) = 0 = \dot{p}_2(t)$ v.e., so $p_2(t)$, being continuous, is constant on $[0, \infty)$.

Since $x_1(\infty)$ and $x_2(\infty)$ are free, we guess that $\varphi_1(t)$ and $\varphi_2(t)$ approach 0 as t approaches 0. Hence $\beta^\infty = 1/2$ and $p_2 \equiv 1/2$. Putting $p_2(t) = 1/2$ into (115(ii)), it follows that $v^*(t) \equiv 1$, and thus from (112), $x_2^*(t) = 1 - t$ in $[0, \infty)$. On (t', ∞), $2x_1^*(t) + x_2^*(t) = 0$, so $x_1^*(t) = \frac{1}{2}t - \frac{1}{2}$, and, in particular, $x_1^*(t') = \frac{1}{2}t' - \frac{1}{2}$. Note also that on (t', ∞), $u^*(t) = (-1/2)v^*(t) = -1/2$.

On $[0, t']$, the conditions above do not determine $u^*(t)$. We suggest putting $u^*(t) = 1$ on $[0, t']$. Then from (111) $x_1^*(t) = 1 - t$ on $[0, t']$ so $x^*(t') = 1 - t'$. Since $x^*(t') = \frac{1}{2}t' - \frac{1}{2}$, we conclude that $t' = 1$ and thus

$$x_1^*(t) = \begin{cases} 1 - t & \text{in } [0, 1] \\ \frac{1}{2}t - \frac{1}{2} & \text{in } (1, \infty) \end{cases}, \qquad x_2^*(t) = 1 - t \text{ in } [0, \infty) \qquad (121)$$

We leave to the reader the easy task of proving that all the conditions (115)–(120) are now satisfied. Actually, in the present problem it is simple to use condition (107) directly. For $t > 1$,

$$p(t) \cdot (x(t) - x^*(t)) = (1, 1/2) \cdot (x_1(t) - \frac{1}{2} \cdot (t - 1), x_2(t) - (1 - t))$$
$$= \frac{1}{2}(2x_1(t) + x_2(t)),$$

which is ≥ 0 for all admissible $(x_1(t), x_2(t))$ on account of (113). Hence (107) is trivially satisfied. Observe, however, that by introducing β^∞ and $(\varphi_1(t), \varphi_2(t))$ as given in note 13, we were able to find suggestions for the appropriate functions $p_1(t)$ and $p_2(t)$.

It was mentioned above that the conditions in theorem 11 did not uniquely determine $u^*(t)$ on $[0, t']$, nor t'. For instance, it is easy to see that $u^*(t) = 0$ on $[0, 3]$, $u^*(t) = -1/2$ on $(3, \infty)$, $v^*(t) \equiv 1$ are also optimal controls.

We consider next a generalization of theorem 3.17.

Theorem 12[12] **(Sufficient conditions. Infinite horizon.)** Let $(x^*(t), u^*(t))$ be an admissible pair in system (84)–(89).

[12]Seierstad (1977a).

Assume that

fₒ, f and g are non-decreasing in x for each (u, t), and concave in (x, u) for each t. (122)

Assume that there exist a vector $p^* = (p_1^*, \ldots, p_n^*)$, a vector $\beta^\infty = (\beta_1^\infty, \ldots, \beta_s^\infty)$, a piecewise continuous function $q(t) = (q_1(t), \ldots, q_s(t))$, a finite number of points $\tau_k \in (t_0, \infty)$, $\tau_1 < \tau_2 < \ldots < \tau_N$, and vectors associated with each τ_k, $\beta_k^+ = (\beta_{k1}^+, \ldots, \beta_{ks}^+)$, $\beta_k^- = (\beta_{k1}^-, \ldots, \beta_{ks}^-)$ such that the following conditions hold: $q(t)$ satisfies the slackness conditions (13), β_k^+ and β_k^- satisfy conditions (22)–(25), β^∞ satisfies (108) and p^* satisfies the following conditions

$$p_i^* \geq 0 \qquad\qquad i = 1, \ldots, l$$

$$p_i^* \geq 0 \,(=0 \text{ if } \overline{\lim_{t \to \infty}} \, x_i(t) > x_i^1) \qquad i = l+1, \ldots, m$$

$$p_i^* = 0 \qquad\qquad i = m+1, \ldots, n. \qquad (123)$$

Assume, finally, that L defined by (8) satisfies

$$\lim_{T \to \infty} \int_{t_0}^T L_u'(x^*(t), u^*(t), p(t, T), q(t), t)[u^*(t) - u(t)] dt \geq 0 \qquad (124)$$

for all admissible control functions $u(t)$, and where for each $T > \tau_N$, $p(t, T)$ as a function of t, is a piecewise continuous and piecewise continuously differentiable function satisfying $p(T, T) = p^* + \beta^\infty \cdot g_x'(x^*(T), u^*(T), T)$, and (with $p(t)$ replaced by $p(t, T)$), (10) and (21). Then the pair $(x^*(t), u^*(t))$ is CU-optimal. ∎

Note 14 If the conditions (1), (2), and (4) described in note 3.23 are satisfied, condition (124) can be replaced by (9), with $p(t) = \lim_{T \to \infty} p(t, T)$.

The conditions on f in (122) can be replaced by the condition in note 3.24.

Exercise 6.6.1 Using theorem 12 show that $u(t) = e^{-t}$ is an optimal solution in the (trivial) problem

$$\max \int_0^\infty u \, dt, \qquad \dot{x} = u, \qquad x(0) = 0, \qquad x \leq 1, \qquad u \in [-2, 2].$$

(Hint: $p(t, T) = p(T, T) = -1$. There are many optimal solutions.)

7. Extensions: Scrap values and more general initial and terminal conditions

In this section we extend the theory presented in this chapter in several directions. Consider the following very general control problem:

$$\max \left\{ \int_{t_0}^{t_1} f_0(x(t), u(t), t) dt + S_0(x(t_0)) + S_1(x(t_1)) \right\}, t_0, t_1 \text{ fixed} \tag{125}$$

$$\dot{x}(t) = f(x(t), u(t), t) \tag{126}$$

$$
\begin{array}{ll}
R_k^0(x(t_0)) = 0 & k = 1, \ldots, r_0' \\
R_k^0(x(t_0)) \geqslant 0 & k = r_0' + 1, \ldots, r_0
\end{array}
\tag{127}
$$

$$
\begin{array}{ll}
R_k^1(x(t_1)) = 0 & k = 1, \ldots, r_1' \\
R_k^1(x(t_1)) \geqslant 0 & k = r_1' + 1, \ldots, r_1
\end{array}
\tag{128}
$$

$$u(t) \in U, \ U \text{ a fixed, convex set in } R^r \tag{129}$$

$$
\begin{array}{ll}
g_j(x(t), u(t), t) \geqslant 0 & j = 1, \ldots, s' \\
g_j(x(t), u(t), t) = \bar{g}_j(x(t), t) \geqslant 0 & j = s' + 1, \ldots, s
\end{array}
\tag{130}
$$

f_0, f, g and their partial derivatives w.r.t. x and u exist and are continuous, S_0, S_1, $R^0 = (R_1^0, \ldots, R_{r_0}^0)$, $R^1 = (R_1^1, \ldots, R_{r_1}^1)$ are C^1-functions. $\qquad (131)$

Several new features are introduced in this problem. Firstly, included in the criterion in (125) are both an evaluation of the initial stock $x(t_0)$ and an evaluation of the final stock $x(t_1)$. Secondly, general terminal conditions on $x(t_0)$ and on $x(t_1)$ are expressed by (127) and (128).

The modifications of our previous results that are needed in order to take account of the new features are partly suggested by the results in Chapter 3. In fact, note 3.5 gives necessary conditions for problem (125)–(131) for the case with no g-constraints. In the following theorem we give sufficient conditions for the problem above, assuming that $p(t)$ is continuous.

Theorem 13 (Sufficient conditions. No jumps in p(t)) Suppose we replace problem (1)–(6) by problem (125)–(131). Then theorem 1 holds with $p(t)$ continuous on $[t_0, t_1]$, provided (11) is replaced by the condition that

there exist numbers $\gamma_1, \ldots, \gamma_{r_1}$ such that for $i = 1, \ldots, n$,

$$p_i(t_1) = \sum_{j=1}^{s} \beta_j \frac{\partial g_j(x^*(t_1), u^*(t_1), t_1)}{\partial x_i} + \frac{\partial S_1(x^*(t_1))}{\partial x_i} + \sum_{k=1}^{r_1} \gamma_k \frac{\partial R_k^1(x^*(t_1))}{\partial x_i}$$

where for $k > r'_1$, $\gamma_k \geq 0$ ($=0$ if $R^1_k(x^*(t_1)) > 0$), (132)

the following assumption is added,

there exist numbers $\beta^0_1, \ldots, \beta^0_s$ and $\delta_1, \ldots, \delta_{r_0}$ such that

$$p_i(t_0) = - \sum_{j=1}^s \beta^0_j \frac{\partial g_j(x^*(t_0), u^*(t_0), t_0)}{\partial x_i}$$

$$- \frac{\partial S_0(x^*(t_0))}{\partial x_i} - \sum_{k=1}^{r_0} \delta_k \frac{\partial R^0_k(x^*(t_0))}{\partial x_i}$$

where for $k > r'_0$, $\delta_k \geq 0$ ($=0$ if $R^0_k(x^*(t_0)) > 0$), (133)

where the numbers $\beta^0_1, \ldots, \beta^0_s$ satisfy

$$\beta^0_j = 0 \qquad\qquad\qquad \text{for } j = 1, \ldots, s' \qquad (134\text{(i)})$$

$$\beta^0_j \geq 0 \,(=0 \text{ if } g_j(x^*(t_0), u^*(t_0), t_0) > 0) \qquad \text{for } j = s'+1, \ldots, s. \qquad (134\text{(ii)})$$

Finally, the following concavity conditions are added:

$S_0(x)$ and $S_1(x)$ are concave in x,

$$\sum_{k=1}^{r_1} \gamma_k R^1_k(x) \text{ and } \sum_{k=1}^{r_0} \delta_k R^0_k(x) \text{ are quasi-concave in } x. \quad \blacksquare \qquad (135)$$

Note 15 Let us check this result against some previous results:

(a) Put $S_0 = S_1 \equiv 0$, $R^0_k(x) = x_k - x^0_k$, $k = 1, \ldots,$ $n = r'_0 = r_0$, $R^1_k(x) = x_k - x^1_k$, $k = 1, \ldots, m$, with $r'_1 = l$, $r_1 = m$. Then (132) is easily seen to reduce to (11). Conditions (133) and (134) place no restriction on $p_i(t_0)$, since $\delta_1, \ldots, \delta_{r_0}$ are all unrestricted in this case.
(b) Put $g_j \equiv 0, j = 1, \ldots, s$. Then we see that (132) reduces to condition (3.16) and (133) reduces to condition (3.33) in Note 3.5.

Proof of theorem 13 Let $(x, u) = (x(t), u(t))$ be any admissible pair. Then we must prove (using the same brief notation as in the proof of theorem 4.5):

$$\bar{\Delta} = \int_{t_0}^{t_1} (f^*_0 - f_0) \mathrm{d}t + S_0(x^*(t_0)) - S_0(x(t_0))$$

$$+ S_1(x^*(t_1)) - S_1(x(t_1)) \geq 0. \qquad (136)$$

Using the same argument as in the proof of theorem 4.5, we get

$$\Delta = \int_{t_0}^{t_1} (f_0^* - f_0)dt \geqslant \int_{t_0}^{t_1} \left(-\frac{\partial H^*}{\partial x}\right) \cdot (x - x^*)dt$$

$$+ \int_{t_0}^{t_1} \left(-\frac{\partial H^*}{\partial u}\right) \cdot (u - u^*)dt + \int_{t_0}^{t_1} p \cdot (\dot{x} - \dot{x}^*)dt. \tag{137}$$

Since $L = H + q \cdot g$, $-\partial H^* / \partial x = -\partial L^* / \partial x + q \partial g^* / \partial x = \dot{p} + q \partial g^* / \partial x$, using (10). Moreover, $-\partial H^* / \partial u = -\partial L^* / \partial u + q \partial g^* / \partial u$, so

$$\Delta \geqslant \int_{t_0}^{t_1} \dot{p} \cdot (x - x^*)dt + \int_{t_0}^{t_1} q \frac{\partial g^*}{\partial x} \cdot (x - x^*)dt$$

$$+ \int_{t_0}^{t_1} \left(-\frac{\partial L^*}{\partial u}\right) \cdot (u - u^*)dt + \int_{t_0}^{t_1} q \frac{\partial g^*}{\partial u} \cdot (u - u^*)dt + \int_{t_0}^{t_1} p \cdot (\dot{x} - \dot{x}^*)dt.$$

The third of these integrals is $\geqslant 0$ by (9), and as in the proof of theorem 2.4 we can add and simplify the first and the last of the integrals, so as to obtain,

$$\Delta \geqslant \int_{t_0}^{t_1} \frac{d}{dt}(p(t) \cdot (x(t) - x^*(t)))dt$$

$$+ \int_{t_0}^{t_1} q\left(\frac{\partial g^*}{\partial x} \cdot (x - x^*) + \frac{\partial g^*}{\partial u} \cdot (u - u^*)\right)dt. \tag{138}$$

The latter integral is

$$\sum_{j=1}^{s} \int_{t_0}^{t_1} q_j[(\partial g_j^* / \partial x) \cdot (x - x^*) + (\partial g_j^* / \partial u) \cdot (u - u^*)]dt,$$

and in the same way as in the proof of theorem 4.5 we see that the quasi-concavity of $(x, u) \to q_j g_j(x, u, t)$ for $j = 1, \ldots, s'$, and the quasi-concavity of $x \to q_j \bar{g}_j(x, t)$ for $j = s' + 1, \ldots, s$ imply that the latter sum is $\geqslant 0$. Using these results and applying (B.1) we obtain:

$$\bar{\Delta} \geqslant p(t_1) \cdot (x(t_1) - x^*(t_1)) - p(t_0) \cdot (x(t_0) - x^*(t_0)) + S_0'(x^*(t_0)) \cdot (x^*(t_0) - x(t_0))$$
$$+ S_1'(x^*(t_1)) \cdot (x^*(t_1) - x(t_1)),$$

that is

$$\bar{\Delta} \geqslant [p(t_1) - S_1'(x^*(t_1))] \cdot (x(t_1) - x^*(t_1))$$
$$+ [-p(t_0) - S_0'(x^*(t_0))] \cdot (x(t_0) - x^*(t_0)). \tag{139}$$

Using (132) and (133) we further get:

$$\bar{\Delta} \geqslant \sum_{j=1}^{s} \beta_j \cdot \frac{\partial g_j^*(t_1)}{\partial x} \cdot (x(t_1) - x^*(t_1))$$

$$+ \sum_{k=1}^{r_1} \gamma_k \frac{\mathrm{d}R_k^1(x^*(t_1))}{\mathrm{d}x} \cdot (x(t_1) - x^*(t_1))$$

$$+ \sum_{j=1}^{s} \beta_j^0 \cdot \frac{\partial g_j^*(t_0)}{\partial x} \cdot (x(t_0) - x^*(t_0))$$

$$+ \sum_{k=1}^{r_0} \delta_k \frac{\mathrm{d}R_k^0(x^*(t_0))}{\mathrm{d}x} \cdot (x(t_0) - x^*(t_0)). \tag{140}$$

Consider the first of these sums. The first s' terms are 0 because of (12). Each of the remaining terms are $\geqslant 0$ by the quasi-concavity of $\beta_j g_j(x, u, t) = \beta_j \bar{g}_j(x, t)$ w.r.t. x ($j = s' + 1, \ldots, s$). In the same way we see that the third sum is $\geqslant 0$. The second and the fourth sums in (140) are $\geqslant 0$ by the same argument as was used in the proof of theorem 3.3. Consequently, $\bar{\Delta} \geqslant 0$, and therefore $(x^*(t), u^*(t))$ is optimal. ∎

Theorem 13 covers cases in which the adjoint function $p(t)$ does not jump. The analogous result when $p(t)$ jumps is given in the next theorem.

Theorem 14 (Sufficient conditions for problem (125)–(131). Jumps in $p(t)$) Suppose we replace problem (1)–(6) by problem (125)–(131). Then theorem 2 still holds provided (11) is replaced by (132), condition (133) is added, with the numbers $\beta_1^0, \ldots, \beta_s^0$ satisfying

$$\beta_j^0 \geqslant 0 (= 0 \text{ if } g_j(x^*(t_0), u^*(t_0), t_0) > 0), \qquad j = 1, \ldots, s$$
$$\beta^0 \cdot \partial g(x^*(t_0), u^*(t_0), t_0)/\partial u \cdot [u - u^*(t_0)] \leqslant 0 \text{ for each } u \in U, \tag{141}$$

and provided the concavity conditions (135) are imposed. ∎

Note 16 The properties described in note 1 also apply to the setting of theorem 14. Note that H need only be concave in the set $\Gamma(t)=\{(x, u): g(x, u, t)\geqslant0, u\in U\}$.

Note 17 Arrow-type sufficient conditions can be derived from theorem 13 and 14 by modifying the assumptions in the way specified in theorem 3.

Proof of theorem 14 The argument in the proof of theorem 13 must be adjusted because of the discontinuities of $p(t)$. The adjustments required are partly suggested by the proof of note 5.1. Defining $r(t)=p(t)\cdot(x(t)-x^*(t))$, and using the appropriate parts of the proof of theorem 13, we obtain

$$\bar{\Delta} \geqslant \int_{t_0}^{t_1} \frac{\mathrm{d}}{\mathrm{d}t} r(t)\mathrm{d}t + S'_0(x^*(t_0))\cdot(x^*(t_0)-x(t_0))$$

$$+ S'_1(x^*(t_1))\cdot(x^*(t_1)-x(t_1)). \tag{142}$$

Now

$$\int_{t_0}^{t_1} \frac{\mathrm{d}}{\mathrm{d}t} r(t)\mathrm{d}t = (r(\tau_1^-)-r(t_0))+(r(\tau_2^-)-r(\tau_1^+))+(r(\tau_3^-)-r(\tau_2^+))+\cdots$$

$$+ (r(\tau_{N-1}^-)-r(\tau_{N-2}^+))+(r(\tau_N^-)-r(\tau_{N-1}^+))+(r(t_1)-r(\tau_N^+))$$

$$= r(t_1)-r(t_0)+ \sum_{k=1}^{N} [r(\tau_k^-)-r(\tau_k^+)]. \tag{143}$$

For $k=1, \ldots, N$, using (21), we get

$$r(\tau_k^-)-r(\tau_k^+)=(p(\tau_k^-)-p(\tau_k^+))\cdot(x(\tau_k)-x^*(\tau_k))$$

$$= \sum_{j=1}^{s} \beta_{kj}^+ \frac{\partial g_j(x^*(\tau_k), u^*(\tau_k^+), \tau_k)}{\partial x}\cdot(x(\tau_k)-x^*(\tau_k))$$

$$+ \sum_{j=1}^{s} \beta_{kj}^- \frac{\partial g_j(x^*(\tau_k), u^*(\tau_k^-), \tau_k)}{\partial x}\cdot(x(\tau_k)-x^*(\tau_k)). \tag{144}$$

Now, $\beta_{kj}^+ g_j(x, u, t)$ is quasi-concave in (x, u). Using (24) we see that

$$\beta_{kj}^+ g_j(x(\tau_k), u(\tau_k^+), \tau_k)\geqslant\beta_{kj}^+ g_j(x^*(\tau_k), u^*(\tau_k^+), \tau_k)=0,$$

so, simplifying the notation,

$$\beta_{kj}^+ \frac{\partial g_j^*(\tau_k^+)}{\partial x} \cdot (x(\tau_k) - x^*(\tau_k)) + \beta_{kj}^+ \frac{\partial g_j^*(\tau_k^+)}{\partial u} \cdot (u(\tau_k^+) - u^*(\tau_k^+)) \geq 0.$$

Repeating this argument we find a similar inequality for each term in the second sum in (144), and therefore

$$r(\tau_k^-) - r(\tau_k^+) \geq - \sum_{j=1}^{s} \beta_{kj}^+ \frac{\partial g_j^*(\tau_k^+)}{\partial u} \cdot (u(\tau_k^+) - u^*(\tau_k^+))$$

$$- \sum_{j=1}^{s} \beta_{kj}^- \frac{\partial g_j^*(\tau_k^-)}{\partial u} \cdot (u(\tau_k^-) - u^*(\tau_k^-))$$

$$= - \beta_k^+ \cdot \frac{\partial g^*(\tau_k^+)}{\partial u} \cdot (u(\tau_k^+) - u^*(\tau_k^+))$$

$$- \beta_k^- \frac{\partial g^*(\tau_k^-)}{\partial u} (u(\tau_k^-) - u^*(\tau_k^+)).$$

Using (22) and (23) we conclude that $r(\tau_k^-) - r(\tau_k^+) \geq 0$, and therefore each term in the sum in (143) is ≥ 0, so

$$\int_{t_0}^{t_1} \frac{d}{dt} r(t) dt \geq r(t_1) - r(t_0)$$

$$= p(t_1) \cdot (x(t_1) - x^*(t_1)) - p(t_0) \cdot (x(t_0) - x^*(t_0)). \qquad (145)$$

Using (145) and (142) we obtain (139) in the proof of theorem 13, which as before implies (140). The second and fourth sum in (140) are ≥ 0 as before. The argument for the non-negativity of two remaining sums in (140) must be changed since the properties of β and β^0 are now different. Consider the first sum in (140). The function $\beta_j g_j(x, u, t)$ is quasi-concave in (x, u). Applying (20(i)), $\beta_j g_j(x(t_1), u(t_1), t_1) \geq \beta_j g_j(x^*(t_1), u^*(t_1), t_1) = \beta_j g_j^*(t_1) = 0$, so $\beta_j \partial g_j^*(t_1)/\partial x \cdot (x(t_1) - x^*(t_1)) + \beta_j \partial g_j^*(t_1)/\partial u \cdot (u(t_1) - u^*(t_1)) \geq 0$. Hence

$$\sum_{j=1}^{s} \beta_j \frac{\partial g_j^*(t_1)}{\partial x} \cdot (x(t_1) - x^*(t_1)) \geq - \sum_{j=1}^{s} \beta_j \frac{\partial g_j^*(t_1)}{\partial u} \cdot (u(t_1) - u^*(t_1)) \geq 0$$

by (20(ii)). In the same way we prove that the third sum in (140) is ≥ 0. Consequently, we have established the required inequality, $\bar{\Delta} \geq 0$. ∎

Necessary conditions

Our next result treats necessary conditions for problem (125)–(131). It generalizes theorem 5.

Theorem 15 (Necessary conditions for problem (125)–(131))[13]   Suppose we replace problem (1)–(6) by problem (125)–(131) assuming in addition that $\bar{g}$ is a C^2-function and that $U = R^r$. Then theorem 5 is still valid with the modifications that there exist numbers $\gamma_1, \ldots, \gamma_{r_1}$ and $\delta_1, \ldots, \delta_{r_0}$ such that the condition

$$(p_0, \gamma_1, \ldots, \gamma_{r_1}, \delta_1, \ldots, \delta_{r_0}, \mu(t_1) - \mu(t_0)) \neq (0, \ldots, 0) \tag{146}$$

holds instead of (63), and such that the conditions

$$p_i(t_1) = p_0 \frac{\partial S_1(x^*(t_1))}{\partial x_i} + \sum_{k=1}^{r_1} \gamma_k \frac{\partial R_k^1(x^*(t_1))}{\partial x_i}$$

where, for $k > r_1'$, $\gamma_k \geq 0$ $(=0$ if $R_k^1(x^*(t_1)) > 0)$ (147)

$$p_i(t_0) = -p_0 \frac{\partial S_0(x^*(t_0))}{\partial x_i} - \sum_{k=1}^{r_0} \delta_k \frac{\partial R_k^0(x^*(t_0))}{\partial x_i}$$

where, for $k > r_0'$, $\delta_k \geq 0$ $(=0$ if $R_k^0(x^*(t_0)) > 0)$ (148)

hold instead of (70). ∎

Note 18 (Non-triviality results)*   A consequence of theorem 15 is the following fact: If (74) holds, the condition

$$(p_0, p(t)) \neq (0, 0) \text{ for some } t \in (t_0, t_1) \tag{149}$$

is satisfied provided (150) and (151) below are satisfied:

    The matrix $\{\partial R_k^0(x^*(t_0))/\partial x_i\}$, $i = 1, \ldots, n$, $k = 1, \ldots, r_0'$
    has a rank equal to r_0',   (150(i))

    There exists a vector $c \in R^n$ such that $(\partial \bar{g}_j(x^*(t_0), t_0)/\partial x) \cdot c > 0$
    for all $j \in G^*(t_0)$, $(\partial R_k^0(x^*(t_0))/\partial x) \cdot c = 0$
    for all $k \leq r_0'$, and $(\partial R_k^0(x^*(t_0))/\partial x) \cdot c > 0$ for all $k \in R_0^* =$
    $\{k: k > r_0', R_k^0(x^*(t_0)) = 0\}$.   (150(ii))

(Condition (ii) holds vacuously if $G^*(t_0) \cup R_0^* = \phi$ and (i) holds vacuously if

[13] Neustadt (1976), Chapter VI, 3.

$r_0' = 0$.)

The matrix $\{\partial R_k^1(x^*(t_1))/\partial x_i\} i = 1, \ldots, n$,
$k = 1, \ldots, r_1'$ has a rank equal to r_1'. $\qquad$ (151(i))

There exists a vector $c \in R^n$ such that
$(\partial \bar{g}_j(x^*(t_1), t_1)/\partial x) \cdot c > 0$ for all $j \in G^*(t_1)$,
$(\partial R_k^1(x^*(t_1))/\partial x) \cdot c = 0$ for $k \leqslant r_1'$, and
$(\partial R_k^1(x^*(t_1))/\partial x) \cdot c > 0$ for all $k \in R_1^*$
$\quad = \{k : k > r_1', R_k^1(x^*(t_1)) = 0\}$. $\qquad$ (151(ii))

(Condition (ii) holds vacuously if $G^*(t_1) \cup R_1^* = \phi$ and (i) holds vacuously if $r_1' = 0$.)

Define the restrictions to be *first-order compatible at* $(x^*(t_0), t_0)$ if (150) holds. (If $x(t_0) = x^0$, $r_0' = r_0$, $R^0 \equiv x - x_0$, then (150) can be replaced by the definition of first-order compatibility for $g = \bar{g}$ stated in connection with (5.120) in note 5.) Define *first-order compatibility at* $(x^*(t_1), t_1)$ by (151).

If we let $g = \bar{g}$ and $U = U^+(t)$ in (5.125), then the properties (5.124), (5.126) and $(p_0, p(t_0^+)) \neq (0, 0)$ hold, provided the relevant conditions in theorem 5.6 and note 5.7 are satisfied. (Replace (5.95) by (74)).

Note that property (5.126) is also a consequence of (150), (151), and (92). (Here (150) can be dropped if $x(t_0) = x^0$.)

Finally, observe that Note 4 remains valid in the present setting.

Free final time problems

Consider problem (125)–(131), assuming that S_1 and R^1 depend explicitly on time and assuming that t_1 is free to vary in some interval $[T_1, T_2]$. Suppose we have a solution to that problem, with optimal time, t_1^*. Then that solution is a solution also to the corresponding problem on the fixed interval $[t_0, t_1^*]$. Thus theorem 15 is valid. As is the case in theorem 2.11, the optimal solution satisfies an extra condition which helps to determine t_1^*. The precise result is as follows:

Theorem 16 (Necessary conditions for (125)–(131). t_1 free)[14] Let $(x^*(t)$,

---

[14]This theorem and (153) can be obtained from the previous results by treating t as a new state variable. If we allow measurable controls the modifications needed are the same as those indicated in footnote 2. In addition, $H^*(t_1^{*-})$ is replaced by $\sup\{H(x^*(t_1^*), u, p(t_1^{*-}), t_1^*) : g_j(x^*(t_1^*), u, t_1^*) > 0, j = 1, \ldots, s'\}$. In (153) a corresponding replacement must be made.

$u^*(t)$, t_1^*) be an admissible triple which solves problem (125)–(131) with $U = R^r$ and $t_1 \in [T_1, T_2]$, $t_0 \leqslant T_1 < T_2$, T_1, T_2 fixed and with $S_1(x)$ replaced by $S_1(x, t)$ and $R^1(x)$ replaced by $R^1(x, t)$. In addition to the standard assumption (131), assume that g_i is C^2 for $i > s'$, and that $\partial f_i / \partial t$, $\partial g_i / \partial t$ ($i \leqslant s'$) exist and are continuous. Then ($x^*(t)$, $u^*(t)$) satisfies the conditions in theorem 15 for $t_1 = t_1^*$. In addition,

$$
H^*(t_1^{*-}) + \sum_{j=s'+1}^{s} (\mu_j(t_1^*) - \mu_j(t_1^{*-})) \cdot \frac{\partial \bar{g}_j(x^*(t_1^*), t_1^*)}{\partial t} + p_0 \frac{\partial S_1(x^*(t_1^*), t_1^*)}{\partial t}
$$

$$
+ \sum_{k=1}^{r_1} \gamma_k \frac{\partial R_k^1(x^*(t_1^*), t_1^*)}{\partial t} \quad \begin{cases} \leqslant 0 \text{ if} & t_1^* = T_1 \\ = 0 \text{ if } T_1 < t_1^* < T_2 \\ \geqslant 0 \text{ if} & t_1^* = T_2 \end{cases} \tag{152}
$$

where $H^*(t) = H(x^*(t), u^*(t), p(t), t)$. (When $t_1^* = T_1 = t_0$, the conditions must be modified in a manner similar to that in note 8.) ∎

Note 19 If we also allow t_0 to vary in a fixed interval $[T_1^0, T_2^0]$, $T_1^0 < T_2^0 \leqslant T_2$, and if we allow S_0 and R^0 to depend explicitly on t, the following condition should be added to the necessary conditions of theorem 16:

$$
-H^*(t_0^{*+}) + \sum_{j=s'+1}^{s} (\mu_j(t_0^{*+}) - \mu_j(t_0^*)) \cdot \frac{\partial \bar{g}_j(x^*(t_0^*), t_0^*)}{\partial t}
$$

$$
+ p_0 \frac{\partial S_0(x^*(t_0^*), t_0^*)}{\partial t} + \sum_{k=1}^{r_0} \delta_k \frac{\partial R_k^0(x^*(t_0^*), t_0^*)}{\partial t} \quad \begin{cases} \leqslant 0 \text{ if } t_0^* = T_1^0 \\ = 0 \text{ if } T_1^0 < t_0^* < T_2^0 \\ \geqslant 0 \text{ if } t_0^* = T_2^0 \end{cases} \tag{153}
$$

Notes 4 and 18 remain valid for $t_0 = t_0^*$, $t_1 = t_1^*$.

The next result generalizes theorem 7.[15]

Theorem 17 (Sufficient conditions for (125)–(131). t_1 free) Consider problem (125)–(131) with $S_1(x)$ replaced by $S_1(x, t)$ and with $R^1(x)$ replaced by $R^1(x, t)$. Suppose that t_1 is free to vary in $[T_1, T_2]$, $t_0 \leqslant T_1 < T_2$. Assume that for each $T \in [T_1, T_2]$ there exists an admissible pair ($x_T(t)$, $u_T(t)$), defined on $[t_0, T]$, with associated multipliers $p_T(t)$, $q_T(t)$, $\beta^T = (\beta_1^T, \ldots, \beta_s^T)$, $\gamma^T = (\gamma_1^T, \ldots, \gamma_{r_1}^T)$, $\beta^{0T} = (\beta_1^{0T}, \ldots, \beta_s^{0T})$ and $\delta^T = (\delta_1^T, \ldots, \delta_{r_0}^T)$ satisfying all the sufficient conditions of theorem 13. Assume, moreover, that $u_T(t)$

---

[15] Seierstad (1984a).

and $q_T(t)$ take values in fixed, bounded subsets of R^r and R^s, respectively, for all $T \in [T_1, T_2]$, that $T \to x_T(T)$ is Lipschitz continuous, that $T \to \beta^T$ and $T \to \gamma^T$ are piecewise continuous, and that $u_T(T)$ for all T belongs to the closure of the set $\{u: u \in U, g_j(x_T(T), u, T) > 0 \text{ for all } j \leqslant s'\}$. Finally, assume that

$$F(T) = H(x_T(T), u_T(T), p_T(T), T) + \beta^T \cdot \frac{\partial g(x_T(T), u_T(T), T)}{\partial t}$$

$$+ \frac{\partial S_1(x_T(T), T)}{\partial t} + \gamma^T \cdot \frac{\partial R^1(x(T), T)}{\partial t} \tag{154}$$

has the property that there exists a $T^* \in [T_1, T_2]$ such that

$$F(T) \geqslant 0 \text{ for } T < T^* \text{ if } T_1 < T^*$$
$$F(T) \leqslant 0 \text{ for } T > T^* \text{ if } T_2 > T^* \tag{155}$$

Then the pair $(x_{T^*}(t), u_{T^*}(t))$ defined on $[t_0, T^*]$, is optimal. ∎

When $T_1 = t_0$ it suffices that the properties of the theorem are satisfied for $T > T_1$.

Note 20 **(Arrow-type sufficient conditions for free time problems)** If $(x_T(t), u_T(t))$ and the associated multipliers satisfy the modifications of the assumptions of theorem 13 mentioned in note 17, then $(x_T(t), u_T(t))$ is still optimal.

Note 21* Except for theorems 7, 16, 17 and notes 4, 7 and 8, the results presented in this chapter hold, if the weaker regularity conditions on f_0, f and h of note 4.9, with $h = (g_1, \dots, g_{s'})$, are satisfied. However, (29), (66(ii)) and (76(ii)) must then be deleted. (Concerning note 5, see note 5.11.)

The same set of results hold even if we assume that the pure state constraints are of the type $g_j \geqslant 0$ in $[a_j, b_j]$ as in (5.131). Each $\mu_j(t)$ is then constant on $[t_0, a_j)$ and $(b_j, t_1]$. (The sets $G^*(t)$, $G^*(t_0^+)$, and $G^*(t_1^-)$ must be redefined in the same manner as in note 5.10).

For the Mangasarian- and Arrow-type sufficiency theorems, the assumptions on f_0, f and $h = (g_1, \dots, g_{s'})$ can even be those described for these cases in note 4.9.

If the set C of discontinuities of f_0, f and $h = (g_1, \dots, g_{s'})$ is empty, and no pure state constraint is of the type (5.131) with $[a_j, b_j] \neq [T_1^0, T_2]$, then even all free initial and free final time results hold.

Exercise 6.7.1 Solve the problem

$$\max \int_0^T (r_1 x_1 + r_2 x_2 - u - v)dt, \qquad \dot{x}_1 = u,$$

$$x_1(0) = 1, \qquad \dot{x}_2 = v, \qquad x_2(0) = 1 \tag{156}$$

$$x_1(T) + x_2(T) \geqslant 3, \qquad \bar{u} \leqslant u, \qquad \bar{v} \leqslant v, \qquad u + v \leqslant r_1 x_1 + r_2 x_2. \tag{157}$$

Here u and v are controls, T, r_1, r_2, $\bar{v}$, and $\bar{u}$ are constants and we assume that $r_1 > r_2 > 0$, $\bar{u} < 0$, $\bar{v} < 0$, and $\exp(r_1 T) > 3$.

(*Interpretation*: x_1 and x_2 are two types of assets bearing interests r_1 and r_2. $r_1 x_1 + r_2 x_2$ is income and $r_1 x_1 + r_2 x_2 - u - v$ is consumption.)

Exercise 6.7.2 Solve the problem

$$\max \left\{ \int_0^T (1-u)x dt - (x(0))^2 + bx(T) \right\}, \quad \dot{x} = ux, \qquad x(0) \text{ free}, \qquad x(T) \text{ free}$$

$$u \in [0, 1], \qquad b \text{ a fixed number}, \qquad 0 < b < 1, \qquad T \text{ fixed}, \qquad T > (1-b).$$

8. Existence theorems

In this final section we present some existence theorems of a rather general kind. The first result is the Filippov–Cesari existence theorem adapted to a generalization of problem (125)–(131).

Theorem 18 (**Filippov–Cesari, existence**)[16] Consider problem (125)–(130), assuming that S_0, S_1, R^0 and R^1 all depend explicitly on t. Assume that t_0 and t_1 are free to vary in $[T_1^0, T_2^0]$ and $[T_1, T_2]$, respectively, allowing $T_1^0 = T_2^0$ and/or $T_1 = T_2$. Assume further that the control region U is closed, but not necessarily convex. Assume that

f_0, f, g, S_0, S_1, R^0, and R^1 are continuous in all arguments at all points

$$(x, u, t) \in R^n \times U \times [T_1^0, T_2]. \tag{158}$$

---

[16]Cesari (1983), Chapter 9.

Suppose that there exists an admissible quadruple $(x(t), u(t), t_0, t_1)$ and that

$$N(x, U, t) = \{(f_0(x, u, t) + \gamma, f(x, u, t)): \gamma \leqslant 0, g_j(x, u, t) \geqslant 0, j \leqslant s', u \in U\}$$
is convex for all $(x, t) \in R^n \times [T_1^0, T_2]$, \hfill (159)

there exists a number b such that $\|x(t)\| < b$ for all admissible
pairs $(x(t), u(t))$ and all t, \hfill (160)

there exists a ball $B(0, b_1)$ in R^r which contains the set
$U(x, t) = \{u: g(x, u, t) \geqslant 0, u \in U\}$ for all $x \in B(0, b)$. \hfill (161)

Then there exists an optimal quadruple $(t_0^*, t_1^*, x^*(t), u^*(t))$, (with $u^*(t)$ measurable). ∎

Note 22 The convexity condition (159) can be dropped in the case where $r_1' = 0$ and U is convex if the following two conditions are imposed:

$f_i(x, u, t) = h_i(x, u, t) + a_i(t)x_i$, $i = 1, \dots, n$ with h_i and a_i as continuous
functions. \hfill (162)

$R_k^1(x, t)$, $S_1(x, t)$, $f_0(x, u, t)$, $g_j(x, u, t)$, $h_i(x, u, t)$ are non-decreasing in
x, and f_0, g_j and h_i are concave in $u \in U$. \hfill (163)

Assumption (161) requires "the interesting set" of values of $u(t)$ to be bounded. That boundedness assumption is dropped in the next theorem. Thus for the first time in this book we introduce the possibility of unbounded control functions in existence theorems.

Theorem 19 (Existence. Unbounded control region U)[17]* Theorem 18 is still valid if (160) and (161) are replaced by the following conditions: There exist a constant $c \in [0, \infty]$ and for each $p \neq 0$, integrable functions $\varphi_p(t)$ and $\psi_p(t)$ such that for all $(x, u, t) \in \Gamma = \{(x, u, t): g(x, u, t) \geqslant 0, u \in U, t \in [T_1^0, T_2], \|x\| \leqslant c\}$,

$$f_0(x, u, t) + p \cdot f(x, u, t) \leqslant \varphi_p(t) + \psi_p(t)\|x\|,$$ \hfill (164)

and, either (i) $c < \infty$ and $\|x(t)\| \leqslant c$ for all t and for all admissible solution $x(t)$, or (ii) $c = \infty$ and

there exists a constant M such that $\|x(t_0)\| \leqslant M$ for all admissible

---

[17]Olech (1969), Rockafellar (1973, 1975). Still, an admissible $x(t)$ is absolutely continuous.

$x(t)$ and either

$f_0(x, u, t) \leqslant \gamma(t)$ for all $(x, u, t) \in \Gamma$, $\gamma(t)$ some integrable

    function, or   (165a)

$\psi_p(t) \leqslant \sigma(t) + \rho(t) \|p\|$, where $\sigma(t)$ and $\rho(t)$ are integrable functions

    independent of p.   (165b)

$N(x, U, t)$ has a closed graph for each t, as a function of

    $x \in \Gamma_t = \{x : (x, u, t) \in \Gamma \text{ for some } u \in U\}$. ∎   (166)

(A *closed graph* means: If $x_n \in \Gamma_t$, $v_n \in N(x_n, U, t)$, $x_n \to x$, $v_n \to v$, then $x \in \Gamma_t$ and $v \in N(x, U, t)$.)

In a number of problems condition (165) is difficult, or impossible, to verify. In the following result, which applies to a less general problem than that of theorem 19, condition (165) is replaced by a different "growth" condition.

Theorem 20 (Existence. Unbounded control region U)[18]* Consider the problem studied in theorem 18, assuming that $r'_1 = 0$, i.e. there are no equality terminal constraints, and that U is closed (but not necessarily convex). In addition to (158) assume that

f_0, f, g, S_1 and R^1 are non-decreasing in x for each

    $(u, t) \in U \times [T_1^0, T_2]$ (167)

Suppose that there exists an admissible quadruple and that the set

$$N(x, U, t) = \{(f_0(x, u, t) + \gamma_0, \ldots, f_n(x, u, t) + \gamma_n) : g(x, u, t) \geqslant 0,$$
$$u \in U, \gamma_i \leqslant 0, i = 0, \ldots, n\} \qquad (168)$$

is convex for $x \in \Gamma_t$ and has a closed graph for each t as a function of $x \in \Gamma_t$, where $\Gamma_t = \{x : g(x, u, t) \geqslant 0 \text{ for some } u \in U\}$. Suppose further that for some fixed vector $q' \in R^{n+1}$, $q' \geqslant 0$, there exists for each $q \geqslant q'$ integrable functions $\varphi_q(t)$ and $\psi_q(t)$ such that

$$\sum_{i=0}^{n} q_i f_i(x, u, t) \leqslant \varphi_q(t) + \psi_q(t) \cdot \max(0, x_1, \ldots, x_n) \text{ for all}$$

    $(x, u, t) \in \Gamma = \{(x, u, t) : g(x, u, t) \geqslant 0, u \in U, t \in [T_1^0, T_2]\}$.   (169)

Suppose also that there exist constants M and M' such that $x_i(t_0) \leqslant M$,

---

[18] Seierstad (1985a). The monotonicity of f can be that of note 3.24, $a_i(t) \leqslant 0$.

$x_i(t_1) \geqslant M'$ for all $i = 1, \ldots, n$ and for all admissible $x(t)$. Suppose, finally, that either

$$\sum_{i=0}^{n} q_i' f_i(x, u, t) \leqslant \gamma(t) \text{ for all } (x, u, t) \in \Gamma \text{ for some integrable}$$

function $\gamma(t)$ or $\qquad (170a)$

$\psi_q(t) \leqslant \sigma(t) + \rho(t) \cdot ||q||$ where $\sigma(t)$ and $\rho(t)$ are integrable

functions, independent of q. $\qquad (170b)$

Then there exists an optimal quadruple $(t_0^*, t_1^*, x^*(t), u^*(t))$ (with $u^*(t)$ measurable).

Note 23 The set $N(x, U, t)$ of (168) *is* convex provided U is convex, g is quasi-concave in $u \in U$ for all (x, t), and f_0 and f are concave in $u \in U$ for all (x, t), $x \in \Gamma_t$.

Note 24 From (169), we obtain the following fact: $N(x, U, t)$ has a closed graph as a function of x if $a_n = (f_0(x_n, u_n, t), \ldots, f_n(x_n, u_n, t))$, $(x_n, u_n, t) \in \Gamma$, $a_n \rightarrow a$ and $x_n \rightarrow x$ as $n \rightarrow \infty$, imply $x \in \Gamma_t$ and $a \in N(x, U, t)$.

Note 25* The theorems 18–20 are valid even with the following modifications of their assumptions:

g is defined on R^{n+r+1} and is continuous in (x, u) and measurable in t, f_0 are f are defined on a subset Γ'' of Γ. On Γ'', f_0 and f are continuous in (x, u) and measurable in t. The set Γ'' has the property that $\Gamma''(t) = \{(x, u): (x, u, t) \in \Gamma''\}$ is contained in the closure $\overline{\Gamma^0(t)}$ of $\Gamma^0(t) = \{(x, u): g(x, u, t) > 0, u \in U\}$. $\qquad (171)$

Require for admissibility of a pair $(x(t), u(t))$ that $(x(t), u(t), t) \in \Gamma''$ a.e. Replace the conditions on U and $N(x, U, t)$ by:

The sets $\Gamma_n'(t) = \{x, u): g(x, u, t) \geqslant 0, u \in U, ||\tilde{f}(x, u, t)|| \leqslant n\}$ are closed in R^{n+r} and contained in $\Gamma''(t)$. The sets $\Gamma_n(t) = \{x: (x, u) \in \Gamma_n'(t) \text{ for some } u\}$ are closed in R^n. $N(x, U, t)$ is convex for each $x \in \Gamma_n(t)$ and has a closed graph as a function of x on this set, $n = 1, 2, \ldots$.[19] $\qquad (172)$

---

[19]Rockafellar (1973) has given conditions related to the results in this note. See also Seierstad (1985a).

Note also the following fact. If for all n and all bounded sets B in R^n, the set $\Gamma'_n(t) \cap (B \times R^r)$ is bounded, then the closed graph property of $N(x, U, t)$ in this note automatically holds, as well as the closedness of the sets $\Gamma_n(t)$.

Example 5 **(Consumption versus investment)** We consider a variant of the optimal growth model studied in several previous examples. (See e.g. the beginning of Chapter 2, section 7 and example 2.7.)

$$\max \int_0^T U(f(K(t)) - u(t))e^{-\rho t} dt,$$

$$\dot{K}(t) = u(t), \qquad K(0) = K_0 > 0, \qquad K(T) \geqslant K_T. \tag{173}$$

$$f(K(t)) - u(t) \geqslant 0, \qquad u(t) \in (-\infty, \infty), \qquad K(t) \geqslant 0. \tag{174}$$

We assume that U and f are defined on $[0, \infty)$, that U' and U'' exist on $(0, \infty)$, with $U' > 0$, $U'' < 0$, and $\lim_{C \to \infty} U(C) = k < \infty$ for some number k, and that $f(0) = 0$, f' and f'' exist on $[0, \infty)$ and $f' > 0$, $f'' < 0$ here. Finally, we assume that $K_T \leqslant \tilde{K}(T)$, where $\tilde{K}(t)$ is the solution of $\dot{K} = f(K)$, $K(0) = K_0$ (which does exist according to the argument in example 2.10). Note that U and f may be extended, as continuous functions to $(-\infty, \infty)$.

We shall prove that an optimal solution exists by using theorem 20 and note 23. Let $f_0 = U(f(K) - u)e^{-\rho t}$, $f_1 = u$, $g_1 = f(K) - u$, $g_2 = K$. These functions are all non-decreasing in K. The functions g_1 and g_2 are concave in u for all K, and f_0 and f_1 are concave in u for all $u \in (-\infty, f(K)]$. By note 23, the convexity condition in (168) holds. It is also easy to test that the closed graph property of N is satisfied by using note 24.

In order to obtain (169), let $q' = (1, 0)$ and let (q_0, q_1) be an arbitrary vector $\geqslant q'$. By concavity of f, $f(K) \leqslant f(1) + f'(1)(K - 1)$, hence by the properties of U and (174), we obtain

$$q_0 f_0 + q_1 f_1 \leqslant q_0 U(f(K) - u) + q_1 u \leqslant q_0 k + q_1 u$$

$$\leqslant q_0 k + q_1 f(K) \leqslant q_0 k + q_1 [f(1) + f'(1)(K - 1)]$$
$$= q_0 k + q_1 (f(1) - f'(1)) + q_1 f'(1)K.$$

We see that the inequality in (169) is satisfied for all $f(K) - u \geqslant 0$, $K \geqslant 0$, if we put $\varphi_q(t) = q_0 k + q_1 (f(1) - f'(1))$, $\psi_q(t) = q_1 f'(1)$, and we observe that (170b) is satisfied. By theorem 20 an optimal pair $(K^*(t), u^*(t))$ exists.

Example 6 (**Consumption versus investment again**) Consider a problem similar to that of example 5:

$$\max \int_0^T (-1/C(t))dt, \qquad \dot{K}(t) = K(t) - C(t),$$

$$K(0) = K_0 > 0, \qquad K(T) \geqslant K_T \tag{175}$$

with $C(t) \in (0, \infty)$ as the control variable. Assume that $K_T \leqslant K_0 e^T/2$, (which ensures that an admissible solution exists).

We shall prove the existence of an optimal solution by using theorem 20. However, since the control region $(0, \infty)$ is not closed, we must also appeal to note 25.

The functions $f_0(K, C, t) = -1/C, f(K, C, t) = K - C$ and $R_1 = K - K_T$ are defined on $\Gamma = R \times (0, \infty) \times [0, T]$, and (158) and (167) are trivially satisfied. The convexity of N is ensured by note 23. The set $\Gamma_n'(t)$ in note 25 is here $\Gamma_n'(t) = \{(K, C): (K, C, t) \in \Gamma, [(1/C)^2 + (K - C)^2]^{1/2} \leqslant n\}$, and the set is easily seen to be closed for each n. If $(K, C) \in \Gamma_n'(t)$, then $|-1/C| \leqslant n$, so $C \geqslant 1/n$, and $|K - C| \leqslant n$, so $C \leqslant K + n$. Hence, for each positive number a, $\Gamma_n'(t) \cap B(0, a) \times R)$ is bounded, $(C \in [1/n, a + n])$. From note 25 we conclude that N has the closed graph property. Property (169) is satisfied with $(q_0, q_1) \geqslant (1, 0)$, $\varphi_q(t) \equiv 0$ and $\psi_q(t) \equiv q_1$. The rest of the requirements in theorem 20 can easily be tested. Thus an optimal solution exists.

If $f_0 = -1/C$ is replaced by $f_0 = \ln C$, again an optimal solution exists. In particular, (169) is satisfied for $(q_0, q_1) \geqslant (0, 1)$: The function $F(C) = q_0 \ln C - q_1 C$ has a maximum value $q_0 \ln q - q_0$ for $C = q = q_0/q_1$ so $q_0 f_0 + q_1 f = q_0 \ln C - q_1 C + q_1 K \leqslant q_0 \ln q - q_0 + q_1 \max (0, K)$. The other conditions are shown to be satisfied in a way similar to that of the case $f_0 = -1/C$.

Infinite horizon*

Finally, we shall state two theorems that extend the two last theorems to the case where t_1 possibly has the value ∞. Thus $T_2 = \infty$, and the optimization problem is now to maximize the criterion

$$\lim_{t \to t_1} \int_{t_0}^t f_0(x(t), u(t), t)dt, \qquad t_1 \in [T_1, \infty] \tag{176}$$

where possibly even $T_1 = \infty$ and the control system is (84)–(86), U not necessarily convex. The initial time point t_0 is now subject to choice in a given interval $[T_1^0, T_2^0]$. The terminal condition is (3) if $t_1 < \infty$ and (87) if $t_1 = \infty$. For this problem, we have the two following theorems:

Theorem 21 (**Existence, unbounded** U, **infinite horizon**)[20]   The following assumptions are made: The functions f_0, f, and g are continuous, U is closed, and (159), (166) hold. There exists an admissible quadruple. Condition (164) holds for $c = \infty$, $\phi_p(t)$ and $\psi_p(t)$ being locally integrable (i.e. integrable on each finite interval).

Finally, there exist integrable functions $v^i(t)$ defined on $[T_1^0, \infty)$ such that for all admissible quadruples $(t_0, t_1, x(t), u(t))$, for all $t \in [t_0, t_1)$,

$$f_i(x(t), u(t), t) \leqslant v^i(t), \qquad i = 0 \text{ and } i = l+1, \ldots, m \tag{177a}$$

$$|f_i(x(t), u(t), t)| \leqslant v^i(t), \qquad i = 1, \ldots, l \tag{177b}$$

Then there exists an optimal quadruple.   ■

Note that in theorem 21 either all admissible quadruples have a criterion value equal to $-\infty$, or some admissible quadruple has a finite criterion value, in which case the optimal quadruple also has a finite criterion value.

Theorem 22 (**Existence, unbounded** U, **infinite horizon**)   The following assumptions are made: There are no equality constraints, $(l=0)$. The functions f_0, f, and g are continuous and non-decreasing in x, and U is closed. There exists an admissible quadruple $(t_0, t_1, x(t), u(t))$. The set $N(x, U, t)$ of (168) has the same properties as in theorem 20, and (169) is satisfied for vectors $q \geqslant q' \geqslant 0$, q' fixed, $\phi_q(t)$ and $\psi_q(t)$ being locally integrable. For each $q \geqslant 0$, $q_i = 0$ for $i > m$, there exists an integrable function $v_q(t)$ defined on $[T_1^0, \infty)$ and for each $s > T_1^0$, a constant θ_s such that for all admissible quadruples $(t_0, t_1, x(t), u(t))$:

$$\sum_{i=0}^{m} (1 + q_i) f_i(x(t), u(t), t) \leqslant v_q(t), \qquad t \in [t_0, t_1) \tag{178a}$$

$$-x^i(t) \geqslant \theta_s \text{ for } t \in [t_0, t_1), \quad t \leqslant s, \ i > m \tag{178b}$$

Then there exists an optimal quadruple.   ■

Even in this theorem we cannot exclude the possibility that all admissible

---

[20]For this and the next theorem, see Baum (1976) and Seierstad (1985a).

quadruples have criterion values equal to $-\infty$. (In *both* theorems the limit in (176) exists in $[-\infty, \infty)$, for any admissible quadruple).

Note 26 Theorems 21 and 22 also hold in the case where f_0 and f are not defined and continuous everywhere and U is non-closed, provided the conditions of note 25, (171), (172), are satisfied.

Moreover, in theorem 22 the condition (178a) can be replaced by the following condition: For each $q \geqslant 0$, $q_i = 0$ for $i > m$, there exists an integrable function $v_q(t)$, continuous functions $\chi_q^i(t)$ and functions $\theta_q^i(s)$, $i = 0, \ldots, n$ defined on $[T_1^0, \infty)$ such that for all admissible quadruples $(t_0, t_1, x(t), u(t))$, for all $t \in [t_0, t_1)$,

$$\sum_{i=0}^m (1 + q_i) f_i(x(t), u(t), t) + \sum_{i=0}^n \chi_q^i(t) f_i(x(t), u(t), t) \leqslant v_q(t). \tag{179}$$

Moreover, for $i = 0, \ldots, n$,

$$-\int_t^T \chi_q^i(t) f_i(x(t), u(t), t) dt \leqslant \theta_q^i(s),$$

$$\text{for } T > t \geqslant s, T, t \in [t_0, t_1) \tag{180}$$

where $\theta_q^i(s)$ has the property that $\theta_q^i(s) \to 0$ when $s \to \infty$.

Note 27 Often an infinite planning horizon is a mathematical idealization of a finite but "very distant" planning horizon. Then the following observations are of interest:

Let t_0 be fixed. Assume first that $m = 0$ (no terminal restrictions, see (87)) and let our control system satisfy the conditions of theorem 21. Let $T_n \to \infty$, and assume that $x_n(t)$ is an optimal solution in the problem where T_n is a fixed finite horizon. (Still $m = 0$, and the existence of such solutions follows from theorem 19.) Then there exists an optimal solution of the infinite horizon problem $x^*(t)$ and a subsequence of $x_n(t)$ converging pointwise (and even uniformly on finite intervals) to $x^*(t)$. Also, $V(T_n) \to V(\infty)$.

If $m > 0$, while $l = 0$, this property may fail if we let the x_i^1's defining the terminal conditions in the infinite horizon problem also define the finite horizon (T_n) terminal conditions. (There may even be no admissible solutions in the latter problems, even if there exist admissible solutions in the infinite horizon problem, as we assume.) But take any infinite horizon optimal solution $\bar{x}(t)$, and define the finite horizon terminal conditions corresponding to T_n by $x_i(T_n) \geqslant \min(x_i^1, \bar{x}_i(T_n))$, $i = 1, \ldots, m$. Then the above result also hold for $m > 0$, $l = 0$. Note that $\lim_n \bar{x}_i(T_n) \geqslant x_i^1$, $i = 1, \ldots, m$.

The case $l > 0$ is more problematic and would require use of conditions as those in theorem 3.16 and normality.

By adding assumptions of uniqueness on $x^*(t)$ and $u^*(t)$ similar to those made in Chapter 3, section 6, C, we could obtain results parallel to those stated there. We do not pursue this matter further.

Note 28 In theorems 19–22 the optimal control known to exist is not necessarily bounded on bounded intervals. The necessary conditions (including those for an infinite horizon, theorems 3.12, 3.16 and 9) generalize in the standard way to measurable control functions with this boundedness property. However, additional assumptions must be made in theorems 3.12 and 9, when the boundedness property fails. For simplicity, we assume no mixed constraints. Then sufficient for the two theorems to hold in the case where $u^*(t)$ is merely known to be measurable, is the existence of a positive, continuous function $n(t)$ and a non-negative locally integrable function $m(t)$ such that $|f_i(x, u^*(t), t)| \leqslant m(t)$ and $||\partial f_i^*(x, u^*(t), t)/\partial x|| \leqslant m(t)$ for all (x, t) such that $||x - x^*(t)|| \leqslant n(t)$.

Sometimes it is possible to show that these properties are satisfied by using the following information concerning the optimal solutions provided by theorems 19–22: f_i^* is integrable for $i = 1, \ldots, m$, and locally integrable for all $i \geqslant 1$. f_0^* is integrable if the criterion has a finite value for some admissible quadruple (automatically satisfied in theorems 19, 20). The same assertions also hold for the notes in this section, except note 26 where the word "integrable" has to be changed to "locally integrable and $\lim_{t \to t_1} f_i^*(t)$ exists (finite)" in these assertions.

Example 7   (Consumption versus investment once more)

$$\max \int_0^\infty -e^{-C} dt, \qquad \dot{K} = aK - C,$$

$$K(0) = K_0 > 0, \qquad K \geqslant 0, \qquad K(T) \text{ free}, \qquad C \in (-\infty, \infty) \qquad (181)$$

(C the control). This model is related to example 2.6. It turns out that theorem 22 is applicable to this problem. We shall not give a detailed verification of all the conditions of that theorem. Notes 23, 24 are used to obtain the convexity and closed graph property of $N(x, U, t)$ in the present case. Property (169) is satisfied for $\psi_q(t) = aq_1$, $\phi_q(t) = \ln(q_1/q_0)$, $q = (q_0, q_1) \geqslant (1, 0) = q'$. Finally, property (178) is satisfied for $v_q(t) \equiv 0$, $\theta_s = 0$.

Example 8 (Consumption versus investment, the last time)

$$\max \int_0^\infty C^a e^{-bt} dt, \qquad \dot{K} = cK - C,$$

$$K(0) = K_0 > 0, \qquad K \geq 0, \qquad K(T) \text{ free}, \qquad C \in [0, \infty) \qquad (182)$$

(*C* the control), $ac < b$, $a \in (0, 1)$, $c > 0$.

Again we shall apply theorem 22 with note 26 and we leave most of the verifications of the required conditions to the reader. The properties of N are obtained from notes 23, 24 again. Property (169) is obtained by choosing $q' = (1, 1)$, and noting that if $q = (q_0, q_1) \geq q'$, then $q_0 C^a e^{-bt} + cKq_1 - Cq_1 \leq q_0 C^a + cKq_1 - Cq_1 \leq q_0 r^a + cKq_1$, where $r = (aq_0/q_1)^{1/(1-a)}$ is a value of C that maximizes $q_0 C^a - q_1 C$.

Let us show that (179) and (180) are satisfied in this problem. In the present case, $m = 0$ and $q = (q_0, 0) \geq 0$, and, evidently, if (179) and (180) hold for $q = (0, 0)$, we can obtain that they hold also for $q = (q_0, 0)$, $q_0 > 0$, by multiplying on both sides of the inequalities by $(1 + q_0)$.

Fix a number d such that $c < d$, $ad < b$. Let $\chi_0^1(t) = e^{-dt}$ and $\chi_0^0(t) \equiv 0$, $(q = 0)$. Note that for any admissible $K(t)$, $K(t) \leq K_0 e^{ct}$. Note also that $C^a e^{-bt} - e^{-dt} C + cKe^{-dt} \leq a^{a/(1-a)} \cdot e^{[a(b-d)/(a-1)-b]t} + cK_0 e^{(c-d)t}$, (the left-hand side has a maximum point as a function of C equal to $a^{1/(1-a)} e^{(b-d)t/(a-1)}$). Since the square bracket on the right-hand side is negative, we are able to obtain (179). To show (180), by partial integration

$$-\int_t^T e^{-d\tau}(cK - C)d\tau = \Big|_t^T -K(\tau)e^{-d\tau}$$

$$-d \int_t^T K(\tau)e^{-d\tau} d\tau \leq K(t)e^{-dt} \leq K_0 e^{(c-d)t},$$

which shows that (180) can be obtained.

Exercise 6.8.1 In (175) of example 6 let $C(t) \in (-\infty, \infty)$, and replace the integrand $-1/C$ by $F(C)$, where $F(C) = 2(1+C)^{1/2} - 2$ for $C_1 \geq 0$, $F(C) = C - C^2$ for $C < 0$. Prove the existence of an optimal control. (Theorem 20 with $q' = (1, 1)$.)

Exercise 6.8.2 (i) Assume in example 5 that U is defined only on $(0, \infty)$ with $U' > 0$, $U'' < 0$ and assume that $U'(C) \to \infty$ when $C \to 0^+$. Using note 25

show that an optimal solution exists. (Cf. example 6 with $f_0 = \ln C$.)
(ii) Assume that U is defined in $(-\infty, \infty)$ with $U' > 0$, $U'' < 0$, $U'(C) \to \infty$ when $C \to -\infty$, $U'(C) \to 0$ when $C \to \infty$. Use note 24 to prove the existence of an optimal solution, when $f(K) - u \geq 0$ is dropped.

Exercise 6.8.3   Show the existence of an optimal control in the problem:

$$\max \int_0^\infty (\ln C) e^{-bt} dt, \qquad \dot{K} = cK - C, \qquad K(0) = K_0 > 0,$$

$$\lim_{t \to \infty} K(t) \geq 0, \qquad C > 0, \quad K \geq 0,$$

(C the control), $b > 0$, $c > 0$.
(Hint: Rewrite the model by defining a new state variable $x(t) = K(t) e^{-ct}$, and a new control $u = C e^{-ct}$ and apply theorem 22 to the rewritten problem. The first paragraph of note 26 is also needed.)

Exercise 6.8.4   Show the existence of an optimal solution in the problem:

$$\max \int_0^\infty [U(C) - U(f(\check{K}))] dt, \qquad \dot{K} = f(K) - C,$$

$$K(0) = K_0 > 0, \qquad K \geq 0, \qquad f(K) - C \geq 0, \qquad C \geq 0,$$

where U and f are C^1 on $[0, \infty)$, $U' > 0$, $U'' < 0$, $f'' < 0$, $f'(K_0) > 0$, $f'(\check{K}) = 0$, where $\check{K}$ is a number $> K_0$.
(Hint: Apply theorem 22 and note 24.)

DIFFERENTIAL EQUATIONS

A.1. Introduction

The dynamic models studied in this book are characterized by ordinary differential equations. In some of the simpler models developed in examples and exercises, the differential equations are of an elementary type. They can often be solved by methods explained in any elementary text on ordinary differential equations. (See e.g. Sydsæter (1981), Chapter 1 and 6 or Gandolfo (1980), Part II.)

However, in a number of other models in this book the differential equations are of a more complicated type. Furthermore, the right-hand sides are often discontinuous w.r.t. the time variable. This appendix contains a number of results covering such systems of differential equations. (For references see section A.8.)

Consider the system

$$\dot{x}_1 = h_1(x_1, x_2, \ldots, x_n, t)$$
$$\cdots\cdots\cdots\cdots\cdots \qquad x_1(t_0) = x_1^0, \ldots, x_n(t_0) = x_n^0 \qquad \text{(A.1)}$$
$$\dot{x}_n = h_n(x_1, x_2, \ldots, x_n, t) ,$$

or, in vector notation, $\dot{x} = h(x, t)$, $x(t_0) = x^0$. Within the context of control theory, the discontinuity of $h(x, t)$ w.r.t. t arises usually from discontinuities in the control functions. Typically, $h(x, u, t)$ equals $f(x, u(t), t)$, $H'_x(x, u(t), p(t), t)$ or similar expressions.

Hence the results in which we are interested concern the vector differential equation

$$\dot{x} = h(x, t), \qquad x(t_0) = x^0 \qquad \text{(A.2)}$$

where we allow $h(x, t)$ to have a finite or countable number of discontinuity points θ_k, $k = 1, 2, \ldots$ as a function of t. More precisely we shall assume that $h(x, t)$ satisfies the following condition:

> $h(x, t)$ is defined in an open set A in R^{n+1} and is continuous
> at all points $(x, t) \in A$, $t \notin \{\theta_k: k = 1, 2, \ldots,\}$. Furthermore, the limits
> of $h(x, t)$ as $(x, t) \to (\bar{x}, \bar{t}^+)$ and as $(x, t) \to (\bar{x}, \bar{t}^-)$ both exist for
> all $(\bar{x}, \bar{t}) \in A$.   (A.3)

We shall need to refer to the case where

(A.3) holds with $A = R^{n+1}$.   (A.4)

As mentioned above, $h(x, t)$ will frequently be of the type $g(x, u(t), t)$ and usually we can assume that

> $u(t) \in U \subset R^r$ and u is piecewise continuous in $(-\infty, \infty)$.   (A.5(i))

> $g(x, u, t)$ is continuous at all points (x, u, t),
> $(x, t) \in A \subset R^{n+1}$, $u \in U'$, where A is an open set in R^{n+1}
> and U' is an open set containing U.   (A.5(ii))

Recall that a function is *piecewise continuous* if it has at most a finite number of discontinuities on each finite interval, with one-sided limits at each point of discontinuity. We assume throughout the book that the value of a control function $u(t)$ at a point of discontinuity t' is equal to the left-hand limit of $u(t)$ at t'. At the ends of an interval of definition, the control functions are assumed to be continuous. Note that (A.5) implies that $g(x, u(t), t)$ satisfies (A.3).

In view of condition (A.3), it is convenient to assume that $u(t)$ is defined on an open interval. If $u(t)$ is defined on a closed interval $[a, b]$, then by putting $u(t) = u(a^+)$ for all $t < a$ and $u(t) = u(b^-)$ for $t > b$, the domain of u can be extended to an interval $(c, d) \supset [a, b]$.

We have to consider the solution concept for (A.2) quite carefully. Since $h(x(t), t)$ might be discontinuous, $\dot{x}(t)$ does not necessarily exist everywhere. If $F(t) = G(t)$ for all t except for a countable numbers of t's in an interval I, we say that $F(t) = G(t)$ *virtually everywhere* in I, and we write "$F(t) = G(t)$ v.e. in I". We also say $F(t) = G(t)$ *for virtually every t in I* (for v.e. t in I).

Let $x = x(t)$ be a function defined on an interval I and with graph in A. Let $(x^0, t_0) \in A$. Then $x(t)$ is called a *solution of* (A.2) *on I through* (x^0, t_0) provided $x(t)$ is continuous, $x(t_0) = x^0$ and $x(t)$ has a derivative v.e. satisfying

$$\dot{x}(t) = h(x(t), t) \qquad \text{v.e. in } I. \tag{A.6}$$

Note that if $h(x, t) = g(x, u(t), t)$ and (A.5) is satisfied, then $\dot{x}(t)$ exists at all continuity points of $u(t)$. At a point of discontinuity of $u(t)$, $\dot{x}(t)$ might not exist. Thus the solution curve traced by $x(t)$ *in* R^n is continuous, but might have kinks.

A.2. Existence and uniqueness of solutions

In the present case the standard local existence theorem is as follows:

Theorem A.1 (**Local existence**) Consider the differential equation $\dot{x} = h(x, t)$, with $h(x, t)$ satisfying (A.4), and let (x^0, t_0) be a given point in R^{n+1}. Then there exists an interval (α, β), $t_0 \in (\alpha, \beta)$, and a solution $x(t)$ defined on (α, β) such that $x(t_0) = x^0$. ∎

In many of the models discussed in this book the function $h(x, t)$ is not defined everywhere. Theorem A.1 then has to be modified somewhat. We include in the formulation of the modified theorem an estimate of the length of the interval of existence. Note that according to standard terminology, $\bar{B}(x^0, r)$ is the closed ball centred at x^0 with radius r, i.e. $\bar{B}(x^0, r) = \{x \in R^n: ||x - x^0|| \leqslant r\}$. (The corresponding open ball is $B(x^0, r) = \{x \in R^n: ||x - x^0|| < r\}$.)

Theorem A.2 (**Local existence**) Consider the differential equation $\dot{x} = h(x, t)$ with $h(x, t)$ satisfying (A.3), and let (x^0, t_0) be a given point in A. Choose positive numbers r and s such that

$$A' = \bar{B}(x^0, r) \times [t_0 - s, t_0 + s] \subset A. \tag{A.7}$$

If M is a number such that

$$||h(x, t)|| \leqslant M \qquad \text{for all } (x, t) \in A', \tag{A.8}$$

then there exists a solution $x(t)$ on the interval $[t_0 - p, t_0 + p]$ with $x(t_0) = x^0$ and $p = \min(s, r/M)$. ∎

Note A.1 The set A' in (A.7) is compact, so by assumption (A.3), there exists a number M satisfying (A.8).

Notice that in order to obtain a solution on a large interval, we must choose s and r as large as possible subject to the condition $A' \subset A$, and M as small as possible subject to (A.8). These ends are partly conflicting.

Theorems A.1 and A.2 guarantee the existence of a solution in an interval around (x^0, t_0). The length of the interval might be very small. In control theory models we are often interested in establishing the existence of a solution on a prescribed interval. The requirements in the existence theorems must then be strengthened to prevent $x(t)$ from "exploding".

Example A.1 Consider the equation $\dot{x} = x^2$, $x(0) = 1$. Its solution is $x(t) = 1/(1-t)$. As t approaches 1, $x(t)$ "disappears to infinity". Hence there is no way to extend this solution to an interval $[0, t_1]$ when $t_1 \geq 1$. (Intuitively, the problem is that $\dot{x}$ grows too fast.)

Before stating our first global result, let us agree that when writing (t_1, t_2), t_1 may be $-\infty$ and t_2 may be ∞. Similarly when writing $[t_1, t_2)$, t_2 may be ∞, and when writing $(t_1, t_2]$, t_1 may be $-\infty$.

Theorem A.3 (Global existence) Consider the differential equation $\dot{x} = h(x, t)$ with $h(x, t)$ satisfying (A.4). Assume moreover that there exist piecewise continuous functions $a(t)$ and $b(t)$ such that

$$||h(x,t)|| \leq a(t)||x|| + b(t)$$
$$\text{for all } x \in R^n \text{ and all } t \in (t_1, t_2). \tag{A.9}$$

If x^0 is any point in R^n and $t_0 \in (t_1, t_2)$, then there exists on (t_1, t_2) a solution $x(t)$ with $x(t_0) = x^0$. ∎

Observe that condition (A.9) is not satisfied for the equation $\dot{x} = x^2$ in example A.1 for any choices of $a(t)$ and $b(t)$.

The next result is for the case in which there are restrictions on the values of x. (By definition, a solution has to lie in A.)

Theorem A.4 (Global existence) Consider the differential equation $\dot{x} = h(x, t)$ with $h(x, t)$ satisfying (A.3). Assume moreover that there exist piecewise continuous functions $a(t)$ and $b(t)$ such that

$$||h(x,t)|| \leq a(t)||x|| + b(t) \qquad \text{for all } (x, t) \in A. \tag{A.10}$$

Let (x^0, t_0) be any point in A and suppose that there exists a closed set A'' contained in A, $(x^0, t_0) \in A''$, and a number $t_1 > t_0$ such that

any solution $x(t)$ defined on an interval $[t_0, p]$, $p \leq t_1$ and
lying in A, with $x(t_0) = x^0$, lies in A'' for all $t \in [t_0, p]$. $\qquad$ (A.11)

Then there exists a solution $x(t)$ on $[t_0, t_1]$ with $x(t_0) = x^0$. If (A.11) is valid for any t_1, then there exists a solution $x(t)$ with $x(t_0) = x^0$ on $[t_0, \infty)$. ∎

Note A.2 (The "backwards version") Let the assumptions be as in theorem A.4 but with $t_1 < t_0$, and with (A.11) replaced by the following condition: Any solution $x(t)$ defined on an interval $[q, t_0]$, $q \geq t_1$, and lying in A, with $x(t_0) = x^0$, also lies in A'' for all $t \in [q, t_0]$. Then there exists a solution $x(t)$ on $[t_1, t_0]$ with $x(t_0) = x^0$.

Note A.3 Suppose $k(t)$ is a solution to the scalar differential equation $\dot{k} = a(t)k + b(t)$, with $k(t_0) = ||x^0||$, and that $h(x, t)$ satisfies (A.3) and (A.10). Then any solution of $\dot{x} = h(x, t)$, $x(t_0) = x^0$, with graph in A, satisfies $||x(t)|| \leqslant k(t)$ for all $t \geqslant t_0$ in the interval of definition of $x(t)$. (Using (A.20) an explicit formula for $k(t)$ can be written down.)

The previous theorems concerned conditions sufficient to guarantee the *existence* of solutions to differential equations. These conditions do not ensure uniqueness. Consider a standard example.

Example A.2 Let $\dot{x} = h(x, t)$, where $h(x, t) = \sqrt{x}$ for $x \geqslant 0$, $h(x, t) = 0$ for $x < 0$. Then $h(x, t)$ is continuous everywhere, so that theorem A.1 applies. But in this case there are two solutions passing through $(t, x) = (0, 0)$: $x(t) \equiv 0$ and $x(t) = 0$ for $t < 0$, $x(t) = \frac{1}{4}t^2$ for $t \geqslant 0$.

In order to obtain uniqueness we impose on the differential equation the following condition, (A is the set occuring in (A.3).)

For each $(x, t) \in A$ there exist a $r > 0$ and an interval (a, b) containing t with $B(x, r) \times (a, b) \subset A$ and a constant L such that for all $y, y' \in B(x, r)$ and all $t \in (a, b)$

$$||h(y, t) - h(y', t)|| \leqslant L||y - y'||. \tag{A.12}$$

The function h is called *locally Lipschitz continuous in x if it satisfies* (A.12). If h and h'_x are continuous, then (A.12) *is* satisfied.

Theorem A.5 (**Uniqueness**) Consider the differential equation $\dot{x} = h(x, t)$ with $h(x, t)$ satisfying (A.3) and (A.12). Let $(x^0, t_0) \in A$ and let $x^1(t)$ and $x^2(t)$ be two solutions defined on the intervals I_1 and I_2 respectively, both passing through (x^0, t_0). Then $x^1(t) = x^2(t)$ on $I_1 \cap I_2$. ∎

A.3. "Bouncing off" results

In this section we study some useful results which are intuitively rather obvious.

Consider a scalar differential equation $\dot{x} = h(x, t)$ with $x(t_0) = x^0$. Assume that t_1 and $\bar{x}$ are given numbers, $t_1 > t_0$, and $x^0 \geqslant \bar{x}$. Suppose we know that $h(\bar{x}, t) \geqslant 0$ for all $t \in [t_0, t_1)$, so that $\dot{x} \geqslant 0$ along the line $x = \bar{x}$. Intuitively, the graph of $x(t)$ cannot pass below the line $x = \bar{x}$ since $\dot{x} \geqslant 0$ along $x = \bar{x}$. (See fig. A.1.) Formally, we have the following result:

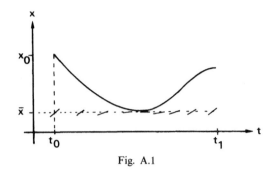

Fig. A.1

Theorem A.6   Consider the scalar differential equation $\dot{x} = h(x, t)$ with $h(x, t)$ satisfying (A.3), and assume that $h(x, t)$ satisfies the Lipschitz condition (A.12). Let $(x^0, t_0) \in A$, and suppose that there exist numbers t_1 and $\bar{x}$, $t_1 > t_0$ and $x^0 \geqslant \bar{x}$ with $\{\bar{x}\} \times [t_0, t_1) \subset A$, such that

$$h(\bar{x}, t) \geqslant 0 \qquad \text{for all } t \in [t_0, t_1). \tag{A.13}$$

Then any solution $x(t)$, $t \in [t_0, t_1)$ with $(x(t), t) \in A$ and $x(t_0) = x^0$ satisfies the inequality $x(t) \geqslant \bar{x}$ for all $t \in [t_0, t_1)$. If the inequality in (A.13) is strict, then $x(t) > \bar{x}$ for all $t \in (t_0, t_1)$, and in this case condition (A.12) is not needed. ■

We call theorem A.6 a "forwards-above bouncing off result". In obvious ways the theorem can be modified to cover three other situations. With $x(t_0) = x^0$, we have in compact notation:

$$x^0 \geqslant \bar{x}, \qquad h(\bar{x}, t) \leqslant 0 \text{ in } (\tau, t_0] \Rightarrow x(t) \geqslant \bar{x} \text{ in } (\tau, t_0] \tag{A.14}$$

$$x^0 \leqslant \bar{x}, \qquad h(\bar{x}, t) \geqslant 0 \text{ in } (\tau, t_0] \Rightarrow x(t) \leqslant \bar{x} \text{ in } (\tau, t_0] \tag{A.15}$$

$$x^0 \leqslant \bar{x}, \qquad h(\bar{x}, t) \leqslant 0 \text{ in } [t_0, t_1) \Rightarrow x(t) \leqslant \bar{x} \text{ in } [t_0, t_1). \tag{A.16}$$

If $h(\bar{x}, t) \leqslant 0$ is replaced by $h(\bar{x}, t) < 0$ in (A.14), then the conclusion changes to $x(t) > \bar{x}$ in (τ, t_0), and in this case we can drop the requirement (A.12). (A.15) and (A.16) can be modified in the same way.

Let us turn to the case in which $\dot{x} = h(x, t)$ is a vector differential equation. Note that if x and y are vectors, $x \geqslant y$ means $x_i \geqslant y_i$ for all i.

Theorem A.7.   Consider the vector differential equation $\dot{x} = h(x, t)$ with $h(x, t)$ satisfying (A.3) and the Lipschitz condition (A.12). Let $(x^0, t_0) \in A$ and suppose that there exist a number $t_1 > t_0$ and a vector $\bar{x} \in R^n$, $x^0 \geqslant \bar{x}$ with $\{\bar{x}\} \times [t_0, t_1) \subset A$, such that for all $t \in [t_0, t_1)$

$$h_i(x, t) \geqslant 0 \text{ if } x \geqslant \bar{x} \qquad \text{and} \qquad x_i = \bar{x}_i, \qquad i = 1, \dots n. \tag{A.17}$$

Then any solution $x(t)$, $t \in [t_0, t_1)$ with $(x(t), t) \in A$ and $x(t_0) = x^0$, satisfies the inequalities

$$x_i(t) \geqslant \bar{x}_i \text{ for all } t \in [t_0, t_1) \text{ and all } i = 1, \ldots, n. \tag{A.18}$$

If the inequality in (A.17) is strict, then $x_i(t) > \bar{x}_i$ for all $t \in (t_0, t_1)$ and all $i = 1, \ldots, n$, and in this case the requirement (A.12) can be dropped.

Note A.4 There are a number of more or less obvious variants of the theorem. For instance, we might look backwards instead of forwards, $\bar{x}_i$ might be $\geqslant x_i^0$ for some i and $\leqslant x_i^0$ for other values of i, etc. We shall not write down any of these simple modifications of the theorem.

A.4. Linear differential equations

Consider first the scalar differential equation

$$\dot{x} = a(t)x + b(t). \quad a(t) \text{ and } b(t) \text{ piecewise continuous}$$
on an interval I. $\tag{A.19}$

According to a standard formula, the solution on I is

$$x(t) = x(t_0) \exp\left(\int_{t_0}^{t} a(\tau)d\tau\right) + \int_{t_0}^{t} b(\tau) \exp\left(\int_{\tau}^{t} a(\zeta)d\zeta\right)d\tau. \tag{A.20}$$

In particular, if $a(t) = a$ and $b(t) = b$ for constants a and b, then it follows from (A.20) that

$$\dot{x} = ax + b, x(t_0) = x_0 \Leftrightarrow x(t) = (x_0 + b/a)e^{a(t - t_0)} - b/a. \tag{A.21}$$

Let us turn to the linear system

$$\dot{x}_1 = a_{11}(t)x_1 + \ldots + a_{1n}(t)x_n + b_1(t)$$

$a_{ij}(t)$ and $b_j(t)$ are piecewise continuous on an interval I.

$$\dot{x}_n = a_{n1}(t)x_1 + \cdots + a_{nn}(t)x_n + b_n(t) \tag{A.22}$$

In vector notation,

$$\dot{x} = A(t)x + B(t),$$

where

$$A(t) = [a_{ij}(t)]_{n \times n}, \qquad x = \begin{bmatrix} x_1 \\ \vdots \\ x_n \end{bmatrix} \qquad B(t) = \begin{bmatrix} b_1(t) \\ \vdots \\ b_n(t) \end{bmatrix}. \tag{A.23}$$

The corresponding homogeneous equation is

$$\dot{x} = A(t)x. \tag{A.24}$$

One can prove that if $t_0 \in I$, there exists on I a unique set of n linearly independent solutions $\varphi_j(t) = [\varphi_{1j}(t), \ldots, \varphi_{nj}(t)]$, $j = 1, \ldots, n$ of (A.24) with $\varphi_j(t_0) = e_j, j = 1, \ldots, n$, where e_j is the jth unit vector in R^n. The resolvent (or *fundamental matrix*) of (A.24) (or A.23) is the matrix

$$\Phi(t, t_0) = \begin{bmatrix} \varphi_{11}(t) & \cdots & \varphi_{1n}(t) \\ \vdots & & \vdots \\ \varphi_{n1}(t) & \cdots & \varphi_{nn}(t) \end{bmatrix}, \qquad \Phi(t_0, t_0) = E, \tag{A.25}$$

with E the $n \times n$ unit matrix. (Thus the jth column of Φ is $\varphi_j(t)$ with $\varphi_j(t_0) = e_j$.) One can now prove that if $x(t; x^0, t_0)$ is the unique solution of (A.24) passing through (x^0, t_0), then $x(t; x^0, t_0) = \Phi(t, t_0)x^0$. Briefly formulated,

$$\dot{x} = A(t)x, x(t_0) = x^0 \iff x = \Phi(t, t_0)x^0 \tag{A.26}$$

with $\Phi(t, t_0)$ given by (A.25). As for the inhomogeneous equation (A.23) we have the following result:

$$\dot{x} = A(t)x + B(t), \; x(t_0) = x^0 \iff x = \Phi(t, t_0)x^0 + \int_{t_0}^{t} \Phi(t, s) \, B(s)ds \tag{A.27}$$

Note A.5. $\Phi(t, s)$ denotes the value of the resolvent at t when the initial point of time is s. Since it can be proved that $\Phi(t, t_0) \cdot \Phi(t_0, t) = E$, $\Phi(t, t_0)$ has an inverse. More generally, $\Phi(t, s) = \Phi(t, \tau) \cdot \Phi(\tau, s)$ for all t, s, and τ. In particular, $\Phi(t, s) = \Phi(t, t_0)(\Phi(s, t_0))^{-1}$. If $A(t)$ in (A.24) is the constant matrix A, then $\Phi(t, s) = \Phi(t - s, 0)$ for all t_0, t and s. Hence:

$$\dot{x} = Ax + B(t), \; x(t_0) = x^0 \iff x = \Phi(t, t_0)x^0 + \int_{t_0}^{t} \Phi(t - s, 0) \, B(s)ds. \tag{A.28}$$

Note finally that if $\Phi(t, s)$ is the resolvent of (A.24), then $(\Phi(s, t))'$ (the transpose of $\Phi(s, t)$) is the resolvent of the equation

$$\dot{z} = -(A(t))'z. \tag{A.29}$$

The solution formulas in (A.26), (A.27) and (A.28) are important from a theoretical point of view. However, using them to find explicit solutions for even very simple problems leads to quite labourious calculations.

A.5. Dependence on initial conditions and on parameters

How does the solution of a differential equation change when the initial conditions change? A precise result which is useful in several of our models is the following:

Theorem A.8 **(Dependence on initial conditions)** Consider the vector differential equation $\dot{x} = h(x, t)$ with $h(x, t)$ satisfying (A.3) and the Lipschitz condition (A.12). Suppose $\bar{x}(t)$ is a solution of the equation on some interval $[a, b]$ with $(x(t), t) \in A$ and let $\bar{t}_0 \in [a, b]$, $\bar{x}^0 = \bar{x}(\bar{t}_0)$.

Then there exists a neighbourhood $N = B(\bar{x}^0, r) \times [\bar{t}_0 - \alpha, \bar{t}_0 + \alpha]$ with $r > 0$, and $\alpha > 0$, such that for all $(x^0, t_0) \in N$ there exists a unique solution through (x^0, t_0) defined on $[a, b]$ with its graph in A. If we denote this solution by $x(t; x^0, t_0)$, then

for each $t \in [a, b]$, the function $(x^0, t_0) \rightarrow x(t; x^0, t_0)$ is continuous in N.

$$(A.30)$$

If $\partial h / \partial x$ exists and satisfies (A.3), and if $t_0 \notin \{\theta_1, \theta_2, \ldots\}$, then $x(t; x^0, t_0)$ is C^1 w.r.t. $(x^0, t_0) \in N$, and

$$\frac{\partial x(t; \bar{x}^0, \bar{t}_0)}{\partial x^0} = \Phi(t, \bar{t}_0), \tag{A.31}$$

$$\frac{\partial x(t; \bar{x}^0, \bar{t}_0)}{\partial t_0} = -\Phi(t, \bar{t}_0) \cdot h(\bar{x}^0, \bar{t}_0), \tag{A.32}$$

where the $n \times n$-matrix $\Phi(t, \bar{t}_0)$ is the resolvent (see section A.4) of the linear differential equation

$$\dot{z} = h'_x(\bar{x}(t), t)z. \tag{A.33}$$

Note A.6 At a point of discontinuity $t_0 \in (\theta_1, \theta_2, \ldots)$, (A.31) is still valid. In this case a right and a left derivative w.r.t. t_0 will exist at $(\bar{x}^0, \bar{t}_0)$ and (A.32) will hold with $h(\bar{x}^0, \bar{t}_0)$ replaced by $h(\bar{x}^0, \bar{t}_0^+)$ and $h(\bar{x}^0, \bar{t}_0^-)$, respectively.

Let us test (A.30), (A.31), and (A.32) on a simple example.

Example A.3 Consider the system $\dot{x}_1 = 2x_2$, $\dot{x}_2 = x_1 + x_2$, which can be written as $\dot{x} = h(x, t)$ if we put

$$x = \begin{bmatrix} x_1 \\ x_2 \end{bmatrix}, \qquad h(x, t) = \begin{bmatrix} 2x_2 \\ x_1 + x_2 \end{bmatrix}. \tag{A.34}$$

The conditions in theorem A.8 are satisfied everywhere for this linear system.

The general solution of $\dot{x} = h(x, t)$ with $x_1(t_0) = x_1^0$ and $x_2(t_0) = x_2^0$ is

$$\begin{aligned}
x_1 &= \frac{2}{3}(x_1^0 - x_2^0)e^{t_0}e^{-t} + \frac{1}{3}(x_1^0 + 2x_2^0)e^{-2t_0}e^{2t} \\[2mm]
x_2 &= -\frac{1}{3}(x_1^0 - x_2^0)e^{t_0}e^{-t} + \frac{1}{3}(x_1^0 + 2x_2^0)e^{-2t_0}e^{2t}
\end{aligned} \tag{A.35}$$

(See e.g. Sydsæter (1981), Example 9, p. 347 for a standard method which can be used to obtain (A.35).) We see from (A.35) that x_1 and x_2 depend continuously on x_1^0, x_2^0, and t_0.

The left-hand side of (A.31) is in the present case the following 2×2 matrix (evaluated at (x_1^0, x_2^0, t_0) rather than at $(\bar{x}_1^0, \bar{x}_2^0, \bar{t}_0)$)

$$\begin{bmatrix} \dfrac{\partial x_1}{\partial x_1^0} & \dfrac{\partial x_1}{\partial x_2^0} \\[3mm] \dfrac{\partial x_2}{\partial x_1^0} & \dfrac{\partial x_2}{\partial x_2^0} \end{bmatrix} = \begin{bmatrix} \frac{2}{3}e^{t_0}e^{-t} + \frac{1}{3}e^{-2t_0}e^{2t} & -\frac{2}{3}e^{t_0}e^{-t} + \frac{2}{3}e^{-2t_0}e^{2t} \\[2mm] -\frac{1}{3}e^{t_0}e^{-t} + \frac{1}{3}e^{-2t_0}e^{2t} & \frac{1}{3}e^{t_0}e^{-t} + \frac{2}{3}e^{-2t_0}e^{2t} \end{bmatrix} \tag{A.36}$$

Notice that since $h(x, t)$ is linear in x, the differential equation in (A.33) is identical to $\dot{x} = h(x, t)$. Thus we see that the right-hand side of (A.31) is the resolvent of $\dot{x} = h(x, t)$. According to (A.25), the two columns of $\Phi(t, t_0)$ are obtained from (A.35) by putting $x_1^0 = x_1(t_0) = 1$, $x_2^0 = x_2(t_0) = 0$ and $x_1^0 = x_1(t_0) = 0$, $x_2^0 = x_2(t_0) = 1$, respectively. Hence

$$\Phi(t, t_0) = \begin{bmatrix} \frac{2}{3}e^{t_0}e^{-t} + \frac{1}{3}e^{-2t_0}e^{2t} & -\frac{2}{3}e^{t_0}e^{-t} + \frac{2}{3}e^{-2t_0}e^{2t} \\[2mm] -\frac{1}{3}e^{t_0}e^{-t} + \frac{1}{3}e^{-2t_0}e^{2t} & \frac{1}{3}e^{t_0}e^{-t} + \frac{2}{3}e^{-2t_0}e^{2t} \end{bmatrix} \tag{A.37}$$

Since the matrices in (A.36) and (A.37) are equal, (A.31) is confirmed.

The left-hand side of (A.32) is in the present case, evaluated at (x_1^0, x_2^0, t_0), using (A.35),

$$\begin{bmatrix} \dfrac{\partial x_1}{\partial t_0} \\[3mm] \dfrac{\partial x_2}{\partial t_0} \end{bmatrix} = \begin{bmatrix} \frac{2}{3}(x_1^0 - x_2^0)e^{t_0}e^{-t} & -\frac{2}{3}(x_1^0 + 2x_2^0)e^{-2t_0}e^{2t} \\[2mm] -\frac{1}{3}(x_1^0 - x_2^0)e^{t_0}e^{-t} & -\frac{2}{3}(x_1^0 + 2x_2^0)e^{-2t_0}e^{2t} \end{bmatrix} \tag{A.38}$$

Furthermore, by using (A.37) we see that

$$-\Phi(t, t_0)\, h(x^0, t_0) = -\Phi(t, t_0) \cdot \begin{bmatrix} 2x_2^0 \\ x_1^0 + x_2^0 \end{bmatrix} \tag{A.39}$$

is equal to the matrix in (A.38). Thus (A.32) is confirmed.

We need also a result on the sensitivity of the solution of a differential equation to changes in parameters.

Theorem A.9 (**Dependence on parameters**) Consider the vector differential equation $\dot{x} = h(x, t, z)$, where z is a vector with s components. We assume that h satisfies (A.3) for $A = A' \times Z$ and with x replaced by (x, z), where A' is an open set in R^{n+1} and Z is an open set in R^s. Suppose $\bar{x}(t)$ is a solution of the equation on some interval $[t_0, t_1]$, with $x(t_0) = x^0$ and $z = \bar{z} \in Z$, x^0 and $\bar{z}$ fixed vectors, and with $(\bar{x}(t), t) \in A'$ for all t.

Then for all z in a neighbourhood of $\bar{z}$ there exists a solution $x(t; z)$ defined on $[t_0, t_1]$ with $x(t_0; \bar{z}) = x^0$, and such that $(x(t; z), t) \in A'$ for all t.

For each $t \in [t_0, t_1]$, $z \to x(t; z)$ is continuously differentiable, and

$$\frac{\partial x(t, \bar{z})}{\partial z} = \int_{t_0}^{t} \Phi(t, s) \cdot h_z'(\bar{x}(s), s, \bar{z})\mathrm{d}s \tag{A.40}$$

with $\Phi(t, s)$ defined as in theorem A.8, provided h_x' and h_z' exist and satisfy (A.3) in $A' \times Z$. ■

A.6. Second-order differential equations

Consider the scalar second-order differential equation

$$\ddot{x} = \varphi(x, \dot{x}, t), \qquad x(t_0) = x^0, \qquad \dot{x}(t_0) = \dot{x}^0, \tag{A.41}$$

where x^0 and $\dot{x}^0$ are given real numbers. Such an equation can always be rewritten as a system of two first-order differential equations in the following way:

$$\begin{aligned} \dot{x} &= y & x(t_0) &= x^0 \\ \dot{y} &= \varphi(x, y, t) & y(t_0) &= \dot{x}^0 \end{aligned} \tag{A.42}$$

This reformulation makes it possible to apply all our results for first-order equations to second-order equations. For easy reference we shall write down some of the most useful results.

First, let us agree that $x(t)$ is a *solution* to (A.41) on an interval I provided: (i) it satisfies the boundary conditions, (ii) $x(t)$ is C^1 on I, (iii) $\ddot{x}(t)$ exists v.e., and, (iv) $\ddot{x}(t) = \varphi(x(t), \dot{x}(t), t)$ v.e. on I.

Theorem A.10 Consider the differential equation $\ddot{x} = \varphi(x, \dot{x}, t)$ with $\varphi(x, y, t)$ continuous at all points $(x, y, t) \in A$ for some given open set A in R^3, and with $t \notin \Gamma$, $\Gamma = \{\theta_1, \theta_2, \ldots\}$, a fixed set of points. Assume moreover that $\varphi(x, y, t)$ has one-sided limits when $(x, y) \to (x', y')$, $t \to t'^{+}$ and $t \to t'^{-}$, for all $(x', y', t') \in A$, $t' \in \Gamma$. Let $(x^0, \dot{x}^0, t_0) \in A$. Then we have the following result:

(A) (Local Existence). Choose positive numbers r and s such that

$$A' = \bar{B}((x^0, \dot{x}^0), r) \times [t_0 - s, t_0 + s] \subset A. \tag{A.43}$$

If M' is a number such that

$$|\varphi(x, y, t)| \leqslant M' \qquad \text{for all } (x, y, t) \in A', \tag{A.44}$$

then there exists a solution $x(t)$ on the interval $[t_0 - p, t_0 + p]$ with $x(t_0) = x^0$, $\dot{x}(t_0) = \dot{x}^0$, and with $p = \min [s, r/M]$, where $M = |\dot{x}^0| + r + M'$.

(B) (Global existence). Suppose $\varphi(x, y, t)$ also satisfies the condition

$$|\varphi(x, y, t)| \leqslant a(t)(|x| + |y|) + b(t) \qquad \text{for all } (x, y, t) \in A, \tag{A.45}$$

with $a(t)$ and $b(t)$ two non-negative piecewise continuous functions. Let $[t_0, t_1]$ be an interval with one of the two following properties: Either $R^2 \times [t_0, t_1] \subset A$, or there exists a closed set $A'' \subset A$ with $(x^0, \dot{x}^0, t_0) \in A''$ such that

> any solution $x(t)$ on an interval $[t_0, p]$, $p \leqslant t_1$ lying in A, with $(x(t_0), \dot{x}(t_0)) = (x^0, \dot{x}^0)$, lies in A'' for all $t \in [t_0, p]$. (A.46)

Then there exists a solution $x(t)$ on $[t_0, t_1]$ lying in A, with $x(t_0) = x^0$ $\dot{x}(t_0) = \dot{x}^0$. If $R^2 \times [t_0, \infty) \subset A$ or if (A.46) is valid for any t_1, $x(t)$ is defined on $[t_0, \infty)$.

(C) (Uniqueness.) Suppose

> $\varphi(x, y, t)$ is locally Lipschitz continuous in (x, y) for all (x, y, t) in A. (A.47)

Then the solutions in (A) and (B) are unique.

(D) (Bouncing off.) Suppose that (A.47) is satisfied and assume that for a given number $\bar{x}$, we have for some $\beta > t_0$ that for all $(x, y, t) \in A$:

$$\varphi(x, y, t) \geqslant 0 \text{ if } y = 0, \qquad x \geqslant \bar{x} \text{ and } t \in [t_0, \beta]. \tag{A.48}$$

Then for any solution $x(t)$ on $[t_0, \beta]$ with graph in A for which

$x(t_0) = x^0 \geqslant \bar{x}$, $\dot{x}(t_0) = \dot{x}^0 \geqslant 0$, we have that $x(t) \geqslant \bar{x}$ and $\dot{x}(t) \geqslant 0$ for all $t \in (t_0, \beta]$. The latter two inequalities are strict if the first inequality in (A.48) is strict, and in this case (A.47) is not needed. ∎

A.7. Maximal interval of existence

Consider the vector differential equation $\dot{x} = h(x, t)$ with initial condition $x(t_0) = x^0$, h satisfying (A.3). Suppose $x = \varphi(t)$ is a solution of the equation defined on an interval I lying in A and passing through (x^0, t_0). We are often interested in determining how far we can extend the solution forwards and backwards. We say that $\hat{\varphi}(t)$ is a *continuation of* $\varphi(t)$ if $\hat{\varphi}(t)$ is a solution to the equation on an interval $\hat{I}$ which properly contains I, $\hat{\varphi}(t) = \varphi(t)$ for all $t \in I$ and $\hat{\varphi}(t)$ lies in A. If such a continuation does not exist, I is called the *maximal interval of existence* of the solution $\varphi(t)$. A useful theorem states that *any solution φ has a maximal continuation $\hat{\varphi}$*, (i.e. $\hat{\varphi}$ is defined on a maximal interval of existence).

Example A.4 The maximal interval of existence of the solution $x(t) = 1/(1-t)$ to $\dot{x} = x^2$, $x(0) = 1$ (see example A.1) is $(-\infty, 1)$. Note that $x(t) \to \infty$ as $t \to 1^-$.

Suppose, in general, that conditions securing existence and uniqueness of solutions to $\dot{x} = h(x, t)$ are satisfied everywhere, and let (x^0, t_0) be given. Suppose the maximal interval of existence is (t_1, t_2). Then the solution cannot be extended to the left of t_1 or to the right of t_2. If $t_2 < \infty$, it is reasonable to expect that as t approaches t_2^-, $||x(t)|| \to \infty$, just like we got in the case of example A.4. Similarly, if $t_1 > -\infty$, we expect that $||x(t)|| \to \infty$ as $t \to t_1^-$. A more general result is given in the next theorem.

Theorem A.11 Consider the vector differential equation $\dot{x} = h(x, t)$, $h(x, t)$ satisfying (A.3). Let $(x^0, t_0) \in A$ and suppose $x(t)$ is a solution passing through (x^0, t_0) and lying in A and with maximal interval of existence (t_1, t_2). If $t_2 < \infty$, then

$$\overline{\lim_{t \to t_2^-}} \, ||x(t)|| = \infty \quad \text{or} \quad \lim_{t \to t_2^-} d[(x(t), t), \complement \, A] = 0, \tag{A.49}$$

with $d[x(t), t), \complement \, A]$ the distance from the vector $(x(t), t)$ to the set $\complement \, A$, i.e. $\inf \{||(x(t), t) - b|| : b \in \complement \, A\}$. If $t_1 > -\infty$, then (A.49) holds when t_2^- is replaced by t_1^+.

Note A.7 If $t_2 < \infty$, $\overline{\lim_{t \to t_2^-}} ||x(t)|| < \infty$, and $h(x, t)$ satisfies (A.3) for A replaced by the closure of A, then $\lim_{t \to t_2^-} x(t)$ exists. A corresponding result holds for $\lim_{t \to t_1^+} x(t)$.

A.8. Comments and references

Generally, proofs of all the results in A.1→A.7 can be found in one or several of the following books: Coddington and Levinson (1955), Hale (1969), Hartman (1964), Hestenes (1966), Pontryagin (1962a), Sansone and Conti (1964). (Coddington and Levinson, Hale, Hestenes and Sansone and Conti contain results involving measurable dependence on t, which covers our case of continuity v.e. in t.) For convenience, a proof of theorem A.6 is given. Assume first that (A.12) holds, and by contradiction that $x(t'') < \bar{x}$ for some $t'' \in [t_0, t_1)$. Choose a point t' in $[t_0, t'']$ such that $x(t') = \bar{x}$, $x(t) < \bar{x}$ for all t in $(t', t'']$. (By necessity, such a point t' exists.) Define $g(x, t) = h(x, t)$ if $x \geqslant \bar{x}$, $g(x, t) = \max(0, h(x, t))$ if $x < \bar{x}$. There exists a solution $y(t)$ of $\dot{y} = g(y, t)$, $y(t') = x(t')$ on a small interval $[t', b]$. It is easily seen that $y(t)$ can never be $< \bar{x}$ in $[t', b]$. Hence, $g(y(t), t) = h(y(t), t)$ and by uniqueness, we must have $x(t) = y(t)$ in $[t', b]$, contradicting the fact that $x(t) < \bar{x}$ in (t', t'').

   If strict inequality in (A.13) holds (and (A.12) is dropped), then if $x(t)$ is near $\bar{x}$, $\dot{x} = h(x(t), t) > 0$, which immediately gives that $x(t) > \bar{x}$ just to the right of t_0, and that this inequality is never violated later on.

A FEW DEFINITIONS AND RESULTS

B.1. Introduction

In this appendix we gather some special definitions and results which are used at several places in the book. We assume a basic familiarity with elementary calculus of functions of several variables and linear algebra.

B.2. Convexity

This brief survey may be supplemented by e.g. Sydsæter (1981) and Rockafellar (1970).

A set S in R^n is convex if x, $y \in S$ and $\lambda \in [0, 1]$ imply $\lambda x + (1 - \lambda)y \in S$. If T is a subset of R^n, $co(T)$, the convex hull of T is the smallest convex set containing T.

A real-valued function f defined on a convex set S in R^n is *concave* provided x, $y \in S$ and $\lambda \in (0, 1)$ imply $f(\lambda x + (1 - \lambda)y) \geq \lambda f(x) + (1 - \lambda)f(y)$. f is *strictly* concave if the inequality is strict for $x \neq y$. f is *convex* if $-f$ is concave.

If f is real-valued and concave on S and differentiable (see below) at $x^0 \in S$, then
$$f(x) - f(x^0) \leq f'(x^0) \cdot (x - x^0) \qquad \text{for all } x \in S. \tag{B.1}$$

Here $f'(x) = (\partial f / \partial x_1, \ldots, \partial f / \partial x_n)$ is the *gradient* of f, also denoted by df/dx. If f is strictly concave on S, the inequality in (B.1) is strict for $x \neq x^0$.

If f is a real-valued function on a convex set S in R^n, then a vector a is a *supergradient* of f at x^0 if

$$f(x) - f(x^0) \leq a \cdot (x - x^0) \qquad \text{for all } x \in S. \tag{B.2}$$

425

If f is differentiable at x^0 and x^0 is an interior point of S, then $a = f'(x^0)$. If f is concave, it has a supergradient at each interior point of S.

A vector a is called a *subgradient* of f at x^0 if (B.2) holds with the inequality sign reversed.

A real-valued function f defined on a convex set S in R^n is *quasi-concave* if

$$f(x) \geqslant f(x^0) \Rightarrow f(\lambda x + (1 - \lambda)x^0) \geqslant f(x^0)$$

for all $x, x^0 \in S$ and all $\lambda \in [0, 1]$. (B.3)

If f is real-valued and quasi-concave on a convex set S in R^n and differentiable at $x^0 \in S$, then

$$f(x) \geqslant f(x^0) \Rightarrow f'(x^0) \cdot (x - x^0) \geqslant 0 \qquad \text{for all } x \in S. \tag{B.4}$$

A vector a in R^n is called *a support functional* to a set S at $x^0 \in S$ if $a \cdot x \geqslant a \cdot x^0$ for all $x \in S$. (A convex set has a nonzero support functional at each boundary point.)

B.3. Differentiability

A vector function $f = (f_1, \ldots, f_m)$ defined on a convex set S in R^n with values in R^m is called *differentiable at* x^0, if there exists an $m \times n$ matrix A such that for all $\varepsilon > 0$ there exists a $\delta > 0$ such that

$$\|f(x^0 + h) - f(x^0) - A \cdot h\| \leqslant \varepsilon \cdot \|h\| \tag{B.5}$$

for all h such that $\|h\| \leqslant \delta$ and $x^0 + h \in S$. If f is differentiable at an interior point x^0 of S, then all partial derivatives of f exist at x^0 and $A = [\partial f_i(x^0)/\partial x_j]$ where $i = 1, \ldots, m, j = 1, \ldots, n$.

A real-valued function f defined on a convex set S in R^n is called *superdifferentiable at* x^0 provided there exists an n-vector a such that for all $\varepsilon > 0$ there exist a $\delta > 0$ such that

$$f(x^0 + h) - f(x^0) \leqslant a \cdot h + \varepsilon \cdot \|h\| \tag{B.6}$$

for all h such that $\|h\| < \delta$ and $x^0 + h \in S$. Then a is called a *superderivative of* f at x^0. If (B.6) holds with the inequality sign reversed, then f is called *subdifferentiable* at x^0, and a is a subderivative.

A C^k- function is a function defined on (or extendable to) an open set, having continuous kth-order partial derivatives w.r.t. all variables.

B.4. Limits

Let f be a real-valued function defined on $[t_0, \infty)$. Define

$$\varliminf_{t \to \infty} f(t) = \lim_{t \to \infty} \inf \{f(s) : s \in [t, \infty)\} \tag{B.7}$$

$$\varlimsup_{t \to \infty} f(t) = \lim_{t \to \infty} \sup \{f(s) : s \in [t, \infty)\}. \tag{B.8}$$

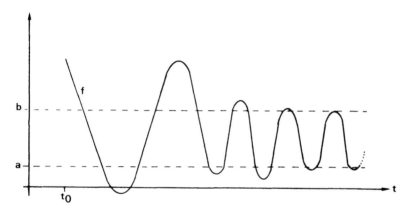

Fig. B.1.

In fig. B.1 we illustrate the definitions in (B.7) and (B.8). In fact, here

$$\varliminf_{t \to \infty} f(t) = a, \qquad \varlimsup_{t \to \infty} f(t) = b.$$

The limits $\varliminf$ and $\varlimsup$ are called "lim inf" and "lim sup", respectively. In general,

$$\varlimsup_{t \to \infty} f(t) \geqslant \varliminf_{t \to \infty} f(t).$$

We note the following basic facts:

$$\varliminf_{t \to \infty} f(t) \geqslant a \Leftrightarrow \begin{cases} \text{For each } \varepsilon > 0 \text{ there exists a } t' \text{ such that} \\ t > t' \text{ implies } f(t) \geqslant a - \varepsilon. \end{cases} \tag{B.9}$$

$$\varlimsup_{t \to \infty} f(t) \geqslant a \Leftrightarrow \begin{cases} \text{For each } \varepsilon > 0 \text{ and for each } t' \text{ there exists} \\ \text{some } t \geqslant t' \text{ where } f(t) \geqslant a - \varepsilon. \end{cases} \tag{B.10}$$

If $\lim_{t \to \infty} f(t)$ exists, then $\lim_{t \to \infty} f(t) = \underline{\lim}_{t \to \infty} f(t) = \overline{\lim}_{t \to \infty} f(t)$ \hfill (B.11)

$$\underline{\lim}_{t \to \infty} (f(t) + g(t)) \geq \underline{\lim}_{t \to \infty} f(t) + \underline{\lim}_{t \to \infty} g(t) \tag{B.12}$$

$$\overline{\lim}_{t \to \infty} (f(t) + g(t)) \leq \overline{\lim}_{t \to \infty} f(t) + \overline{\lim}_{t \to \infty} g(t) \tag{B.13}$$

$$\underline{\lim}_{t \to \infty} f(t) = -\overline{\lim}_{t \to \infty} (-f(t)). \tag{B.14}$$

The expression $f(t^-)$ means the left limit at t, while $f(t^+)$ is the right limit. Similarly $(d^-/dt) f(t)$ means the left derivative at t, equal to the limit of $(f(t+h) - f(t))/h$ when $h \to 0^- (h < 0, h \to 0)$. The right derivative is denoted $(d^+/dt) f(t)$.

ANSWERS TO EXERCISES

Chapter 1

1.1.1 (a) (i) $J(x) = (4/3)(e^2 - 1)^2$, (ii) $J(x) = \frac{1}{2}(e^2 - 1)^2(1 + (\pi/2)^2)$,
(iii) $J(x) = e^4 - 1$, (iv) $J(x) = (11/30)a^2 - (1/6)(e^2 - 1)a + (4/3)(e^2 - 1)^2$.

1.2.1 (i) $\ddot{x} + \frac{1}{2} = 0$, $x = -\frac{1}{4}t^2 + At + B$. (ii) $x = -\frac{1}{4}t^2 - \frac{3}{4}t + 1$, (iii) $x = -\frac{1}{4}t^2 + (x_1 - x_0 + \frac{1}{4})t + x_0$ (iv) $V(x_0, x_1) = (x_1 - x_0 + \frac{1}{4})^2 - \frac{1}{12}$, $\partial V/\partial x_0 = 2x_0 - 2x_1 - \frac{1}{2}$, $\partial V/\partial x_1 = -\partial V/\partial x_0$

1.2.2 $\ddot{x} - x = e^t$, $x = Ae^t + Be^{-t} + \frac{1}{2}te^t$.

1.2.3 (i) $x = (1/7)(t^3 + 6)$. (ii) $J(x) = (17/2 - 12\ln 2)a^2 + (4\ln 2 - 3)a + 1/2 \to \infty$ as $a \to \infty$. (iii) No. A priori, the maximum as well as the minimum might fail to exist. However, see exercise 1.6.1.

1.2.5 (i) $\ddot{K} - (a + b)\dot{K} + abK = 0$, (ii) $C(t) = B(b - a)e^{at}$, (iii) $\dot{C}/C = a > 0 \Leftrightarrow b > \rho$, (iv) $C(t)$ as in (ii) with $B = (K_0 e^{bT} - K_T)/(e^{bT} - e^{aT}) > 0$, so $C(t) > 0$ for all t.

1.2.7 Euler: $U'_C \cdot f'_K + (d/dt)U'_C = 0$, $\dot{C}/C = ((-U''_{Ct}/U'_C) - f'_K)/\check{\omega}$.

1.2.8 (i) $pD'_p + D - b'(D)D'_p - (d/dt)(pD'_p - b'(D)D'_p) = 0$
(ii) $p = C_1 e^{\lambda t} + C_2 e^{-\lambda t} + k$, where $\lambda = [(\alpha A - 1)A/\alpha B^2]^{1/2}$,
$k = (\beta A + 2\alpha AC - C)/2A(1 - \alpha A)$.

1.3.1 $F''_{\dot{x}\dot{x}}$ is 2, 2 and $2/t^2$ respectively, so the Legendre necessary condition for minimum is satisfied in all three cases.

1.3.2 $F''_{\dot{p}\dot{p}} = -2\alpha B^2 < 0$, so the Legendre condition is satisfied.

1.4.1 Solution of the Euler equation: $x = -t$. It solves the problem if $a = -1$.

1.4.2 (i) $x = -\frac{1}{2}t$, (ii) $x = At^4 + B$, (iii) $x = Ae^t + Be^{-t} + \frac{1}{2}\sin t$.

1.4.3 $p\phi'(K(t)) = q(\rho + \gamma)$.

1.4.4 (i) $\ddot{y} - (2/\sigma)\dot{y} + (1/\sigma^2)y = (\bar{z}/\sigma^2)(l(t) - \sigma l(t))$,
(ii) $y = Ae^{t/\sigma} + Bte^{t/\sigma} + \bar{z}l_0 e^{\alpha t}/(1 - \alpha\sigma)$.

1.4.6 $[y(A - y)]^{k+1}(w - y')^{-(k+1)}[w - (k+1)y'] = C$.

1.5.1 (i) $x = -\frac{1}{4}t^2 + 1$, (ii) $x = -\frac{1}{4}t^2 + \frac{1}{4}t + 1$.

1.5.2 $x = 16^{1/3}(t^3 - 1)/3$, $t_1 = 4^{1/3}$.

1.5.3 (i) $\ddot{A} - r\dot{A} + (\rho - r)U'/U'' = 0$, $A(0) = A_0$, $A(T) = A_T$.

(ii) $A = Ke^{rt} + (r - \rho)t/br + L$. K and L are determined from
$A(0) = A_0$, $A(T) = A_T$.

1.5.4 (i) $\ddot{K} - b(1 + 1/v)\dot{K} + (b^2/v)K = 0$, and $K = Ae^{bt} + Be^{bt/v}$.

(ii) (1) $K_0 = A + B$, (2) $kL_0e^{nt_1} = Ae^{bt_1} + Be^{bt_1/v}$, (3) $knL_0e^{nt_1} = bAe^{bt_1}$.

1.6.3 (i) $\ddot{\bar{Y}} = m^2\bar{Y}$ where $m^2 = (\alpha_2 r_1^2 + \alpha_1 r_2^2)/a_2$.

(ii) $\bar{Y} = Ae^{mt} + Be^{-mt}$, where $A = (r_1 + m)\bar{Y}_0/[e^{2mT}(m - r_1) + (m + r_1)]$,
$B = Y_0 - A$.

1.6.4 $z(t) = A(e^{bt} - 1)/(e^{bT} - 1)$, $\qquad b = ra/(1 - a)$.

1.7.1 $x(t) = -\frac{1}{4}t^2 - t + 1$.

1.7.2 $A(t)$ satisfies the Euler equation (see the answer to exercise 1.5.3),
$A(0) = A_0$ and $\phi'(A(T)) = U'(C(T))$.

1.7.3 (i) $[(K_0 - A)(1 - 1/v)b]^{-v}e^{-bT} = \alpha\beta[Ae^{bT} + (K_0 - A)e^{bT/v}]^{\beta - 1}$.

(ii) $K(t) = Ae^{2t} + (10 - A)e^{4t}$ where $A = 190e^{20}/(19e^{20} - 1)$.

1.8.1 $x_1(t) = x_2(t) = t$.

1.8.2 $(\ddot{K} - \mu\dot{K})\phi''(\dot{K} - \mu K) + (\mu - \delta)\phi'(\dot{K} - \mu K) + F_K'(K, L) = 0$, $F_L'(K, L) = \omega$.

1.8.3 $\lambda_1 g_1 h_1 \dfrac{\dot{C}_1}{C_1^2} + \lambda_2 g_1 h_1 \dfrac{\dot{C}_2}{C_2^2} = \dfrac{1}{C_1}$, $\lambda_1 g_2 h_2 \dfrac{\dot{C}_1}{C_1^2} + \lambda_2 g_2 h_2 \dfrac{\dot{C}_2}{C_2^2} = \dfrac{1}{C_2}$.

1.8.4 (ii) $(\partial F/\partial \dot{x}_i)_{t = t_1} + \partial S(x(t_1))/\partial x_i = 0$, $\qquad i = 1, \ldots, n$.

1.8.7 (i) $x(t) = (e^t - e^{-t})/(e^2 - e^{-2})$, $T^* = 2$ (iii) $x(t) = (e^t - e^{-t})/(3^{1/2} - 3^{-1/2})$,
$T^* = \frac{1}{2}\ln 3$.

1.9.2 $a \neq 0$.

1.10.1 $d^4x/dt^4 - x = 0$, $x = \cos t$.

1.10.2 $d^4y/dt^4 = -\rho/\mu$, $y = (-\rho/24\mu)(x^2 - 1)^2$.

Chapter 2

2.3.1 (i) $J = \frac{1}{2}e - 1 + \frac{1}{2}(\cos 1 - \sin 1) \approx 0.209$, (ii) $J = \frac{1}{2}e - 1 \approx 0.359$. (The optimal value of J is $e - 2 \approx 0.718$.)

2.3.2 (i) With $H = p_0(x + u) + p(-x + u + t)$, there exist a number p_0 and a continuous function $p(t)$ such that $p_0 = 0$ or 1, $(p_0, p(t)) \neq (0, 0)$ for all t and

(1) $u^*(t)$ maximizes $p_0(x^*(t) + u) + p(t)(-x^*(t) + u + t)$ for $u \in [0, 1]$,

(2) $\dot{p}(t) = -\partial H^*/\partial x = -p_0 + p(t)$, $p(1) = 0$

(ii) $p(t) = -e^{t - 1} + 1$. Note that $p(t) \geq 0$ in $[0, 1]$.

(iii) $u^*(t) \equiv 1$

(iv) $\dot{x}^*(t) = -x^*(t) + 1 + t$, $x^*(0) = 1$ gives $x^*(t) = e^{-t} + t$.

(v) $x(t) = e^{-t} + e^{-t}\int_0^t e^s(u(s) + s)ds$. $u(t) = 1$ gives the highest values to $x(t)$ and to the criterion functional.

2.3.3 (ii) $p(t) = T - t$ on $[t^*, T]$.

(iv) $s^*(t) = 1$ in $[0, t_*]$, $s^*(t) = 0$ in $(t_*, T]$, where $t_* = \max(T - 1, \ln(k_T/k_0))$.

2.3.4 (i) $\dot{p} = -p_0 f'(k^*)e^{(\rho - \delta)t} + p\lambda - f'(k^*)e^{\rho t}(p - p_0 e^{-\delta t})s^*(t)$, $p(T) \geq 0$ $(= 0$ if $k^*(T) > k_T)$.

(ii) $s^*(t) = 1$ or 0 according as $p(t)$ is $>$ or $< p_0 e^{-\delta t}$. If $p(t) = p_0 e^{-\delta t}$, $s^*(t)$ is not determined by the maximum condition.

2.3.5 $u^*(t) \equiv 1$ is the only admissible control, and is hence optimal.

2.3.6 $u^*(t) = 1$ in $[0, T/2]$, $u^*(t) = 0$ in $(T/2, T]$.

2.3.7 (i) Put $\tau = at$, $k_1(\tau) = k(\tau/a)$, $s_1(\tau) = s(\tau/a)$, $T_1 = aT$.

(ii) $s^*(t) = 1$ in $[0, T - 1/a]$, $s^*(t) = 0$ in $(T - 1/a, T]$.

2.4.1 (i) In $[0, T - 2/a]$: $x_1^*(t) = x_1^0 e^{at}$, $x_2^*(t) = x_2^0$. In $(T - 2/a, T]$: $x_1^*(t) = x_1^0 e^{aT - 2}$, $x_2^*(t) = x_2^0 + ax_1^0 e^{aT - 2}(t - (T - 2/a))$.

2.4.2 $u^*(t) = 1$ in $[0, T - 1/a]$, $u^*(t) = 0$ in $(T - 1/a, T]$. (In $[0, T - 1/a]$: $x_1^*(t) = x_1^0 e^{at}$, $x_2^*(t) = x_2^0$. In $(T - 1/a, T]$: $x_1^*(t) = x_1^0 e^{aT - 1}$. $x_2^*(t) = x_2^0 + ax_1^0 e^{aT - 1}[t - (T - 1/a)]$.)

2.4.7 $H^c = p_0(1 - s)e^{\rho t}f(k) + q(se^{\rho t}f(k) - \lambda k)$.
$\dot{q} = -p_0(1 - s^*)e^{\rho t}f'(k^*) - q(s^*e^{\rho t}f'(k^*) - (\lambda + \delta))$
$s^*(t) = 1$ or 0 according as $q(t)$ is $>$ or $< p_0$.

2.5.1 (i) $H = p_0(U(x) - b(x) - gz) + pax$, (ii) $\dot{p}(t) = p_0 g$, (iii) $p(T) = 0$.
(vi) $\dot{x} = -ag/(U'' - b'') > 0$. (vii) $(d/dt)[U'(\dot{z}/a) - b'(\dot{z}/a)] = -ag$.

2.7.2 $u^*(t) = 0$ in $[0, T - 1]$, $u^*(t) = (1 - e^{t + 1 - T})/(e^{1 - T} - e^{t + 1 - T} + K_0 + t)$ in $(T - 1, T]$.

2.8.1 In exercise 2.3.2, $N(x, U, t) = \{(x + u + \gamma, -x + u + t): \gamma \leq 0, u \in [0, 1]\}$, and this is an infinite rectangular set as in fig. 2.11, hence convex. In example 3, $N(x_1, x_2, U, t) = \{(x_2 + \gamma, aux_1, a(1 - u)x_1): \gamma \leq 0, u \in [0, 1]\}$, and this is an infinite rectangular box, hence convex.

2.8.3 (b) $u^*(t) = 1$ in $[0, 1/2]$, $u^*(t) = 0$ in $(1/2, 1]$.

2.8.5 (i) The necessary conditions have the following solutions:
For $T < 4$, only $u_1^*(t) \equiv 0$.

For $T \geq 4$, $u_1^*(t)$, $u_2^*(t) \equiv 1$ and $u_3^*(t) = \begin{cases} 0 & \text{in } [0, T - 4] \\ 1 & \text{in } (T - 4, T] \end{cases}$

(ii) For $T \leq 6$, u_1^* is optimal. For $T \geq 6$, $u_2^*(t)$ is optimal.

2.8.7 (b) $u^*(t) = 1$ in $[0, t']$, $u^*(t) = 0$ in $(t', 1]$, where $t' = 1/x_0 - (1/x_0 - 1)^{1/2}$. If $t' \geq 1$, then $u^*(t) \equiv 1$.

2.9.1 (v) $T^* = (c - x_2^0)/ax_1^0$

(vi) $u^*(t)$ switches from 1 to 0 at $t^* = T^* - 1/a$, $T^* = (1/a)(1 + \ln[(c - x_2^0)/x_1^0])$

2.9.2   (ii) $x = ae^{-\delta(y-b)/u_i} + \dfrac{\alpha}{\delta}\left(y - \dfrac{u_i}{\delta}\right)$, $i = 0, 1$.

2.9.3   (i) $q(t_1) = 0$

      (ii)   Optimal path: $x(t) = \alpha y^2/2u_1 + (x(0) - \alpha(y(0))^2/2u_1)$

2.9.5   (i)   Define $x(t) = \bar{x} - \int_0^t u(s)ds$. Then $\dot{x}(t) = -u(t)$, $x(0) = \bar{x}$ and $x(T) \geqslant 0$.

      (ii)   Define $u = u^*(t; \bar{p})$ as a solution to $U'(u) = \bar{p}e^{rt}$, where $\bar{p}$ is a positive parameter, $\bar{p} > \lim_{u \to \infty} U'(u)$. Define $T = T(\bar{p})$ as the solution of $\int_0^T u^*(t; \bar{p})dt = \bar{x}$. (Then $T(\bar{p}) \in (0, \bar{x}/u_0)$.) Let $\bar{p}^*$ be the solution to $U(u^*(T(\bar{p}); \bar{p}))e^{-rT(\bar{p})} = \bar{p}u^*(T(\bar{p}); \bar{p})$. Then $u^*(t) = u^*(t, \bar{p}^*)$ is the unique solution candidate with $p(t) = \bar{p}^*$ as the adjoint variable, and $T^* = T(\bar{p}^*)$.

2.9.7   (ii) $u^*(t) = 1$ in $[0, 4]$, $u^*(t) = 0$ in $(4, 8]$, $T^* = 8$.

2.9.8   (i) $u^*(t) = 1$ in $[0, 7]$, $u^*(t) = 0$ in $(7, 8]$, $T^* = 8$.

      (ii)   Two optimal solutions: $u_1^*(t) = 1$, $T_1^* = 1$ and $u_2^*(t) = 1$ in $[1, T_2^* - 1]$, $u_2^*(t) = 0$ in $(T_2^* - 1, T_2^*]$, $T_2^* = 4 + 2\sqrt{2}$.

2.9.9   (ii) $u^*(t) \equiv 0$, $T^* = 1$.

Chapter 3

3.1.1   (a) $p_1(T) = \gamma c$, $p_2(T) = \gamma d$, where $\gamma \geqslant 0$ ($= 0$ if $cx^*(T) + dy^*(T) > A$)

      (b)   Put $F(t) = (1/a)[c(ax_0 + by_0)e^{at} + (ad - bc)y_0]$. Then $u^*(t) = 1$, $v^*(t) = 1$ in $[0, t']$, $u^*(t) = 0$, $v^*(t)$ arbitrary in $(t', T]$. Here $t' = T - 1/a$ if $cx^*(T) + dy^*(T) = F(T - 1/a) \geqslant A$. If $F(T - 1/a) < A$, $t' \geqslant T - 1/a$, and t' is determined by $F(t') = A$.

      (c)   If $F(T - 1/a) \geqslant A$, we get the first of the two solutions in (b). If $F(T - 1/a) \leqslant A$, there are two possibilities. (I) The solution is of the latter of the two types in (b) if $p_1(t') \geqslant p_2(t')$, i.e. $d(1 - a(T - t')) + bc(T - t') \leqslant c$. (II) If the last inequality fails, $u^*(t) = 1$ in $[0, t^*]$, $u^*(t) = 0$ in $(t^*, T]$ where $p_2(t^*) = 1$, $t^* \in (0, T)$ and $v^*(t) = 1$ in $[0, t'']$, $v^*(t) = 0$ in $(t'', t^*]$, and $v^*(t)$ arbitrary in $(t^*, T]$, where $t'' < t^*$. If $t'' > 0$, then $p_1(t'') = p_2(t'')$.

      (d)   If $a + f > b + g$, the solution is as in (c). If $a + f \leqslant b + g$, $u^*(t)$ switches from 1 to 0 at $t^* = \max[0, T - 1/b, t']$ where t' satisfies $cx^*(t') + dy^*(t') = A$.

3.1.2 There exist numbers $\gamma_1, \ldots, \gamma_n$ such that $p_i(t_1) = \gamma_i$, $i = 1, \ldots, m$, $p_i(t_1) = 0$, $i = m+1, \ldots, n$. Hence the only restrictions on $p_i(t_1)$ are $p_i(t_1) = 0$, $i = m+1, \ldots, n$, in accordance with theorem 2.2.

3.1.3 Put $\dot{x} = u$, $R(x, t) = x - g(t)$. Then $p(t) = -\partial F/\partial \dot{x}$, $H = F(t, x, \dot{x}) + p\dot{x}$, so (3.7) and (3.9) imply (1.32c)

3.2.1 u^* switches from 1 to 0 at $t' = \max(0, T+a-1)$. For $T > 1$, u^* switches in this case later than in the corresponding problem in exercise 2.3.3.

3.2.2 Put $S_1(x_1, \ldots, x_n) = x_1^2$, $R_k(x_1, \ldots, x_n) = x_1/\lambda_1 - x_k/\lambda_k$, $k = 2, \ldots, n$. Transv.cond: $p_1(T) = 2p_0 x_1(T) - (1/\lambda_1) \sum_{k=2}^{n} \lambda_k p_k(T)$.

3.3.1 (i) $\tau^* = (1/r)\ln(cr/q)$ (ii) $p(t) = -qt + (q/r)(1 + \ln(cr/q))$
(iii) $x^*(t) \equiv 0$ in $[0, \tau^*]$, $x^*(t) = x_T$ in $(\tau^*, T]$.

3.5.1 (ii) $\partial V/\partial x^0 = x^1/x^0 = p(t_0)$, $\partial V/\partial x^1 = t_1 - t_0 - \ln(x^1/x^0) - 1 = -p(t_1)$
$\partial V/\partial t_0 = -x^1 = -H^*(t_0)$, $\partial V/\partial t_1 = x^1 = H^*(t_1)$.

3.5.2 (i) $\hat{H}$ is not concave in x
(ii) $V(x_0) = \frac{1}{2}ex_0^2$.

3.6.1 $k_r(t) = k_0 e^t$ in $[0, \tau+r]$, $k_r(t) = k_0 e^{\tau+r}$ in $[\tau+r, T]$, $p(t) = (T-\tau)e^{\tau-t}$ in $[0, \tau]$, $p(t) = T-t$ in $(\tau, T]$.

3.6.3 $dV^+(0)/ds = K_0(1-a)(1-b)^{-1} \exp[(b-a)t'' + (1-b)t']$

3.9.3 $p(t, T) = e^{(1/2)(t-T)} - 1$.

Chapter 4

4.1.1 (i) $\tilde{x}(t) = e^t - 1$ on $[0, \ln 2]$, $\tilde{x}(t) = \frac{1}{2}(u_0 + 1)e^t - u_0$ on $(\ln 2, 1]$
(ii) $u_0 \in [-1, 4e^{-1} - 1]$
(iii) $J = (\ln 2 - 2 + \frac{1}{2}e)u_0 + \frac{1}{2}e - \ln 2$
$J_{\max} = 4e^{-1} \ln 2 - 8e^{-1} + 4 - 2 \ln 2 \approx 0.6906$ for $u_0 = 4e^{-1} - 1$.
(iv) On $(\ln 2, 1]$, $q_1 = q_2 = q_3 = 0$. By (18), $p(t) = 0$ which contradicts (22).

4.2.2 $u^*(t) = c$ in $[0, t']$, $u^*(t) = ax^*(t)$ in $(t', T]$, with $t' = \max(T - 1/a, t'')$, where t'' satisfies $(x_0 - c/a)e^{at''} + c/a = x_T$.

4.2.3 $u^*(t) = x^*(t) = e^{-t}$ in $[0, T-1]$, $u^*(t) = 0$ in $(T-1, T]$.

4.3.1 (i) The criterion is 1 for $(\bar{x}, \bar{u})$ and 2 for (x^*, u^*),

4.3.2 $u^* = -\frac{1}{2}t - \frac{1}{4}(2 \ln 3 - 3)$ in $[-1 - \ln 3, -\ln 3]$,
$u^* = \frac{1}{4}e^{-t}$ in $(-\ln 3, 0]$, $u^* = \frac{1}{4} - \frac{1}{2}t$ in $(0, 1/2]$.

4.4.2 (i) $t'' = 20$, $u^*(t) = 0$ in $[0, 20]$, $u^*(t) = 1$ in $(20, 50]$, $x^*(t) = 30$ in $[0, 20]$, $x^*(t) = 50 - t$ in $(20, 50]$.
(ii) $y_1(t) = 0$ on $[0, 20]$, $y_1(t) = (1/a)e^{2at} - (1/a)e^{a(t+20)}$ on $(20, 50]$

(iii) $c(t) = \bar{c}e^{2a(t-t')}$ on $(t', 50]$.

(iv) On $(t'\ 20]$, $y^*(t) = (\bar{c}/a)(2e^{a(t-t')} - 1 - e^{2a(t-t')})$

(v) $z = e^{-50a} + [(1 - e^{-30a})/\bar{c}]^{1/2}$. For $a = 0.1$, $\bar{c} = 10$, $t' = -10\ln z$ ≈ 11.55.

(vi) $c^*(0) = (e^2 + e - 1 - e^{7/5})/(e^2 - e)$.

4.4.3 $\dot{y}^*(t''^+) - \dot{y}^*(t''^-) = \bar{u}e^{rt''}$.

4.4.4 $u^*(t) = 1$ and $v^*(t) = (t + 1/8)^{1/2}$ in $[0, 1/8]$, $u^*(t) = 0$ and $v^*(t) = 1/2$ in $(1/8, 1]$, $(x^*(t) = t + 1/8$ and $p(t) = t - (t + 1/8)^{1/2} + 3/8$ in $[0, 1/8]$, $x^*(t) = 1/4$ and $p(t) = 0$ in $(1/8, 1]$.)

Chapter 5

5.2.1 (i) $u^*(t) = -1$ on $[0, 1]$, $u^*(t) = 0$ on $(1, 2]$.

5.2.2 (i) $u^*(t) = 1$ in $[0, \pi/3]$, $u^*(t) = -1$ in $(\pi/3, 2\pi/3]$, $u^* = 0$ in $(2\pi/3, \pi]$

(ii) $u^*(t) = 2$ in $[0, 1]$, $u^*(t) = 1$ in $(1, 2]$, $u^*(t) = 0$ in $(2, 3]$

5.2.3 (i) If $\alpha T \leqslant \ln a - \ln(a - \alpha)$: $u^*(t) \equiv 0$.

If $\alpha T > \ln a - \ln(a - \alpha)$: $u^*(t) = \bar{u}$ in $[0, t^*]$, $u^*(t) = 0$ in $(t^*, T]$ where $t^* = T - (1/\alpha)[\ln a - \ln(a - \alpha)]$

(ii) (1) $u^*(t)$ maximizes $(p(t) - 1)u + ax^*(t) - \alpha p(t)x^*(t)$ for $u \in [0, \bar{u}]$

(2) $\dot{p}(t) = -a + \alpha p(t) + \lambda(t)$ (3) $\lambda(t) \geqslant 0$ ($= 0$ if $x^*(t) < \bar{x}$)

(4) $p(T^-) - p(T) = -\beta$ (5) $\beta \geqslant 0$ ($= 0$ if $x^*(T) < \bar{x}$)

(6) $p(T) = 0$

(H is concave in (x, u) and $g(x, t) = \bar{x} - x$ is quasi−concave in (x, u).)

(iii) If $\alpha T \leqslant \ln a - \ln(a - \alpha)$: $u^*(t) = 0$. If $\alpha T > \ln a - \ln(a - \alpha)$: $u^*(t) = \bar{u}$ in $[0, t_*]$, $u^*(t) = \alpha\bar{x}$ in $(t_*, t^*]$, $u^*(t) = 0$ in $(t^*, T]$ where t^* is given in (i) and $t_* = (1/\alpha)[\ln(\bar{u} - \alpha x_0) - \ln(\bar{u} - \alpha\bar{x})]$

5.2.5 $u^*(t) = 1 - e^{t-3}$ in $[0, 3]$, $u^*(t) = 0$ in $(3, 4]$.

5.2.6 (a) (i) $u^*(t) = 0$ on $[0, t'']$, $u^*(t) = [(2ab - h)t' + ht]/2b$ on $(t'', t']$, $u^*(t) = at$ on $(t', T]$. $x^*(t) > 0$ on $[0, t')$, $x^*(t) \equiv 0$ in $[t', T]$. $t'' = (1 - 2ab/h)t'$ and t'' is determined by

$$1 + \int_0^{t'} (-at)\,dt + \int_{t''}^{t'} [((2ab - h)t' + ht)/2b]\,dt = 0$$

The left-hand side of this equation is 1 for $t'=0$ and <0 for $t'=T$ due to the assumption $aT^2(1-2ab/h)>2$.

(ii) $u^*(t)=0$ on $[0, t'']$,
$u^*(t)=(h(t-T)+K-c)/2b$ on $(t'', T]$, $x^*(t)>0$ on $(0, T)$ with $x^*(T)=0$. $t''=T$ is impossible due to $aT^2>2$. If $T^2<4b/(2ab-h)$, then $t''=T-(K-c)/h>0$ with $K>c$ determined by

$$1+\int_0^T(-at)dt+\int_{t''}^T[(h(t-T)+K-c)/2b]dt=0.$$

If $T^2\geqslant4b/(2ab-h)$, then $t''=0$ with $K>c$ determined by this equation for $t''=0$.

5.3.1 (i) If $u^*(t)\equiv0$, $V=4$. If $u^*(t)\equiv1$, $V=6$.
(ii) $t^*=3$ maximizes $V(t^*)$. (ii) Example 4 partially confirmed.
5.3.2 $u^*(t)=1$ in $[0, 1]$, $u^*(t)=0$ in $(1, 3]$ and $u^*(t)=-1$ in $(3, 4]$ is optimal. (There is another (non-optimal) candidate.)

Chapter 6

6.2.1 $u^*(t)=1$, $v^*(t)=(t+1/16)^{1/2}$ in $[0, 1/16]$, $u^*(t)=0$, $v^*(t)=\sqrt{2}/4$ in $(1/16, 1]$.
6.2.2 $u^*(t)=e^{1-t}$ in $[0, 1]$, $u^*(t)=-1$ on $(1, 2]$
6.2.3 $u^*(t)=-\frac14(2\ln3-3)-\frac12t$ in $[-1-\ln3, -\ln3]$, $u^*(t)=\frac14e^{-t}$ in $(-\ln3, 0]$, $u^*(t)=\frac14-\frac12t$ in $(0, 1/4]$, $u^*(t)=-1/8$ in $(1/4, 1]$.
6.2.4 $u^*(t)=0$, $v^*(t)=k^*(t)$ in $[0, t'-1]$, $u^*(t)=k^*(t)$, $v^*(t)=0$, in $(t'-1, t']$, $u^*(t)=1<k^*(t)$, $v^*(t)=0$ in $(t', 2]$, where t' is the solution to $2e^{t'-1}=2+t'$. $(t'\in(1, 2).)$
6.2.6 For $T\leqslant3$, $u^*(t)=x^*(t)=e^{-t}$ in $[0, T-1]$, $u^*(t)=0$ in $(T-1, T]$.
For $T>3$, $u^*(t)=x^*(t)=e^{-t}$ in $[0, 2]$, $u^*(t)=0$ in $(2, T]$.
6.3.2 (i) $u^*(t)\equiv-t-2$ (ii) $u^*(t)=-t+2$ on $[0, 1/2]$, $u^*(t)=t+1$ on $(1/2, 1]$.
6.4.1 (i) $u^*(t)=1/(1-t)^2$ in $[0, 1/2]$, $u^*(t)=0$ in $(1/2, 3-a]$, $u^*(t)=-1$ in $(3-a, 3]$
(ii) $u^*(t)=x^*(t)=e^t$ in $[0, t']$, $u^*(t)=-1$ on $(t', T]$.
Here $t'=\max(T-1, t'')$, where t'' is the solution to $e^{t''}=T-t''$.
(iii) One solution $u_1^*\equiv1$ when $\varphi(T)=-3e^{-T}+4-e^T>0$. When $\varphi(T)<0$, three candidates: $u_1^*(t)$, $u_2^*(t)\equiv e^t$ and $u_3^*(t)=-1$ on

$[0, t']$, $0 < t' < 1$, $u_3^*(t) = (1 - t')$ e^t on (t', T), t' a solution of $(-4e^{t''} + (1 - t')$ e$^{2t''})$ e$^{-T} + 4 - (1 - t')$ e$^T = 0$.

6.7.1 $u^*(t) = r_1 x_1^*(t) + r_2 x_2^*(t)$ in $[0, t]$, $u^*(t) = \bar{u}$ in $(t', T]$, $v^*(t) \equiv \bar{v}$. Here $t' = \max(T - 1/r_1, \; t'')$, where t'' is defined by $x^*(t'') = 3$. $(x^*(t) = (r_2/r_1^2)(\bar{v} + r_1)(e^{r_1 t} - 1) + e^{r_1 t} - (r_2/r_1)\bar{v}t$ in $[0, t']$.)

6.7.2 $u^*(t) = 1$ in $[0, T - (1 - b)]$, $u^*(t) = 0$ in $(T - (1 - b), T]$, $x^*(0) = \frac{1}{2}e^{T - (1 - b)}$.

REFERENCES

Aarrestad, J., 1979, "Resource extraction, financial transactions and consumption in an open economy", Scandinavian Journal of Economics, *81*, 552–565.

Atkinson, A.B., 1971, "Capital taxes, the redistribution of wealth and individual savings", Review of Economic Studies, 38, 209–227.

Arrow, K.J. and Kurz, M., 1970, Public Investment, the Rate of Return, and Optimal Fiscal Policy, John Hopkins Press, Baltimore.

Baum, R.F., 1976, "Existence theorems for Lagrange control problems with unbounded time domain", Journal of Optimization Theory and Applications, *19*, 89–116.

Bensoussan, A., Hurst Jr., E.G., and Näslund, B., 1974, Management Applications of Modern Control Theory, North-Holland, Amsterdam.

Bryson Jr., A.E. and Ho, Y.-C., 1975, Applied Optimal Control, Hemisphere Publ. Co., Washington.

Cesari, L., 1983, Optimization – Theory and Applications, Springer, New York.

Chow, G.C., 1975, Analysis and Control of Dynamic Economic Systems, Wiley, New York.

Clark, C.W., 1976, Mathematical Bioeconomics: The Optimal Management of Renewable Resources, Wiley, New York.

Clark, C.W., 1982, (A review of Kamien and Schwartz, 1981) American Journal of Agricultural Economics, *64*, 167.

Clarke, F.H., 1976, "The maximum principle with minimal hypothesis", Siam Journal on Control and Optimization, *14*, 1078–1091.

Coddington, E.A. and Levinson, N., 1955, Theory of Ordinary Differential Equations, McGraw-Hill, New York.

Dorfman, R., 1969, "An economic interpretation of optimal control theory", American Economic Review, *59*, 817–831.

Elsgolc, L.E., 1962, Calculus of Variations, Addison-Wesley, Reading, Mass.

Feichtinger, G. and Hartl, R.F., 1986, Optimale Kontrolle Ökonomischer Prozesse, De Gruyter. To appear.

Fleming, W.H. and Rishel, R.W., 1975, Deterministic and Stochastic Optimal Control, Springer, Berlin.

Fourgeaud, C., Lenclud, B. and Michel, P., 1982, "Technological renewal of natural resource stocks", Journal of Economic Dynamics and Control, *4*, 1–36.

Gandolfo, G., 1980, Economic Dynamics: Methods and Models, North-Holland, Amsterdam, 2nd edn.

Gelfand, I.M., and Fomin, S.V., 1969, Calculus of Variations, Prentice-Hall, Englewood Cliffs, NJ.

Getz, W.M. and Martin, D.H., 1980, "Optimal control systems with state variable jump discontinuities", Journal of Optimization Theory and Applications, *31*, 195–205.

Hadley, G. and Kemp, M.C., 1971, Variational Methods in Economics, North-Holland, Amsterdam.

Hale, J.K., 1969, Ordinary Differential Equations, Wiley-Interscience, New York.

Halkin, H., 1974, "Necessary conditions for optimal control problems with infinite horizons", Econometrica, *42*, 267–272.

Hartman, P., 1964, Ordinary Differential Equations, Wiley, New York.

Hestenes, M.R., 1966, Calculus of Variations and Optimal Control Theory, Wiley, New York.

Hoel, M., 1978, "Distribution and growth as a differential game between workers and capitalists", International Economic Review, *19*, 335–350.

Hotelling, H., 1931, "The Economics of Exhaustible Resources", Journal of Political Economy, *39*, 137–175.

Ioffe, A.D., and Tihomirov, V.M., 1979, Theory of Extremal Problems, North-Holland, Amsterdam.

Kamien, M.I. and Schwartz, N.L., 1981, Dynamic Optimization. The Calculus of Variations and Optimal Control In Economics and Management, North-Holland, New York.

Koopmans, T.C., 1974, "Proof of a case when discounting advances doomsday", Review of Economic Studies, *41*, 117–120.

Lancaster, K., 1973, "The dynamic inefficiency of capitalism", Journal of Political Economy, *81*, 1092–1109.

Lee, E.B. and Markus, L., 1967, Foundations of Optimal Control Theory, Wiley, New York.

Makowski, K. and Neustadt, L., 1974, Optimal control problems with mixed control-phase variable equality and inequality constraints, Siam Journal of Control, *12*, 184–228.

Malanowski, K., 1984, "On differentiability with respect to parameters of solutions to convex optimal control problems subject to state space constraints", Applied Mathematics and Optimization, *12*, 231–247.

Mereau, P.M. and Powers, W.F., 1976, "A direct sufficient condition for free final time optimal control problems", SIAM Journal on Control and Optimization, *14*, 613–623.

Michel, P., 1977, "Une demonstration élémentaire du principe du maximum de Pontryagin", Bulletin de Mathematiques Economiques, *14*, 9–23.

Michel, P., 1981, "Choice of projects and their starting dates: An extension of Pontraygin's maximum principle to a case which allows choice among different possible evolution equations", Journal of Economic Dynamics and Control, *3*, 97–118.

Michel, P., 1982, "On the transversality condition in infinite horizon optimal control problems", Econometrica, *50*, 975–985.

Neustadt, L.W., 1963, "The existence of optimal controls in the absence of convexity conditions", Journal of Mathematical Analysis and Applications, *7*, 110–117.

Neustadt, L.W., 1976, Optimization. A Theory of Necessary Conditions, Princeton University Press, New Jersey.

Olech, C., 1969, "Existence theorems for optimal control problems with vector-valued cost functions", Transactions of the American Mathematical Society, *136*, 159–180.

Oniki, H., 1973, "Comparative dynamics (sensitivity analysis) in optimal control theory", Journal of Economic Theory, *6*, 265–283.

Pallu de la Barrière, R., 1980, Optimal Control Theory (English edn.) Dover, New York.

Petersson, D.W. and Zalkin J.H., 1978, "A review of direct sufficient conditions in optimal control theory", International Journal on Control, *28*, 589–610.

Pontryagin, L.S., (1962a), Ordinary Differential Equations, Addison Wesley, Reading, Mass.

Pontryagin, L.S., Boltyanskii, V.G., Gamkrelidze, R.V. and Mishchenko, E.F., 1962b. The Mathematical Theory of Optimal Processes, Interscience, New York.

Ramsey, F.P., 1928, "A mathematical theory of saving", Economic Journal, *38*, 543–549.

Rockafellar, R.T., 1970a, Convex Analysis, Princeton University Press, Princeton.

Rockafellar, R.T., 1970b, "Conjugate convex functions in optimal control and the calculus of variations", Journal of Mathematical Analysis and Applications, *32*, 174–222.

Rockafellar, R.T., 1973, "Optimal arcs and the minimum value functions in problems of Lagrange", Transactions of the American Mathematical Society, *180*, 53–83.

Rockafellar, R.T., 1975, "Existence theorems for general control problems of Bolza and Lagrange", Advances in Mathematics, *15*, 312–333.

Russel, D.L. (ed.), 1976, Calculus of Variations and Control Theory, Academic Press, New York.

Sansone, G. and Conti, R., 1964, Nonlinear Differential Equations, (Revised edn.), Pergamon, Oxford.

Seierstad, A., 1975, "Existence of a control satisfying Mangasarian's sufficiency conditions for optimality makes these conditions necessary, Memorandum from Institute of Economics, University of Oslo, Dec. 5.

Seierstad, A., 1977a, "A sufficient condition for control problems with infinite horizons", Memorandum from Institute of Economics, University of Oslo, Jan. 26.

Seierstad, A., 1977b, "Transversality conditions for control problems with infinite horizons". Memorandum from Institute of Economics, University of Oslo, Jan. 27.

Seierstad, A., 1981a, "Derivatives and subderivatives of the optimal value function in control theory", Memorandum from Institute of Economics, University of Oslo, Feb. 26.

Seierstad, A., 1981b, "Necessary conditions and sufficient conditions for optimal control with jumps in the state variables", Memorandum from Institute of Economics, University of Oslo, June 15.

Seierstad, A., 1982, "Differentiability properties of the optimal value function in control theory", Journal of Economic Dynamics and Control, *4*, 303–310.

Seierstad, A., 1984a, "Sufficient conditions in free final time optimal control problems", Memorandum from Department of Economics, University of Oslo, Jan. 15.

Seierstad, A., 1984b, "Sufficient conditions in free final time optimal control problems. A comment", Journal of Economic Theory, *32*, 367–370.

Seierstad, A., 1985a, "Existence of optimal unbounded control functions in infinite horizon problems with a monotonicity property", Memorandum from Department of Economics, University of Oslo.

Seierstad, A., 1985b, "Sufficient conditions, uniqueness and dependence on parameters in optimal control problems", Memorandum from Department of Economics, University of Oslo.

Seierstad, A., 1985c, "Existence of an optimal control with sparse jumps in the state variable". Journal of Optimization Theory and Applications, *45*, 265–293.

Seierstad, A., 1986a, "Nontrivial multipliers and necessary conditions for optimal control with infinite horizons and time path restrictions", Memorandum from Department of Economics, University of Oslo.

Seierstad, A., 1986b, "Nontrivial necessary conditions in optimal control problems with state restrictions", Memorandum from Department of Economics, University of Oslo.

Seierstad, A. and Sydsæter, K., 1977, "Sufficient conditions in optimal control theory", International Economic Review, *18*, 367–391.

Seierstad, A. and Sydsæter, K., 1981. "Sufficient conditions applied to an optimal control problem of resource management. A full analysis", Memorandum from Institute of Economics, University of Oslo, Oct. 5.

Seierstad, A. and Sydsæter, K., 1983, "Sufficient conditions applied to an optimal control problem of resource management", Journal of Economic Theory, *31*, 375–382.

Sethi, S.P. and Thompson, G.L., 1981, Optimal Control Theory Applications to Management Science, Martinus Nijhoff, Boston, Mass.

Shell, K., 1967, "Optimal program of capital accumulation for an economy in which there is exogenous technical change", in Essays on the Theory of Optimal Economic Growth (ed. K. Shell) 1–30, M.I.T. Press, Cambridge, Mass.

Solow, R.M. and Vickery, W.S., 1971, "Land use in a long narrow city", Journal of Economic Theory, *3*, 430–477.

Sydsæter, K., 1978, Optimal control theory and economics. Some critical remarks on the literature. Scandinavian Journal of Economics, *80*, 113–117.

Sydsæter, K., 1981, Topics in Mathematical Analysis for Economists, Academic Press, London.

Takayama, A., 1974, Mathematical Economics, The Dryden Press, Hinsdale Ill.

Vind, K., 1967, "Control systems with jumps in the state variables", Econometrica, *35*, 273–277.

Warga, J., 1972, Optimal Control of Differential and Functional Equations, Academic Press, New York.

Young, L.C., 1969, Lectures on the Calculus of Variations and Optimal Control Theory, Saunders, New York.

LIST OF THEOREMS, NOTES AND EXAMPLES

Theorems

Atle Seierstad and Knut Sydsæter

Notes

1.1 , 22	2.7 , 87	3.1 , 179	3.24, 246	5.6 , 344	6.15, 391
1.2 , 30	2.8 , 106	3.2 , 183	3.25, 246	5.7 , 346	6.16, 394
1.3 , 36	2.9 , 106	3.3 , 186	3.26, 257	5.8 , 346	6.17, 394
1.4 , 36	2.10, 108	3.4 , 186	3.27, 257	5.9 , 347	6.18, 396
1.5 , 44	2.11, 109	3.5 , 186		5.10, 348	6.19, 398
1.6 , 45	2.12, 109	3.6 , 194	4.1 , 277	5.11, 348	6.20, 399
1.7 , 45	2.13, 109	3.7 , 197	4.2 , 277	5.12, 352	6.21, 399
1.8 , 47	2.14, 111	3.8 , 197	4.3 , 277	5.13, 353	6.22, 401
1.9 , 48	2.15, 133	3.9 , 201	4.4 , 278		6.23, 403
1.10, 51	2.16, 133	3.10, 208	4.5 , 279	6.1 , 362	6.24, 403
1.11, 52	2.17, 133	3.11, 214	4.6 , 280	6.2 , 363	6.25, 403
1.12, 54	2.18, 136	3.12, 214	4.7 , 280	6.3 , 369	6.26, 407
1.13, 55	2.19, 136	3.13, 215	4.8 , 290	6.4 , 373	6.27, 407
1.14, 56	2.20, 137	3.14, 222	4.9 , 290	6.5 , 374	6.28, 408
1.15, 64	2.21, 137	3.15, 228	4.10, 292	6.6 , 374	
1.16, 64	2.22, 137	3.16, 235	4.11, 292	6.7 , 376	A.1, 413
	2.23, 138	3.17, 236	4.12, 302	6.8 , 377	A.2, 414
2.1 , 76	2.24, 138	3.18, 236	4.13, 302	6.9 , 377	A.3, 415
2.2 , 86	2.25, 143	3.19, 237		6.10, 377	A.4, 417
2.3 , 86	2.26, 146	3.20, 245	5.1 , 318	6.11, 381	A.5, 418
2.4 , 86	2.27, 146	3.21, 245	5.2 , 320	6.12, 386	A.6, 420
2.5 , 86	2.28, 147	3.22, 245	5.3 , 333	6.13, 386	A.7, 424
2.6 , 87		3.23, 246	5.4 , 334	6.14, 389	
			5.5 , 337		

Examples

1.1 , 13	1.14, 48	2.7 , 121	3.5 , 202	4.1 , 271	6.1 , 360
1.2 , 20	1.15, 52	2.8 , 133	3.6 , 212	4.2 , 279	6.2 , 364
1.3 , 20	1.16, 54	2.9 , 134	3.7 , 213	4.3 , 283	6.3 , 382
1.4 , 26	1.17, 61	2.10, 138	3.8 , 214	4.4 , 292	6.4 , 386
1.5 , 27	1.18, 62	2.11, 147	3.9 , 217	4.5 , 302	6.5 , 404
1.6 , 29	1.19, 63	2.12, 151	3.10, 222		6.6 , 405
1.7 , 37		2.13, 155	3.11, 238	5.1 , 314	6.7 , 408
1.8 , 37	2.1 , 76	2.14, 166	3.12, 239	5.2 , 321	6.8 , 409
1.9 , 38	2.2 , 78		3.13, 241	5.3 , 321	
1.10, 38	2.3 , 89	3.1 , 179	3.14, 242	5.4 , 337	A.1, 414
1.11, 39	2.4 , 92	3.2 , 184	3.15, 247	5.5 , 340	A.2, 415
1.12, 45	2.5 , 109	3.3 , 187	3.16, 260	5.6 , 348	A.3, 419
1.13, 45	2.6 , 113	3.4, 199	3.17, 261		A.4, 423

INDEX

Printed and bound by CPI Group (UK) Ltd, Croydon, CR0 4YY

08/05/2025

01864826-0001